Progress in Probability

Volume 55

High Dimensional Probability III

Jørgen Hoffmann-Jørgensen
Michael B. Marcus
Jon A. Wellner
Editors

Birkhäuser Verlag
Basel · Boston · Berlin

Editors' addresses:

Jørgen Hoffmann-Jørgensen
Department of Mathematical Sciences
University of Aarhus
Building 530
Ny Munkegade
8000 Arhus C
Denmark
hoff@imf.au.dk

Michael B. Marcus
Department of Mathematics
City College
New York, NY 10031
USA
mbmarcus@earthlink.net

Jon A. Wellner
University of Washington
Department of Statistics
Box 354322
Seattle, WA 98195-4322
USA
jaw@stat.washington.edu

2000 Mathematics Subject Classification 60B11, 60B12, 60E05, 60E07, 60E15, 60F05, 60F10, 60F15, 60F17, 60G15, 60G17, 60G40, 60G44, 60G50, 60H25, 60J25, 60J30, 60J60, 60J65, 62E17, 62E20, 62G07, 62G20

A CIP catalogue record for this book is available from the Library of Congress, Washington D.C., USA

Bibliographic information published by Die Deutsche Bibliothek
Die Deutsche Bibliothek lists this publication in the Deutsche Nationalbibliografie; detailed bibliographic data is available in the Internet at <http://dnb.ddb.de>.

ISBN 3-7643-2187-3 Birkhäuser Verlag, Basel · Boston · Berlin

Part of Springer Science+Business Media
Printed on acid-free paper produced from chlorine-free pulp. TCF ∞
Printed in Germany
ISBN 3-7643-2187-3

9 8 7 6 5 4 3 2 1

www.birkhauser.ch

Table of Contents

Preface

The title High Dimensional Probability is an attempt to describe the many tributaries of research on Gaussian processes and probability in Banach spaces that started in the early 1970's. In each of these fields it is necessary to consider large classes of stochastic processes under minimal conditions. There are rewards in research of this sort. One can often gain deep insights, even about familiar processes, by stripping away details that in hindsight turn out to be extraneous.

Many of the problems that motivated researchers in the 1970's were solved. But the powerful new tools created for their solution, such as randomization, isoperimetry, concentration of measure, moment and exponential inequalities, chaining, series representations and decoupling turned out to be applicable to other important areas of probability. They led to significant advances in the study of empirical processes and other topics in theoretical statistics and to a new approach to the study of aspects of Lévy processes and Markov processes in general. Papers on these topics as well as on the continuing study of Gaussian processes and probability in Banach are included in this volume.

The papers in this volume are by many of the participants in the Third International Conference on High Dimensional Probability, held at Sandjberg, Denmark, June 24–28, 2002. Most of them are based on the author's talks at the conference. This conference was the thirteenth in a series of conferences that began with the Colloque International sur les Processus Gaussians et les Distributions Aléatoires held at Strasbourg in 1973 and continued in nine conferences on Probability in Banach Spaces and three on High Dimensional Probability. On page 5 of the book, *Probability in Banach Spaces*, Springer–Verlag, by M. Ledoux and M. Talagrand, there is a list of the first seven Probability in Banach Spaces conferences, (as well as many related conferences), up to 1989. The subsequent conferences, of which this is the thirteenth, are:

Probability in Banach Spaces 8, Bowdoin (Maine) 1990. *Progress in Probability, vol.* 30. Birkhäuser, Boston 1992.

Probability in Banach Spaces 9, Sandjberg (Denmark) 1993. *Progress in Probability, vol.* 35. Birkhäuser, Boston 1994.

High Dimensional Probability I, Oberwolfach 1996. *Progress in Probability, vol.* 43. Birkhäuser, Basel 1998.

High Dimensional Probability II, Seattle (Washington) 1999. *Progress in Probability vol.* 47. Birkhäuser, Boston 2000.

The Third International Conference on High Dimensional Probability was held at Arhus University's beautiful conference center in southern Jutland. It was our third conference at this center. We are grateful to MaPhySto (Center for Mathematical Physics and Statistics) for funding the conference and providing secretarial assistance. MaPhySto is funded by a grant from the Danish National Research Foundation.

Jorgen Hoffmann–Jørgensen
Michael B. Marcus
Jon A. Wellner

Part 1

Measures on General Spaces and Inequalities

Progress in Probability, Vol. 55, 3–34

Stochastic inequalities and perfect independence

J. Hoffmann-Jørgensen

Abstract. The objective of this work is to extend the classical inequalities for sums of independent Bochner measurable random vectors to arbitrary "independent" random elements taking values in a linear space. Note that the usual definition of independence do not apply to arbitrary functions. In the literature this problem has been surpassed by assuming that the functions are "independently defined". However, this impose an unpleasant restriction on the form of the underlying probability space and below I shall introduce two new concepts of "independence" which are weaker than "independently defined" and applies to arbitrary probability spaces. The main inequality proved in this paper is a version of the Ottaviani inequality; see Thm. 3.1, with a constant which improves the usual constant and which is strong enough to imply Lévy's inequality under asumptions which are much weaker than the assumptions of the known results and where the usual proof do not apply.

1. Introduction

In the theory of empirical processes and uniform convergence of stochastic processes, non-measurable norms and non-measurable "random elements" occur naturally and a crucial part of the theory consists in extending the classical inequalities (Lévy's inequality, Ottaviani's inequality, Jensen's inequality, the symmetrization inequalities, the exponential inequality and the subgaussian inequalities) to vector valued non-measurable functions under various independence and regularity conditions. However, the usual definition of independence do not apply to non-measurable functions. In the literature, this problem has been bypassed by introducing the concept "independently defined" (see [2], [9] and (2.3) and (2.4) below). The notion of "independently defined" imposes an unpleasant restriction on the underlying probability space and – as we shall see below – is unnecessarily restrictive for the validity of the inequalities.

Received by the editors December 1, 2002.
Supported by MaPhySto. Center for Mathematical Phycics and Stochastics, funded by a grant from the Danish National Research Foundation.

Let us consider *Lévy's inequality:* If $X_1, \dots, X_n$ are independent, symmetric, Bochner measurable random vectors with values in a Banach space $(\Lambda, ||\cdot||)$ and $S_i := X_1 + \cdots + X_i$ denotes the partial sums, then we have

(1.1) $P(\max_{1\le i\le n} \|S_i\|>t)\le 2\,P(\|S_n\|>t) \ \ \forall t\ge 0$

The non-measurable version of Lévy's inequality states that if $X_1, \dots, X_n$ are "independent" and "symmetric" functions with values in a Banach space $(\Lambda, ||\cdot||)$, then we have

(1.2) $P^*(\max_{1\le i\le n} \|S_i\|>t)\le 2\,P^*(\|S_n\|>t) \ \ \forall t\in \mathbf{R}$

where "independent" and "symmetric" are defined in various ways; see for instance [3; p. 289–290] and [9; p. 431]. The current proofs of (1.2) all require (at the least) that $P^*(||X_1 + X_2|| > t) = P^*(||X_1 - X_2|| > t)$ for all $t > 0$. Let me give a simple example of "independent" and "symmetric" functions $X_1, \dots, X_n$ for which this condition fails (but the Lévy inequality prevails):

Example 1. Let $T = [0,1]$ denote the unit interval and let λ denote the Lebesgue measure on the Borel σ-algebra $\mathcal{B}$ on T . Let ξ be a random variable on $(T, \mathcal{B}, \lambda)$ which assumes the values $+1$ and -1 with probability $\frac{1}{2}$. Let us define $X(t,\omega) := \xi(\omega)$ if $t, \omega \in T$ and $t \neq \omega$, and $X(t,\omega) := 1$ if $t = \omega \in T$. Let $\ell^\infty(T)$ denote the set of all bounded functions $f : T \to \mathbf{R}$ and let $||f||_T := \sup_{t\in T} |f(t)|$ denote the sup-norm on $\ell^\infty(T)$. Let $n \ge 2$ be a given integer and let $(\Omega, \mathcal{F}, P) = (T^n, \mathcal{B}^n, \lambda^n)$ denote the n-fold product space of $(T, \mathcal{B}, \lambda)$. If we define $X_i(t,\omega) := X(t, \omega_i)$ for $\omega = (\omega_1, \dots, \omega_n) \in \Omega$ and $i = 1, \dots, n$, then $X_1, \dots, X_n$ are independently defined; see (2.3) and (2.4) below, symmetric, stochastic processes taking values in the Banach space $(\ell^\infty(T), ||\cdot||_T)$. Let $\omega = (\omega_1, \dots, \omega_n) \in \Omega$ be given. Then a straightforward computation shows that $||X_1(\omega) + X_2(\omega)||_T = 2$ and

$$\begin{aligned} ||X_1(\omega) - X_2(\omega)||_T &= 0 \text{ if } (\xi(\omega_1), \xi(\omega_2)) = (1,1) \\ ||X_1(\omega) - X_2(\omega)||_T &= 2 \text{ if } (\xi(\omega_1), \xi(\omega_2)) \neq (1,1) \end{aligned}$$

In particular, we see that norms $||X_1 + X_2||_T$ and $||X_1 - X_2||_T$ are measurable functions satisfying $P(||X_1 + X_2||_T = 2) = 1$ and $P(||X_1 - X_2||_T = 2) = \frac{3}{4}$. This means that the classical proof of Lévy's inequality does *not* apply to the "independent" and "symmetric" functions $X_1, \dots, X_n$. However, in Remark (f) to Thm. 3.1 we shall see that $X_1, \dots, X_n$ indeed satisfy (1.2) but with a different proof.

Recall that *Ottaviani's inequality* states that if $X_1, \dots, X_n$ are "independent" functions taking values in an Banach space $(\Lambda, ||\cdot||)$, then we have

(1.3) $\beta(s)\,P^*(\max_{1\le i\le n} \|S_i\|>s+t)\le P^*(\|S_n\|>t) \ \forall\, s,t\ge 0$

for some specified *Ottaviani constant* $\beta(s)$. In the current literature the Ottaviani inequality is proved with $\beta(s) := \min_{1\le i<n} P_*(||S_i - S_n|| \le s)$; see [9; p. 430] and [3; p. 17–18]. Note that the Ottaviani inequality (1.3) implies the Lévy

inequality (1.2) if $\beta(0) \geq \frac{1}{2}$. However, with the classical Ottaviani constant this condition is by far too restrictive to be of any use, but in Thm. 3.1, I shall provide an Ottaviani constant $\beta(s)$ satisfying (1.3) which improves the classical constant and I shall show that the new Ottaviani constant implies the Lévy inequality far beyond the classical assumptions (in particular, we obtain the Lévy inequality for the example above).

Throughout this paper we let $(\Omega, \mathcal{F}, P)$ denote a fixed probability space and we let $(\Omega, \mathcal{F}_P, \bar{P})$ denote the completion of $(\Omega, \mathcal{F}, P)$. We let $\mathbf{R}$ denote *the real line*, we let $\mathbf{N} := \{1, 2, \ldots\}$ denote the positive integers and we let $\bar{\mathbf{R}} := [-\infty, \infty]$ denote *the extended real line.* If $x, y \in \bar{\mathbf{R}}$, we let $x \cdot y$ denote the usual extension of the multiplication with the convention $0 \cdot (\pm\infty) := 0$. Let $(a_i \mid i \in I) \subseteq \bar{\mathbf{R}}$ be given numbers. Then sums $\sum_{i\in I} a_i^+$ and $\sum_{i\in I} a_i^-$ are well-defined and non-negative but possibly $+\infty$, and we define *the upper sum* as follows: $\sum_{i\in I} a_i := \sum_{i\in I} a_i^+ - \sum_{i\in I} a_i^-$ with the convention $\infty - \infty := \infty$. If $a_1, \ldots, a_n \in \bar{\mathbf{R}}$, we define $a_1 + \cdots + a_n := \sum_{1\leq i\leq n} a_i$ and $a_1 - a_2 := a_1 + (-a_2)$. Then $+$ is an associative and commutative extension of the addition on $\mathbf{R}$ to the extended real line $\bar{\mathbf{R}}$.

Upper and lower expectations and envelopes. If $A \subseteq \Omega$, we let $P^*(A)$ and $P_*(A)$ denote *the inner* and *outer probabilities* of A and we let A_* and A^* denote the (a.s. unique) *P-kernel* and *P-hull* of A . If X is a function from Ω into the extended real line $\bar{\mathbf{R}}$, we let E^*X and E_*X denote *the upper* and *lower expectations* of X and we let X^* and X_* denote the (a.s. unique) *upper* and *lower P-envelopes* of X (see [9] and [1]). Then E^* is *subadditive*, i.e. $E^*(X+Y) \leq E^*X + E^*Y$ with equality if $X \in L^1(P)$, and *positively homogeneous*; i.e. $E^*(cX) = c\,E^*X$ for all $0 \leq c < \infty$ and if $X \geq 0$ a.s., then $E^*(cX) = c\,E^*X$ for all $0 \leq c \leq \infty$. Moreover, we have the following well-known rules of calculus; see [9] and [1]:

(1.4) $\{X>t\}^* = \{X^*>t\}$ a.s. and $\{X\leq t\}_* = \{X^*\leq t\}$ a.s. $\forall t \in \mathbf{R}$

(1.5) $E^*X = \lim_{a\to-\infty} E^*(X \vee a)$, $(\sup_{n\geq 1} X_n)^* = \sup_{n\geq 1} X_n^*$ a.s.

(1.6) $E^*X = E^*X^* = E^*X^+ - E_*X^- = \int_0^\infty P^*(X^+>t)\,dt - \int_0^\infty P_*(X^->t)\,dt$

(1.7) $(YX)^* = Y X^*$ a.s. if $P^*(Y=\infty\ ,\ X^*=0\ ,\ X<0) = 0$ and Y is non-negative and P-measurable

(1.8) $\sum_{i=1}^\infty E^*X_i^- < \infty \ \Rightarrow\ E^*\Big(\sum_{i=1}^\infty X_i\Big) \leq \sum_{i=1}^\infty E^*X_i$

(1.9) $E^*\Big(\sum_{i=1}^\infty 1_{F_i} X_i\Big) \geq \sum_{i=1}^\infty E^*(1_{F_i} X_i)$ if $F_1, F_2, \ldots \in \mathcal{F}_P$ are mutually disjoint where all the sums and differences in (1.4)–(1.9) are the upper sums and differences as defined above.

Measurable spaces and functions. Let T be given set. Then 2^T will denote the set of all subsets of T and a *paving* on T is subset of 2^T. If S is a set, then S^T will denote the set of all functions $f : T \to S$. If $\mathcal{A}$ is a paving on T, we let $\sigma(\mathcal{A})$ denote the smallest σ-algebra on T containing $\mathcal{A}$ and if $A \subseteq T$, we let $A^c := T \setminus A$ denote the complement of A. Let $(M, \mathcal{B})$ be a measurable space. Then a measurable function $X : (\Omega, \mathcal{F}_P) \to (M, \mathcal{B})$ will be called a *random function.* If $X : \Omega \to (M, \mathcal{B})$ is a random function, we let P_X denote its *distribution*; that is, the probability measure P_X on $(M, \mathcal{B})$ given by $P_X(B) := \bar{P}(X \in B)$ for all $B \in \mathcal{B}$, and we let $\mathcal{B}_X$ denote the σ-algebra of all P_X-measurable subsets of M. Recall that $(M, \mathcal{B})$ is *countably separated* if there exists an injective measurable function $f : (M, \mathcal{B}) \to \mathbf{R}$ (see [4; Exc. 1.83–89 p. 91–93]), that $(M, \mathcal{B})$ is *discrete* if $\mathcal{B} = \sigma(B_1, B_2, \ldots)$ for some sequence $B_1, B_2, \ldots$ of mutually disjoint sets, and that $(M, \mathcal{B})$ is a *Blackwell space* if $f(M)$ is an analytic subset of $\mathbf{R}$ for every measurable function $f : (M, \mathcal{B}) \to \mathbf{R}$.

Measurable linear spaces. Let Λ be a vector space. Then we let Λ^* denote the set of all linear functionals $\xi : \Lambda \to \mathbf{R}$. Then Λ^* is a linear space and if $\Xi \subseteq \Lambda^*$, we let $\operatorname{span} \Xi$ denote *the linear span* of Ξ and we let $\operatorname{aff} \Xi$ denote *the affine span* of Ξ; that is, the set of all functions of the form $\psi(x) = \xi(x) + a$ for some function $\xi \in \operatorname{span} \Xi$ and some number $a \in \mathbf{R}$. Recall that $(\Lambda, \mathcal{L})$ is a *measurable linear space* if $\mathcal{L}$ is a σ-algebra on Λ such that the addition map $(u, v) \to u + v$ is measurable from $(\Lambda \times \Lambda, \mathcal{L} \otimes \mathcal{L})$ into $(\Lambda, \mathcal{L})$ and the scalar multiplication $(\alpha, u) \to \alpha u$ is measurable from $(\mathbf{R} \times \Lambda, \mathcal{B}(\mathbf{R}) \otimes \mathcal{L})$ into $(\Lambda, \mathcal{L})$. If $\Xi \subseteq \Lambda^*$ is a set of linear functions on Λ and $\sigma(\Xi)$ denotes the smallest σ-algebra on Λ making all functions in Ξ measurable, then $(\Lambda, \sigma(\Xi))$ is a measurable linear space. If $(\Lambda_i, \mathcal{L}_i)$ is a measurable linear space for all $i \in I$, then the product space $(\prod_{i \in I} \Lambda_i, \otimes_{i \in I} \mathcal{L}_i)$ is a measurable linear space.

Convex and subadditive functions. Let Λ be a real vector space. If $Q : \Lambda \to \bar{\mathbf{R}}$ is a given function, we say that Q is *subadditive* if $Q(u + v) \leq Q(u) + Q(v)$ for all $u, v \in \Lambda$ (with our convention $\infty - \infty := \infty$), we say that Q is *convex* if $Q(u) > -\infty$ for all $u \in \Lambda$ and $Q(\alpha u + (1 - \alpha) v) \leq \alpha Q(u) + (1 - \alpha) Q(v)$ for all $u, v \in \Lambda$ and all $\alpha \in [0, 1]$, and we say that Q is *quasi-convex* if $Q((x + y)/2) \leq \max(Q(x), Q(y))$ for all $x, y \in \Lambda$. If $\Xi \subseteq \Lambda^*$ is a non-empty set, we say that Q is Ξ*-convex* if $Q(u) = \sup_{\psi \in \Psi} \psi(u)$ for some non-empty set $\Psi \subseteq \operatorname{aff} \Xi$. If $\Xi \subseteq \Lambda^*$, then *the orthogonal complement* $\Xi^{\perp} := \{u \in \Lambda \mid \xi(u) = 0 \ \forall \xi \in \Xi\}$ is a linear subspace of Λ and we have the following well-known facts:

(1.10) If Q is Ξ-convex, then Q is a convex function such that $Q(u) = Q(v)$ for all $u, v \in \Lambda$ satisfying $u - v \in \Xi^{\perp}$

(1.11) Let $[\alpha, \beta] \subseteq \bar{\mathbf{R}}$ be given (possibly infinite) interval and let $f : [\alpha, \beta] \to \bar{\mathbf{R}}$ be a nondecreasing continuous function such that f is finite and convex on the open interval (α, β). If Q is Ξ-convex and $C \subseteq \Lambda$ is a given set satisfying $Q(u) \in [\alpha, \beta]$ for all $u \in C$, then there exists a Ξ-convex function $Q_0 : \Lambda \to \bar{\mathbf{R}}$ satisfying $Q_0(u) = f(Q(u))$ for all $u \in C$

(1.12) If Λ is a locally convex space, then Q is lower semicontinuous and convex if and only if Q is Λ'-convex where Λ' denotes the set of all continuous linear functionals

The Pettis mean. Let Λ be a vector space and let $\Xi \subseteq \Lambda^*$ be a non-empty set. If $X : \Omega \to \Lambda$ is a given function, we say that X is *Pettis Ξ-integrable* if $\xi(X) \in L^1(P)$ for all $\xi \in \Xi$ and there exists a vector $a \in \Lambda$ such that $\xi(a) = E\xi(X)$ for all $\xi \in \Xi$. If X is Pettis Ξ-integrable, we let

(1.13) $\mathcal{E}_\Xi(X) := \{ a \in \Lambda \mid E\xi(X) = \xi(a) \ \forall \xi \in \Xi \}$

denote the set of all Ξ-*means* of X. If X is Pettis Ξ-integrable, then we have

(1.14) $\Xi^\perp = \mathcal{E}_\Xi(X) - \mathcal{E}_\Xi(X)$ and $\mathcal{E}_\Xi(X) = a + \Xi^\perp \quad \forall a \in \mathcal{E}_\Xi(X)$

(1.15) $\psi(X) \in L^1(P)$ and $E\psi(X) = \psi(a) \quad \forall \psi \in \operatorname{aff} \Xi \ \forall a \in \mathcal{E}_\Xi(X)$

In particular, we see that the set of Ξ-means of X is an affine subspace of Λ and that X admits a unique Ξ-mean if and only if $\Xi^\perp = \{0\}$.

Stochastic processes. Let T be a non-empty set. Then $\mathbf{R}^T$ is a vector space and we let $\mathcal{B}^T := \otimes_{t \in T} \mathcal{B}(\mathbf{R})$ denote the product σ-algebra on $\mathbf{R}^T$. Then $(\mathbf{R}^T, \mathcal{B}^T)$ is a measurable linear space and the projections $\pi_t(f) := f(t)$ are linear functionals on $\mathbf{R}^T$ for all $t \in T$. Let $\sup_T(f) := \sup_{t \in T} f(t)$ denote *the sup-functional* and let $\|f\|_T := \sup_{t \in T} |f(t)|$ denote *the sup-norm* on $\mathbf{R}^T$. Then $\sup_T$ and $\|\cdot\|_T$ are subadditive and Π-convex where $\Pi := \{\pi_t \mid t \in T\}$ denotes the set of all projections. Recall that $(X(t) \mid t \in T)$ is a *stochastic process* if and only if $X(\omega) := X(\omega, \cdot)$ is a random function from Ω into $(\mathbf{R}^T, \mathcal{B}^T)$. Let $X : \Omega \to \mathbf{R}^T$ be a stochastic process. Then X is Pettis Π-integrable if and only if $X(t) \in L^1(P)$ for all $t \in T$ and if so, then we have $\mathcal{E}_\Pi(X) = \{\mu\}$ where $\mu(t) := EX(t)$ is *the mean value function* of X.

Saturated and measurable sets. Let M be a given set and let $X : \Omega \to M$ be a given function without any specified measurability properties. Then we let $\mathcal{S}_P(X)$ denote the paving of all P-a.s. X-*saturated* subsets of Ω; that is, $F \in \mathcal{S}_P(X)$ if and only if there exists $A \subseteq M$ such that $F = X^{-1}(A)$ P-a.s. Then $\mathcal{S}_P(X)$ is a σ-algebra on Ω such that $X : (\Omega, \mathcal{S}_P(X)) \to (M, 2^M)$ is measurable. We let $\mathcal{M}_P(X)$ denote the largest σ-algebra on M making X P-measurable; that is, $A \in \mathcal{M}_P(X)$ if and only if $X^{-1}(A) \in \mathcal{F}_P$. We let $\mathcal{K}_P(X)$ denote the set of all P-kernels of sets in $\mathcal{S}_P(X)$, we let $\mathcal{H}_P(X)$ denote the set of all P-hulls of sets in $\mathcal{S}_P(X)$ and we define the following pavings:

$$\mathcal{L}_P(X) := \sigma(\mathcal{H}_P(X)) = \sigma(\mathcal{K}_P(X)), \quad \mathcal{C}_P(X) := \sigma(\mathcal{L}_P(X) \cup \mathcal{S}_P(X))$$

$$\mathcal{G}_P(X) := \left\{ \bigcup_{i=1}^{\infty} (A_i \cap B_i) \,\middle|\, A_i \in \mathcal{L}_P(X), \;\; B_i \in \mathcal{S}_P(X) \; \forall i \geq 1 \right\}$$

We let $\mathcal{A}_P(X)$ denote the set of all sets $A \in \mathcal{G}_P(X)$ satisfying $A^c \in \mathcal{G}_P(X)$. If $X : \Omega \to M$ and $Y : \Omega \to L$ are given functions, then we have the following trivial facts:

(1.16) $\mathcal{K}_P(X)$ is stable under countable intersections, $\mathcal{H}_P(X)$ is stable under countable unions and $K \in \mathcal{K}_P(X)$ if and only if $K^c \in \mathcal{H}_P(X)$

(1.17) $X^{-1}(\mathcal{M}_P(X)) \subseteq \mathcal{H}_P(X) \subseteq \mathcal{L}_P(X) \subseteq \mathcal{F}_P$

(1.18) If $(L, \mathcal{A})$ is countably separated space such that $Y : (\Omega, \mathcal{S}_p(X)) \to (L, \mathcal{A})$ is measurable, then there exists a function $f : M \to L$ such that $Y = f(X)$ a.s.

(1.19) If $f : M \to L$ is a given function such that $Y = f(X)$ a.s., then we have

$$\mathcal{S}_P(Y) \subseteq \mathcal{S}_P(X) \;,\;\; \mathcal{H}_P(Y) \subseteq \mathcal{H}_P(X) \;,\;\; \mathcal{M}_P(Y) = f^{-1}(\mathcal{M}_P(X))$$

(1.20) $\mathcal{G}_P(X)$ is stable under countable unions and finite intersections and $\mathcal{A}_P(X)$ is an algebra containing $\mathcal{L}_P(X) \cup \mathcal{S}_P(X)$. If $A_1, A_2, \ldots \in \mathcal{L}_P(X)$ and $S_1, S_2, \ldots \in \mathcal{S}_P(X)$ are given sets such that either $A_1, A_2, \ldots$ are mutually disjoint or $S_1, S_2, \ldots$ are mutually disjoint, then we have $\cup_{i=1}^{\infty} (A_i \cap S_i) \in \mathcal{A}_P(X)$

Perfect functions. Let $(M, \mathcal{B})$ be a measurable space and let $X : \Omega \to (M, \mathcal{B})$ be random functions with distribution P_X. Then we let P_X^* and $(P_X)_*$ denote the outer and inner P_X-probabilities and we let $(E_X)_*$ and E_X^* denotes the lower and upper P_X-expectations. Then we have (see [1] and [9])

(1.21) $(P_X)_*(A) \leq P_*(X^{-1}(A)) \leq P^*(X^{-1}(A)) \leq P_X^*(A) \quad \forall A \subseteq M$

(1.22) $(E_X)_*(f) \leq E_* f(X) \leq E^* f(X) \leq E_X^* f \quad \forall f \in \bar{\mathbf{R}}^M$

If $X : \Omega \to M$ is a given function, we say that X is $(P, \mathcal{B})$-*perfect* if $X : \Omega \to (M, \mathcal{B})$ is P-measurable and $P^*(X^{-1}(A)) = P_X^*(A)$ for all $A \subseteq M$. Let $(L, \mathcal{A})$ and $(M, \mathcal{B})$ be measurable spaces and let $X : \Omega \to M$ and $f : M \to L$ be given functions. Then we have

(1.23) If X is $(P, \mathcal{B})$-perfect and f is $(P_X, \mathcal{A})$-perfect, then $f(X)$ is $(P, \mathcal{A})$-perfect

(1.24) If $(M, \mathcal{B})$ is discrete, then X is $(P, \mathcal{B})$-perfect if and only if $X : \Omega \to (M, \mathcal{B})$ is P-measurable and $P^*(X = u) = P_X(\Lambda_u)$ for all $u \in M$ where Λ_u denotes the unique $\mathcal{B}$-atom containing u

(1.25) If X is $(P,\mathcal{B})$-perfect, then we have

$$\mathcal{L}_P(X) = \mathcal{F}_P \cap \mathcal{S}_P(X) \ , \ \mathcal{S}_P(X) = \mathcal{C}_P(X) \text{ and } \mathcal{M}_P(X) = \mathcal{B}_X$$

$$(E_X)_*(f) = E_*f(X) \leq E^*f(X) = E^*_X f \quad \forall f \in \bar{\mathbf{R}}^M$$

(1.26) If $(\Omega,\mathcal{F})$ is a Blackwell space and $(M,\mathcal{B})$ is countably separated, then every measurable function $Y : (\Omega,\mathcal{F}) \to (M,\mathcal{B})$ is $(P,\mathcal{B})$-perfect

(1.27) If $X : \Omega \to (M,\mathcal{B})$ is P-measurable and injective, then X is $(P,\mathcal{B})$-perfect $\Leftrightarrow \mathcal{M}_P(X) \subseteq \mathcal{B}_X \Leftrightarrow X(F) \in \mathcal{B}_X \ \forall F \in \mathcal{F} \Leftrightarrow X(\Omega) \in \mathcal{B}_X$ and $\mathcal{F} \subseteq X^{-1}(\mathcal{B}_X)$

Theorem 1.1 (M. Talagrand; see [3]). *Let $(\Omega_i,\mathcal{F}_i,P_i)$ be a given probability space for all i in the non-empty set I . Let $\alpha \subseteq I$ be a non-empty set, let $\Omega^\alpha := \prod_{i\in\alpha} \Omega_i$ denote the product space, let $\mathcal{F}^\alpha := \otimes_{i\in\alpha} \mathcal{F}_i$ denote the product σ-algebra on Ω^α and let $P^\alpha := \otimes_{i\in\alpha} P_i$ denote the product probability on $(\Omega^\alpha,\mathcal{F}^\alpha)$. If $\pi^\alpha : \Omega \to \Omega^\alpha$ denotes the projection map $\pi^\alpha(\omega) := (\omega_i)_{i\in\alpha}$, then π^α is $(P^I,\mathcal{F}^\alpha)$-perfect with distribution P^α and if α is countable and $A_i \subseteq \Omega_i$ are given sets for all $i \in \alpha$, then we have*

$$(P^\alpha)^*\Big(\prod_{i\in\alpha} A_i\Big) = \prod_{i\in\alpha} P_i^*(A_i) \ , \ (P^\alpha)_*\Big(\prod_{i\in\alpha} A_i\Big) = \prod_{i\in\alpha} (P_i)_*(A_i)$$

2. Independence notions

Let $(M_i \mid i \in I)$ be a non-empty family of non-empty sets and let $(X_i : \Omega \to M_i \mid i \in I)$ be a family of functions. If $\mathcal{B}_i$ is a σ-algebra on M_i for all $i \in I$, we say that $(X_i : \Omega \to (M_i,\mathcal{B}_i) \mid i \in I)$ are *independent* if $X_i : \Omega \to (M_i,\mathcal{B}_i)$ is P-measurable for all $i \in I$ and the random functions (X_i) are independent in the usual sense; that is, for every non-empty finite set $\alpha \subseteq I$ we have

(2.1) $\bar{P}(X_i \in B_i \ \forall i \in \alpha) = \prod_{i\in\alpha} \bar{P}(X_i \in B_i)$ if $B_i \in \mathcal{B}_i \ \forall i \in \alpha$

Suppose that X_i is an arbitrary function without any specified measurability properties. Then it is tempting to say that (X_i) is "independent" if (2.1) holds with $\bar{P}$ replaced by P^* whenever $B_i \subseteq M_i$ for all $i \in \alpha$. However, this definition of "independence" doesn't seem to be useful our context and below you find three definitions of "independence" which are much more appropriate for extending the classical stochastic inequalities to arbitrary functions.

Independence notions. Suppose that we have defined a notion of "independence" for non-empty families of functions $(X_i : \Omega \to M_i \mid\in I)$. Then we say that the notion of "independence" is *consistent* if it satisfies the following consistency condition:

(2.2) Let $(I_\theta \mid \theta \in \Theta)$ be a non-empty family of mutually disjoint non-empty subsets of the set I. Let $(L_\theta \mid \theta \in \Theta)$ be a family of non-empty sets and let $f_\theta : M^\theta \to L_\theta$ be given function for all $\theta \in \Theta$ where $M^\theta := \prod_{i \in I_\theta} M_i$. If $(X_i : \Omega \to M_i \mid i \in I)$ are "independent", then the functions $(f_\theta(X^\theta) : \Omega \to L_\theta \mid \theta \in \Theta)$ are "independent" where $X^\theta : \Omega \to M^\theta$ is the function given by $X^\theta(\omega) := (X_i(\omega))_{i \in I_\theta}$ for all $\omega \in \Omega$ and all $\theta \in \Theta$

If $G, H \subseteq \Omega$ are given sets, we say that G and H are "independent" if the indicator functions $1_G : \Omega \to \{0,1\}$ and $1_H : \Omega \to \{0,1\}$ are "independent" and if $\mathcal{G}$ and $\mathcal{H}$ are pavings on Ω, we say $\mathcal{G}$ and $\mathcal{H}$ are "independent" if G and H are "independent" for all $G \in \mathcal{G}$ and all $H \in \mathcal{H}$.

Independently defined functions. Let M_i be a given non-empty set and let $X_i : \Omega \to M_i$ be a given function for all i in the non-empty set I. Then we say that the functions $(X_i : \Omega \to M_i \mid i \in I)$ are *independently defined* if there exist probability spaces $(\tilde{\Omega}, \tilde{\mathcal{F}}, \tilde{P})$ and $(\Omega_i, \mathcal{F}_i, P_i)$ and functions $\hat{X}_i : \Omega_i \to M_i$ for $i \in I$ satisfying

(2.3) $\Omega = \tilde{\Omega} \times \prod_{i \in I} \Omega_i \ , \quad \mathcal{F} = \tilde{\mathcal{F}} \otimes \bigotimes_{i \in I} \mathcal{F}_i \ , \quad P = \tilde{P} \otimes \bigotimes_{i \in I} P_i$

(2.4) $X_i(\omega) = \hat{X}_i(\omega_i) \quad \forall \omega = (\tilde{\omega}, (\omega_\nu)_{\nu \in I}) \in \Omega \, \forall i \in I$

Observe that the concept "independently defined" imposes an unpleasant restriction on the form of the underlying probability space $(\Omega, \mathcal{F}, P)$ and below we shall we shall remove this restriction.

Perfect independence. Let M_i be a given non-empty set and let $X_i : \Omega \to M_i$ be a given function for all i in the non-empty set I. Then we say that the functions $(X_i : \Omega \to M_i \mid i \in I)$ are *perfectly independent* if there exist measurable spaces $(\Omega_i, \mathcal{F}_i)$ and functions $\xi_i : \Omega \to \Omega_i$ and $\hat{X}_i : \Omega_i \to M_i$ satisfying:

(2.5) The functions $(\xi_i : \Omega \to (\Omega_i, \mathcal{F}_i) \mid i \in I)$ are measurable and independent and we have $X_i(\omega) = \hat{X}_i(\xi_i(\omega))$ for all $\omega \in \Omega$ and all $i \in I$

(2.6) The function $\xi(\omega) := (\xi_i(\omega))_{i \in I}$ from Ω into the product space $(\Omega^I, \mathcal{F}^I)$ is $(P, \mathcal{F}^I)$-perfect where $\Omega^I := \prod_{i \in I} \Omega_i$ and $\mathcal{F}^I := \otimes_{i \in I} \mathcal{F}_i$

Fubini independence. If $A, B \subseteq \Omega$ are given sets, we say that A and B are *non-negatively correlated* if $P^*(A \cap B) \geq P^*(A)\, P^*(B)$. If $A, B \in \mathcal{F}_P$ then A and B are non-negatively correlated if and only if $\mathrm{cov}(1_A, 1_B) \geq 0$. Let L and M be two given sets and let $X : \Omega \to L$ and $Y : \Omega \to M$ be given functions. If $\psi : L \times M \to \bar{\mathbf{R}}$ is a given function, we say that X and Y are *non-negatively ψ-correlated* if the following Fubini inequality holds:

(2.7) $E^*\psi(X,Y) \geq E^*\phi(X)$ where $\phi(x) := E^*\psi(x,Y) \ \forall x \in L$

If $A \subseteq L \times M$, we say that X and Y are non-negatively A-correlated if X and Y are non-negatively 1_A-correlated; that is, if

$$(2.8) \quad P^*((X,Y) \in A) \geq E^*\phi(X) \text{ where } \phi(x) := P^*((x,Y) \in A) \ \forall x \in L$$

We say that X and Y are *Fubini independent* if X and Y are non-negatively ψ-correlated for every function $\psi : L \times M \to \bar{\mathbf{R}}$. Note that if $X : \Omega \to (L, \mathcal{A})$ and $Y : \Omega \to (M, \mathcal{B})$ are independent random functions, then X and Y are non-negatively ψ-correlated for every $P_{X,Y}$-measurable function $\psi : L \times M \to \bar{\mathbf{R}}$.

Note that the notion "Fubini independent" is non-symmetric in (X,Y) and I don't know if X and Y are Fubini independent implies that Y and X are Fubini independent in general. However, if L and M are countable sets, then Thm. 2.5 shows that this indeed so.

Let M_i be a given set and let $X_i : \Omega \to M_i$ be a given function for all i in the non-empty set I . If $\alpha \subseteq I$ is a non-empty subset of I , we let $M^\alpha := \prod_{i \in \alpha} M_i$ denote the product space and we let $X^\alpha(\omega) := (X_i(\omega))_{i \in \alpha}$ denote *the α-marginal* of X . Then X^α is a function from Ω into the product space M^α and we say that the functions $(X_i : \Omega \to M_i \mid i \in I)$ are *Fubini independent* if X^α and X^β are Fubini independent whenever α and β are disjoint non-empty subsets of I . Note that if $I = \{1,2\}$, then (X_1, X_2) are Fubini independent if and only if X_1 and X_2 are Fubini independent and X_2 and X_1 are Fubini independent.

Theorem 2.1. *Let $X : \Omega \to L$ and $Y : \Omega \to M$ be given functions, let $\mathcal{G} \subseteq 2^{L \times M}$ be paving and let $W(\mathcal{G})$ denote the set of all functions $\psi : L \times M \to \bar{\mathbf{R}}$ satisfying*

$$\forall -\infty < s < t < \infty \ \exists G \in \mathcal{G}^* \text{ so that } \{\psi > t\} \subseteq G \subseteq \{\psi > s\}$$

where $\mathcal{G}^* := \mathcal{G} \cup \{\emptyset, L \times M\}$.

(A) If X and Y are non-negatively G-correlated for all $G \in \mathcal{G}$, then X and Y are non-negatively ψ-correlated for all $\psi \in W(\mathcal{G})$.

(B) X and Y are Fubini independent if and only if X and Y are non-negatively D-correlated for all $D \subseteq L \times M$.

(C) If $f : L \to M$ is a given function, then the following three statements are equivalent:

(1) *X and $f(X)$ are Fubini independent*

(2) *$f(X)$ and X are Fubini independent*

(3) $\exists b \in M$ so that $f(X) = b$ a.s.

(D) The independence notions "independently defined" and "perfectly independent" are consistent independence notions and if (X_i) are independently defined, then (X_i) are perfectly independent.

Proof. (A) Note that X and Y are non-negatively $\emptyset$-correlated and $L \times M$-correlated. Hence, without loss of generality, we may assume that $\mathcal{G} = \mathcal{G}^*$. Let $\psi \in W(\mathcal{G})$ be a given function such that $\psi(x,y) \geq 0$ for all $(x,y) \in L \times M$ and let us define $\phi(x) := E^*\psi(x,Y)$ for $x \in L$. Let $n \geq 1$ be a given integer.

Since $\psi \in W(\mathcal{G})$ and $\mathcal{G} = \mathcal{G}^*$, there exists $G_i \in \mathcal{G}$ such that $\{\psi > \frac{i+1}{n}\} \subseteq G_i \subseteq \{\psi > \frac{i}{n}\}$ for all $i = 0, 1, \dots$. Let us define $\phi_i(x) := P^*((x,Y) \in G_i)$ for $x \in L$ and $i = 0, 1, 2, \dots$. Since $t \to P^*(\psi(x,Y) > t)$ is nonincreasing and $\{\psi > \frac{i+1}{n}\} \subseteq G_i$, we have by (1.6)

$$\phi(x) = \int_0^\infty P^*(\psi(x,Y) > t)\, dt \le \tfrac{1}{n} \sum_{i=0}^\infty P^*(\psi(x,Y) > \tfrac{i}{n})$$

$$\le \tfrac{2}{n} + \tfrac{1}{n} \sum_{i=1}^\infty P^*(\psi(x,Y) > \tfrac{i+1}{n}) \le \tfrac{2}{n} + \tfrac{1}{n} \sum_{i=1}^\infty \phi_{i+1}(x)$$

Since $G_{i+1} \in \mathcal{G}$, we have that X and Y are non-negatively G_{i+1}-correlated and since $G_{i+1} \subseteq \{\psi > \frac{i+1}{n}\}$, we have

$$E^* \phi_{i+1}(X) \le P^*((X,Y) \in G_{i+1}) \le P^*(\psi(X,Y) > \tfrac{i+1}{n})$$

Since $t \to P^*(1_H \psi(X,Y) > t)$ is nondecreasing, we have by (1.6)

$$E^* \phi(X) \le \tfrac{2}{n} + \tfrac{1}{n} \sum_{i=1}^\infty E^* \phi_{i+1}(X) \le \tfrac{2}{n} + \tfrac{1}{n} \sum_{i=1}^\infty P^*(\psi(X,Y) > \tfrac{i+1}{n})$$

$$\le \tfrac{2}{n} + \int_0^\infty P^*(\psi(X,Y) > t)\, dt = \tfrac{2}{n} + E^* \psi(X,Y)$$

Letting $n \to \infty$, we see that $E^*\phi(X) \le E^*\psi(X,Y)$ for every non-negative function $\psi \in W(\mathcal{G})$. Let ψ be any function in $W(\mathcal{G})$, let $a \in \mathbf{R}$ be a given number and let us define $\phi(x) := E^*\psi(x,Y)$, $\psi_a(x,y) := (\psi(x,y) \vee a) - a$ and $\phi_a(x) := E^*\psi_a(x,Y)$ for all $x \in L$ and all $y \in M$. Then we have $\phi_a(x) + a = E^*(\psi(x,Y) \vee a) \ge \phi(x)$ and since $\psi \in W(\mathcal{G})$, we have $\psi_a \in W(\mathcal{G})$. Since $\psi_a \ge 0$, then by the argument above we have

$$E^*(\phi(X,Y) \vee a) = E^*(\psi_a(X,Y) + a) = a + E^*\psi_a(X,Y)$$

$$\ge a + E^*\phi_a(X) = E^*(a + \psi_a(X)) \ge E^*\phi(X)$$

for all $a \in \mathbf{R}$. Hence, by (1.5) we see that $E^*\phi(X) \le E^*\psi(X,Y)$ for all $\psi \in W(\mathcal{G})$ which proves (A).

(B) follows from (A).

(C) Suppose that (1) holds and let us define $D := \{(x,y) \in L \times M \mid y \neq f(x)\}$. Since X and $f(X)$ are Fubini independent and $\{(X, f(X)) \in D\} = \emptyset$, we have $E^*\phi(X) = 0$ where $\phi(x) := P^*((x, f(X)) \in D)$. Since $\phi \ge 0$, we have $\phi(X) = 0$ a.s. In particular, there exists at least one point $a \in L$ such that $\phi(a) = 0$ and since $\phi(a) = P^*(f(X) \neq f(a))$, we see that $f(X) = f(a)$ a.s. Hence, we see that (1) implies (3). Suppose that (2) holds and let us define $D := \{(y,x) \in M \times L \mid y \neq f(x)\}$. As above we see that there exists $b \in M$ such that $P^*(f(X) \neq b) = P^*((b,X) \in D) = 0$; that is, (2) implies (3). The implications "(3) ⇒ (1)" and "(3) ⇒ (2)" are evident.

(D) is an easy consequence of Thm. 1.1. □

Theorem 2.2. *Let* $X : \Omega \to L$ *and* $Y : \Omega \to M$ *be given functions and let us define* $\psi^\diamond(x) := E^*\psi(x, Y)$ *for all* $x \in L$ *whenever* $\psi : L \times M \to \bar{\mathbf{R}}$ *is a given function.*

(A) If $H \subseteq \Omega$ *is a given set such that* $(1_H, X)$ *and* Y *are Fubini independent, then we have*

$$(1) \qquad E^*\{1_H\psi(X,Y)\} \geq E^*\{1_H\psi^\diamond(X)\} \ , \ E^*\{1_{H^c}\psi(X,Y)\} \geq E^*\{1_{H^c}\psi^\diamond(X)\}$$

for all $\psi \in \bar{\mathbf{R}}^{L\times M}$. *Conversely, if* $H \in \mathcal{F}_P$ *and (1) holds for every indicator function* $\psi = 1_D$, *then* $(1_H, X)$ *and* Y *are Fubini independent.*

(B) If $\mathcal{L} \subseteq \mathcal{F}_P$ *is an algebra such that* $(1_L, X)$ *and* Y *are Fubini independent for all* $L \in \mathcal{L}$, *then* (ξ, X) *and* Y *are Fubini independent for every* $\sigma(\mathcal{L})$*-measurable, positive integer-valued function* $\xi : \Omega \to \mathbf{N}$.

(C) If X *and* Y *are Fubini independent and* $U : \Omega \to L_0$ *and* $V : \Omega \to M_0$ *are given functions such that* $U = f(X)$ *a.s. and* $V = h(Y)$ *a.s. for some functions* $f : M \to M_0$ *and* $h : L \to L_0$, *then* U *and* V *are Fubini independent.*

(D) Let $(S, \mathcal{A})$ *be a countably separated measurable space and let* $\xi : (\Omega, \mathcal{S}_P(X)) \to (S, \mathcal{A})$ *be a measurable function. If* X *and* Y *are Fubini independent, then* (ξ, X) *and* Y *are Fubini independent.*

Proof. (A): If $(1_H, X)$ and Y are Fubini independent, then (1) follows from the (2.7) applied to the functions $\psi_0(\delta, x, y) := \delta\,\psi(x,y)$ and $\psi_1(\delta,x,y) := (1-\delta)\,\psi(x,y)$ for $(\delta,x,y) \in \{0,1\} \times L \times M$. So suppose that $H \in \mathcal{F}_P$ and that (1) holds for all indicators functions. Let $D \subseteq \{0,1\} \times L \times M$ be given and let us define $D_\delta := \{(x,y) \mid (\delta,x,y) \in D\}$ and $\phi(\delta,x) := P^*((\delta,x,Y) \in D)$ for $\delta \in \{0,1\}$ and $x \in L$. Then we have $D = (\{0\} \times D_0) \cup (\{1\} \times D_1)$ and since

$$\{(1_H, X, Y) \in D\} = \big(H \cap \{(X,Y) \in D_1\}\big) \cup \big(H^c \cap \{(X,Y) \in D_0\}\big)$$

and $H \in \mathcal{F}_P$, we have by (1), (1.8) and (1.9)

$$\begin{aligned} P^*((1_H, X, Y) \in D) &= P^*(H \cap \{(X,Y) \in D_1\}) + P^*(H^c \cap \{(X,Y) \in D_0\}) \\ &\geq E^*\{1_H\phi(1,X)\} + E^*\{1_{H^c}\phi(0,X)\} \\ &= E^*\{1_H\phi(1,X) + 1_{H^c}\phi(0,X)\} = E^*\phi(1_H, X) \end{aligned}$$

Hence, we see that $(1_H, X)$ and Y are non-negatively D-correlated for all $D \subseteq \{0,1\} \times L \times M$. So by Thm. 2.1 we see that $(1_H, X)$ and Y are Fubini independent.

(B) Let $D \subseteq L \times M$ be a given set and let us define $C := \{(X,Y) \in D\}$ and $\phi(x) := P^*((x,Y) \in D)$. If we define $\mu(F) := P^*(F \cap C)$ and $\nu(F) := E^*1_F\,\phi(X)$ for all $F \in \mathcal{F}_P$. then by (1.8) and (1.9) we see that μ and ν are finite measures on $(\Omega, \mathcal{F}_P)$. Since $\mathcal{L} \subseteq \mathcal{F}_P$ and $(1_L, X)$ and Y are Fubini independent, then by (1) we have $\mu(L) \geq \nu(L)$ for all $L \in \mathcal{L}$ and since $\mathcal{L}$ is an algebra, then by [4; Exc. 1.48 p. 77] we have $P^*(L \cap C) = \mu(L) \geq \nu(L) = E^*(1_L\,\phi(X))$ for all $L \in \sigma(\mathcal{L})$ and all $D \subseteq L \times M$. Let $\xi : (\Omega, \sigma(\mathcal{L})) \to \mathbf{N}$ be a given measurable function and let $D \subseteq \mathbf{N} \times L \times M$ be a given set. Let us define $\phi(n,x) := P^*((n,x,Y) \in D)$

for $(n,x) \in \mathbf{N} \times L$, $D_n := \{(x,y) \mid (n,x,y) \in D\}$ and $L_n := \{\xi = n\}$ for $n \in \mathbf{N}$. Since $L_n \in \sigma(\mathcal{L})$, we have $P^*(L_n \cap \{(X,Y) \in D_n) \geq E^*(1_{L_n}\,\phi(n,X))$ for all $n \geq 1$ and since $L_1, L_2, \ldots \in \mathcal{F}_P$ is disjoint partition of Ω , we have by (1.8) and (1.9)

$$P^*((\xi,X,Y) \in D) = \sum_{n=1}^{\infty} P^*(L_n \cap \{(X,Y) \in D_n\}) \geq \sum_{n=1}^{\infty} E^*\{1_{L_n}\,\phi(n,X)\}$$
$$= E^*\Big(\sum_{n=1}^{\infty} 1_{L_n}\,\phi(n,X)\Big) = E^*\phi(\xi,X)$$

Hence, we see that (ξ,X) and Y are non-negatively D-correlated for all $D \subseteq \mathbf{N} \times L \times M$. So by Thm. 2.1 we see that (ξ,X) and Y are Fubini independent.

(C) Let $\psi_0 : L_0 \times M_0 \to \bar{\mathbf{R}}$ be a given function and let us define $\phi_0(u) := E^*\psi_0(u,V)$ for $u \in L_0$ and $\phi(x) := E^*\psi(x,Y)$ for $x \in L$ where $\psi(x,y) := \psi_0(f(x),h(y))$ for all $(x,y) \in L \times M$. Since $V = h(Y)$ a.s. we have $\phi_0(f(x)) = \phi(x)$ for all $x \in L$ and since $U = f(X)$ a.s., we have $\phi_0(U) = \phi(X)$ a.s. Since X and Y are Fubini independent and $(U,V) = (f(X),h(Y))$ a.s., we have $E^*\psi_0(U,V) = E^*\psi(X,Y) \geq E^*\phi(X) = E^*\phi_0(U)$. Hence, we see that U and V are Fubini independent.

(D) Since $(S,\mathcal{A})$ is countably separated and $\xi : (\Omega, \mathcal{S}_P(X)) \to (S,\mathcal{A})$ is measurable, then by (1.18) there exists a function $f : L \to S$ such that $\xi = f(X)$ a.s. Hence, by (C) we see that (ξ,X) and Y are Fubini independent. □

Theorem 2.3. *Let $X : \Omega \to L$ and $Y : \Omega \to M$ be functions and let $(L_0,\mathcal{A}_0)$ and $(M_0,\mathcal{B}_0)$ be countably separated measurable spaces. Let I be a non-empty set and let $(X_i, Y_i : \Omega \to M_i \mid i \in I)$ be given functions such that $X_i = Y_i$ a.s. for all $i \in I$.*

(A) If X and Y are perfectly independent, then X and Y are Fubini independent and Y and X are Fubini independent.

(B) The independence notion "Fubini independent" is consistent and if (X_i) are perfectly independent, then (X_i) are Fubini independent.

(C) If I is countable and the functions (X_i) are perfectly independent or Fubini independent, then so are the functions (Y_i) .

(D) If X and Y are perfectly independent and $X_0 : (\Omega, \mathcal{C}_P(X)) \to (L_0,\mathcal{A}_0)$ and $Y_0 : (\Omega, \mathcal{C}_P(Y)) \to (M_0,\mathcal{B}_0)$ are measurable functions, then (X_0,X) and (Y_0,Y) are perfectly independent.

Remark. By (B) and Thm. 2.1 we have: "independently defined" ⇒ "perfectly independent" ⇒ "Fubini independent". In Thm. 2.6, we shall see that Fubini independence implies ordinary independence of the random functions $X_i : \Omega \to (M_i, \mathcal{M}_P(X_i))$. It is easy to construct examples of perfectly independent functions which are *not* independently defined and, using (C) in Thm. 2.1, it is easy to construct examples of ordinarily independent random functions which are *not* Fubini independent. I don't know if the independence notions "perfectly independent" and "Fubini independent" coincide. If X_1 and X_2 only assumes finitely many values, then in Thm. 2.5 we shall see that X_1 and X_2 are Fubini independent if and only if X_1 and X_2 are perfectly independent but I conjecture that this is not so if X_1 and X_2 are arbitrary.

Proof. (A) Since X and Y are perfectly independent, there exist measurable spaces $(\Gamma, \mathcal{G})$ and $(\Upsilon, \mathcal{H})$ and independent random functions $\xi : \Omega \to (\Gamma, \mathcal{G})$ and $\eta : \Omega \to (\Upsilon, \mathcal{H})$ such that $\zeta(\omega) := (\xi(\omega), \eta(\omega))$ is $(P, \mathcal{G} \otimes \mathcal{H})$-perfect and $X(\omega) = U(\xi(\omega))$ and $Y(\omega) = V(\eta(\omega))$ for some functions $U : \Gamma \to L$ and $V : \Upsilon \to L$. Let $\psi : L \times M \to \bar{\mathbf{R}}$ be given and let us define $\phi(x) := E^*\psi(x, Y)$ for $x \in L$ and $\psi_0(\gamma, v) := \psi(U(\gamma), V(v))$ for $(\gamma, v) \in \Gamma \times \Upsilon$. Since ξ is perfect, we have $E^*_\xi \phi(U) = E^*\phi(X)$ and since η is perfect, we have $\phi(U(\gamma)) = E^*_\eta \psi_0(\gamma, \cdot)$ for all $\gamma \in \Gamma$. Since $\zeta = (\xi, \eta)$ is perfect with $P_\zeta = P_\xi \otimes P_\eta$, then by [9; Lem. 1.2.6 p. 11] we have

$$E^*\psi(X, Y) = E^*\psi_0(\xi, \eta) = E^*_\zeta \psi_0 \geq E^*_\xi \phi(U) = E^*\phi(X)$$

which proves (A).

(B) By Thm. 2.2, we see that "Fubini independent" is a consistent independence notion. So suppose that (X_i) are perfectly independent and let α and β be disjoint non-empty subsets of I. Since "perfectly independent" is consistent, we have that the marginals X^α and X^β are perfectly independent. Hence, by (A) we see that X^α and X^β are Fubini independent, that is, (X_i) are Fubini independent.

(C) Suppose that (X_i) are perfectly independent and let $\xi_i : \Omega \to (\Omega_i, \mathcal{F}_i)$ and $\hat{X}_i : \Omega_i \to M_i$ be chosen according to (2.5) and (2.6). Since I is countable there exists a P-null set $N \in \mathcal{F}$ such that $X_i(\omega) = Y_i(\omega)$ for all $\omega \in \Omega \setminus N$ and all $i \in I$. Let M_i^* be chosen such that $\Omega_i \cap M_i^* = \emptyset$ and M_i^* has the same cardinality as M_i for all $i \in I$. Let $f_i : M_i \to M_i^*$ be a bijection and let us define $\tilde{\Omega}_i := \Omega_i \cup M_i^*$ and $\tilde{\mathcal{F}}_i := \{F \subseteq \tilde{\Omega}_i \mid F \cap \Omega_i \in \mathcal{F}_i\}$. Then $(\tilde{\Omega}_i, \tilde{\mathcal{F}}_i)$ is a measurable space and if we define $\tilde{\xi}_i(\omega) := \xi_i(\omega)$ for $\omega \in \Omega \setminus N$ and $\tilde{\xi}_i(\omega) := f_i(Y_i(\omega))$ for $\omega \in N$, then $\tilde{\xi}_i$ is a function from Ω into $\tilde{\Omega}_i$ such that $\tilde{\xi}_i^{-1}(A) \setminus N = \xi_i^{-1}(A \cap \Omega_i)$ for all $A \subseteq \tilde{\Omega}_i$. Since N is a P-null set and $\xi_i \to (\Omega_i, \mathcal{F}_i)$ is P-measurable, we see that $\tilde{\xi}_i : \Omega \to (\tilde{\Omega}_i, \tilde{\mathcal{F}}_i)$ is P-measurable for all $i \in I$. Let us consider the product spaces $(\Omega^I, \mathcal{F}^I) := (\prod_{i \in I} \Omega_i, \otimes_{i \in I} \mathcal{F}_i)$ and $(\tilde{\Omega}^I, \tilde{\mathcal{F}}^I) := (\prod_{i \in I} \tilde{\Omega}_i, \otimes_{i \in I} \tilde{\mathcal{F}}_i)$ and let $\xi(\omega) := (\xi_i(\omega))$ and $\tilde{\xi}(\omega) := (\tilde{\xi}_i(\omega))$ denote the associated random "vectors". Since I is countable and $\Omega_i \in \mathcal{F}_i$, we have $\Omega^I \in \mathcal{F}^I \subseteq \tilde{\mathcal{F}}^I$ and since $\tilde{\xi}_i^{-1}(\Omega_i) = \Omega \setminus N$, we have $P_{\tilde{\xi}_i}(\Omega_i) = 1$ for all $i \in I$. Since I is countable we have $P_{\tilde{\xi}}(\Omega^I) = 1$ and $P_{\tilde{\xi}}(F) = P_\xi(F \cap \Omega^I)$ for all $F \in \tilde{\mathcal{F}}^I$. Since (ξ_i) are independent, we see that $(\tilde{\xi}_i)$ are independent and since $\mathcal{F}^I \subseteq \tilde{\mathcal{F}}^I$, then by (1.27) we see that the injection map $\eta(v) := v$ from Ω^I into $\tilde{\Omega}^I$ is $(P_\xi, \tilde{\mathcal{F}}^I)$-perfect. Since ξ is $(P, \mathcal{F}^I)$-perfect, then by (1.23) we have that $\eta \circ \xi$ is $(P, \tilde{\mathcal{F}}^I)$-perfect and since $\tilde{\xi} = \eta \circ \xi$ a.s. we conclude that $\tilde{\xi}$ is $(P, \tilde{\mathcal{F}}^I)$-perfect. Let us define $\hat{Y}_i(v) := \hat{X}_i(v)$ if $v \in \Omega_i$ and $\hat{Y}_i(v) := f_i^{-1}(v)$

if $v \in M_i^*$. Then $\hat{Y}_i$ maps $\tilde{\Omega}_i$ into M_i and we have

$$Y_i(\omega) = X_i(\omega) = \hat{X}_i(\xi_i(\omega)) = \hat{Y}_i(\tilde{\xi}_i(\omega)) \ \forall \omega \in \Omega \setminus N$$
$$Y_i(\omega) = f_i^{-1}(f_i(Y_i(\omega))) = \hat{Y}_i(\tilde{\xi}_i(\omega)) \ \ \forall \omega \in N$$

Thus, we conclude that (Y_i) are perfectly independent. So suppose that (X_i) are Fubini independent and let X^α and Y^α denote the α-marginals of (X_i) and (Y_i) whenever $\alpha \subseteq I$ is non-empty. Let $\alpha, \beta \subseteq I$ be given non-empty disjoint sets. Since I is countable, we have $X^\alpha = Y^\alpha$ a.s. and $X^\beta = Y^\beta$ a.s. and since X^α and X^β are Fubini independent, then by Thm. 2.2 we see that Y^α and Y^β are Fubini independent. Thus, we conclude that the functions (Y_i) are Fubini independent.

(D) Let us adopt the notation from the proof of (A). Since $X = U(\xi)$, we have $\mathcal{C}_P(X) \subseteq \mathcal{C}_P(\xi)$ and since ξ is perfect, then by (1.25) we have $\mathcal{C}_P(\xi) = \mathcal{S}_P(\xi)$. Since $X_0 : (\Omega, \mathcal{C}_P(X)) \to (L_0, \mathcal{A}_0)$ is measurable, we see that $X_0 : (\Omega, \mathcal{S}_P(\xi)) \to (L_0, \mathcal{A}_0)$ is measurable and since $(L_0, \mathcal{A}_0)$ is countably separated, there exists a function $f : \Gamma \to L_0$ such that $X_0 = f(\xi)$ a.s. For the same reasons there exists a function $h : \Upsilon \to M_0$ such that $Y_0 = h(\eta)$ a.s. Since $(f(\xi(\omega)), X(\omega)) = (f(\xi(\omega)), U(\xi(\omega)))$ and $(h(\eta(\omega)), Y(\omega)) = (h(\eta(\omega)), V(\eta(\omega)))$ for all $\omega \in \Omega$, we see that $(f(\xi), X)$ and $(h(\eta), Y)$ are perfectly independent and since $(X_0, X) = (f(\xi), X)$ a.s. and $(Y_0, Y) = (h(\eta), Y)$ a.s., then by (C) we conclude that (X_0, X) and (Y_0, Y) are perfectly independent. □

Lemma 2.4. *Let $G \subseteq \Omega$ and $H \subseteq \Omega$ be Fubini independent sets. Then we have*

(1) $\quad P(G^*)\,P(H^*) = P^*(G)\,P^*(H) = P^*(G \cap H) = P(G^* \cap H^*)$

(2) $\quad P(G_*)\,P(H_*) = P_*(G)\,P_*(H) = P_*(G \cap H) = P(G_* \cap H_*)$

(3) $\quad (G \cap H)^* = G^* \cap H^*$ a.s. and $(G \cup H)_* = G_* \cup H_*$ a.s.

Proof. (1) The first equality in (1) is evident and applying (2.7) on the function $\psi(x,y) := xy$, we see that $P^*(G \cap H) \geq P^*(G)\,P^*(H)$. Since $1_G \vee P^*(H) = P^*(H) + (1 - P^*(H))\,1_G$, then applying (2.7) on the function $\psi(x,y) := x \vee y$, we have

$$P^*(G \cup H) = E^*(1_G \vee 1_H) \geq E^*(1_G \vee P^*(H)) = P^*(H) + (1 - P^*(H))\,P^*(G)$$

Hence, by [4; Exc. 1.32 p. 70–71] we have

$$\begin{aligned} P^*(G \cup H) + P^*(G \cap H) &\geq P^*(G \cup H) + P^*(G)\,P^*(H) \\ &\geq P^*(G) + P^*(H) \geq P^*(G \cup H) + P^*(G \cap H) \end{aligned}$$

In particular, we see that $P^*(G \cup H) + P^*(G \cap H) = P^*(G) + P^*(H)$ and that $P^*(G \cap H) = P^*(G)\,P^*(H)$ and by (1.5), we have $(G \cup H)^* = G^* \cup H^*$ a.s. Hence, we have

$$\begin{aligned} P^*(G) + P^*(H) &= P^*(G \cup H) + P^*(G \cap H) = P(G^*) + P(H^*) \\ &= P(G^* \cup H^*) + P(G^* \cap H^*) = P^*(G \cup H) + P(G^* \cap H^*) \end{aligned}$$

Thus, we see that $P^*(G \cap H) = P(G^* \cap H^*)$ which completes the proof of (1).

(2) The first equality in (2) is evident and the last equality follows from [1; Prop. 2.5 p. I.12]. By Thm. 2.2, we have that G^c and H^c are Fubini independent. Hence, as in the proof of (1) we have $P^*(G^c \cup H^c) + P^*(G^c \cap H^c) = P^*(G^c) + P^*(H^c)$ and since $P_*(A) = 1 - P^*(A^c)$ for all $A \subseteq \Omega$, we have by (1)

$$\begin{aligned} P_*(G \cap H) &= 1 - P^*(G^c \cup H^c) = 1 - \{P^*(G^c) + P^*(H^c) - P^*(G^c \cap H^c)\} \\ &= 1 + P^*(G^c)\, P^*(H^c) - P^*(G^c) - P^*(H^c) \\ &= P_*(G) - P^*(H^c)\, P_*(G) = P_*(G)\, P_*(H) \end{aligned}$$

which completes the proof of (2).

(3) Since $G \cap H \subseteq G^* \cap H^*$ a.s., we see that the first equality in (3) follows from the last equality in (1) and since $(A_*)^c = (A^c)^*$ a.s. for all $A \subseteq \Omega$, we see that the last equality in (3) follows from the first equality. □

Theorem 2.5. *Let* $X : \Omega \to L$ *and* $Y : \Omega \to M$ *be given functions and let* $\mathcal{A}$ *and* $\mathcal{B}$ *be pavings on* L *and* M *satisfying*

$$(1) \qquad \forall\, A \subseteq M\ \exists\, A_1, A_2, \ldots \in \mathcal{A} \text{ so that } X^{-1}(A) = \bigcup_{i=1}^{\infty} X^{-1}(A_i) \text{ a.s.}$$

$$(2) \qquad \forall\, B \subseteq M\ \exists\, B_1, B_2, \ldots \in \mathcal{B} \text{ so that } Y^{-1}(B) = \bigcup_{i=1}^{\infty} Y^{-1}(B_i) \text{ a.s.}$$

(A) If X *and* Y *are Fubini independent, then the* σ*-algebras* $\mathcal{L}_P(X)$ *and* $\mathcal{L}_P(Y)$ *are independent and* $\{X \in A\ ,\ Y \in B\}^* = \{X \in A\}^* \cap \{Y \in A\}^*$ *a.s. for all* $A \subseteq L$ *and all* $B \subseteq M$ *. Conversely, if* L *is countable,* $\mathcal{H}_P(X)$ *and* $\mathcal{H}_P(Y)$ *are independent and* $\{X \in A\ ,\ Y \in B\}^* = \{X \in A\}^* \cap \{Y \in A\}^*$ *a.s. for all* $A \in \mathcal{A}$ *and all* $B \in \mathcal{B}$ *, then* X *and* Y *are Fubini independent.*

(B) If L *and* M *are countable, then* X *and* Y *are Fubini independent if and only if* Y *and* X *are Fubini independent and if and only if* $\mathcal{H}_P(X)$ *and* $\mathcal{H}_P(Y)$ *are independent and* $\{X = x\ ,\ Y = y\}^* = \{X = x\}^* \cap \{Y = y\}^*$ *a.s.* $\forall\, (x, y) \in L \times M$ *.*

(C) If L *and* M *are finite, then* X *and* Y *are Fubini independent if and only if* X *and* Y *are perfectly independent.*

Proof. (A) Suppose that X and Y are Fubini independent and let $A \subseteq L$ and $B \subseteq M$ be given. By Thm. 2.2, we have that $X^{-1}(A)$ and $Y^{-1}(B)$ are Fubini independent. So by Lem. 2.4 we see that $X^{-1}(A)^*$ and $Y^{-1}(B)^*$ are independent and $\{X \in A\ ,\ Y \in B\}^* = \{X \in A\}^* \cap \{Y \in B\}^*$ a.s. In particular, we see that $\mathcal{H}_P(X)$ and $\mathcal{H}_P(Y)$ are independent and since $\mathcal{K}_P(Z)$ is the set of complements in $\mathcal{K}_P(Z)$, we see that $\mathcal{K}_P(X)$ and $\mathcal{K}_P(Y)$ are independent. Since $\mathcal{K}_P(Z)$ is stable under countable intersections and $\mathcal{L}_P(Z) = \sigma(\mathcal{K}_P(Z))$, we have that the σ-algebras $\mathcal{L}_P(X)$ and $\mathcal{L}_P(Y)$ are independent (see [4; (2.5) p. 106]) which completes the proof of the first part of (A). Suppose that L is countable, that $\mathcal{H}_P(X)$ and $\mathcal{H}_P(Y)$ are independent and that $\{X \in A\ ,\ Y \in B\}^* = \{X \in A\}^* \cap \{Y \in A\}^*$ a.s. for all $A \in \mathcal{A}$ and all $B \in \mathcal{B}$. As above, we see that $\mathcal{L}_P(X)$ and $\mathcal{L}_P(Y)$ are independent and by (1), (2) and (1.5), we see that $(X^{-1}(A) \cap Y^{-1}(B))^* = X^{-1}(A)^* \cap Y^{-1}(B)^*$

a.s. for all $A \subseteq L$ and all $B \subseteq M$. Let $D \subseteq L \times M$ be a given set and let us define $\phi(x) := P^*(Y \in D_x)$ where $D_x := \{y \in M \mid (x,y) \in D\}$. If $C := \{(X,Y) \in D\}$, we have $C = \cup_{x\in L}\{X = x \, , \, Y \in D_x\}$ and since L is countable and $(X^{-1}(A) \cap Y^{-1}(B))^* = X^{-1}(A)^* \cap Y^{-1}(B)^*$ a.s. for all $A \subseteq L$ and all $B \subseteq M$, we have by (1.5)

$$C^* = \bigcup_{x\in L} \{X = x \, , \, Y \in D_x\}^* = \bigcup_{x\in L} A_x \cap B_x \text{ a.s.} \qquad (*)$$

where $A_x := \{X = x\}^*$ and $B_x := \{Y \in D_x\}^*$. Let $[0,1]^L$ be equipped with the product σ-algebra $\mathcal{B}^L := \otimes_{x\in L}\mathcal{B}[0,1]$ and let us define $\xi(\omega) := (1_{A_x}(\omega))_{x\in L}$ and $\eta(\omega) := (1_{B_x}(\omega))_{x\in L}$. Since $A_x \in \mathcal{L}_P(X)$ and $B_x \in \mathcal{L}_P(Y)$, we have $\xi : (\Omega, \mathcal{L}_P(X)) \to ([0,1]^L, \mathcal{B}^L)$ and $\eta : (\Omega, \mathcal{L}_P(Y)) \to ([0,1]^L, \mathcal{B}^L)$ are measurable and since $\mathcal{L}_P(X)$ and $\mathcal{L}_P(Y)$ are independent, we see that ξ and η are independent random functions. Let us define $H(u,v) := \sup_{x\in L} u_x v_x$ for all $u = (u_x) \in [0,1]^L$ and all $v = (v_x) \in [0,1]^L$. Since L is countable, we have that $H : [0,1]^L \times [0,1]^L \to [0,1]$ is $(\mathcal{B}^L \otimes \mathcal{B}^L)$-measurable and by (*) we have $1_{C^*} = H(\xi,\eta)$ a.s. Since ξ and η are independent random function, we have $P^*(C) = E1_{C^*} = EF(\xi)$ where $F(u) := EH(u,\eta)$ for all $u \in [0,1]$ and since

$$F(u) = E\Big(\sup_{x\in L} u_x 1_{B_x}\Big) \geq \sup_{x\in L} u_x P(B_x) = \sup_{x\in L} u_x \phi(x) \quad \forall u = (u_x) \in [0,1]^L$$

and $\{X = x\} \subseteq A_x$, we have $F(\xi) \geq \sup_{x\in L} 1_{A_x}\phi(x) \geq \phi(X)$. Hence, we have $P^*((X,Y) \in D) = P(C^*) = EF(\xi) \geq E^*\phi(X)$ for all $D \subseteq L \times M$. So by Thm. 2.1 we see that X and Y are Fubini independent which proves the last part of (A).

(B) Since L and M are countable, we see that the pavings $\mathcal{A} := \{\{x\} \mid x \in L\}$ and $\mathcal{B} := \{\{y\} \mid y \in M\}$ satisfy the conditions (1) and (2). Hence, we see that (B) follows from (A) applied to (X,Y) and (Y,X).

(C) Suppose that X and Y are Fubini independent. Let $u \in L$ be given and let D_u be a given P-hull of $X^{-1}(u)$ satisfying $D_u \in \mathcal{F}$ and $D_u \supseteq X^{-1}(u)$. Let $\mathcal{L}$ denote the σ-algebra generated by the sets $(D_u \mid u \in L)$ and let $(F_i \mid i \in I)$ denote the non-empty atoms of the σ-algebra $\mathcal{L}$. Since L is finite, we have that I is a non-empty finite set and that $F_i \in \mathcal{L} \subseteq \mathcal{F}$ for all $i \in I$. Let $(\Gamma, \mathcal{G})$ denote the measurable space given by

$$\Gamma := \{\, (i,u) \in I \times L \mid X^{-1}(u) \cap F_i \neq \emptyset\} \ , \ \mathcal{G} := \{\Gamma \cap (\pi \times L) \mid \pi \subseteq I\}$$

Then $(\Gamma, \mathcal{G})$ is a discrete measurable space and if we define $\Gamma_i := \Gamma \cap (\{i\} \times L)$, then $(\Gamma_i \mid i \in I)$ is the set of all atoms of $(\Gamma, \mathcal{G})$. Let $\omega \in \Omega$ be given. Then there exists a unique $i(\omega) \in I$ such that $\omega \in F_{i(\omega)}$ and we define $\xi(\omega) := (i(\omega), X(\omega))$. Since $\omega \in F_{i(\omega)} \cap X^{-1}(X(\omega))$, we see that ξ maps Ω into Γ and if $\hat{X}(i,u) := u$ for all $(i,u) \in \Gamma$, then we have $X(\omega) = \hat{X}(\xi(\omega))$ for all $\omega \in \Omega$ and $\xi^{-1}(\Gamma_i) = F_i$ for all $i \in I$. In particular, we see that $\xi : (\Omega, \mathcal{L}) \to (\Gamma, \mathcal{G})$ is measurable. Let $(i,u) \in \Gamma$ be given. Then Γ_i is the unique $\mathcal{G}$-atom containing (i,u) and since $(i,u) \in \Gamma$ and $X^{-1}(u) \subseteq D_u$, we have $D_u \cap F_i \neq \emptyset$. Since $D_u \in \mathcal{L}$ and F_i is a

$\mathcal{G}$-atom, we have $D_u \cap F_i = F_i$ and since $F_i \in \mathcal{F}_P$ and $\xi^{-1}(i,u) = F_i \cap X^{-1}(u)$, we have $\xi^{-1}(i,u)^* = F_i \cap D_u = F_i$ a.s. Hence, we see that $\mathcal{L} \subseteq \mathcal{L}_P(X)$ is a σ-algebra on Ω, $(\Gamma, \mathcal{G})$ is a discrete measurable space and $\xi : \Omega \to \Gamma$ and $\hat{X} : \Gamma \to L$ are functions such that $\xi : (\Omega, \mathcal{L}) \to (\Gamma, \mathcal{G})$ is measurable and

(i) $\xi^{-1}(\gamma) = \xi^{-1}(\Gamma^\gamma) \cap X^{-1}(\hat{X}(\gamma))$ and $X(\omega) = \hat{X}(\xi(\omega))$ $\forall \gamma \in \Gamma \, \forall \omega \in \Omega$

(ii) $\xi^{-1}(\gamma)^* = \xi^{-1}(\Gamma^\gamma) \subseteq X^{-1}(\hat{X}(\gamma))^*$ a.s. $\forall \gamma \in \Gamma$

where Γ^γ is the unique $\mathcal{G}$-atom containing γ. In the same manner, we see that there exists a σ-algebra $\mathcal{K} \subseteq \mathcal{L}_P(Y)$, a discrete measurable space $(\Upsilon, \mathcal{H})$ and functions $\eta : \Omega \to \Upsilon$ and $\hat{Y} : \Upsilon \to M$ such that $\eta : (\Omega, \mathcal{K}) \to (\Upsilon, \mathcal{H})$ is measurable and

(iii) $\eta^{-1}(\gamma) = \eta^{-1}(\Upsilon^v) \cap Y^{-1}(\hat{Y}(v))$ and $Y(\omega) = \hat{Y}(\eta(\omega))$ $\forall v \in \Upsilon \, \forall \omega \in \Omega$

(iv) $\eta^{-1}(v)^* = \eta^{-1}(\Upsilon^v) \subseteq Y^{-1}(\hat{Y}(v))$ a.s. $\forall v \in \Upsilon$

where Υ^v is the unique $\mathcal{H}$-atom containing v.

Since X and Y are Fubini independent, then by (A) we see that $\mathcal{L}_P(X)$ and $\mathcal{L}_P(Y)$ are independent and since $\mathcal{L} \subseteq \mathcal{L}_P(X)$ and $\mathcal{K} \subseteq \mathcal{L}_P(Y)$, we see that $\xi : \Omega \to (\Gamma, \mathcal{G})$ and $\eta : \Omega \to (\Upsilon, \mathcal{H})$ are independent random functions. Let us consider the product space $(\Gamma \times \Upsilon, \mathcal{G} \otimes \mathcal{H})$ and let us define $\zeta(\omega) := (\xi(\omega), \eta(\omega))$ for all $\omega \in \Omega$. Then $(\Gamma \times \Upsilon, \mathcal{G} \otimes \mathcal{H})$ is a discrete measurable space and $\zeta : \Omega \to (\Gamma \times \Upsilon, \mathcal{G} \otimes \mathcal{H})$ is a random function. Let $(\gamma, v) \in \Gamma \times \Upsilon$ be given. Then $\Gamma^\gamma \times \Upsilon^v$ is the unique $(\mathcal{G} \otimes \mathcal{H})$-atom containing (γ, v) and if $(u, v) := (\hat{X}(\gamma), \hat{Y}(v))$, then by (i) and (iii) we have

$$\zeta^{-1}(\gamma, v) = \xi^{-1}(\Gamma^\gamma) \cap X^{-1}(u) \cap \eta^{-1}(\Upsilon^v) \cap Y^{-1}(v)$$

Since X and Y are Fubini independent, then by (A) we have $(X^{-1}(u) \cap Y^{-1}(v))^* = X^{-1}(u)^* \cap Y^{-1}(v)^*$ and since $\xi^{-1}(\Gamma^\gamma)$ and $\eta^{-1}(\Upsilon^v)$ belong to $\mathcal{F}_P$, then by (ii) and (iv) we have

$$\begin{aligned} \zeta^{-1}(\gamma, v)^* &= \xi^{-1}(\Gamma^\gamma) \cap \eta^{-1}(\Upsilon^v) \cap \{X^{-1}(u) \cap Y^{-1}(v)\}^* \\ &= \xi^{-1}(\Gamma^\gamma) \cap \eta^{-1}(\Upsilon^v) \cap X^{-1}(u)^* \cap Y^{-1}(v)^* = \zeta^{-1}(\Gamma^\gamma \cap \Upsilon^v) \text{ a.s.} \end{aligned}$$

Since $(\Gamma \times \Upsilon, \mathcal{G} \otimes \mathcal{H})$ is a discrete measurable space, then by (1.24) we see that ζ is $(P, \mathcal{G} \otimes \mathcal{H})$-perfect and since ξ and η are independent, then by (i) and (iii), we see that X and Y are perfectly independent. The converse implication follows from Thm. 2.3. □

Theorem 2.6. *Let $X : \Omega \to L$ and $Y : \Omega \to M$ be Fubini independent functions, let $f : L \to L_0$ and $h : M \to M_0$ be given functions and let $\xi, \eta : \Omega \to \mathbf{N}$ be integer valued functions.*

(A) The σ-algebras $\mathcal{L}_P(X)$ and $\mathcal{L}_P(Y)$ are independent and the algebras $\mathcal{A}_P(X)$ and $\mathcal{A}_P(Y)$ are perfectly independent.

(B) If $\{\xi = n\} \in \mathcal{G}_P(X)$ and $\{\eta = n\} \in \mathcal{G}_P(Y)$ for all $n \geq 1$, then ξ and η are Fubini independent.

(C) If L_0 *and* M_0 *are countable, then* $h(Y)$ *and* $f(X)$ *are Fubini independent, and if* L_0 *and* M_0 *are finite, then* $f(X)$ *and* $h(Y)$ *are perfectly independent.*

(D) If $\phi : L \to \bar{\mathbf{R}}$ *and* $\psi : M \to \bar{\mathbf{R}}$ *are given functions, then* $\phi(X)^*$ *and* $\psi(Y)^*$ *are independent random variables and if* ϕ *and* ψ *are non-negative, then we have*

$$(E^*\phi(X)) \cdot (E^*\psi(Y)) = E\{\phi(X)^*\,\psi(Y)^*\} = E^*\{\phi(X)\,\psi(Y)\}$$

(E) If $(X_i : \Omega \to M_i \mid i \in I)$ *are Fubini independent functions, then the* σ*-algebras* $(\mathcal{L}_P(X_i) \mid i \in I)$ *are independent and the random functions* $(X_i : \Omega \to (M_i, \mathcal{M}_P(X_i)) \mid i \in I)$ *are independent.*

Proof. (A) By Thm. 2.5, we have that $\mathcal{L}_P(X)$ and $\mathcal{L}_P(Y)$ are independent. Let $G \in \mathcal{G}_P(X)$ and $H \in \mathcal{G}_P(Y)$ be given. Then there exist sets $A_1, A_2, \ldots \in \mathcal{L}_P(X)$ and $S_1, S_2, \ldots \in \mathcal{S}_P(X)$ such that $G = \cup_{i=1}^{\infty} (A_i \cap S_i)$ and there exist sets $B_1, B_2, \ldots \in \mathcal{L}_P(Y)$ and $T_1, T_2, \ldots \in \mathcal{S}_P(Y)$ such that $H = \cup_{i=1}^{\infty} (B_i \cap T_i)$. Since $A_i \in \mathcal{L}_P(X) \subseteq \mathcal{F}_P$, then by (1.5) and (1.7) we have $G^* = \cup_{i=1}^{\infty} (A_i \cap S_i^*) \in \mathcal{L}_P(X)$ and for the same reason we have $H^* = \cup_{j=1}^{\infty} (B_j \cap T_j^*) \in \mathcal{L}_P(Y)$. Since X and Y are Fubini independent, then by Thm. 2.2 we have that S_i and T_j are Fubini independent. Hence, by Lem. 2.4, we have $(S_i \cap T_j)^* = S_i^* \cap T_j^*$ a.s. and since $A_i, B_j \in \mathcal{F}_P$, we have by (1.5) and (1.7)

$$\begin{aligned}(G \cap H)^* &= \bigcup_{i=1}^{\infty} \bigcup_{j=1}^{\infty} (A_i \cap S_i \cap B_j \cap T_j)^* = \bigcup_{i=1}^{\infty} \bigcup_{j=1}^{\infty} \{(A_i \cap B_j) \cap (S_i \cap T_j)^*\} \\ &= \bigcup_{i=1}^{\infty} \bigcup_{j=1}^{\infty} (A_i \cap S_i^* \cap B_j \cap T_j^*) = G^* \cap H^* \text{ a.s.}\end{aligned}$$

Hence, we see that $G^* \in \mathcal{L}_P(X)$, $H^* \in \mathcal{L}_P(Y)$ and $(G \cap H)^* = G^* \cap H^*$ for all $G \in \mathcal{G}_P(X)$ and all $H \in \mathcal{G}_P(Y)$. Let $G \in \mathcal{A}_P(X)$ and $H \in \mathcal{A}_P(Y)$ be given. Then 1_G and 1_H takes values in the finite set $\{0, 1\}$. Since G and its complement belong to $\mathcal{G}_P(X)$, we see that $\mathcal{H}_P(1_G) \subseteq \mathcal{L}_P(X)$ and for the same reason we have $\mathcal{H}_P(1_H) \subseteq \mathcal{L}_P(Y)$. Hence, we see that $\mathcal{H}_P(1_G)$ and $\mathcal{H}_P(1_H)$ are independent and since $\{1_G = x\} \in \mathcal{G}_P(X)$ and $\{1_H = y\} \in \mathcal{G}_P(Y)$ for $x, y = 0, 1$, we see that the functions 1_G and 1_H satisfy the hypotheses of (B) in Thm. 2.5. So by (B) and (C) in Thm. 2.5 we concluded that 1_G and 1_H are perfectly independent which completes the proof of (A)

(B) Since $\{\xi = n\} \in \mathcal{G}_P(X)$ for all $n \geq 1$ and $\mathcal{G}_P(X)$ is stable under countable unions, we have $\{\xi = n\} \in \mathcal{A}_P(X)$ for all $n \geq 1$ and $\{\xi \in B\} \in \mathcal{G}_P(X)$ for all $B \subseteq N$. So by the argument in the proof of (A), we see that $\mathcal{H}_P(\xi) \subseteq \mathcal{L}_P(X)$. For the same reasons, we $\{\eta = n\} \in \mathcal{A}_P(Y)$ for all $n \geq 1$ and $\mathcal{H}_P(\eta) \subseteq \mathcal{L}_P(Y)$. Hence, by (A) we see that $\{\xi = n\}$ and $\{\eta = k\}$ are perfectly independent for all $n, k \geq 1$ and by (A), we have that $\mathcal{L}_P(X)$ and $\mathcal{L}_P(Y)$ are independent. So by (A) and (B) in Thm. 2.5 we see that ξ and η are Fubini independent.

(C) is an immediate consequences of Thm.2.2 and Thm. 2.5.

(D) By (1.4), we have that $\phi(X)^*$ is $\mathcal{L}_P(X)$-measurable and that $\psi(Y)^*$ is $\mathcal{L}_P(Y)$-measurable. So by (A) we see that $\phi(X)^*$ and $\psi(Y)^*$ are independent random variables. Suppose that ϕ and ψ are non-negative and let us define $\xi(x,y) := \phi(x)\,\psi(y)$ and $\eta(x) := E^*\xi(x,Y)$ for all $x \in L$ and all $y \in M$. Since X and Y and Fubini independent, we have $E^*\xi(X,Y) \geq E^*\eta(X)$ and since $\phi, \psi \geq 0$, we have by (1.7) $\eta(x) = \phi(x)\,E^*\psi(Y)$ and $E^*\eta(X) = (E^*\phi(X))\,(E^*\psi(Y))$. Since $\phi, \psi \geq 0$, we have $\phi(X)^*\,\psi(Y)^* \geq \phi(X)\,\psi(Y)$ and since $\phi(X)^*$ and $\psi(Y)^*$ are independent, we have by (1.6)

$$\begin{aligned} E^*\{\phi(X)\,\psi(Y)\} &= E^*\xi(X,Y) \geq E^*\eta(X) = (E^*\phi(X)) \cdot (E^*\psi(Y)) \\ &= (E\phi(X)^*) \cdot (E\psi(Y)^*) = E\{\phi(X)^*\,\psi(Y)^*\} \geq E^*\{\phi(X)\,\psi(Y)\} \end{aligned}$$

which completes the proof of (D).

(E) Let $\alpha, \beta \subseteq I$ be non-empty disjoint sets. Since the marginals X^α and X^β are Fubini independent, then by (A) we have that $\mathcal{L}_P(X^\alpha)$ and $\mathcal{L}_P(X^\beta)$ are independent. By (1.19), we have $\mathcal{L}_P(X_i) \subseteq \mathcal{L}_P(X^\gamma)$ for all $i \in \gamma \subseteq I$. Hence, we see that $\sigma(\cup_{i\in\alpha}\mathcal{L}_P(X_i))$ and $\sigma(\cup_{i\in\beta}\mathcal{L}_P(X_i))$ are independent whenever $\alpha, \beta \subseteq I$ are non-empty disjoint sets but then it follows easily that the σ-algebras $(\mathcal{L}_P(X_i))$ are independent and since $X_i^{-1}(\mathcal{M}_P(X_i)) \subseteq \mathcal{L}_P(X_i)$, we see that the random functions $(X_i : \Omega \to (M_i, \mathcal{M}_P(X_i)) \mid i \in I)$ are independent. □

Theorem 2.7. *Let $Y : \Omega \to M$ be a given function, let $(L, \mathcal{A})$ be a measurable space, let $X : \Omega \to (L, \mathcal{A})$ be a random function and let $Q \in \mathcal{A}$ be a countable set such that $\bar{P}(X \in Q) = 1$ and $\{q\} \in \mathcal{A}$ for all $q \in Q$. Let $\psi : L \times M \to \bar{\mathbf{R}}$ be a given function. If X and Y are Fubini independent, or more generally, if $\{X = q\}$ and $\psi(q,Y)^*$ are independent for all $q \in Q$, then we have*

$$E^*\psi(X,Y) \geq \sum_{q\in Q} P(X = q)\,E^*\psi(q,Y)$$

with equality if either Q is finite or if there exists $U \in L^1(P)$ such that $\psi(q,Y) \geq U$ a.s. for all $q \in Q$.

Proof. Suppose that X and Y are Fubini independent. By Thm.2.6, we see that $\{X = q\}$ and $\psi(q,Y)^*$ are independent for all $q \in Q$. So suppose that $\{X = q\}$ and $\psi(q,Y)^*$ are independent for all $q \in Q$. Let $q \in Q$ be given and let us define $F_q := \{X = q\}$ and $Y_q := \psi(q,Y)^*$. Since $F_q \in \mathcal{F}_P$, then by (1.6) and (1.7), we have $(1_{F_q}\,\psi(q,Y))^* = 1_{F_q}\,Y_q$ a.s. and $E^*(1_{F_q}\,\psi(q,Y)) = E^*(1_{F_q}\,Y_q)$ and since F_q and Y_q are independent, we have $E(1_{F_q}\,Y_q^\pm) = P(F_q)\,EY_q^\pm$. So by (1.6) we have

$$\begin{aligned} E^*(1_{F_q}\,\psi(q,Y)) &= E^*(1_{F_q}\,Y_q) = E(1_{F_q}\,Y_q^+) - E(1_{F_q}\,Y_q^-) \\ &= P(F_q)\,EY_q^+ - P(F_q)\,EY_q^- = P(F_q)\,(EY_q^+ - EY_q^-) \\ &= P(F_q)\,E^*Y_q = P(F_q)\,E^*\psi(q,Y) \end{aligned}$$

Since $\psi(X,Y) = \sum_{q\in Q} 1_{F_q}\,\psi(q,Y)$ a.s. and (F_q) are P-measurable, mutually disjoint sets, then by (1.9) we have $E^*\psi(X,Y) \geq \sum_{q\in Q} P(F_q)\,E^*\psi(q,Y)$. If

Q is finite, then the converse inequality holds by subadditivity of E^* and if there exists $U \in L^1(P)$ satisfying $\psi(q,Y) \geq U$ a.s. for all $q \in Q$, then we have $1_{F_q}\psi(q,Y)^- \leq 1_{F_q} U^-$ and so by (1.8) we see that the converse inequality holds. □

3. Stochastic inequalities

Throughout the this section, we let Λ denote a fixed vector space with algebraic dual Λ^* and we let $X_1,\ldots,X_n : \Omega \to \Lambda$ be given functions where $n \geq 2$ is a given integer. We let $S_i := X_1+\cdots+X_i$ denote the partial sums for $i = 1,\ldots,n$ and we let $Q : \Lambda \to \bar{\mathbf{R}}$ denote a given function. The objective is to prove the classical inequalities of the introduction where the norm is replaced by the function Q ; that is, we shall the study the "norms" $Q(S_1),\ldots,Q(S_n)$ of the partial sums and the maximal "norm" $M_n := \max(Q(S_1),\ldots,Q(S_n))$.

It is well-known that symmetrizations plays an important role in establishing inequalities for sums of random elements. Let $E^n := \{-1,1\}^n$ denote the set of all ± 1-vectors. Then a random function $\varepsilon = (\varepsilon_1,\ldots,\varepsilon_n) : \Omega \to E^n$ will be called a *random sign vector* and we let $S_n^\varepsilon(\omega) := \sum_{i=1}^n \varepsilon_i(\omega)\, X_i(\omega)$ denote *the sign symmetrization* of S_n . In particular, we define $S_n^\delta(\omega) = \sum_{i=1}^n \delta_i\, X_i(\omega)$ if $\delta = (\delta_1,\ldots,\delta_n) \in E^n$. In the literature, the sign symmetrizations is usually only considered, when ε is a *Bernoulli vector* (that is, $\varepsilon_1,\ldots,\varepsilon_n$ are independent random variables taking the values $+1$ and -1 with probability $\frac{1}{2}$) but the symmetrization equalities derived below do *not* require this.

Theorem 3.1 (Ottaviani and Lévy's inequalities). *Let* $Q : \Lambda \to \bar{\mathbf{R}}$ *be a given function and let* Q^i *be a given upper* P*-envelope of* $Q_i := Q(S_i)$ *for all* $1 \leq i \leq n$. *Let us define*

$$\tau_s(\omega) = \min\{\, 1{\leq}i{\leq}n \mid Q^i(\omega){>}s\} \quad \forall\, \omega \in \Omega \,\forall\, s \in \mathbf{R}$$
$$\beta(s) := \inf_{t\in\mathbf{R}} \min_{1\leq i<n} P(Q^n{>}t \mid \tau_{s+t} = i) \quad \forall\, s \in \mathbf{R}$$

with the conventions $\min\emptyset := \infty$ *and* $P(F \mid G) := 1$ *if* $\bar{P}(G) = 0$. *Then we have*

(1) $$\beta(s)\, P^*\Big(\max_{1\leq i<n} Q(S_i){>}s+t\Big){\leq}P^*(Q(S_n){>}t) \quad \forall\, s,t \in \mathbf{R}$$

(2) $$\beta(s)\, P^*\Big(\max_{1\leq i\leq n} Q(S_i){>}s+t\Big){\leq}P^*(Q(S_n){>}t) \quad \forall\, s{\geq}0\,\forall\, t \in \mathbf{R}$$

Let Q^{in} *and* R^{in} *be given upper* P*-envelopes of* $Q_{in} := Q(S_i - S_n)$ *and* $R_{in} := Q(2S_i - S_n)$, *respectively, let* $s \in \mathbf{R}$ *be given real number and let us define*

$$\beta_*(s) := \min_{1\leq i<n} P_*(Q_{in}{\leq}s) \;\;,\;\; \beta^*(s) := \min_{1\leq i<n} P^*(Q_{in}{\leq}s)$$

Then we have:

(A) If Q *is quasi-convex and* $(Q^1,\dots,Q^i,Q^n)$ *and* $(Q^1,\dots,Q^i,R^{in})$ *have the same distribution* $\forall 1 \le i < n$ *, then* $\beta(0) \ge \frac{1}{2}$.

(B) If Q *is subadditive and* $\{\tau_{s+t} = i\}$ *and* $\{Q^{in} \le s\}$ *are non-negatively correlated* $\forall t \in \mathbf{R}\ \forall 1 \le i < n$ *, then* $\beta(s) \ge \beta_*(s)$.

(C) If Q *is subadditive and* $\{Q_{in} \le s\}$ *and* $\{\tau_{s+t} \ge i\ ,\ Q_i > s+t\}$ *are non-negatively correlated* $\forall t \in \mathbf{R}\ \forall 1 \le i < n$ *, then* $\beta(s) \ge \beta^*(s)$.

(D) Suppose that $Q(x) = \sup_{\psi\in\Psi} \psi(x)$ *for some non-empty set* $\Psi \subseteq \bar{\mathbf{R}}^\Lambda$ *of subadditive functions and let us define*

$$D_{st} := \{\, (\delta,x,y) \in \{0,1\}\times\Lambda\times\Lambda \mid \delta = 1\ ,\ Q(x) > s+t\ ,\ Q(x-y) > t\}$$
$$\beta^\circ(s) := \min_{1\le i<n}\ \inf_{\psi\in\Psi}\ P^*(\psi(S_i - S_n) \le s)$$

for all $t \in \mathbf{R}$ *. If* $(1_{\{\tau_{s+t}\ge i\}}, S_i)$ *and* $S_i - S_n$ *are non-negatively* D_{st}*-correlated for all* $t \in \mathbf{R}$ *and all* $1 \le i < n$ *, then* $\beta(s) \ge \beta^\circ(s)$

Remarks. Note that the Ottaviani inequalities (1) and (2) hold without any assumptions whatsoever on the random elements $X_1,\dots,X_n$ and the "norm" Q and that (2) implies the Lévy inequality (1.2) of the introduction provided that $\beta(0) \ge \frac{1}{2}$. So the game is to find lower estimates of *the Ottaviani constant* $\beta(s)$ defined in the theorem. Note that $\beta_*(s)$ is the classical Ottaviani constant and that $\beta^\circ(s) \ge \beta^*(s) \ge \beta_*(s)$ for all $s \in \mathbf{R}$. Hence, if Q is subadditive, we see that (B) implies the classical Ottaviani inequality under the independence hypotheses of (B) and that (C) and (D) improve the Ottaviani inequality under the independence hypotheses of (C) and (D). Let $t,u \in \mathbf{R}$ and $1 \le i \le n$ be given. Since $\{\tau_t \ge i\} = \{Q^\nu \le t\ \forall 1 \le \nu < i\}$ and $\{\tau_t = i\} = \{\tau_t \ge i\ ,\ Q^i > t\}$, we have by (1.4) and (1.5)

(3) $$\{Q_{in} \le t\} \in \mathcal{S}_P(X_{i+1},\dots,X_n)\ ,\ \{Q^{in} \le t\} \in \mathcal{K}_P(X_{i+1},\dots,X_n)$$

(4) $$\{\tau_t = i\} \in \mathcal{L}_P(X_1,\dots,X_i) \cap \sigma(Q^1,\dots,Q^i)\ ,\ \{\tau_t \ge i\} \in \mathcal{L}_P(X_1,\dots,X_{i-1})$$

(5) $$\{\tau_t \ge i\ ,\ Q(S_i) > u\} \in \mathcal{A}_P(X_1,\dots,X_i)$$

Let $\mathcal{B}$ be a given σ-algebra on Λ such that $(\Lambda,\mathcal{B})$ is a measurable linear space. Then by Thm. 2.3 and Thm. 2.6 we have:

(a) If Q is subadditive, then the Ottaviani inequalities (1) and (2) hold with $\beta(s)$ replaced by $\beta_*(s)$ in either of the following three cases (i): $(Q^1,\dots,Q^i)$ and Q^{in} are independent for all $1 \le i < n$; (ii): $X_1,\dots,X_n$ are Fubini independent; (iii): Q is $\mathcal{B}$-measurable and $X_1,\dots,X_n : \Omega \to (\Lambda,\mathcal{B})$ are independent random functions.

(b) If Q is subadditive, then the Ottaviani inequalities (1) and (2) hold with $\beta(s)$ replaced by $\beta^*(s)$ in either of the following three cases (i): $(Q_1\dots,Q_i)$ and Q_{in} are Fubini independent for all $1 \le i < n$; (ii): $X_1,\dots,X_n$ are Fubini independent; (iii): Q is $\mathcal{B}$-measurable and $X_1,\dots,X_n : \Omega \to (\Lambda,\mathcal{B})$ are independent random functions.

(c) If $Q(x) = \sup_{\psi\in\Psi} \psi(x)$ for some non-empty set $\Psi \subseteq \bar{\mathbf{R}}^\Lambda$ of subadditive functions, then the Ottaviani inequalities (1) and (2) hold with $\beta(s)$ replaced by $\beta^\circ(s)$ in either of the following three cases (i): $(1_{\{\tau_{s+t}\geq i\}}, S_i)$ and $S_i - S_n$ are Fubini independent for all $t \in \mathbf{R}$ and all $1 \leq i < n$; (ii): $X_1, \ldots, X_n$ are perfectly independent; (iii): Q is $\mathcal{B}$-measurable and $X_1, \ldots, X_n : \Omega \to (\Lambda, \mathcal{B})$ are independent random functions.

(d) Let $X = (X_1, \ldots, X_n)$ and $X^i = (X_1, \ldots, X_i, -X_{i+1}, \ldots, -X_n)$ for $1 \leq i < n$. Since $2S_i - S_n = X_1 + \cdots + X_i - X_{i+1} - \cdots - X_n$, then by (A) we see that the Lévy inequality (1.2) holds if (i): $P^*(X \in A) = P^*(X^i \in A)$ for all $A \subseteq \Lambda^n$ and all $1 \leq i < n$; (ii): Q is $\mathcal{B}$-measurable, $X_1, \ldots, X_n : \Omega \to (\Lambda, \mathcal{B})$ are random functions such that X and X^i have the same distribution on the n-fold product space $(\Lambda^n, \mathcal{B}^n)$ for all $1 \leq i < n$. In particular, we see that (A) implies the Lévy inequalities in [9; p. 431] and [3; p. 289–290].

(e) Suppose that $X_1, \ldots, X_n$ are perfectly independent and $Q(x) = \sup_{\psi\in\Psi} \psi(x)$ for some non-empty set $\Psi \subseteq \bar{\mathbf{R}}^\Lambda$ of subadditive functions. If $\psi(S_i - S_n)$ admits a non-positive median (i.e. $P^*(\psi(S_i - S_n) \leq 0) \geq \frac{1}{2}$), for all $\psi \in \Psi$ and all $1 \leq i < n$, then $\beta^\circ(0) \geq \frac{1}{2}$ and the Lévy inequality (1.2) holds.

(f) Let T be a non-empty set and let $X_1, \ldots, X_n : \Omega \to \mathbf{R}^T$ be perfectly independent stochastic processes. If $Q := \|\cdot\|_T$ is the sup-norm on $\mathbf{R}^T$ and $\Psi = \{\pi_t \mid t \in T\} \cup \{-\pi_t \mid t \in T\}$ where $\pi_t(f) := f(t)$, then $Q = \sup_{\psi\in\Psi} \psi$ and we have

$$\beta^\circ(s) = \min_{1\leq i<n} \inf_{t\in T} \{P(S_i(t) - S_n(t) \leq s) \wedge P(S_n(t) - S_i(t) \leq s)\} \quad \forall\, s \in \mathbf{R}$$

Hence, if 0 is a median of $S_i(t) - S_n(t)$ for all $t \in T$ and all $1 \leq i < n$, then $\beta^\circ(0) \geq \frac{1}{2}$ and the Lévy inequality (1.2) holds for the sup-norm. In particular, we that the Lévy inequality holds for stochastic processes in Example 1 of the introduction – but the classical proof fails. If $Q = \sup_T$ is the sup-functional, then $Q = \sup_{t\in T} \pi_t$ and we have

$$\beta^\circ(s) = \min_{1\leq i<n} \inf_{t\in T} P(S_i(t) - S_n(t) \leq s) \quad \forall\, s \in \mathbf{R}$$

Hence, if $S_i(t) - S_n(t)$ admits a non-positive median for all $t \in T$ and all $1 \leq i < n$, then $\beta^\circ(0) \geq \frac{1}{2}$ the Lévy inequality (1.2) holds for the sup-functional.

Proof. Let $M_i := \max_{1\leq\nu\leq i} Q_\nu$ and $M^i := \max_{1\leq\nu\leq i} Q^\nu$ denote the partial maxima for $i = 1, \ldots, n$.

(1+2): Let $s, t \in \mathbf{R}$ be given. Then we have $\{\tau_{s+t} \leq i\} = \{M^i > s + t\}$ for all $i = 1, \ldots, n$ and by (1.5) we have $M^i = M_i^*$ a.s. for all $i = 1, \ldots, n$. By the definition of $\beta(s)$, we have $\beta(s)\, P(\tau_{s+t} = i) \leq P(Q^n > t\, ,\ \tau_{s+t} = i)$ for all $i = 1, \ldots, n-1$. Summing these inequalities, we have by (1.4)

$$\beta(s)\, P^*(M_{n-1} > s + t) = \beta(s)\, P(M^{n-1} > s + t) = \beta(s)\, P(\tau_{s+t} \leq n - 1)$$

$$= \sum_{i=1}^{n-1} \beta(s)\, P(\tau_{s+t} = i) \leq \sum_{i=1}^{n-1} P(Q^n > t\, ,\ \tau_{s+t} = i) \leq P(Q^n > t) = P^*(Q_n > t)$$

which proves relation (1). So suppose that $s \geq 0$. Since $\beta(s) < 1$ and $s \geq 0$, we have $\{\tau_{s+t} = n\} \subseteq \{Q^n > s+t\} \subseteq \{Q^n > t\}$ and $\beta(s)\, P(\tau_{s+t} = n) \leq P(Q^n > t, \tau_{s+t} = n)$. Hence, we see that (2) follows in the same manner as above.

(A) Let $t \in \mathbf{R}$ and $1 \leq i < n$ be given. Since $S_i = \frac{1}{2}(S_n + (2S_i - S_n))$ and Q is quasi-convex, we have $Q_i \leq Q_n \vee R_{in}$ and consequently we have $Q^i \leq Q^n \vee R^{in}$ a.s. In particular, we have $\{Q^i > t\,,\ Q^n \leq t\} \subseteq \{Q^i > t\,, R^{in} > t\}$ and since $\{\tau_t = i\} \subseteq \{Q^i > t\}$, we have

$$\begin{aligned} P(\tau_t = i) &= P(\tau_t = i\,,\ Q^n > t) + P(\tau_t = i\,,\ Q^i > t\,,\ Q^n \leq t) \\ &\leq P(\tau_t = i\,,\ Q^n > t) + P(\tau_t = i\,,\ R^{in} > t) \end{aligned}$$

Since $\{\tau_t = i\} = \{Q^i > t\,, M^{i-1} \leq t\}$ and $(Q^1, \ldots, Q^i, Q^n)$ and $(Q^1, \ldots, Q^i, R^{in})$ have the same distribution, we have $P(\tau_t = i\,,\ R^{in} > t) = P(\tau_t = i\,,\ Q^n > t)$. Hence, we see that $P(\tau_t = i) \leq 2\, P(\tau_t = i\,,\ Q^n > t)$ and so we have $\beta(0) \geq \frac{1}{2}$.

(B) Let $1 \leq i < n$ and $t \in \mathbf{R}$ be given and let us define $F := \{\tau_{s+t} = i\}$. Then $F \in \mathcal{F}_P$ and since Q is subadditive and $S_i = S_n + (S_i - S_n)$, we have $Q_i \leq Q_{in} + Q_n$ and consequently $Q^i \leq Q^{in} + Q^n$ a.s. By (1.4), we have $\beta_*(s) \leq P_*(Q_{in} \leq s) = P(Q^{in} \leq s)$ and since $F \subseteq \{Q^i > s+t\}$, we have $F \cap \{Q^{in} \leq s\} \subseteq F \cap \{Q^n > t\}$. Since F and $\{Q^{in} \leq s\}$ are non-negatively correlated, we have

$$\begin{aligned} \beta_*(s)\, P(\tau_{s+t} = i) &\leq P(Q^{in} \leq s)\, P(F) \leq P(F \cap \{Q^{in} \leq s\}) \\ &\leq P(F \cap \{Q^n > t\}) = P(\tau_{s+t} = i\,,\ Q^n > t) \end{aligned}$$

for all $t \in \mathbf{R}$ and all $1 \leq i < n$ which proves (B).

(C) Let $1 \leq i < n$ and $t \in \mathbf{R}$ be given and let us define $H := \{\tau_{s+t} \geq i,\ Q_i > s+t\}$. By (1.4) and (1.7), we have $H^* = \{\tau_{s+t} \geq i, Q^i > s+t\} = \{\tau_{s+t} = i\}$ a.s. and by subadditivity of Q we have $Q_i \leq Q_n + Q_{in}$. Since $Q^n \geq Q_n$ a.s., we have

$$\begin{aligned} H \cap \{Q_{in} \leq s\} &= H \cap \{Q_i > s+t\,,\ Q_{in} \leq s\} \\ &\subseteq H \cap \{Q_n > t\} \subseteq \{\tau_{s+t} = i\,,\ Q^n > t\} \end{aligned}$$

and since $\beta^*(s) \leq P^*(Q_{in} \leq s)$ and H and $\{Q_{in} \leq s\}$ are non-negatively correlated, we have

$$\begin{aligned} \beta^*(s)\, P(\tau_{s+t} = i) &\leq P^*(Q_{in} \leq s)\, P^*(H) \leq P^*(H \cap \{Q_{in} \leq s\}) \\ &\leq P(\tau_{s+t} = i\,,\ Q^n > t) \end{aligned}$$

for all $t \in \mathbf{R}$ and all $1 \leq i < n$ which proves (C).

(D) Suppose that $Q = \sup_{\psi \in \Psi} \psi$ for some non-empty set Ψ of subadditive functions and let $t \in \mathbf{R}$ and $1 \leq i < n$ be given. Then Q is subadditive and we define $F := \{\tau_{s+t} \geq i\}$ and $f(\delta, x) := P^*(S_i - S_n \in D_{st}^{\delta x})$ for all $(\delta, x) \in \{0,1\} \times \Lambda$ where $D_{st}^{\delta x} := \{y \in \Lambda \mid (\delta, x, y) \in D_{st}\}$ is the (δ, x)-section of D_{st}. Since $(1_F, S_i)$ and $S_i - S_n$ are non-negatively D_{st}-correlated and $Q^n \geq Q_n$ a.s, we have

$$\begin{aligned} E^* f(1_F, S_i) &\leq P^*(\,(1_F, S_i, S_i - S_n) \in D_{st}\}) = P^*(F \cap \{Q_i > s+t\,,\ Q_n > t\}) \\ &\leq P(\tau_{s+t} = i\,,\ Q^n > t) \end{aligned}$$

Let $x \in A := \{x \in \Lambda \mid Q(x) > s+t\}$ be given. Since $Q(x) = \sup_{\psi\in\Psi} \psi(x) > s+t$, there exists a function $\psi_x \in \Psi$ such that $\psi_x(x) > s+t$. Let $y \in \Lambda$ be a given vector satisfying $\psi_x(y) \le s$. Since $x \in A$, we have $D^{1x}_{st} = \{z \in \Lambda \mid Q(x-z) > t\}$ and since ψ_x is subadditive, we have

$$s + t < \psi_x(x) \le \psi_x(y) + \psi_x(x-y) \le s + Q(x-y)$$

So we have $\{y \in \Lambda \mid \psi_x(y) \le s\} \subseteq D^{1x}_{st}$ for all $x \in A$ and since $D^{0x}_{st} = \emptyset$, we have $f(0,x) = 0$. Since $\beta^\circ(s) \le P^*(\psi_x(S_i - S_n) \le s)$, we have $f(\delta, x) \ge \delta\,\beta^\circ(s)\,1_A(x)$ for all $(\delta, x) \in \{0,1\} \times \Lambda$. Hence, if $H := F \cap \{S_i \in A\} = \{\tau_{s+t} \ge i, Q_i > s+t\}$, then we have $f(1_F, S_i) \ge \beta^\circ 1_H$ and as in the proof of (C), we see that $P^*(H) = P(\tau_{s+t} = i)$. Since $f(1_F, S_i) \ge \beta^\circ(s)\,1_F 1_A(S_i) = \beta^\circ(s)\,1_H$, we have by the inequality derived above

$$\beta^\circ(s)\,P(\tau_{s+t} = i) = \beta^\circ(s)\,P^*(H) \le E^* f(1_F, S_i) \le P(\tau_{s+t} = i\,,\ Q^n > t)$$

for all $s, t \in \mathbf{R}$ and all $1 \le i < n$ which proves (D). □

Theorem 3.2 (The symmetrization inequalities). *Let $\mu_1, \dots, \mu_n \in \Lambda$ be given vectors and let us define $U_n := \sum_{i=1}^n (X_i - \mu_i)$. Let $\Psi \subseteq \bar{\mathbf{R}}^\Lambda$ be a non-empty set of subadditive functions and let us define $Q(x) := \sup_{\psi\in\Psi} \psi(x)$ for all $x \in \Lambda$. Let Π denote the set of all sets $\pi \subseteq \{1, \dots, n\}$ with $\pi \ne \emptyset$ and $\pi \ne \{1, \dots, n\}$ and let us define*

$$\kappa(s) = \inf_{\psi\in\Psi} P^*(\psi(U_n) \le s)\ ,\quad \sigma(s) = \kappa(s) \wedge \min_{\pi\in\Pi}\ \inf_{\psi\in\Psi} P^*\Big(\psi\Big(\sum_{i\in\pi}(X_i - \mu_i)\Big) \le \tfrac{s}{2}\Big)$$

for all $s \in \mathbf{R}$. Let $S^\circ_n(\omega_1, \omega_2) := S_n(\omega_1) - S_n(\omega_2)$ for $(\omega_1, \omega_2) \in \Omega \times \Omega$ denote the usual product symmetrization of S_n. Then we have

(1) $\quad \kappa(s)\,P^*(Q(U_n) > s+t) \le (P \otimes P)^*(Q(S^\circ_n) > t) \quad \forall\, s, t \in \mathbf{R}$

Let $\varepsilon : \Omega \to E^n$ be a random sign vector such that ε and $(X_1, \dots, X_n) : \Omega \to \Lambda^n$ are Fubini independent. If $X_1, \dots, X_n$ are Fubini independent, then we have

(2) $\quad \sigma(s)\,P^*(Q(U_n) > s+t+u) \le P^*(Q(S^\varepsilon_n) > t) + P^*(Q(-S^\varepsilon_n) > u) \quad \forall\, s, t, u \in \mathbf{R}$

and if $X_1, \dots, X_n$ are perfectly independent, then we have

(3) $\quad \kappa(s)\,P^*(Q(U_n) > s+t+u) \le P^*(Q(S^\varepsilon_n) > t) + P^*(Q(-S^\varepsilon_n) > u) \quad \forall\, s, t, u \in \mathbf{R}$

Proof. Let $s, t, u \in \mathbf{R}$ be given. Let $S_\pi := \sum_{i\in\pi} X_i$ and $\mu_\pi := \sum_{i\in\pi} \mu_i$ denote the partial sums whenever $\pi \in \Pi$ and let us define $\mu := \mu_1 + \cdots + \mu_n$.

(1) Let us define $V_1(\omega_1, \omega_2) := U_n(\omega_1)$ and $V_2(\omega_1, \omega_2) := -U_n(\omega_2)$. Then V_1 and V_2 are independently defined on the product space $(\Omega \times \Omega, \mathcal{F} \otimes \mathcal{F}, P \otimes P)$ and we have $S^\circ_n = V_1 + V_2$. By Thm. 1.1 we have

$$(P \otimes P)^*(\psi(V_1 - (V_1 + V_2)) \le s) = (P \otimes P)^*(\psi(-V_2) \le s) = P^*(\psi(U_n) \le s)$$

for all $s \in \mathbf{R}$ and all $\psi \in \Psi$. Hence, we see that (1) follows from (1) and (D) in Thm. 3.1 applied to (V_1, V_2).

(2) Let $s, t, u \in \mathbf{R}$ be given numbers and let me first prove (2) when $\varepsilon \in E^n$ is a given non-random vector.

Suppose that $\varepsilon = (1,\dots,1)$. If $Q(-\mu) \le s+u$, then by subadditivity of Q , we have $Q(U_n) \le Q(S_n) + Q(-\mu) \le Q(S_n) + s + u$ and consequently, we have $\{Q(U_n) > s+t+u\} \subseteq \{Q(S_n) > t\}$ and since $\sigma(s) \le 1$, we see that (2) holds. Suppose that $Q(-\mu) > s+u$. Since $Q = \sup_{\psi\in\Psi} \psi$, there exists $\psi \in \Psi$ such that $\psi(-\mu) > s+u$. Since ψ is subadditive and $\psi \le Q$, we have $s+u < \psi(-\mu) \le \psi(U_n) + Q(-S_n)$ and consequently, we have $\{\psi(U_n) \le s\} \subseteq \{Q(-S_n) > u\}$. Since $\sigma(s) \le \kappa(s) \le P^*(\psi(U_n) \le s)$, we see that (2) holds.

Suppose that $\varepsilon = -(1,\dots,1)$. By the argument above with $s+u$ replaced by $s+t$, we see that (2) holds.

Suppose that $\varepsilon \ne \pm(1,\dots,1)$ and let us define $\pi := \{1 \le i \le n \mid \varepsilon_i = 1\}$ and $v := \{1 \le i \le n \mid \varepsilon_i = -1\}$. Since $\varepsilon \ne \pm(1,\dots,1)$, we see that $\pi, v \in \Pi$ and since $X_1,\dots,X_n$ are Fubini independent and $\pi \cap v = \emptyset$, we see that the pairs $(S_\pi - \mu_v, \mu_v - S_v)$ and $(S_v - \mu_\pi, \mu_\pi - S_\pi)$ satisfy the hypotheses of (D) in Thm. 3.1 (note that $n=2$ and $\{\tau_{s+t} \ge 1\} = \Omega$). Since

$$S_n^\varepsilon = (S_\pi - \mu_v) + (\mu_v - S_v) \ , \ (S_\pi - \mu_v) - S_n^\varepsilon = S_v - \mu_v$$
$$- S_n^\varepsilon = (S_v - \mu_\pi) + (\mu_\pi - S_\pi) \ , \ (S_v - \mu_\pi) + S_n^\varepsilon = S_\pi - \mu_\pi$$

and $\sigma(s) \le \kappa(s) \le P^*(\psi(S_\alpha - \mu_\alpha) \le \frac{s}{2})$ for all $\alpha \in \Pi$ and all $\psi \in \Psi$, then by (1) and (D) in Thm3.1 we have

$$\sigma(s)\, P^*(Q(S_\pi - \mu_v) > \tfrac{s}{2} + t) \le P^*(Q(S_n^\delta) > t)$$
$$\sigma(s)\, P^*(Q(S_v - \mu_\pi) > \tfrac{s}{2} + u) \le P^*(Q(-S_n^\delta) > t)$$

Since $U_n = (S_\pi - \mu_v) + (S_v - \mu_\pi)$, then by subadditivity of Q we have

$$P^*(Q(U_n) > s+t+u) \le P^*(Q(S_\pi - \mu_v) > \tfrac{s}{2} + t) + P^*(Q(S_v - \mu_\pi) > \tfrac{s}{2} + u)$$

multiplying by $\sigma(s)$ and inserting the inequalities derived above we see that (2) holds whenever ε is a fixed non-random sign vector. Hence, if $\varepsilon : \Omega \to E^n$ is a random sign vector such that ε and $(X_1,\dots,X_n)$ are Fubini independent, we see (2) follows from Thm. 2.7.

(3) As in the proof of (2), it suffices to prove (3) when $\varepsilon \in E^n$ is a fixed non-random vector. Since $X_1,\dots,X_n$ are perfectly independent we have that $X_i(\omega) = Y_i(\xi_i(\omega))$ for some functions $Y_i : \Gamma_i \to \Lambda$ and some independent random functions $\xi_i : \Omega \to (\Gamma_i, \mathcal{F}_i)$ such that $\xi := (\xi_1,\dots,\xi_n)$ is $(P,\mathcal{F}^n)$-perfect where $\mathcal{F}^n := \otimes_{i=1}^n \mathcal{F}_i$ denotes the product σ-algebra on the product space $\Gamma^n := \prod_{i=1}^n \Gamma_i$. In particular, we have $P_\xi = P_{\xi_1} \otimes \cdots \otimes P_{\xi_n}$ and $\mu := P_\xi \otimes P_\xi$ is a probability measure on $(\Gamma^n \times \Gamma^n, \mathcal{F}^n \otimes \mathcal{F}^n)$. Let $u := (u_1,\dots,u_n) \in \Gamma^n$ and $v := (v_1,\dots,v_n) \in \Gamma^n$ and let us define $J_\varepsilon(u,v) := (u',v')$ where $(u_i',v_i') = (u_i,v_i)$ if $\varepsilon_i = 1$ and $(u_i',v_i') = (v_i,u_i)$ if $\varepsilon_i = -1$. Then $J_\varepsilon : \Gamma^n \times \Gamma^n \to \Gamma^n \times \Gamma^n$ is the map which interchange the coordinates u_i and v_i if $\varepsilon_i = -1$ and leaves the coordinates u_i and v_i at their original positions if $\varepsilon_i = 1$. In particular, we see that J_ε is a measurable bijection of $(\Gamma^n \times \Gamma^n, \mathcal{F}^n \otimes \mathcal{F}^n)$ onto itself such that $J_\varepsilon = J_\varepsilon^{-1}$. So by (1.27) we see that J_ε is $(\mu, \mathcal{F}^n \otimes \mathcal{F}^n)$-perfect and since $\mu = P_{\xi_1} \otimes \cdots \otimes P_{\xi_n} \otimes P_{\xi_1} \otimes \cdots \otimes P_{\xi_n}$, then by the definition of J_ε , we have

$\mu(J_\varepsilon^{-1}(F)) = \mu(F)$ for all $F \in \mathcal{F}^n \otimes \mathcal{F}^n$ and since J_ε is $(\mu, \mathcal{F}^n \otimes \mathcal{F}^n)$-perfect, we have $\mu^*(J_\varepsilon^{-1}(D)) = \mu^*(D)$ for all $D \subseteq \Gamma^n \times \Gamma^n$.

Let us define $\Sigma(u,v) := \sum_{i=1}^n (Y_i(u_i) - Y_i(v_i))$ and $\Sigma^\varepsilon(u) := \sum_{i=1}^n \varepsilon_i Y_i(u_i)$ for all $(u,v) \in \Omega^n \times \Omega^n$ and let $\eta(\omega_1,\omega_2) := (\xi(\omega_1), \xi(\omega_2))$ for $(\omega_1,\omega_2) \in \Omega \times \Omega$ denote the product map from $\Omega \times \Omega$ into $\Gamma^n \times \Gamma^n$. Then $\Sigma(\eta) = S_n^\circ$ and if we define $D := \{(u,v) \in \Gamma^n \times \Gamma^n \mid Q(\Sigma(u,v)) > t + u\}$, then we have $\{Q(S_n^\circ) > t + u\} = \eta^{-1}(D)$. Since μ is the distribution of η, then by (1.21) we have

$$(P \otimes P)^*(Q(S_n^\circ) > t + u) = (P \otimes P)^*(\eta^{-1}(D)) \le \mu^*(D) = \mu^*(J_\varepsilon^{-1}(D))$$

Since J_ε interchange the coordinates u_i and v_i if $\varepsilon_i = -1$ and leaves the coordinates u_i and v_i at their original positions if $\varepsilon_i = 1$, then by subadditivity of Q we have

$$\begin{aligned} J_\varepsilon^{-1}(D) &= \{(u,v) \mid Q(\Sigma^\varepsilon(u) - \Sigma^\varepsilon(v)) > t + u\} \\ &\subseteq (\{u \mid Q(\Sigma^\varepsilon(u)) > t\} \times \Omega^n) \cup (\Omega^n \times \{v \mid Q(-\Sigma^\varepsilon(v)) > u\}) \end{aligned}$$

Since $S_n^\varepsilon(\omega) = \Sigma^\varepsilon(\xi(\omega))$, then by (1.21) and perfectness of ξ we have

$$\begin{aligned} (P \otimes P)^*(Q(S_n^\circ) > t + u) &\le \mu^*(J_\varepsilon^{-1}(D)) \le P_\xi^*(Q(\Sigma^\varepsilon) > t) + P_\xi^*(Q(-\Sigma^\varepsilon) > u) \\ &= P^*(Q(S_n^\varepsilon) > t) + P^*(Q(-S_n^\varepsilon) > u) \end{aligned}$$

and by (1) we have $\kappa(s)\, P^*(Q(U_n) > s + t + u) \le (P \otimes P)^*(Q(S_n^\circ) > t + u)$. Hence, we see that (3) holds. □

Theorem 3.3 (Subgaussian inequalities). *Let $X_1, \dots, X_n : \Omega \to (\mathbf{R}^T, \mathcal{B}^T)$ be stochastic processes stochastic processes. Let $\mu_1, \dots, \mu_n \in \mathbf{R}^T$ be given functions and let us define $U_n(t) : \sum_{i=1}^n (X_i(t) - \mu_i(t))$. Let $s, \rho \ge 0$ be given non-negative numbers satisfying*

(1) $\|S_n^\delta\|_T$ *is P-measurable for all sign vectors $\delta \in E^n$*

(2) $\rho\, P^*(\|U_n\|_T > s + 2y) \le P(\|S_n^\delta\|_T > y) \quad \forall\, y \ge 0\ \forall\, \delta \in E^n$

Let $\varepsilon : \Gamma \to E^n$ be a random sign vector defined on the probability space $(\Gamma, \mathcal{G}, P_0)$ and let us define

$$R^\varepsilon(\omega, y) := P_0\big(\sup_{t \in T} |\langle \varepsilon, X(\omega,t)\rangle| > y\big) \quad \forall\, y \ge 0\ \forall\, \omega \in \Omega$$

where $X(\omega,t) = (X_1(\omega,t), \dots, X_n(\omega,t))$ is the associated vector process and $\langle x, y\rangle := x_1 y_1 + \cdots + x_n y_n$ denotes the usual inner product on $\mathbf{R}^n$. Then $R^\varepsilon(\,\cdot\,, y)$ is P-measurable for all $y \ge 0$ and we have

(3) $\rho\, P^*(\|U_n\|_T > s + 2y) \le E\, R^\varepsilon(y) \quad \forall\, y \ge 0$

Let $C, r \ge 0$ be a given non-negative numbers and let $V : \Omega \to [0, \infty]$ be a given function satisfying

(4) $R^\varepsilon(\omega, r + y) \le C \exp\Big(-\frac{y^2}{2\, V(\omega)}\Big) \quad \forall\, y > 0\ \forall\, \omega \in \Omega$

with the conventions $\exp(-\frac{a}{0}) := 0$ *and* $\exp(-\frac{a}{\infty}) := 1$ *for all* $a \geq 0$. *If* $f : [0,\infty) \to [0,\infty)$ *is a nondecreasing right continuous function such that* $f(0) = 0$, *then we have*

$$(5) \qquad \rho\, P^*(\, \|U_n\|_T > s + 2r + 2y) \leq C\, E_* \exp\Big(-\tfrac{y^2}{2\,V}\Big) = C\, E \exp\Big(-\tfrac{y^2}{2\,V_*}\Big) \ \ \forall\, y \geq 0$$

$$(6) \qquad \rho\, P^*(\, \|U_n\|_T > s + 2r + 2y) \leq C\, P(V_* > x) + C\, \exp\Big(-\tfrac{y^2}{2x}\Big) \ \ \forall\, x, y \geq 0$$

$$(7) \qquad \rho\, E^* f\Big(\tfrac{1}{2}(\|U_n\|_T - s - 2r)^+\Big) \leq C\, E f^\diamond(V_*)$$

where $f(\infty) := \lim_{x\to\infty} f(x)$ *and* $f^\diamond(x)$ *is given by*

$$(8) \qquad f^\diamond(x) := \int_0^\infty e^{-\frac{t^2}{2x}}\, f(dt) \ \ \forall\, 0 \leq x \leq \infty$$

Remarks. (a) The measurability condition (1) goes under several names in the literature; for instance, *permissible*, *admissible* or *measurable*. The condition is indispensable in general but is valid in most cases of interest (see [5; Appendix F p. 262–276]).

(b) Note that condition (2) is a symmetrization inequality and since $||f||_T = \sup_{\psi\in\Psi} \psi(f)$ where $\Psi := \{\pi_t \mid t \in T\} \cup \{-\pi_t \mid t \in T\}$ and $\pi_t(f) := f(t)$ is the projection map, then the symmetrization inequalities of the previous theorem show that (2) holds with $\rho := \frac{1}{2}\,\sigma(s)$ or $\rho := \frac{1}{2}\,\kappa(s)$ under the appropriate independence assumptions on the processes $X_1, \ldots, X_n$.

(c) Note that $B(x) := \langle \varepsilon, x\rangle$ is a stochastic process with index set $\mathbf{R}^n$ defined on the probability space $(\Gamma, \mathcal{G}, P_0)$. Let us define $B_A := \sup_{x\in A} |B(x)|$ whenever $A \subseteq \mathbf{R}^n$ is a non-empty set. Since $\sigma_A(y) := \sup_{x\in A} |\langle y, x\rangle|$ is a lower semicontinuous seminorm on $\mathbf{R}^n$ and $B_A = \sigma_A(\varepsilon)$, we see that B_A is a measurable function of ε . If A is bounded, then $P_0(B_A > \rho) = 0$ where $\rho := \sup_{x\in A} ||x||_1$ and $||x||_1 := |x_1| + \cdots + |x_n|$ is the ℓ^1-norm. If A is unbounded and $P_0(\varepsilon = \delta) > 0$ for all $\delta \in E^n$, then the reader easily verifies that $P_0(B_A = \infty) > 0$. The stochastic process $(B(x))$ is a very regular process which has been studied thoroughly in the literature and there exist a series of subgaussian inequalities of the form:

$$(9) \qquad P_0(B_A > r + y) \leq C\, \exp\Big(-\tfrac{y^2}{2\, v(r,A)}\Big) \ \ \forall\, y \geq 0$$

where $C > 0$ is an absolute constant and the "variance" $v(r, A)$ is a specified constant in $[0, \infty]$. Let $A_\omega := \{X(\omega, t) \mid t \in T\} \subseteq \mathbf{R}^n$ denote the (random) range of the vector process $X(\omega, t)$. Since $R^\varepsilon(\omega, s) = P_0(B_{A_\omega} > s)$, we see (9) that implies condition (4) with $V(\omega) := v(r, A_\omega)$.

(d) Let $||x||_2$ denote the Euclidian norm on $\mathbf{R}^n$, let $\Delta_0(A)$ denote the Euclidian diameter of $A \cup \{0\}$ and let us define $\Delta(A) := \sup_{x\in A} ||x||_2$. Let $N^1(r, A)$ denote the smallest number of closed $||\cdot||_1$-balls with centers in A and radii r needed to cover A , and let $D^2(r, A)$ denote the largest number of points in A of mutual Euclidian distance $> r$. If ε is a Bernoulli vector, then

(9) holds with the following choices of the constants (see [7], [8], [5; Thm. 6.4 p. 185–186] and [9; Cor. 2.2.8 p. 101]):

$$C := 2e \ , \ r \geq 0 \ \text{ and } \ v(r,A) := (1 \vee \log N^1(r,A)) \cdot \Delta(A)^2$$

$$C := 2e \ , \ r := 0 \ \text{ and } \ v(0,A) := 8\sqrt{6} \int_0^{\Delta_0(A)} \sqrt{1 + \log D^2(s,A)}\, ds$$

(e) Let $f : [0,\infty) \to [0,\infty)$ be a nondecreasing right continuous function such that $f(0) = 0$. If $f^\diamond(x)$ denote *the Gaussian transform* of f as defined in (8), then $f^\diamond$ is a nondecreasing function and with our conventions, we have $f^\diamond(0) = 0$ and $f^\diamond(\infty) = f(\infty)$. Note that we have the following Gaussian transforms:

$$f(x) = 1_{(\beta,\infty)}(x) \ \Rightarrow \ f^\diamond(x) = \exp\Big(-\tfrac{\beta^2}{2x}\Big)$$

$$f(x) = x^\beta \ \Rightarrow \ f^\diamond(x) = (2x)^{\beta/2}\,\Gamma(\tfrac{\beta}{2}+1)$$

$$f(x) = e^{\beta x} - 1 \ \Rightarrow \ f^\diamond(x) = \beta\sqrt{2\pi x}\; e^{\frac{1}{2}\beta^2 x}\,\Phi(\beta\sqrt{x})$$

where $\Gamma(x)$ is the gamma function and $\Phi(x)$ is the standard normal distribution function. Hence, we see that (7) delivers various upper bounds for the moments and exponential moments of $||U_n||_T$.

Proof. (3) By (1), we see that $Y(\gamma,\omega,t) := \langle \varepsilon(\gamma), X(\omega,t)\rangle$ is a stochastic process on the product space $(\Gamma \times \Omega, \mathcal{G} \otimes \mathcal{F}, P_0 \otimes P)$ such that $||Y(\gamma,\omega)||_T$ is $(\mathcal{G} \otimes \mathcal{F}_P)$-measurable. So by Fubini's theorem, we have that $R^\varepsilon(\,\cdot\,,y)$ is P-measurable and

$$(P_0 \otimes P)(\,\|Y\|_T > y) = \int_\Omega P_0(\,\|Y(\,\cdot\,,\omega)\|_T > y)\, P(d\omega) = ER^\varepsilon(y)$$

Let us define $\tilde{X}(\gamma,\omega) = X(\omega,\,\cdot\,)$ $\tilde{S}_n^\delta(\gamma,\omega) := S_n^\delta(\omega)$ and $\tilde{\varepsilon}(\gamma,\omega) := \varepsilon(\gamma)$ for all $(\gamma,\omega) \in \Gamma \times \Omega$ and all $\delta \in E^n$. By (2) we have

$$\rho\, P^*(\,\|U_n\|_T > s + 2y) \leq P(\,\|S_n^\delta\|_T > y) = (P_0 \otimes P)(\,\|\tilde{S}_n^\delta\|_T > y)$$

for all $y \geq 0$ and all $\delta \in E^n$. Since $\tilde{\varepsilon}$ and $\tilde{X}$ are independently defined on the product space $(\Gamma \times \Omega, \mathcal{G} \otimes \mathcal{F}, P_0 \otimes P)$ and $Y(\gamma,\omega,t) = S_n^{\varepsilon(\gamma)}(\omega,t)$, then by Thm. 2.7 we have

$$\rho\, P^*(\|U_n\|_T > s + 2y) \leq (P_0 \otimes P)^*\,(\|Y\|_T > y) = ER^\varepsilon(y)$$

which proves (3).

(5,6): The first inequality in (5) follows directly from (3) and (4) and since $v \to \exp(-\frac{y^2}{2v})$ is nondecreasing and continuous on $[0,\infty]$, we see that the last equality in (5) follows from [9; Exc. 8 p. 13]. Since $x \to \exp(-\frac{y^2}{2x})$ is nondecreasing on $[0,\infty]$, we see that (6) follows from (5).

(7) Let us define $Q := ||U_n||_T$ and $W := \frac{1}{2}\,(Q - s - 2r)^+$. By (5) and (1.4), we have

$$\rho\, P(W^* > u) = \rho\, P^*(W > u) = \rho\, P^*(Q > s + 2r + 2u) \leq C\, E\exp\Big(-\tfrac{u^2}{2V_*}\Big)$$

for all $u \geq 0$ and since $u \to \exp(-\frac{u^2}{2v})$ is bounded and continuous on $[0,\infty)$ for all $0 \leq v \leq \infty$, we that the last expression is continuous in u on $[0,\infty)$. Hence, we have $\rho P(W^* \geq u) \leq C\,E\exp(-\frac{u^2}{2V_*})$ for all $u > 0$ and since $f(0+) = f(0) = 0$, we have $\lambda_f(\{0\}) = 0$ where λ_f is the Lebesgue Stieltjes f-measure. So by Fubini's theorem, we have

$$C\,Ef^\circ(V_*) = C\int_0^\infty E\Big\{\exp(-\tfrac{u^2}{2V_*})\Big\}\,f(du) \geq \rho\int_0^\infty P(W^* \geq u)\,f(du)$$
$$\geq \rho\int_\Omega P(d\omega)\int_{[0,W^*(\omega)]} f(du) = \rho\,Ef(W^*) \geq \rho\,E^*f(W)$$

which proves (7). □

Theorem 3.4 (The exponential inequality). *Let $Q : \Lambda \to [0,\infty]$ be a non-negative subadditive function such that $Q(x) = Q(-x)$ for all $x \in \Lambda$. Let $Q^1,\dots,Q^n$ be given upper P-envelopes of $Q(S_1),\dots,Q(S_n)$, let τ_s be defined as in Thm. 3.1 and let us we define*

$$L_n = \max_{1\leq\nu\leq n} Q(X_\nu)\,,\ M_n = \max_{1\leq\nu\leq n} Q(S_\nu)\,,\ R_i = \max_{i<\nu\leq n} Q(S_\nu - S_i)\ \text{for}\ 1\leq i<n$$
$$\gamma(s,t) = \max_{1\leq i<n} P(R_i^*>s \mid \tau_t = i)\,,\ \mu(s,t) = \max_{1\leq i<n} P(Q(S_n - S_i)^*>s \mid \tau_t = i)$$

for all $s,t \in \mathbf{R}$ with the convention $P(F \mid G) := 0$ if $F,G \in \mathcal{F}_P$ and $P(G) = 0$. If $r,s,t \geq 0$ are given non-negative numbers, then we have

(1) $\quad P^*(M_n>r+s+t) \leq P^*(L_n>r) + \gamma(s,t)\,P^*(M_n>t)$

(2) $\quad P^*(Q(S_n)>r+s+t) \leq P^*(L_n>r) + \mu(s,t)\,P^*(M_n>t)$

If $(Q^1,\dots,Q^i)$ and R_i^* are independent for all $1 \leq i < n$, then we have

(3) $\quad P^*(M_n>r+2s+t) \leq P^*(L_n>r) + P^*(M_n>s)\,P^*(M_n>t)$

and if $(Q^1,\dots,Q^i)$ and $Q(S_n - S_i)^*$ are independent for all $1 \leq i < n$, then we have

(4) $\quad P^*(Q(S_n)>r+2s+t) \leq P^*(L_n>r) + P^*(Q(S_n)>s)\,P^*(M_n>t)$

Remark. If $(\Lambda,\mathcal{B})$ is a measurable linear space, then by Thm. 2.6, we see that the exponential inequalities (3) and (4) hold in either of the following two cases: (i) $X_1,\dots,X_n$ are Fubini independent; (ii): Q is $\mathcal{B}$-measurable and $X_1,\dots,X_n : \Omega \to (\Lambda,\mathcal{B})$ are independent random functions.

Proof. Let us define $Q_i := Q(S_i)$ for $i = 1,\dots,n$, let $M_i := \max_{1\leq\nu\leq i} Q_\nu$ and $M^i := \max_{1\leq\nu\leq i} Q^\nu$ denote the partial maxima for $i = 1,\dots,n$ and let us define $M^0 = Q^0 = R_n := 0$. Since $S_\nu = X_1 + (S_\nu - S_1)$ and Q is non-negative and subadditive, we have $Q_\nu \leq L_n + R_1$ for all $1 \leq \nu \leq n$ and since $Q^0 = 0$, we have $Q^\nu \leq Q^0 + L_n^* + R_1^*$ a.s. for all $1 \leq \nu \leq n$. Let $1 < i \leq n$ be given. Since $R_n = 0$, $S_\nu = S_{i-1} + X_i + (S_\nu - S_i)$ and Q is non-negative subadditive, we have $Q_\nu \leq Q_{i-1} + L_n + R_i$ for all $1 \leq i \leq \nu \leq n$. Hence, we have $Q^\nu \leq Q^{i-1} + L_n^* + R_i^*$ a.s. for all $1 \leq i \leq \nu \leq n$ and since $M^0 := 0$, we have

(i) $M^n \le M^{i-1} \vee (Q^{i-1} + L_n^* + R_i^*)$ a.s. $\forall i = 1, \dots, n$

Let us define $U^n := 0$ and $U^i := Q(S_n - S_i)^*$ for all $1 \le i < n$. Since $Q^0 = 0$ and Q is non-negative and subadditive, then as above we obtain the following inequalities:

(ii) $Q^n \le Q^{i-1} + L_n^* + U^i$ a.s. $\forall i = 1, \dots, n$

Let $r, s, t \ge 0$ be given numbers.

(1) Let $1 \le i \le n$ be a given integer. Since $Q^{i-1}(\omega) \le M^{i-1}(\omega) \le t \le r + s + t$ for all $\omega \in \{\tau_t = i\}$, then by (i) we have

$$\{\tau_t = i\,,\, M^n > r + s + t\} \subseteq \{\tau_t = i\,,\, L_n^* > r\} \cup \{\tau_t = i\,,\, R_i^* > s\}$$

Hence, if $1 \le i < n$, then we have

$$\begin{aligned} P(\tau_t = i\,,\, M^n > r + s + t) &\le P(\tau_t = i\,,\, L_n^* > r) + P(\tau_t = i\,,\, R_i^* > s) \\ &\le P(\tau_t = i\,,\, L_n^* > r) + \gamma(s,t)\, P(\tau_t = i) \end{aligned}$$

and since $R_n^* = 0$, we see that this inequality also for $i = n$. Since $\{\tau_t \le n\} = \{M^n > t\} \supseteq \{M^n > r + s + t\}$, then summing this inequality over $i = 1, \dots, n$, we see that $P(M^n > r + s + t) \le P(L_n^* > r) + \gamma(s,t)\, P(M^n > t)$. Hence, we see that (1) follows from (1.4).

(2) Let $1 \le i \le n$ be a given integer. Since $U^n = 0$, then applying (ii) in the same manner as in the proof of (1), we have

$$\begin{aligned} P(\tau_t = i\,,\, Q^n > r + s + t) &\le P(\tau_t = i\,,\, L_n^* > r) + P(\tau_t = i\,,\, U_i > s) \\ &\le P(\tau_t = i\,,\, L_n^* > r) + \mu(s,t)\, P(\tau_t = i) \end{aligned}$$

and since $\{\tau_t \le n\} = \{M^n > t\} \supseteq \{Q_n > r + s + t\}$, we see that (2) follows in the same manner as in the proof of (1).

(3) Let $1 \le i < n$ be given. Since Q is subadditive and even, we have $R_i^* \le 2\, M^n$ and by the definition of τ_t , we have $\{\tau_t = i\} \in \sigma(Q^1, \dots, Q^i)$. Since $(Q^1, \dots, Q^i)$ and R_i^* are independent, we have

$$P(\tau_t = i\,,\, R_i^* > 2s) = P(\tau_t = i)\, P(R_i^* > 2s) \le P(\tau_t = i) \cdot P(M^n > s)$$

Hence, we have $P(M^n > s) \ge \gamma(2s, t)$ and so we see that (3) follows from (1) and (1.4).

(4) Follows from (2) in the same manner as in the proof of (3). □

Theorem 3.5 (Jensen's inequality). *Let* $\Xi \subseteq \Lambda^*$ *be a non-empty set of linear functionals. Let* $Q : \Lambda \to \bar{\mathbf{R}}$ *be a* Ξ*-convex function and let* $U : \Omega \to \Lambda$ *be a Pettis* Ξ*-integrable function. Then we have*

(1) $Q(a) = Q(b) \le E_* Q(U) \quad \forall\, a, b \in \mathcal{E}_\Xi(U)$

Let $V : \Omega \to \Lambda$ *be a given function, let* $r, q \in \mathbf{R}$ *be given numbers and let us define* $Q_{rq}(x,y) := Q(rx + qy)$ *for all* $(x,y) \in \Lambda \times \Lambda$ *. If* V *and* U *are non-negatively* Q_{rq}*-correlated, then we have*

(2) $E^* Q(qa + rV)\} \le E^* Q(qU + rV) \quad \forall\, a \in \mathcal{E}_\Xi(U)$

Proof. (1) Let $a, b \in \mathcal{E}_\Xi(U)$ be given. Since Q is Ξ-convex, there exists a non-empty set $\Psi \subseteq \operatorname{aff}\Xi$ such that $Q = \sup_{\psi\in\Psi} \psi$ and since $a, b \in \mathcal{E}_\Xi(X)$, then by (1.10) and (1.14) we have $Q(a) = Q(b)$. Let $\psi \in \Psi$ be given. Since $\psi(u) \leq Q(u)$ and $\psi(U) \in L^1(P)$, then by (1.15) we have $\psi(a) = E\psi(U) \leq E_* Q(U)$. Taking supremum over $\psi \in \Psi$, we obtain (1).

(2) Let $x \in \Lambda$ and $a \in \mathcal{E}_\Xi(U)$ be given. Since $rx + qa \in \mathcal{E}_\Xi(rx + qU)$, then by (1) we have $Q(rx + qa) \leq E^* Q(rx + qU) = E^* Q_{rq}(x, U)$ and since V and U are non-negatively Q_{rq}-correlated, we see that (2) holds. □

Theorem 3.6 (Convex symmetrization inequalities). *Let $\Xi \subseteq \Lambda^*$ be a non-empty set of linear functionals such that $X_1, \ldots, X_n : \Omega \to \Lambda$ are Pettis Ξ-integrable with Ξ-means $\mu_1, \ldots, \mu_n$ and let us define $S_+^\delta := \sum_{i=1}^n \delta_i^+ X_i$ and $S_-^\delta := \sum_{i=1}^n \delta_i^- X_i$ for all $\delta = (\delta_1, \ldots, \delta_n) \in E^n$. Let $\varepsilon = (\varepsilon_1, \ldots, \varepsilon_n) : \Omega \to E^n$ be a random sign vector and let $Q : \Lambda \to \bar{\mathbf{R}}$ be a Ξ-convex function. Let us consider the following functions on $\Lambda \times \Lambda$:*

$$Q_1(x,y) := Q(2x + 2y) \ , \quad Q_2(x,y) := Q(-2x - 2y) \ , \quad Q_3(x,y) := Q(2x - 2y)$$

If ε and $Q(S_n^\delta)^*$ are independent and S_+^δ and S_-^δ are non-negatively Q_i-correlated for all sign vectors $\delta \in E^n$ and for $i = 1, 2$, then we have

$$\text{(1)} \qquad E^* Q\Big(\sum_{i=1}^n \varepsilon_i (X_i - \mu_i)\Big) \leq \tfrac{1}{2} E^* Q\Big(2 \sum_{i=1}^n X_i\Big) + \tfrac{1}{2} E^* Q\Big(-2 \sum_{i=1}^n X_i\Big)$$

and if ε and $Q(2S^\delta)^*$ are independent and S_+^δ and S_-^δ are non-negatively Q_3-correlated for all sign vectors $\delta \in E^n$, then we have

$$\text{(2)} \qquad E^* Q\Big(\sum_{i=1}^n (X_i - \mu_i)\Big) \leq \tfrac{1}{2} E^* Q\Big(2 \sum_{i=1}^n \varepsilon_i X_i\Big) + \tfrac{1}{2} E^* Q\Big(-2 \sum_{i=1}^n \varepsilon_i X_i\Big)$$

Remark. Let $\mathcal{B}$ be a σ-algebra on Λ such that $(\Lambda, \mathcal{B})$ is a measurable linear space. Then (1) and (2) hold in either of the following two cases (i): $X_1, \ldots, X_n$ are Fubini independent and ε and $(X_1, \ldots, X_n)$ are Fubini independent; (ii): Q is $\mathcal{B}$-measurable, $X_1, \ldots, X_n : \Omega \to (\Lambda, \mathcal{L})$ are independent random functions and ε and $(X_1, \ldots, X_n)$ are independent.

Proof. Let $\delta = (\delta_1, \ldots, \delta_n) \in E^n$ be a given and let us define $\mu^\delta := \sum_{i=1}^n \delta_i \mu_i$ and $\mu_\pm^\delta := \sum_{i=1}^n \delta_i^\pm \mu_i$. Then we have $S_n = S_+^\delta + S_-^\delta$ and $S_n^\delta = S_+^\delta - S_-^\delta$.

Suppose that ε and $Q(S_n^\delta)^*$ are independent and that S_+^δ and S_-^δ are non-negatively Q_i-correlated for all $\delta \in E^n$ and for $i = 1, 2$. Since $Q(\pm 2x)$ is Ξ-convex and $\mu_\pm^\delta$ is a Ξ-mean of $S_\pm^\delta$, then by Thm. 3.5 we have $E^* Q(2S_+^\delta + 2\mu_-^\delta) \leq E^* Q(2S_n)$ and $E^* Q(-2S_-^\delta - 2\mu_+^\delta) \leq E^* Q(-2S_n)$. Since $S_n^\delta - \mu_\delta = (S_+^\delta + \mu_-^\delta) - (S_-^\delta + \mu_+^\delta)$ and Q is convex, we have

$$\begin{aligned} E^* Q(S_n^\delta) &\leq E^* \{\tfrac{1}{2} Q(2S_+^\delta + 2\mu_-^\delta) + \tfrac{1}{2} Q(-2S_-^\delta - 2\mu_+^\delta)\} \\ &\leq \tfrac{1}{2} E^* Q(2S_+^\delta + 2\mu_-^\delta) + \tfrac{1}{2} E^* Q(-2S_-^\delta - 2\mu_+^\delta)\} \\ &\leq \tfrac{1}{2} E^* Q(2S_n) + \tfrac{1}{2} E^* Q(-2S_n) \end{aligned}$$

for all $\delta \in E^n$. Since ε and $Q(S_n^\delta)^*$ are independent for all $\delta \in E^n$, we see that (1) follows from Thm. 2.7.

Suppose that ε and $Q(2S^\delta)^*$ are independent and that S_+^δ and S_-^δ are non-negatively Q_3-correlated for all $\delta \in E^n$. Let us define $\mu := \mu_1 + \cdots + \mu_n$. Since $S_n - \mu = (S_+^\delta - \mu_-^\delta) + (S_-^\delta - \mu_+^\delta)$, then by convexity of Q , we have

$$E^*Q(S_n - \mu) \le \tfrac{1}{2} E^*Q(2S_+^\delta - 2\mu_-^\delta) + \tfrac{1}{2} E^*Q(2S_-^\delta - 2\mu_+^\delta)$$

and since $S^\delta = S_+^\delta - S_-^\delta$ and $-S_\delta = S_-^\delta - S_+^\delta$, then by Thm. 3.5 we have $E^*Q(2S_+^\delta - 2\mu_-^\delta) \le E^*Q(2S_n^\delta)$ and $E^*Q(2S_-^\delta - 2\mu_+^\delta) \le E^*Q(2(-S_n^\delta))$. Hence, we have

$$E^*Q(S_n - \mu) \le \tfrac{1}{2} E^*Q(2S_n^\delta) + \frac{1}{2} E^*Q(-2S_n^\delta) \quad \forall \delta \in E^n$$

for all $\delta \in E^n$ and since ε and $Q(2S_n^\delta)^*$ are independent for all $\delta \in E^n$, we see that (2) follows in from Thm. 2.7. □

References:

[1] Andersen, N.T. (1985): *The calculus of non-measurable functions and sets*, Var. Publ. Ser. No. 36, Institute of Mathematics, University of Aarhus

[2] Dudley, R.M. (1984): *A course on empirical processes*, LNS 1097 p. 2–141, Springer Verlag, Berlin, New York and London

[3] Dudley, R.M. (1999): *Uniform central limits theorems*, Cambridge University Press, Cambridge and New York

[4] Hoffmann-Jørgensen, J. (1994): *Probability with a view toward statistics*, Vol.I, Chapman & Hall, New York and London

[5] Hoffmann-Jørgensen, J. (1999): *The Glivenko-Cantelli theorem and the Ranga Rao theorem*, in *Functional Analysis VI* (edited by S. Kurepa, H. Kraljević & D. Butković), Var. Publ. Ser. No. 43 p. 65–290, Institute of Mathematics, University of Aarhus

[6] Ledoux, M. & Talagrand, M. (1991): *Probability in Banach spaces*, Springer Verlag, Berlin, New York and London

[7] Pollard, D. (1984): *Convergence of stochastic processes*, Springer Verlag, Berlin, New York and London

[8] Pollard, D. (1990): *Empirical processes: Theory and application*, NSF-CBMS Regional Conference Ser. in Prob. and Stat., Vol. 2, Institute of Mathematical Statistics, Hayward California

[9] van der Vaart & A.W & Wellner, J.A. (2000): *Weak convergence and empirical processes*, Springer Verlag, Berlin, New York and London (2nd printing)

Department of Mathematical Sciences, University of Aarhus, Ny Munkegade, DK 8000 Aarhus C, Denmark

E-mail address: hoff@imf.au.dk

Progress in Probability, Vol. 55, 35–42

Prokhorov–LeCam–Varadarajan's Compactness Criteria for Vector Measures on Metric Spaces

Jun Kawabe

Abstract. The purpose of the paper is to give compactness and sequential compactness criteria for a set of vector measures on a complete separable metric space with values in a certain semi-Montel space. Among others it is shown that a set of such vector measures is uniformly bounded and uniformly tight if and only if the corresponding set of real measures is relatively sequentially compact with respect to the weak convergence of measures.

1. Introduction

In 1956, Yu. V. Prokhorov [14, Theorem 1.12] gave a sequential compactness criterion for the weak topology of measures in the space of all positive, finite measures on a complete separable metric space. This criterion was extended by L. LeCam [11, Theorems 6 and 7] and V. S. Varadarajan [16, Theorem 30, Part II] to real Radon measures. These results are called *Prokhorov–LeCam–Varadarajan's compactness criteria*, and play an important role in the study of stochastic convergence in probability theory and statistics.

The attempt to extend these compactness criteria to vector measures was first made by M. März and R. M. Shortt [13, Theorem 1.5] in the case of Banach space-valued vector measures on a metric space. This result has been extended to locally convex space-valued vector measures on an arbitrary completely regular space; see [8, 9]. However, in the process of these extensions, only one-sided implication has been discussed; that is, the implication that every uniformly bounded and uniformly tight set of vector measures is relatively (sequentially) compact with respect to the weak topology of vector measures. The purpose of the paper is to show that the converse of the above assertion is valid for vector measures on a complete separable metric space with values in a certain semi-Montel space.

Received by the editors December 1, 2002.
2000 *Mathematics Subject Classification.* Primary 28B05, 28C15; Secondary 46G10.
Key words and phrases. weak convergence of vector measures, uniform tightness, compactness criterion, semi-Montel space.
Research supported by Grant-in-Aid for General Scientific Research No. 13640162, Ministry of Education, Culture, Sports, Science and Technology, Japan.

In Section 2, we prepare notation and definitions, and recall some necessary results concerning vector measures and an integral of a real function with respect to a vector measure.

In Section 3, we give compactness and sequential compactness criteria for a set of vector measures on a complete separable metric space with values in a certain semi-Montel space. Among others it is shown that a set of such vector measures is uniformly bounded and uniformly tight if and only if the corresponding set of real measures is relatively sequentially compact with respect to the weak convergence of measures; see Theorem 3.

In this paper, all the topological vector spaces are Hausdorff, and the scalar fields of topological vector spaces are taken to be the field $\mathbb{R}$ of all real numbers. We also denote by $\mathbb{N}$ the set of all natural numbers.

2. Notation and Preliminaries

Let X be a locally convex topological vector space (for short, LCTVS). Denote by X^* the topological dual of X and by $\langle x, x^* \rangle$ the natural duality between X and X^*. The weak topology of X means the $\sigma(X, X^*)$-topology on X. If $x^* \in X^*$ and p is a seminorm on X, we write $x^* \leq p$ whenever $|\langle x, x^* \rangle| \leq p(x)$ for all $x \in X$.

Let S be a metric space. Denote by $C(S)$ the Banach space of all bounded, continuous real functions f on S with norm $\|f\|_\infty := \sup_{s \in S} |f(s)|$. Denote by $\mathcal{B}(S)$ the σ-field of all Borel subsets of S, that is, the σ-field generated by the open subsets of S. Let $\mu : \mathcal{B}(S) \to X$ be a finitely additive set function. We say that μ is a *vector measure* if it is σ-additive, that is, for any sequence $\{A_n\}_{n \in \mathbb{N}}$ of pairwise disjoint sets in $\mathcal{B}(S)$, we have $\mu\left(\bigcup_{n=1}^\infty A_n\right) = \sum_{n=1}^\infty \mu(A_n)$, where the series is unconditionally convergent for the initial topology of X. If μ is a vector measure, then $x^*\mu := \langle \mu, x^* \rangle$ is a real measure for each $x^* \in X^*$. Conversely, a theorem of Orlicz and Pettis ensures that a finitely additive set function $\mu : \mathcal{B}(S) \to X$ is σ-additive if so is $x^*\mu$ for each $x^* \in X^*$; see [12, Theorem 1.1].

Let $\mu : \mathcal{B}(S) \to X$ be a vector measure and p a continuous seminorm on X. The *p-semivariation* of μ is the set function $\|\mu\|_p : \mathcal{B}(S) \to [0, \infty]$ defined by $\|\mu\|_p(A) := \sup_{x^* \leq p} |x^*\mu|(A)$ for all $A \in \mathcal{B}(S)$, where $|x^*\mu|(\cdot)$ is the total variation of the real measure $x^*\mu$. Then $\|\mu\|_p(A) < \infty$ for all $A \in \mathcal{B}(S)$ [12, page 158]. For further properties of p-semivariations see [10] and [12]. See also [3, 4].

We say that a vector measure $\mu : \mathcal{B}(S) \to X$ is *Radon* if for each $\varepsilon > 0$, $A \in \mathcal{B}(S)$, and continuous seminorm p on X, there exists a compact subset K of A such that $\|\mu\|_p(A - K) < \varepsilon$, and it is *tight* if the condition is satisfied for $A = S$. By [12, Theorem 1.6] a vector measure $\mu : \mathcal{B}(S) \to X$ is Radon if and only if for each $x^* \in X^*$, the real measure $x^*\mu$ is Radon; that is, for each $\varepsilon > 0$ and $A \in \mathcal{B}(S)$, there exists a compact subset K of A such that $|x^*\mu|(A - K) < \varepsilon$. Thus, every X-valued vector measure on a complete separable metric space is Radon; see [5, Theorem IV.1.1]. Denote by $\mathcal{M}_t(S, X)$ the set of all Radon vector measures $\mu : \mathcal{B}(S) \to X$. We write $\mathcal{M}_t(S) := \mathcal{M}_t(S, \mathbb{R})$.

A subset $\mathcal{V}$ of $\mathcal{M}_t(S,X)$ is said to be *uniformly bounded* if $\sup_{\mu\in\mathcal{V}} \|\mu\|_p(S) < \infty$ for every continuous seminorm p on X. For any subset $\mathcal{V}$ of $\mathcal{M}_t(S,X)$, put $x^*(\mathcal{V}) := \{x^*\mu : \mu \in \mathcal{V}\}$. Then it follows from [9, page 496] that a subset $\mathcal{V}$ of $\mathcal{M}_t(S,X)$ is uniformly bounded if and only if $x^*(\mathcal{V})$ is uniformly bounded for each $x^* \in X^*$, that is, $\sup_{\mu\in\mathcal{V}} |x^*\mu|(S) < \infty$, and this is the case that $\sup_{\mu\in\mathcal{V}} \left|\int_S f d(x^*\mu)\right| < \infty$ for each $f \in C(S)$.

In this paper, we need an integral of a real function with respect to a vector measure. Let $\mu : \mathcal{B}(S) \to X$ be a vector measure. A Borel measurable real function f on S is said to be *μ-integrable* if (a) f is $x^*\mu$-integrable for each $x^* \in X^*$, and (b) for each $A \in \mathcal{B}(S)$, there exists an element of X, denoted by $\int_A f d\mu$, such that $\langle \int_A f d\mu, x^*\rangle = \int_A f d(x^*\mu)$ for each $x^* \in X^*$. This type of integrals is defined in [12, Definition 2.1], and we refer the reader to [10] and [12] for the properties of the integral. We only note here that if X is sequentially complete, then every bounded, Borel measurable real function on S is μ-integrable.

3. Compactness Criteria for Vector Measures

We introduce the notion of the weak convergence of vector measures. Let S be a metric space. Let X be a sequentially complete LCTVS. Let $\{\mu_\alpha\}_{\alpha\in\Gamma}$ be a net in $\mathcal{M}_t(S,X)$ and $\mu \in \mathcal{M}_t(S,X)$.

Definition 1. We say that $\{\mu_\alpha\}_{\alpha\in\Gamma}$ *converges weakly to* μ, and write $\mu_\alpha \xrightarrow{w} \mu$ if for each $f \in C(S)$ we have $\int_S f d\mu_\alpha \to \int_S f d\mu$ for the initial topology of X.

In the following, we always equip $\mathcal{M}_t(S,X)$ with the topology determined by this weak convergence and call it the *weak topology of vector measures.* It is easy to prove that the neighborhood base of $\mu_0 \in \mathcal{M}_t(S,X)$ for this topology is given by the family of sets of the form

$$W(\mu_0; f_1,\dots,f_n,U) := \bigcap_{i=1}^{n} \left\{\mu \in \mathcal{M}_t(S,X) : \int_S f_i d\mu - \int_S f_i d\mu_0 \in U\right\},$$

where $n \in \mathbb{N}$, $f_1,\dots,f_n \in C(S)$, and U is a neighborhood of the origin in X. This topology is a natural analogy of that defined by Dekiert [2] for Banach space-valued vector measures, and coincides with the usual weak topology of real measures in the case that $X = \mathbb{R}$; see [1, 11, 14, 15, 16].

Definition 2. A subset $\mathcal{V}$ of $\mathcal{M}_t(S,X)$ is said to be *uniformly tight* if for each $\varepsilon > 0$ and continuous seminorm p on X, there exists a compact subset K of S such that $\|\mu\|_p(S-K) < \varepsilon$ for all $\mu \in \mathcal{V}$.

Recall that a subset M of $\mathcal{M}_t(S)$ is *uniformly tight* if for each $\varepsilon > 0$, there is a compact subset K of S such that $|m|(S-K) < \varepsilon$ for all $m \in M$. Then it is well-known that if S is a metric space, then every uniformly bounded and uniformly tight subset of $\mathcal{M}_t(S)$ is relatively compact and relatively sequentially compact in $\mathcal{M}_t(S)$; see [11, Proposition 1 and Theorem 6] and [16, Corollary III, Part II]. It is also well-known that the converse of the above result is valid when the metric space S is complete and separable; see [16, Theorem 30, Part II].

The following example shows that a subset $\mathcal{V}$ of $\mathcal{M}_t(S,X)$ is not necessarily uniformly tight even if $x^*(\mathcal{V})$ is uniformly bounded and uniformly tight for every $x^* \in X^*$.

Example. Let H be a separable Hilbert space with inner product $(\cdot,\cdot)$. Let $\{e_n\}_{n\in\mathbb{N}}$ be a complete orthonormal basis in H. Let $\{\delta_n\}_{n\in\mathbb{N}}$ be a sequence of Dirac measures on $\mathbb{R}$ with mass n. For each $n \in \mathbb{N}$, define a vector measure $\mu_n : \mathcal{B}(\mathbb{R}) \to H$ by $\mu_n(A) := \delta_n(A)e_n, A \in \mathcal{B}(\mathbb{R})$. Then $\mu_n \in \mathcal{M}_t(\mathbb{R},H)$. For each $x \in H$ and $\mu \in \mathcal{M}_t(\mathbb{R},H)$, put $(x\mu)(A) := (x,\mu(A)), A \in \mathcal{B}(\mathbb{R})$. Then $|x\mu_n| = |(x,e_n)|\delta_n$, $\|\mu_n\| = \delta_n$ for all $n \in \mathbb{N}$, where $\|\mu_n\|$ denotes the semivariation of μ_n with respect to the norm of H.

Put $\mathcal{V} := \{\mu_n\}_{n\in\mathbb{N}}$ and fix $x \in H$. Then $|x\mu_n|(\mathbb{R}) = |(x,e_n)| \le \|x\|$ for all $n \in \mathbb{N}$, and hence $x(\mathcal{V})$ is uniformly bounded.

Let $\varepsilon > 0$. Since (x,e_n) converges to 0, there is $n_0 \in \mathbb{N}$ such that $n \ge n_0$ implies $|(x,e_n)| < \varepsilon$. Hence we have $\sup_{n\ge n_0} |x\mu_n|(\mathbb{R}) = \sup_{n\ge n_0} |(x,e_n)| \le \varepsilon$.

On the other hand, since each $x\mu_n$ is Radon, the finite set $\{x\mu_n : 1 \le n < n_0\}$ is uniformly tight, so that there is a compact subset K of $\mathbb{R}$ such that $\sup_{1\le n<n_0} |x\mu_n|(\mathbb{R}-K) < \varepsilon$. Consequently, we have

$$\sup_{n\in\mathbb{N}} |x\mu_n|(\mathbb{R}-K) \le \max\left(\sup_{1\le n<n_0} |x\mu_n|(\mathbb{R}-K), \sup_{n\ge n_0} |x\mu_n|(\mathbb{R}) \right) = \varepsilon,$$

and this implies that $x(\mathcal{V})$ is uniformly tight for each $x \in H$.

However, $\mathcal{V}$ is not uniformly tight. Indeed, since any compact subset K of $\mathbb{R}$ is contained in some bounded interval $[-N_0,N_0]$ $(N_0 \in \mathbb{N})$, we have

$$\|\mu_{N_0+1}\|(\mathbb{R}-K) = \delta_{N_0+1}(\mathbb{R}-K) \ge \delta_{N_0+1}(\mathbb{R}-[-N_0,N_0]) = 1.$$

Recall that a LCTVS X is a *semi-Montel space* if every bounded subset of X is relatively compact. Every semi-Montel space is semi-reflexive and sequentially complete [6, Propositions 11.5.1 and 11.5.2].

The following theorem extends Prokhorov–LeCam–Varadarajan's compactness criteria to vector measures with values in the space $\mathcal{S}$ of all rapidly decreasing, infinitely differentiable functions, the space $\mathcal{D}$ of all test functions, and the strong duals of those spaces. These spaces are important examples of semi-Montel spaces, each of whose duals has a countable set separating points of X.

Theorem 3. *Let S be a complete separable metric space. Let X be a semi-Montel space whose dual X^* has a countable set which separates points of X. Let $\mathcal{V}$ be a subset of $\mathcal{M}_t(S,X)$. Then the following conditions are equivalent:*

(1) *For each $x^* \in X^*$, $x^*(\mathcal{V})$ is uniformly bounded and uniformly tight.*
(2) *For each $x^* \in X^*$, $x^*(\mathcal{V})$ is relatively compact in $\mathcal{M}_t(S)$.*
(3) *For each $x^* \in X^*$, $x^*(\mathcal{V})$ is relatively sequentially compact in $\mathcal{M}_t(S)$.*
(4) *$\mathcal{V}$ is uniformly bounded and uniformly tight.*
(5) *$\mathcal{V}$ is relatively compact in $\mathcal{M}_t(S,X)$.*
(6) *$\mathcal{V}$ is relatively sequentially compact in $\mathcal{M}_t(S,X)$.*

Remark 4. It is routine to check that the condition that X^* has a countable set which separates points of X is equivalent to X^* being separable for the weak topology $\sigma(X^*, X)$.

To prove the theorem we need several technical results which may essentially be given in [16], where the space of all regular Baire real measures defined on a general topological space is studied. However, since the concepts of Baire set and Borel set coincide in a metric space, many results in [16] are applied to the study of the space of all regular Borel real measures on a metric space. We shall use them in the proofs of the following results in a positive manner.

Lemma 5. *Let S be a metric space. Let X be a sequentially complete LCTVS. Let $\mu \in \mathcal{M}_t(S, X)$. Let $\delta > 0$, F a closed subset of S and p a continuous seminorm on X. Assume that $p(\int_S g d\mu) \le \delta$ whenever $g \in C(S)$, $\|g\|_\infty \le 1$, and $g(s) = 0$ for all $s \in F$. Then $\|\mu\|_p(S - F) \le \delta$.*

Proof. Fix $x^* \in X^*$ with $x^* \le p$. Assume that $g \in C(S)$, $\|g\|_\infty \le 1$, $g(s) = 0$ for all $s \in F$ imply $p\left(\int_S g d\mu\right) \le \delta$. Then, we have $\left|\int_S g d(x^*\mu)\right| = \left|\langle \int_S g d\mu, x^*\rangle\right| \le p\left(\int_S g d\mu\right) \le \delta$. It follows from [16, Lemma 3, Part I] that $|x^*\mu|(S - F) \le \delta$. Thus $\|\mu\|_p(S - F) = \sup_{x^* \le p} |x^*\mu|(S - F) \le \delta$ and the proof is complete. □

Lemma 6. *Let S be a metric space. Let X be a sequentially complete LCTVS. Let $\mu \in \mathcal{M}_t(S, X)$. Let $\delta > 0$, $\alpha_0 > 0$, and $g_0 \in C(S)$ with $\|g_0\|_\infty \le 1$. Let F_0 be a closed subset of S and p a continuous seminorm on X. Assume that $p(\int_S g d\mu - \int_S g_0 d\mu) \le \delta$ whenever $g \in C(S)$, $\|g\|_\infty \le 1$ and $|g(s) - g_0(s)| < \alpha_0$ for all $s \in F_0$. Then $p(\int_S g d\mu) \le 2\delta$ whenever $g \in C(S)$, $\|g\|_\infty \le 1$ and $g(s) = 0$ for all $s \in F_0$.*

Proof. As is seen in the proof of Lemma 5, it suffices to prove the lemma for real measures. However, it can be proved by the same argument of the proof of [11, Lemma 6] as is indicated in [16, page 196]. So we omit the proof. □

Let S be a metric space. Recall that a sequence $\{F_k\}_{k\in\mathbb{N}}$ of closed subsets of S is *regular* if it satisfies the following two conditions: (1) F_k increases to S, (2) for each $k \in \mathbb{N}$ there exists an open subset U_k of S such that $F_k \subset U_k \subset F_{k+1}$; see [16, page 168]. Since a real function on S is continuous if and only if it is continuous on each set F_k in a regular sequence $\{F_k\}_{k\in\mathbb{N}}$ of closed subsets of S [16, Theorem 12, Part I], every regular sequence is a k'-sequence in the sense of [16, page 195].

The proof of the following proposition is an easy modification of the proof of [16, Theorem 19, Part II], but we shall give the sketch of the proof for the completeness of the paper.

Proposition 7. *Let S be a metric space. Let X be a sequentially complete LCTVS. Let $\{\mu_n\}_{n\in\mathbb{N}}$ be a sequence in $\mathcal{M}_t(S, X)$. Assume that for each $f \in C(S)$, the sequence $\{\int_S f d\mu_n\}_{n\in\mathbb{N}}$ converges in X. Then $\sup_{n\in\mathbb{N}} \|\mu_n\|_p(S - F_k) \to 0$ as $k \to \infty$ for each continuous seminorm p on X and regular sequence $\{F_k\}_{k\in\mathbb{N}}$ of closed subsets of S.*

Proof. Let $\{F_k\}_{k\in\mathbb{N}}$ be a regular sequence of closed subsets of S. Let I be a unit ball of $C(S)$. For $f, g \in I$, put $\rho_k(f,g) := \sup_{s\in F_k} |f(s) - g(s)|$ $(k \in \mathbb{N})$ and $\rho(f,g) := \sum_{k=1}^{\infty} \rho_k(f,g)/2^k$. Then, ρ is a metric on I, and the metric space (I,ρ) is complete. It is routine to prove that for each $\mu \in \mathcal{M}_t(S,X)$, the mapping $g \in (I,\rho) \mapsto \int_S g d\mu \in X$ is continuous.

Let p be a continuous seminorm on X. Fix $\varepsilon > 0$. For each $l \in \mathbb{N}$, put

$$Q_l := \left\{ g \in I : p\left(\int_S g d\mu_n - \int_S g d\mu_m \right) \le \varepsilon \quad \text{for all } m, n \ge l \right\}.$$

Then it follows from the continuity of the mapping $g \in (I,\rho) \to \int_S g d\mu \in X$ that each Q_l is a closed subset of (I,ρ). Further, since the sequence $\{\int_S f d\mu_n\}_{n\in\mathbb{N}}$ converges in X, we have $I = \bigcup_{l=1}^{\infty} Q_l$. By Baire's category theorem, some Q_{l_0} has an interior point g_0. Hence, there exist $k_0 \in \mathbb{N}$ and $r_0 > 0$ such that

$$p\left(\int_S g d\mu_n - \int_S g d\mu_m \right) \le \varepsilon,$$

so that

$$p\left(\int_S g d(\mu_n - \mu_m) - \int_S g_0 d(\mu_n - \mu_m) \right) \le 2\varepsilon$$

whenever $m, n \ge l_0$, $g \in I$ and $|g(s) - g_0(s)| < r_0$ for all $s \in F_{k_0}$. It follows from Lemma 6 that

$$p\left(\int_S g d\mu_n - \int_S g d\mu_m \right) \le 4\varepsilon$$

whenever $m, n \ge l_0$, $g \in I$ and $g(s) = 0$ for all $s \in F_{k_0}$. Hence by Lemma 5 $\|\mu_n - \mu_m\|_p(S - F_{k_0}) \le 4\varepsilon$ for all $m, n \ge l_0$. Take $k_1 \ge k_0$ so that $\|\mu_{l_0}\|_p(S - F_{k_1}) \le \varepsilon$. Then for all $n \ge l_0$,

$$\|\mu_n\|_p(S - F_{k_1}) \le \|\mu_n - \mu_{l_0}\|_p(S - F_{k_1}) + \|\mu_{l_0}\|_p(S - F_{k_1}) \le 4\varepsilon + \varepsilon = 5\varepsilon.$$

Finally we take $k_2 \ge k_1$ so that $\|\mu_n\|_p(S - F_{k_2}) \le 5\varepsilon$ for $n = 1, 2, \ldots, l_0 - 1$. □

Proposition 8. *Let S be a metric space. Let X be a sequentially complete LCTVS. Assume that a subset $\mathcal{V}$ of $\mathcal{M}_t(S,X)$ is relatively sequentially compact in $\mathcal{M}_t(S,X)$. Then $\mathcal{V}$ has the following properties:*

(1) *$\mathcal{V}$ is uniformly bounded.*

(2) $\sup_{\mu\in\mathcal{V}} \|\mu\|_p(S - F_k) \to 0$ *as $k \to \infty$ for every continuous seminorm p on X and regular sequence $\{F_k\}_{k\in\mathbb{N}}$ of closed subsets of S.*

Proof. (1) Since $\mathcal{V}$ is relatively sequentially compact, for each $f \in C(S)$ the set $\{\int_S f d\mu : \mu \in \mathcal{V}\}$ is a weakly bounded subset of X. It follows from the Banach–Steinhaus theorem that for every $x^* \in X^*$ we have

$$\sup_{\mu\in\mathcal{V}} |x^*\mu|(S) = \sup \left\{ \left| \int_S f d(x^*\mu) \right| : \mu \in \mathcal{V}, f \in C(S), \|f\|_\infty \le 1 \right\} < \infty,$$

which implies the uniform boundedness of $\mathcal{V}$ by [7, Proposition 1].

(2) Suppose (2) fails. Then there exist $\delta > 0$, a continuous seminorm p on X, a regular sequence $\{F_n\}_{n\in\mathbb{N}}$ of closed subsets of S and a sequence $\{\mu_n\}_{n\in\mathbb{N}}$ in

$\mathcal{V}$ such that $\|\mu_n\|_p(S - F_n) > \delta$ for all $n \in \mathbb{N}$. Since $\mathcal{V}$ is relatively sequentially compact, there exist a subsequence $\{\mu_{n_k}\}_{k\in\mathbb{N}}$ of $\{\mu_n\}_{n\in\mathbb{N}}$ and $\mu_0 \in \mathcal{M}_t(S, X)$ such that $\mu_{n_k} \xrightarrow{w} \mu_0$. Thus it follows from Proposition 7 that there exists $i_0 \in \mathbb{N}$ such that $\sup_{k\in\mathbb{N}} \|\mu_{n_k}\|_p(S - F_{i_0}) < \delta$. Choose $k_0 \in \mathbb{N}$ so that $n_{k_0} \geq i_0$. Then

$$\delta \leq \|\mu_{n_{k_0}}\|_p(S - F_{n_{k_0}}) \leq \|\mu_{n_{k_0}}\|_p(S - F_{i_0}) \leq \sup_{k\in\mathbb{N}} \|\mu_{n_k}\|_p(S - F_{i_0}) < \delta,$$

which is a contradiction. □

The following result extends [11, Theorem 7] and a one-sided implication of [16, Theorem 30, Part II] to vector measures.

Proposition 9. *Let S be a complete separable metric space. Let X be a sequentially complete LCTVS. If a subset $\mathcal{V}$ of $\mathcal{M}_t(S, X)$ is relatively sequentially compact in $\mathcal{M}_t(S, X)$, then it is uniformly bounded and uniformly tight.*

Proof. Assume that $\mathcal{V}$ is relatively sequentially compact. By Proposition 8 it suffices to prove the uniform tightness of $\mathcal{V}$.

Let p be a continuous seminorm on X. Fix $n \in \mathbb{N}$ and $\varepsilon > 0$. The separability of S implies that there exists a sequence $\{S_{nj}\}_{j\in\mathbb{N}}$ of open balls with center 0 and radius $1/n$ such that $S = \bigcup_{j=1}^{\infty} S_{nj}$. Put $G_k := \bigcup_{j=1}^{k} S_{nj}$ for all $k \in \mathbb{N}$. Then, by [16, Theorem 15, Part I] there exists a regular sequence $\{F_k\}_{k\in\mathbb{N}}$ of closed subsets of S such that $S - G_k \subset S - F_k$ for all $k \in \mathbb{N}$. It follows from Proposition 8 that $\sup_{\mu\in\mathcal{V}} \|\mu\|_p(S - G_k) \to 0$ as $k \to \infty$, so that there exists $k_n \in \mathbb{N}$ such that $\sup_{\mu\in\mathcal{V}} \|\mu\|_p(S - \bigcup_{j=1}^{k_n} S_{nj}) < \varepsilon/2^n$. Put $Q_n := \bigcup_{j=1}^{k_n} \bar{S}_{nj}$ for all $n \in \mathbb{N}$ and $K_\varepsilon := \bigcap_{n=1}^{\infty} Q_n$, where $\bar{S}_{nj}$ denotes the closure of the set S_{nj}. Then, for all $\mu \in \mathcal{V}$ we have

$$\|\mu\|_p(S - K_\varepsilon) \leq \sum_{n=1}^{\infty} \|\mu\|_p(S - Q_n) \leq \sum_{n=1}^{\infty} \|\mu\|_p\Big(S - \bigcup_{j=1}^{k_n} S_{nj}\Big) \leq \sum_{n=1}^{\infty} \frac{\varepsilon}{2^n} = \varepsilon,$$

and this implies that $\sup_{\mu\in\mathcal{V}} \|\mu\|_p(S - K_\varepsilon) \leq \varepsilon$.

It is routine to prove that K_ε is a totally bounded, closed subset of S, so that it is compact since the metric space S is complete. Thus, the proof of the uniform tightness of $\mathcal{V}$ is complete. □

Remark 10. The converse of the assertion in Proposition 9 has been discussed in [8, Theorem 2] and [9, Theorem 2] for the weaker notion of the weak convergence of vector measures, which is defined by using the weak topology on the range space of the vector measures.

To prove Theorem 3, we need still another result in [8, Theorem 2].

Theorem 11. *Let S be a completely regular space whose compact subsets are all metrizable. Let X be a semi-Montel space whose topological dual X^* has a countable set which separates points of X. Let $\mathcal{V}$ be a subset of $\mathcal{M}_t(S, X)$ and assume that for each $x^* \in X^*$, $x^*(\mathcal{V})$ is uniformly bounded and uniformly tight. Then, $\mathcal{V}$ is relatively compact and metrizable, so that it is relatively sequentially compact in $\mathcal{M}_t(S, X)$.*

We are now ready to prove Theorem 3.

Proof of Theorem 3. The implications that (4) implies (1), that (5) implies (2), and that (6) implies (3) are trivial. The implication that (3) implies (1) and the implication that (2) implies (1) follow from [16, Theorem 30, Part II]. The implication that (1) implies (5) and the implication that (1) implies (6) follow from Theorem 11. Finally, the implication that (6) implies (4) follows from Proposition 9, and the proof is complete.

References

[1] P. Billingsley, *Convergence of Probability Measures.* John Wiley & Sons, New York, 1968.

[2] M. Dekiert, *Kompaktheit, Fortsetzbarkeit und Konvergenz von Vectormassen.* Dissertation, University of Essen, 1991.

[3] J. Diestel and J. J. Uhl, Jr., *Vector Measures.* Amer. Math. Soc. Surveys No. 15, Providence, 1977.

[4] N. Dinculeanu, *Vector Integration and Stochastic Integration in Banach Spaces.* John Wiley & Sons, New York, 2000.

[5] J. Hoffmann-Jørgensen, *The Theory of Analytic Spaces.* Matematisk Institut, Aarhus Universitet, Various Publication Series No. 10, Aarhus, 1970.

[6] H. Jarchow, *Locally Convex Spaces.* B. G. Teubner, Stuttgart, 1981.

[7] J. Kawabe, *Weak convergence of tensor products of vector measures with values in nuclear spaces.* Bull. Austral. Math. Soc. **59** (1999), 449–458.

[8] ———, *Compactness criteria for the weak convergence of vector measures in locally convex spaces.* Publ. Math. Debrecen **60** (2002), 115–130.

[9] ———, *Compactness and metrizability in the space of vector measures in locally convex spaces.* Sci. Math. Japonicae **55** (2002), 493–503.

[10] I. Kluvánek and G. Knowles, *Vector Measures and Control Systems.* North-Holland, 1976.

[11] L. LeCam, *Convergence in distribution of stochastic processes.* Univ. California Publ. Statist. **2** (1957), 207–236.

[12] D. R. Lewis, *Integration with respect to vector measures.* Pacific J. Math. **33** (1970), 157–165.

[13] M. März and R. M. Shortt, *Weak convergence of vector measures.* Publ. Math. Debrecen **45** (1994), 71–92.

[14] Yu. V. Prokhorov, *Convergence of random processes and limit theorems in probability theory.* Theory Probab. Appl. **1** (1956), 157–214.

[15] F. Topsøe, *Topology and Measure.* Lecture Notes in Math. **133**, Springer, Berlin, 1970.

[16] V. S. Varadarajan, *Measures on topological spaces.* Amer. Math. Soc. Transl. Ser. II **48** (1965), 161–228.

Department of Mathematics, Faculty of Engineering
Shinshu University, Wakasato, Nagano 380-8553, Japan
E-mail address: `jkawabe@gipwc.shinshu-u.ac.jp`

Progress in Probability, Vol. 55, 43–54

On Measures in Locally Convex Spaces

A.V. Uglanov

Abstract. The deep and close connection between absolutely summing (or, more generally, p-absolutely summing) mappings and measure theory is well known (see, e.g., [1–3]). However, the related considerations have only been concerned with Banach spaces. (In general, both absolutely summing mappings and measures in nonnormed spaces are studied very poorly.) The basic result of this paper, Theorem 3.1, appears purely topological, for no measures are mentioned in its statement. However, in reality, this result is promising in the study of the relation mentioned above in the general-topological case. This statement is not unfounded, because Theorem 3.1 has already made it possible to obtain a new generalization of the Sazonov theorem, to prove the existence of a Radon–Nikodym density for vector measures in a fairly general situation, (and, as a corollary, the existence of a logarithmic gradient of a differentiable measure), and to write the Gauss–Ostrogradskii formula in an efficient (for applications; see [3]) scalar form (see Theorems 4.1, 6.4, 7.3, and 7.8, respectively). Without going into details, we also mention that Theorem 3.1 is important for the calculus of variations, Lagrange problem, and boundary value problems on nonmetrizable spaces (these problems are not considered in this paper, but some results are given in [4, 5]).

1. General notation

We shall use the following notation: E is an abstract locally convex space (LCS); p_A is the gauge function of a subset $A \subset E$; $\mathbb{U}(E)$ is a fundamental system of closed absolutely convex neighborhoods of zero of the space E; $L(E, F)$ is the space of linear continuous mappings from E to an LCS F; $(\cdot)^*$ is the object topologically conjugate to an object $(\cdot)$; (Ω, Σ) is an abstract measurable space ($\Omega \in \Sigma$); $M(\Omega, E)$ is the space of E-valued measures (i.e., countably additive functions $\Sigma \to E$); $M(\Omega) = M(\Omega, \mathbb{R}^1)$; X, Y, and Z are Hausdorff LCS's; and H is a Hilbert space. All linear spaces under consideration are real. The subsets of topological spaces

---

Received by the editors November 30, 2002.
2000 *Mathematics Subject Classification.* Primary 46G12; Secondary 46G10, 58C35.
Key words and phrases. locally convex space, absolutely summing map, measure, vector integration.
This work was financially supported by the Russian Foundation for Basic Research (project 01-01-00701).

are assumed to be endowed with the induced topology, unless otherwise specified. Any topological space T is considered measurable with Borel σ-algebra Σ_T.

2. Preliminaries

Definition 2.1. The sequence $\{e_n\} \in E$ is called: *weakly summable*, if the inequality $\sum_n |(e_n, a)| < \infty$ is valid for any $a \in E^*$; *absolutely summable*, if the inequality $\sum_n p_U(e_n) < \infty$ is valid for any set $U \in \mathbb{U}(E)$.

We denote by $l^1[E]$ and $l^1\{E\}$ the sets of all weakly summable and absolutely summable sequences from E. These sets are linear spaces, and the systems of seminorms

$$\sigma_U\{e_n\} \stackrel{\text{def}}{=} \sup\left\{\sum_n |(e_n, a)| : a \in U^0\right\}, \quad q_U\{e_n\} \stackrel{\text{def}}{=} \sum_n p_U(e_n),$$

where U runs through the set $\mathbb{U}(E)$ define the locally convex topologies in the spaces $l^1[E]$ and $l^1\{E\}$ respectively (here $U^0 \subset E^*$ is the polar of U; see [6], 1.2, 1.4 for details).

Definition 2.2. The operator $S \in L(E,X)$ is called *absolutely summing* if $S\left(l^1[E]\right) \subset l^1\{X\}$.

We denote the set of all absolutely summing operators from $L(E,X)$ by $L_1(E,X)$.

Definition 2.3. The space X has B-property, if for any bounded set $A \subset l^1\{X\}$ there exists a bounded set $C \subset X$, such that

$$\sum_n p_C(x_n) \leq 1 \quad \forall \{x_n\} \in A. \tag{2.1}$$

Throughout, we assume that the space X is sequentially complete and has the B-property. It should be mentioned that virtually all spaces arising in analysis, such as metric, dual-metric, nuclear, dual-nuclear, etc., have the B-property (see [6], 1.4 for details).

3. A result on absolutely summing operators

Theorem 3.1. *The inclusion $S \in L_1(H,X)$ holds if and only if there exist a Hilbert space G, a Hilbert–Schmidt operator $T \in L(H,G)$, and an operator $R \in L(G,X)$ such that $S = RT$.*

Proof. As is known ([6], 2.5), if G is a Hilbert space, then

$$L_1(H,G) = \text{the set of all Hilbert–Schmidt operators from } L(H,G). \tag{3.1}$$

Now sufficiency of the theorem's conditions is evident. Let us prove necessity of the conditions. Note, that if X is a Banach space, then this necessity was proved

([3], 1.2). Let us consider the case of general LCS X. We denote by V the unit ball in H; then the space's $l^1[H]$ topology is normalized with the norm $\|\cdot\| = \sigma_V\{\cdot\}$. If V^1 is the unit ball in $l^1[H]$, then the set $A = \{\{S(h_n)\} : \{h_n\} \in V^1\}$ is bounded in $l^1\{X\}$ ([6], 3.1), and let C be the set, which appeared in the B-property's definition. Since the closed absolutely convex hull of a bounded set is the bounded set, we can regard that the set C is closed absolutely convex one. We put $F = \{tC : t \in \mathbb{R}^1\}$; then F is the normed space with the norm $\|\cdot\|_F \stackrel{\text{def}}{=} p_C(\cdot)$ ([6], 0.11), and by virtue of the space's X sequential completeness, F is the Banach space. Since C is bounded set, F is continuously enclosed in X. The inequality (2.1) gives the inclusion $S(V) \subset F$ and, further, the inclusion $S \in L_1(H, F)$. □

4. Sazonov theorem

The notion of (real-valued sign-indefinite) cylindrical measure in an LCS used below can be found in, e.g., [2, 7]. A cylindrical measure m in E is called continuous if, for any $\varepsilon > 0$ there exists a neighborhood U of zero in E such that, for any half-space $L \subset E$ disjoint from U and cylindrical set $K \subset L$, the inequality $|m(K)| < \varepsilon$ holds. An operator $S \in L(E, Y)$ is called radonifying if, for any continuous bounded cylindrical measure m in E, the cylindrical measure $m \circ S^{-1}$ induced in Y is countably additive.

Theorem 4.1. *An operator $S \in L(H, X)$ is radonifying if and only if $S \in L_1(H, X)$.*

Proof. 1. Sufficiency. Let G, T, R be the same that in Theorem 3.1. Since T is a Hilbert–Schmidt operator, by Sazonov theorem (see, e.g., [2], VI.5.2), the cylindrical measure $m \circ T^{-1}$ induced in G is countably additive. Since R is linear continuous operator, the cylindrical measure $\left(m \circ T^{-1}\right) \circ R^{-1}$ induced in X is countably additive. But evidently $m \circ S^{-1} = \left(m \circ T^{-1}\right) \circ R^{-1}$.
2. Necessity. For $U \in \mathbb{U}(X)$ let $N(U) = \{x \in X : p_U(x) = 0\}$; $N(U)$ is linear subspace of X, and we consider the factor space $X/N(U)$. It is the normed space with respect to canonical norm $\|\cdot\|_U$. Since the operator S is radonifying, and the canonical map $J_U : X \to X/N(U)$ is linear and continuous operator, the operator $J_U S : H \to X/N(U)$ is radonifying one. But then $J_U S \in L_1(H, X/N(U))$ ([2], VI.5.2) i.e., $\sum_n \|J_U S(h_n)\|_U < \infty$ for any sequence $\{h_n\} \in l^1[H]$. But $\|J_U S(\cdot)\|_U = \|J_U(S(\cdot))\|_U$, and $\|J_U(\cdot)\|_U = p_U(\cdot)$ □

Remark 4.2. By (3.1), the result stated indeed generalizes the Sazonov theorem (to be more precise, its most important corollary; see [2], VI.5.2). Let us also mention that the result is possibly new even for a Banach space X (this case is thoroughly examined in [1, 2]).

5. Vector integrals

In this section we assume that the spaces X and Y are vector-dual in the sense that an X-hypocontinuous bilinear form $\langle\cdot,\cdot\rangle : Y \times X \to Z$ is defined. (The X-hypocontinuity of the form means that, for any bounded set $B \subset X$ and any neighborhood $U \in \mathbb{U}(Z)$, there exists a neighborhood $V \in \mathbb{U}(Y)$ such that $\langle V, B\rangle \subset U$. X-hypocontinuity follows from joint continuity in all variables and, as a rule, implies separate continuity. See [8], 7.7.1 for more details.) Further, we assume that the LCS Z is sequentially complete space. We denote by π_Q the totality of all countable measurable partitions of the set $Q \in \Sigma$. For measure $\mu \in M(\Omega, X)$ and set $U \in \mathbb{U}(X)$ we put

$$|\mu|_U(Q) = \left\{ \sup_{\{Q_n\} \in \pi_Q} \sum_n p_U(\mu(Q_n)) \right\}.$$

(If X is a Banach space and U is the unit ball in X, then $|\mu|_U$ is ordinary total variation of the measure μ; in these cases we shall omit the lower index U.) As is known, the function $|\mu|_U : \Sigma \to [0,\infty]$ is countably additive [9]. We shall call the measure μ *bounded*, if $|\mu|_U(\Omega) < \infty$ for any $U \in \mathbb{U}(X)$. Let $MB(\Omega, X)$ be the collection of all bounded measures from $M(\Omega, X)$. For $\mu \in MB(\Omega, X)$ the set $A = \{\{\mu(\Omega_n)\} : \{\Omega_n\} \in \pi_\Omega\}$ is bounded in the space $l^1\{X\}$. Let the set $C \subset X$ and Banach space F be the same that in the Theorem's 3.1 proof. The inequality (2.1) ($x_n = \mu(\Omega_n)$, $\{\Omega_n\} \in \pi_\Omega$) and measure's μ additivity give the inclusion $\{\mu(\Sigma)\} \subset F$ and, further, countable additivity of the function $\mu_F : \Sigma \to F : \mu_F(Q) = \mu(Q)$; moreover, the measure μ_F is bounded. We call a function $f : \Omega \to Y$:
a) *simple* if there is a partition $\{\Omega_n\} \in \pi_\Omega$ such that $f = l_n =$ const on Ω_n;
b) *measurable* if there exists a sequence $f_n : \Omega \to Y$ of simple functions uniformly converging to f;
c) *integrable* with respect to a measure $\mu \in MB(\Omega, X)$ or *μ-integrable* if f is measurable and the integral $\int_\Omega p_U(f)\, d|\mu|$ is finite for any $U \in \mathbb{U}(Y)$ (hereafter $|\mu| \overset{\text{def}}{=} |\mu_F|$).
For a simple integrable function f we put

$$\int_Q f\, d\mu = \sum_{n=1}^{\infty} \langle l_n, \mu(Q \cap \Omega_n)\rangle \qquad (Q \in \Sigma), \tag{5.1}$$

and for an integrable function f we put

$$\int_Q f\, d\mu = \lim_{n\to\infty} \int_Q f_n\, d\mu. \tag{5.2}$$

(a *vector integral*).
It is easy to check that the definition of the integral is correct (i.e., the series in (5.1) converges and the sum doesn't depends of a partition's Ω_n choice; the limit

in (5.2) exists and doesn't depends of a sequence's f_n choice), and the entered integral is linear with respect to f and μ.

Remark 5.1. The integrals of Banach-valued functions with respect to Banach-valued measures were introduced in [10] and thoroughly studied in [11, 12, 3]. The case of general LCS's was for the first time considered in [13, 14]. (True, in the papers [10 - 13, 3], it was assumed that $Y = L(X, Z)$ with the topology of uniform convergence on all bounded sets and $\langle y, x\rangle = y(x)$). Present definition is a descendant of [10, 13].

In what follows, the Banach space $F = F(\mu)$ is the same that has been constructed in this section.

Lemma 5.2. *For any $U \in \mathbb{U}(Z)$ there exists $V \in \mathbb{U}(Y)$ such that, for any μ-integrable $f : \Omega \to Y$ and $Q \in \Sigma$ the inequality*

$$p_U\left(\int_Q f\,d\mu\right) \leq \int_Q p_V(f)\,d|\mu| \tag{5.3}$$

holds.

Proof. Let a function f be simple. It follows from form's $\langle\cdot,\cdot\rangle : Y \times X \to Z$ X-hypocontinuous that

$$\forall U \in \mathbb{U}(Z)\ \exists V \in \mathbb{U}(Y):\ p_U(\langle y, x\rangle) \leq p_V(y)\,\|x\|_F \quad \forall x \in F,\ y \in Y. \tag{5.4}$$

The inequality (5.3) now follows from definition (5.1). Inequality (5.3) in a general case is obtained from the proved and definition (5.2) with the help of passage to limit. □

Theorem 5.3. *Let a function $f : \Omega \to Y$ be μ-integrable. Then the function $\Sigma \to Z : Q \mapsto \int_Q f\,d\mu$ is countably additive.*

Proof. – Follows from vector integral's definition and Lemma 5.2. □

Theorem 5.4. *Let $\{f_n\}$ be a sequence of measurable functions from Ω into Y, $\mu \in MB(\Omega, X)$ and the following conditions are fulfilled: 1) for any $\omega \in \Omega$ $\lim_{n\to\infty} f_n(\omega) \overset{\text{def}}{=} f(\omega)$ exists; 2) for any $U \in \mathbb{U}(Y)$ there exists a $|\mu|$-integrable function $f_U : \Omega \to \mathbb{R}^1$ such, that $\forall \omega \in \Omega$ $\sup_n p_U(f_n(\omega)) \leq f_U(\omega)$; 3) the function f is measurable. Then the equality*

$$\lim_{n\to\infty} \int_Q f_n\,d\mu = \int_Q f\,d\mu$$

is well posed and valid.

Proof. μ-integrability of functions f_n, f follows from definition and theorem's conditions. By virtue of (5.3) for any $U \in \mathbb{U}(Z)$ there is $V \in \mathbb{Y}$ such that

$$p_U\left(\int\limits_Q f\,d\mu - \int\limits_Q f_n\,d\mu\right) = p_U\left(\int\limits_Q (f - f_n)\,d\mu\right) \le \int\limits_Q p_V(f - f_n)\,d|\mu|. \quad (5.5)$$

The classical Lebesgue theorem gives that right side of (5.5) converges to zero when $n \to \infty$. □

Remark 5.5. If Y is a metrizable space, then the Theorem's 5.4 condition 3 is unnecessary (it follows from the condition 1).

6. Reduction: vector integral to classical integrals

How is the introduced integral related to the classical integrals in the case of $X = \mathbb{R}^1$? We examine a more general situation. Suppose that $\mu \in MB(\Omega, X)$, $\nu \in M(\Omega)$. First of all we note that according to vector integral's definition (5.2) and construction of the Banach space F and the measure $\mu_F \in MB(\Omega, F)$ the equality

$$\int\limits_Q f\,d\mu = \int\limits_Q f\,d\mu_F,$$

holds, and bilinear form $\langle\cdot,\cdot\rangle : Y \times F \to Z$ is continuous in all variables (see the inequality (5.4)). In this connection we remind that Bochner ν-integrable function $\frac{d\mu}{d\nu} : \Omega \to F$ is called a Radon–Nikodym density (of the measure μ_F with respect to the measure ν) if $\forall Q \in \Sigma$ $\mu(Q) = (\mu_F(Q)) = \int\limits_Q \frac{d\mu}{d\nu}\,d\nu$. Below we assume that the spaces $\mathbb{R}^1$ and E are in the natural vector duality determined by $\langle e, t\rangle = te$.

Lemma 6.1. *Let $X = \mathbb{R}^1, Y = Z$ be a Banach space. A function $g : \Omega \to Y$ is Bochner ν-integrable, if and only if there exists a ν-integrable in our sense function $g_0 : \Omega \to Y$ such, that $g = g_0$ almost everywhere (with respect to ν). In this connection the corresponding integrals coincide.*

Proof. As was proved in [15], Bochner ν-integrability of the function g is equivalent to its Bochner measurability (= strong ν-measurability) together with the validity of the inequality $\int\limits_\Omega \|g\|_Y d|\nu| < \infty$. The Bochner measurability of g is equivalent to the existence of a Borel separable-valued function $g_0 : \Omega \to Y$, which is equal to g almost everywhere ([2], II.1). The function Borel measurability together with the separability of the set of the function values are equivalent to the strong measurability (i.e., the measurability which figured in section 4, see [2], I.1); the lemma statement follows now from the definitions of the Bochner integrals and section 4. □

This lemma allows us not to specify later on what ν-integral we consider: the Bochner one (in particular the Lebesgue one) or which had been introduced above.

Theorem 6.2. *Suppose that the Radon–Nikodym density* $\frac{d\mu}{d\nu}$ *exists. Then, for any* μ*-integrable function* $f : \Omega \to Y$*, the equality*

$$\int_\Omega f\, d\mu = \int_\Omega \left\langle f, \frac{d\mu}{d\nu} \right\rangle d\nu\,, \tag{6.1}$$

is well posed and valid; the integral in the right-hand side can also be read as an integral in the sense of Pettis, and if the space Z *is Banach, then this integral can be understood in the sense of Bochner, too.*

Proof. The ν-integrability of the function $\frac{d\mu}{d\nu}$ delivers the ν-integrability of the function $\left\| \frac{d\mu}{d\nu} \right\|_F$, therefor for any $Q \in \Sigma$ we have

$$|\mu|(Q) = \int_Q \left\| \frac{d\mu}{d\nu} \right\|_F d|\nu|$$

([2], II.2). Hence, by inequality (5.4) and μ-integrability of the function f we obtain

$$\int_\Omega p_U\left(\left\langle f, \frac{d\mu}{d\nu} \right\rangle\right) d|\nu| \le \int_\Omega p_V(f) \left\| \frac{d\mu}{d\nu} \right\|_F d|\nu| =$$

$$= \int_\Omega p_V(f) \frac{|d\mu|}{|d\nu|} d|\nu| = \int_\Omega p_V(f)\, d|\mu| < \infty\,,$$

which proves the existence of the integral in the right-hand side of (6.1). Let f be a simple function. Then

$$\int_\Omega f\, d\mu = \sum_{n=1}^{\infty} \langle l_n, \mu(Q_n) \rangle = \sum_{n=1}^{\infty} \left\langle l_n, \int_{Q_n} \frac{d\mu}{d\nu} d\nu \right\rangle. \tag{6.2}$$

The bilinear form's $\langle \cdot, \cdot \rangle : Y \times F \to Z$ continuity (5.4) and the vector integral's definition imply the equality

$$\left\langle l_n, \int_{Q_n} \frac{d\mu}{d\nu} d\nu \right\rangle = \int_{Q_n} \left\langle l_n, \frac{d\mu}{d\nu} \right\rangle d\nu$$

(by virtue of Lemma 6.1 we consider that function $\frac{d\mu}{d\nu}$ is measurable). Hence, by (6.2) we obtain

$$\int_\Omega f\, d\mu = \sum_{n=1}^{\infty} \int_{Q_n} \left\langle l_n, \frac{d\mu}{d\nu} \right\rangle d\nu = \sum_{n=1}^{\infty} \int_\Omega 1_{Q_n} \left\langle l_n, \frac{d\mu}{d\nu} \right\rangle d\nu. \tag{6.3}$$

Now let us note the following.

1. For any $\omega \in \Omega$ $\lim\limits_{n\to\infty} f_n(\omega) \stackrel{\text{def}}{=} \sum\limits_{j=1}^{n} 1_{Q_j} \left\langle l_j, \frac{d\mu}{d\nu} \right\rangle = \left\langle f(\omega), \frac{d\mu_F}{d\nu}(\omega) \right\rangle.$

2. $\forall\, U \in \mathbb{U}(Z)$, $\exists\, V \in \mathbb{U}(Y)$, $\forall\, \omega \in \Omega$ the inequality $\sup\limits_n p_U \left\langle f_n(\omega), \frac{d\mu}{d\nu}(\omega) \right\rangle \le$

$p_V(f(\omega)) \left\| \frac{d\mu}{d\nu}(\omega) \right\|_F \stackrel{\text{def}}{=} f_U(\omega)$, and the function f_U is ν-integrable.

3. The function $\left\langle f, \frac{d\mu}{d\nu} \right\rangle : \Omega \to Z$ is measurable on any set Q_n, and therefore is measurable.

Let us put in theorem 5.4 $X = \mathbb{R}^1, Y = Z, (\langle y, x \rangle = xy), \mu = \nu$. Then all conditions of the theorem are fulfilled, and, using (6.3), we obtain the equality (6.1). The coincidence of the integral in the right-hand side of (6.1) and Pettis (Bochner, if Z is a Banach space) follows from definitions of corresponding integrals. □

Note that, if $\mu \in M(\Omega, E)$, $A \in L(E, X)$, then the measure $A\mu \in M(\Omega, X)$ specified by the equality $A\mu(Q) = A(\mu(Q)) \quad (\forall Q \in \Sigma)$ is well defined; this observation is used in what follows.

Definition 6.3. Let E, G be Banach spaces. An operator $A \in L(E, G)$ is called *a Radon–Nikodym operator*, if for any $\mu \in MB(\Omega, E)$, $\nu \in M(\Omega)$ such, that μ is absolutely continuous with respect to $\nu\,(\mu \ll \nu)$, the Radon–Nikodym density $\frac{dA\mu}{d\nu}$ exists.

Theorem 6.4. *Let $\mu \in M(\Omega, H)$, $S \in L_1(H, X)$. Then:*

1) $S\mu \in MB(\Omega, X)$;
2) *there exists a measure $\nu \in M(\Omega)$ such that $S\mu \ll \nu$;*
3) *for any measure $\lambda \in M(\Omega)$, such that $S\mu \ll \lambda$, the Radon–Nikodym density $\frac{dS\mu}{d\lambda}$ exists.*

Proof. Let us note firstly, that we may consider the case $\ker S = 0$ (the general case is easily reduced to this case). Let G, T, R be the same that in Theorem 3.1. As $T \in L_1(H, G)$ (see (3.1)), then $T\mu \in MB(\Omega, G)$ ([3], 1.2), and all the more $S\mu \in MB(\Omega, X)$. As $T\mu \in MB(\Omega, G)$, then $\nu \stackrel{\text{def}}{=} |T\mu| \in M(\Omega)$. It is obviously that $T\mu \ll \nu$, and all the more $S\mu \ll \nu$. Let $\lambda \in M(\Omega)$ and $S\mu \ll \lambda$. Let G, T, R be the same that in Theorem 3.1, with X is replaced by $F = F(S\mu)$ (see section 4; we remind, that $S \in L_1(H, F)$). As $\ker S = 0$, then $T\mu \ll \lambda$. As G is a Hilbert space, then the identical mapping $J : G \to G$ is a Radon–Nikodym operator ([1], 24.2.4, 24.3.6). But the Radon–Nikodym operators class is an operator ideal ([1], 24.2.7), that is why $R = JR$ is a Radon–Nikodym operator. Therefor the Radon–Nikodym density $\frac{d(RT\mu)}{d\lambda}$ exists (we remind, that $T\mu \in MB(\Omega, G)$). But $RT\mu = S\mu$. □

Remark 6.5. Much attention has been paid to the problem of the existence of Radon–Nikodym densities for Banach-valued measures (see, e.g., [1]). Some considerations of nonnormed case are present in [7].

7. Smooth and surface measures

In this section X_0 is a linear subset of X, $V \in \Sigma_X$, $V^0, \overline{V}$, and ∂V are, respectively, the interior, closure, and boundary of V.

Definition 7.1. A measure $\nu \in M(X)$ is called *differentiable* (with respect to X_0), if for all $h \in X_0$ and $C \in \Sigma_Z$, there exists a limit

$$\lim_{t\to 0} \frac{1}{t}[\nu(C+th) - \nu(C)] \stackrel{\text{def}}{=} D_h\nu(C).$$

Definition 7.2. A measure $\nu \in M(V)$ is called differentiable, if a differentiable measure $\bar{\nu} \in M(X, \mathbb{R}^1))$ exists such that $\bar{\mu} \equiv \mu$ on V.

We denote by $M^1(V) \subset M(V)$ the set of all differentiable Radon measures. If $\nu \in M^1(V)$, then for any $h \in X_0$, the function $D_h\nu : \Sigma_V \to \mathbb{R}^1 : C \mapsto D_h\bar{\nu}(C)$ is countably additive (Nikodym theorem); i.e., $D_h\nu \in M(V)$. Therefore, higher order derivatives of ν can be defined inductively. It is well known ([7], IV.4) that, if X_0 is a Fréchet space, and the inclusion map $X_0 \to X$ is continuous, then for any $\nu \in M^1(V)$, $\nu' : X_0 \to M(V) : h \mapsto D_h\nu$ is a linear and continuous mapping (the space $M(V)$ is considered with the norm $\|\cdot\| = \text{var}(\cdot)$). This result and the Pettis theorem ([8], 9.4.14) imply that, if X_0 is reflexive, then the measure $\nu' \in M(V, X_0^*)$ is well defined by the equality $(\nu'(C), h) = D_h\mu(C) \quad (\forall h \in X_0,\, C \in \Sigma_V)$. In what follows, $Z_0 = H$, and we suppose that the embedding operator $I : H \to X$ is absolutely summing; $V \in \Sigma_X$, $\nu \in M^1(\overline{V})$. We note, that by Theorem 6.4, $I\nu' \in MB(V, X)$. In the following theorem $F = F(I\nu\prime)$.

Theorem 7.3. *The Radon–Nikodym density* $\frac{dI\nu'}{d\nu}$ *exists.*

Proof. As is known, $\nu' \ll \nu$ ([7], IV.4), and all the more $I\nu' \ll \nu$. The result of the theorem follows now from Theorem 6.4 ($\mu = \nu', S = I, \lambda = \nu$). □

Remark 7.4. The density $\frac{dI\nu'}{d\nu}$ is called the logarithmic gradient of the measure ν. The existence of the logarithmic gradient of a differentiable measure is very important (see, e.g., [16 - 18]). Until paper [18] appeared, only partial results on this topic were known. In [18], Theorem 7.3 is proved for a Banach X.

Definition 7.5. A set $S \in \Sigma_X$ is called a *H-smooth surface* if for any point $s \in S$, there exist an open neighborhood $U \ni s$ and a continuous bounded function $f : U \to \mathbb{R}^1$, continuously differentiable with respect to H such that $f'(s) \neq 0$ and $S \cap U = f^{-1}(0)$.

Let $S \subset V$ be a H-smooth surface. A locally finite Borel measure ν_S supported by S (*surface measure*) was constructed in [3]. The following natural properties of the surface measure were proved: localization (i.e., if $\nu = \lambda$ in a neighborhood of S, then $\nu_S = \lambda_S$); the iterated integration, integration-by-parts, Gauss–Ostrogradskii, and (both) Green's formulas; surface-layer theorems, etc. If $X = H = \mathbb{R}^n$ and ν is a Lebesgue measure, then ν_S coincides with the classical (geometric) surface measure on S.

Remark 7.6. In fact the surface measure is constructed on surfaces much more general than H-smooth surfaces [3]. In particular, the surface S can possess singular points (roughly speaking, the set of singular points must be a countable union of surfaces of codimension 2). Accordingly, the following Theorem 7.8 can be stated for a wider class of sets V.

Lemma 7.7. *For any $A \in L(X, X^*)$, the operator $I^*AI : H \to H$ is nuclear.*

Proof. As is known (see, e.g., [19], VI.6), the following statements are valid: 1) the Hilbert–Schmidt operators class is an operator ideal; 2) an operator conjugate to Hilbert–Schmidt operator is an Hilbert–Schmidt one; 3) the composition of two Hilbert–Schmidt operators is nuclear operator. Let G, T, R be the same that in Theorem 3.1 ($S = I$). Thus, we have the chain $X^* \overset{R^*}{\to} G^* \overset{T^*}{\to} H \overset{T}{\to} G \overset{R}{\to} X$, and also $I = RT$, $I^* = T^*R^*$. The result of the lemma follows now from the statements 1) – 3). □

Hereafter: it is supposed, that ∂V is H-smooth surface; ∂V_n is the set of all points $s \in \partial V$ where the outer (with respect to V) unit normal $n = n(s) \in H$ to the surface ∂V exists (a normal $n(s)$ at a point s is called outer if, for any vector $h \in H$ satisfying condition $(n(s), h) > 0$, there exists an $\varepsilon > 0$ such that $s - \delta h \in V^0$, and $s + \delta h \in X \backslash \overline{V}$ for all $\delta \in (0, \varepsilon)$); $\nu_{\partial V}$ is a surface measure on ∂V; $b : \overline{V} \to X^*$ is a bounded continuous function (X^* is considered with the strong topology) continuously and boundedly differentiable in the sense of Fréchet (i.e., with respect to a system of bounded sets in X) on the set V^0. We continue the function $n : \partial V_n \to H$ on ∂V assuming that $n(s) = 0$ if $s \in \partial V \backslash \partial V_n$. By Lemma 7.7, the function $\operatorname{div} b : V^0 \to \mathbb{R}^1$ such that $x \mapsto \operatorname{Tr} I^* b'(x) I$ is well defined; as a matter of fact, it is even bounded. Note that $|\nu|(\partial V) = 0$, because ∂V is a surface.

Theorem 7.8. *If the function b is measurable, and the function $\|I^*b\| : \partial V \to \mathbb{R}^1$ is $\mu_{\partial V}$-integrable, then the equality*

$$\int_V \left[\operatorname{div} b + \left(b, \frac{dI\mu'}{d\mu} \right) \right] d\nu = \int_{\partial V} (b, In)\, d\nu_{\partial V} \tag{7.1}$$

(the Gauss–Ostrogradskii formula) is well posed and valid.

Proof. Under the theorem's conditions the equality

$$\int_V \operatorname{div} b\, d\nu + \int_V b\, dI\nu' = \int_{\partial V} (b, In)\, d\nu_{\partial V} \tag{7.2}$$

was established in [13], where $\int_V b\, dI\nu'$ is the vector integral ($\Omega = V, Y = X^*, Z = \mathbb{R}^1, \mu = I\nu', f = b$). By Theorem 6.2 we have the equality

$$\int_V b\, dI\nu' = \int_V \left(b, \frac{dI\nu'}{d\nu} \right) d\nu,$$

which proves the theorem. □

Remark 7.9. The Gauss–Ostrogradskii formula (in an LCS) in an inefficient (containing limit expressions) scalar form is given in [3] and in an efficient vector form (7.2), in [13]. Form (7.1) is most efficient, for it only involves usual Lebesgue integrals. We note in addition that Gauss–Ostrogradskii formula is extremely important for applications (see [3]).

References

[1] A. Pietsch, *Operator Ideals*, VEB, Berlin, 1978.

[2] N.N. Vakhaniya, V.I. Tarieladze, and S.A. Chobanyan, *Probabilistic Distributions in Banach Spaces*, Nauka, Moscow, 1985; English transl., Reidel, New York, 1987.

[3] A.V. Uglanov, *Integration on Infinite-Dimensional Surfaces and Its Applications*, Kluwer Academic Publishers, Dordrecht, 2000.

[4] A.V. Uglanov, Control of Systems, Evolving in "Infinite-Dimensional Time", *Intern. Conf. "New Direction in Dynamical Systems". Abstracts.* Kyoto Univ. Publ., Kyoto, 2002, 470–474.

[5] A.V. Uglanov, Potentials and Boundary Value Problems in Locally Convex Spaces, *Dokl. Akad. Nauk Russia*, 387 (2002), 1–5; English transl. in *Dokl. Math. Sci.*, 66 (2002).

[6] A. Pietsch, *Nukleare lokalkonvexe räume*, Academie-Verlag, Berlin, 1965.

[7] Yu.L. Daletskii and S.V. Fomin, *Measures and Differential Equations in Infinite-Dimensional Spaces*, Nauka, Moscow, 1983; English transl., Kluwer Academic Publishers, Dordrecht, 1991.

[8] R. Edwards, *Functional Analysis. Theory and Applications*, Holt, Rinehart, and Winston, New York, 1965.

[9] N. Dinculeanu, Vector Measures, VEB Deutscher Verlag, Berlin, 1966.

[10] E.I. Efimova and A.V. Uglanov, Formulas of Vector Analysis on a Banach Space, *Dokl. Akad. Nauk SSSR*, 271 (1983), 1302–1307; English transl. in *Soviet Math. Dokl.*, 28 (1983).

[11] A.V. Uglanov, Fubini Theorem for Vector Measures, Matem. Sbornik, 181 (1990), 423–432; English transl. in *Math. USSR Sbornik*, 69 (1991), 453–463.

[12] A.V. Uglanov, Integrals with Respect to Vector Measures: Theoretical Problems and Applications, *Amer. Math. Soc. Transl.*, ser. 2, 163 (1995), 171–184.

[13] A.V. Uglanov, Vector Integrals, Dokl. Akad. Nauk Russia, 373 (2000), 737–740; English transl. in *Russian Acad. of Sci. Dokl.* (2000).

[14] A.V. Uglanov, Absolutely Summing Mappings of Locally Convex Spaces in Measure Theory, *Dokl. Acad. Nauk Russia*, 380 (2001), 319–322; English transl. in *Russian Acad. of Sci. Dokl.* (2000).

[15] J. Diestel and J.J. Uhl, *The Theory of Vector Measures.* Providence, 1977.

[16] N.V. Norin, Stochastic Integrals and Differentiable Measures, *Teor. Ver. i Prim.*, 32 (1987), 114–124; English transl. in *Theory Prob. Appl.*, 32 (1987).

[17] Yu. L. Daletskii Yu.L. and V.R. Steblovskaya, On Infinite Dimensional Variational Problems. *Stochastic Analysis and Applications*, 14 (1996), 47–71.

[18] A.V. Uglanov, Variational Calculus on Banach Spaces, *Matem. Sbornik*, 191 (2000), 105–118; English transl. in *Sbornik Math.*, 191 (2000), 1527–1540.

[19] M. Reed and B. Simon, *Methods of Modern Mathematical Physics. 1. Functional Analysis*, Academic Press, New York, 1972.

Yaroslavl State University, Yaroslavl, 150000, Russia
E-mail address: uglanov@univ.uniyar.ac.ru

Part 2

Gaussian Processes

Progress in Probability, Vol. 55, 57–93

Karhunen–Loève Expansions for Weighted Wiener Processes and Brownian Bridges via Bessel Functions

Paul Deheuvels and Guennady Martynov

Abstract. We provide Karhunen–Loève expansions on $(0,1)$ for the processes $t^{\theta}B(t^{\rho})$ and $t^{\theta}W(t^{\rho})$, where $B(\cdot)$ a Brownian bridge, $W(\cdot)$ is a Wiener process, and ρ and θ are arbitrary real numbers such that $\theta > -(\rho+1)/2$. The eigenfunctions of these expansions have simple expressions in terms of Bessel functions $J_{\nu}(\cdot)$ and $J_{\nu-1}(\cdot)$ of the first kind with indexes $\nu = \rho/(2\theta+\rho+1)$ and $\nu - 1 = -(2\theta+1)/(2\theta+\rho+1)$. The corresponding eigenvalues have simple expressions in terms of the positive zeros $z_{\nu,1} < z_{\nu,2} < \ldots$ of $J_{\nu}(\cdot)$ and $z_{\nu-1,1} < z_{\nu-1,2} < \ldots$ of $J_{\nu-1}(\cdot)$. These results allow us, in particular, to obtain explicitly, for each $\nu > 0$, a Fourier-Bessel expansion on $(0,1)$ of $t^{-\nu}B(t^{2\nu})$, and a Dini expansion on $(0,1)$ of $t^{-\nu}W(t^{2\nu})$. As a consequence, we obtain explicit forms for the distribution functions of the $L^2(0,1)$ norms of $t^{\theta}W(t^{\rho})$ and $t^{\theta}B(t^{\rho})$, whose upper quantiles are tabulated in part. These distributions are also shown to be identical up to a multiplicative constant after the change of ν into $\nu+1$. We provide sharp evaluations for small deviations as well as for large deviations for these $L^2(0,1)$ norms. This allows us to obtain simple proofs of the Rayleigh extensions of Euler's formula involving the positive zeros of $J_{\nu}(\cdot)$, an example of which is given by the identity $\sum_{k=1}^{\infty} 1/z_{\nu,k}^2 = 1/(4(\nu+1))$ for $\nu > -1$.

1. Introduction – Main Results

1.1. Introduction

The aim of this paper is to describe some of the intimate connections existing between Bessel functions and weighted Wiener processes and Brownian bridges. The fact that Bessel functions have close relationships with Wiener processes is

Received by the editors : October 10, 2002.
2000 *Mathematics Subject Classification.* Primary: 62G10 ; Secondary: 60F05, 60F10, 60F25, 60J65, 62E20, 62G20, 62G30.
Key words and phrases. Karhunen–Loève expansions, Wiener process, Brownian bridge, Cramér-von Mises tests of fit, Tests of goodness of fit, Statistical tables.

not new. As early as in 1953, Paul Lévy ([23]) established the following property for the $\mathbb{R}^d$-valued Brownian motion $W_d(\cdot)$ ($d = 1, 2, \ldots$). Denoting by $|\cdot|$ the Euclidian norm in $\mathbb{R}^d$, it holds that, for each $x > 0$,

$$P\Big(\sup_{0\leq t\leq 1} |W_d(t)| \leq x\Big) = \sum_{k=1}^{\infty} \xi_{d,k} \exp\Big(-\frac{z_{\nu,k}^2}{2x^2}\Big),$$

where $\{z_{\nu,k} : k \geq 1\}$ denote the positive zeros of the Bessel function $J_\nu(\cdot)$ (see, e.g., §2.1 below), taken with $\nu = \frac{1}{2}d - 1$, and $\{\xi_{d,k} : k \geq 1\}$ are constants, whose explicit form were given later by Ciesielski and Taylor (see, e.g., Th. 2, p. 444 in [7]) as follows. For $k = 1, 2, \ldots$,

$$\xi_{d,k} = \Big\{\frac{1}{2^{d-1}\Gamma(\nu+1)}\Big\} \frac{z_{\nu,k}^{\nu-1}}{J_{\nu+1}(z_{\nu,k})}.$$

Further results of the kind have been obtained by Kiefer ([20]), Gikhman ([13]), Yor ([45], [46], [47]), Donati-Martin and Yor ([9], [10], [11]) and Pitman and Yor ([32], [34], [33], [35]). The connection of some of Yor's and his co-authors' results with our work will be discussed later. More recently, Shi ([38]) has shown that, for $d = 1, 2, \ldots$ and $\nu = \frac{1}{2}d - 1$, it holds that, for an arbitrary $0 \leq p < \frac{1}{2}$,

$$\lim_{x\downarrow 0} x^2 \log \mathbb{P}\left(\sup_{0<t\leq 1} t^{-p}|W_d(t)| < x\right) = -\frac{z_{\nu,1}^2}{2(1-2p)}.$$

We will provide later on other examples of the kind in connection with our forthcoming theorems, including the work of Li ([24], [25], [26]) and Scott ([37]). Below, we shall be primarily concerned with Karhunen–Loève expansions for $t^\theta W(t^\rho)$ and $t^\theta B(t^\rho)$, where $\rho > 0$ and $\theta > -\frac{1}{2}(\rho+1)$ are arbitrary constants. Our exposition is organized as follows. We start in §1.2 by a brief summary of the main results we will need on Karhunen–Loève [KL] expansions for a general Gaussian process $Z(\cdot)$ on $(0, 1)$. We are specially interested in applications of KL expansions to evaluate the distribution of $D^2 = \int_0^1 Z^2(t)dt$. In particular, we recall how the Smirnov formula for $\mathbb{P}(D^2 > t)$ may be used in order to describe the asymptotic behavior of $\mathbb{P}(D^2 > t)$ as $t \to \infty$. In §1.3, we show that the derivation of the KL expansions for general weighted Wiener processes and weighted Brownian bridges reduces to solving a second order differential equation with appropriate limit conditions. Our main results are stated in Theorems 1.1–1.4 of §1.3. In §1.4, we establish exact distribution of the $L^2(0, 1)$ norms of $t^\theta W(t^\rho)$ and $t^\theta B(t^\rho)$ and tabulate the corresponding quantiles for various choices of the parameters. In §1.5, we provide estimates for small deviations of these distributions, the case of large deviation being treated in §1.6. In §1.7, we apply these results to obtain simple proofs of the Rayleigh extension of Euler's formula (see, e.g., [44]). In §1.8, we prove simple identities relating the $L^2(0, 1)$ norms of the weighted processes $t^\beta W(t)$. In §1.9 we describe the limiting behavior of these $L^2(0, 1)$ norms when, either $\beta \downarrow -1$, or $\beta \uparrow \infty$. In §2, we recall some important facts on Bessel functions which are needed in our work. Finally §3 provides details on the proofs of Theorems 1.1–1.4.

1.2. Preliminaries on Karhunen–Loève expansions

We recall the following well-known facts about Karhunen–Loève [KL] expansions, (refer to [2], [4], [15], [18]). Let $\{Z(t) : 0 < t < 1\}$ denote a centered Gaussian process with covariance function $\{R(s,t) = \mathbb{E}(Z(s)Z(t)) : 0 < s,t < 1\}$, fulfilling

$$0 < \int_0^1 R(t,t)dt < \infty. \tag{1.1}$$

Then, there exist nonnegative constants $\{\lambda_k : k \geq 1\}$, $\lambda_k \downarrow$, together with functions $\{e_k(t) : k \geq 1\} \subset L^2(0,1)$ of $t \in (0,1)$ such that $(K.1\text{–}2\text{–}3\text{–}4)$ below hold.

$(K.1)$ For all $i \geq 1$ and $k \geq 1$,

$$\int_0^1 e_i(t)e_k(t)dt = \begin{cases} 1 & \text{if } i = k, \\ 0 & \text{if } i \neq k. \end{cases}$$

$(K.2)$ The $\{\lambda_k, e_k(\cdot) : k \geq 1\}$ form a complete set of solutions of the Fredholm equation in $(\lambda, e(\cdot))$, $\lambda \neq 0$.

$$\lambda e(t) = \int_0^1 R(s,t)e(s)ds \quad \text{for} \quad 0 < t < 1 \quad \text{and} \quad \int_0^1 e^2(t)dt = 1. \tag{1.2}$$

The λ_k's (resp. $e_k(\cdot)$) will be referred to as *eigenvalues* (resp. *eigenfunctions*) of the Fredholm transformation

$$f \in L^2(0,1) \to Tf \in L^2(0,1) : Tf(t) = \int_0^1 R(s,t)f(s)ds, \quad t \in (0,1).$$

$(K.3)$ The series expansion

$$R(s,t) = \sum_{k\geq 1} \lambda_k e_k(s)e_k(t) \quad \text{for} \quad 0 < s,t < 1, \tag{1.3}$$

is convergent in $L^2\big((0,1)^2\big)$.

$(K.4)$ There exist independent and identically distributed [i.i.d.] $N(0,1)$ random variables $\{\omega_k : k \geq 1\}$ such that the Karhunen–Loève [KL] expansion

$$Z(t) = \sum_{k\geq 1} \sqrt{\lambda_k}\, \omega_k e_k(t) \quad 0 < t < 1, \tag{1.4}$$

of $Z(\cdot)$ holds, with the series (1.4) converging almost surely [a.s.] and in integrated mean square on $(0,1)$.

Remark 1.1. 1°) The sequence $\{\lambda_k, e_k(\cdot) : k \geq 1\}$ in $(K.1\text{–}2\text{–}3\text{–}4)$ may very well be finite. Below, we will implicitly exclude this case and specialize in infinite KL expansions with k ranging through $\mathbb{N}^* = \{1, 2, \ldots\}$, and $\lambda_1 > \lambda_2 > \ldots > 0$.

2°) If, in addition to (1.1), $Z(\cdot)$ is a.s. continuous on $[0,1]$ with covariance function $R(\cdot,\cdot)$ continuous on $[0,1]^2$, then, we may choose the functions $\{e_k(\cdot) : k \geq 1\}$ in the KL expansion (1.4) to be continuous on $[0,1]$. The series (1.3) is then absolutely and uniformly convergent on $[0,1]^2$, and the series (1.4) is a.s. uniformly convergent on $[0,1]$ (see, e.g., [2]).

There are very few Gaussian processes of interest with respect to statistics for which the KL expansion is known through explicit values of $\{\lambda_k : k \geq 1\}$ and with simple forms of the functions $\{e_k(\cdot) : k \geq 1\}$ (see, e.g., [30] for a review). It is most interesting to have a precise knowledge of the λ_k's, since we infer from (1.4) that

$$D^2 = \int_0^1 Z^2(t)dt = \sum_{k\geq 1} \lambda_k \omega_k^2. \tag{1.5}$$

This readily implies (see, e.g. (6.23), p. 200 in [15]), that the moment-generating function of the distribution of D^2 is given by

$$\psi_{D^2}(z) = \mathbb{E}(\exp(zD^2)) = \prod_{k=1}^{\infty} \left\{\frac{1}{1-2z\lambda_k}\right\}^{1/2} \quad \text{for} \quad \mathrm{Re}(z) < \frac{1}{2\lambda_1}. \tag{1.6}$$

We note that $|E(\exp(zD^2))| < \infty$ for all $z \in \mathbb{C}$ with $\mathrm{Re}(z) < \frac{1}{2\lambda_1}$, subject to the additional conditions that

$$\begin{array}{lll} (i) & \lambda_1 > \lambda_2 > \ldots > 0, \quad \text{and} & \\ (ii) & \displaystyle\sum_{k=1}^{\infty} \lambda_k = \sum_{k=1}^{\infty} \frac{1}{\gamma_k} < \infty, \quad \text{where} \quad \gamma_k = \frac{1}{\lambda_k} \quad \text{for} \quad k \geq 1. \end{array} \tag{1.7}$$

Since D^2 is a weighted sum of independent χ_1^2 components, its distribution is easy to compute under (1.7) via the *Smirnov formula* (see, e.g., [8], [28], [29], [41], [43]). For $t > 0$,

$$\begin{aligned} \mathbb{P}\Big(D^2 > t\Big) &= \frac{1}{\pi}\sum_{k=1}^{\infty}(-1)^{k+1} \int_{\gamma_{2k-1}}^{\gamma_{2k}} \frac{e^{-tu/2}du}{u\sqrt{|\mathbb{F}(u)|}} \\ &= \frac{1+o(1)}{\pi} \int_{\gamma_1}^{\gamma_2} \frac{e^{-tu/2}du}{u\sqrt{|\mathbb{F}(u)|}} \quad \text{as} \quad t \to \infty, \end{aligned} \tag{1.8}$$

where $\mathbb{F}(u)$ is the *Fredholm determinant* defined, under (1.7), by

$$\mathbb{F}(u) = \prod_{k=1}^{\infty} \Big\{1 - u\lambda_k\Big\} = \prod_{k=1}^{\infty} \Big\{1 - \frac{u}{\gamma_k}\Big\} \quad \text{for} \quad u \geq 0. \tag{1.9}$$

In view of (1.6)–(1.9), we note that

$$\psi_{D^2}(z) = \{\mathbb{F}(2z)\}^{-1/2} \quad \text{for} \quad \mathrm{Re}(z) < \frac{1}{2\lambda_1}. \tag{1.10}$$

We refer to Martynov ([30], [28]) for a study of the convergence of the series (1.8), together with versions of this formula holding when some the consecutive terms of the sequence $\lambda_1 \geq \lambda_2 \geq \ldots > 0$ are equal. Under (1.7) and making use of (1.8)–(1.9), it is not too difficult to derive the following lemma (see, e.g., [14], [48], and Lemma 1 in Li ([25])).

Lemma 1.1. *Assume that* (1.7) *holds and let* $\mathbb{F}(\cdot)$ *be as in* (1.9). *Then, as* $t \to \infty$,

$$\mathbb{P}\Big(D^2 > t\Big) = (1 + o(1))\sqrt{\frac{2}{\pi}}\frac{1}{\gamma_1\sqrt{-\mathbb{F}'(\gamma_1)}} \times \frac{\exp\big(-\frac{1}{2}t\gamma_1\big)}{\sqrt{t}}. \tag{1.11}$$

Proof. Making use of (1.8), we will show that, as $t \to \infty$,

$$\begin{aligned}\mathbb{P}\Big(D^2 > t\Big) &= \frac{1+o(1)}{\pi}\int_{\gamma_1}^{\infty}\frac{e^{-tu/2}du}{u\sqrt{|\mathbb{F}(u)|}} = \frac{(1+o(1))e^{-t\gamma_1/2}}{\pi\gamma_1\sqrt{-\mathbb{F}'(\gamma_1)}}\int_0^{\infty}\frac{e^{-tv/2}dv}{\sqrt{v}} \\ &= \frac{(1+o(1))e^{-t\gamma_1/2}}{\pi\gamma_1\sqrt{-\mathbb{F}'(\gamma_1)}} \times \frac{2}{\sqrt{t}}\int_0^{\infty} e^{-w^2/2}dw, \end{aligned} \tag{1.12}$$

(by setting $u = v + \gamma_1$ and $v = w^2/t$) from where (1.11) is straightforward. To establish (1.12), we use the following general argument. We select a function $m(t)$ such that $m(t) \to \infty$ as $t \to \infty$, in such a way that we may approximate the distribution of D^2 by a sum of $\lfloor m(t) \rfloor$ terms, with

$$\mathbb{P}\Big(D^2 > t\Big) = \frac{1+o(1)}{\pi}\sum_{k=1}^{\lfloor m(t)\rfloor}(-1)^{k+1}\int_{\gamma_{2k-1}}^{\gamma_{2k}}\frac{e^{-tu/2}du}{u\sqrt{|\mathbb{F}(u)|}} \quad \text{as} \quad t \to \infty.$$

By setting

$$a_k = \frac{1}{\pi}\int_{\gamma_{2k-1}}^{\gamma_{2k}}\frac{du}{u\sqrt{|\mathbb{F}(u)|}}, \qquad k = 1, 2, \ldots,$$

we get

$$\mathbb{P}\Big(D^2 > t\Big) = \frac{1+o(1)}{\pi}\int_{\gamma_1}^{\gamma_2}\frac{e^{-tu/2}du}{u\sqrt{|\mathbb{F}(u)|}}(1 + b(t)) \quad \text{as} \quad t \to \infty,$$

where

$$|b(t)| \le \sum_{k=2}^{\lfloor m(t)\rfloor}\frac{a_k e^{-t\lambda_{2k-1}}}{a_1 e^{-t\lambda_2}} \le \frac{1}{a_1}\Big(\lfloor m(t)\rfloor - 1\Big)\max_{2\le k\le\lfloor m(t)\rfloor} a_k e^{-t(\lambda_3-\lambda_2)}.$$

The conclusion follows by letting the function $m(t)$ converge to infinity at a sufficiently slow rate to ensure that $b(t) \to 0$. □

Remark 1.2. By setting $\gamma_1 = 1/\lambda_1$ in (1.11), and since, via (1.9),

$$\frac{1}{\gamma_1\sqrt{-\mathbb{F}'(\gamma_1)}} = \sqrt{\lambda_1}\prod_{k=2}^{\infty}\Big\{1 - \frac{\lambda_k}{\lambda_1}\Big\}^{-1/2},$$

the conclusion of Lemma 1.1 is in agreement with Lemma 1 in Li ([25]).

1.3. KL expansions for weighted Wiener processes and Brownian bridges

In this section, we provide KL expansions for weighted Wiener processes and Brownian bridges under the following assumptions. Throughout, $\{W(t) : t \geq 0\}$ will denote a standard Wiener process, and $\{B(t) : 0 \leq t \leq 1\}$ a Brownian bridge. These processes are centered with covariance functions

$$\mathbb{E}(W(s)W(t)) = s \wedge t \quad \text{for} \quad s, t \geq 0, \tag{1.13}$$
$$\mathbb{E}(B(s)B(t)) = s \wedge t - st \quad \text{for} \quad 0 \leq s, t \leq 1. \tag{1.14}$$

Denote by $\{\psi(t) : 0 < t < 1\}$ a positive and continuous function on $(0, 1)$, whose definition will, at times, be extended by continuity to $(0, 1]$ or $[0, 1]$. Below, we will work under additional conditions taken among the following.

$(C.1)$ $\psi(\cdot)$ is continuous on $(0, 1]$;

$(C.2)$ (i) $\lim_{t \downarrow 0} t\psi(t) = 0;$ (ii) $\int_0^1 t\psi^2(t)dt < \infty;$

$(C.3)$ (i) $\lim_{t \downarrow 0} t\psi(t) = \lim_{t \uparrow 0}(1 - t)\psi(t) = 0;$ (ii) $\int_0^1 t(1 - t)\psi^2(t)dt < \infty.$

It is readily checked from (1.13) (resp. (1.14)) that $(C.2)(ii)$ (resp. $(C.3)(ii)$) is the version of (1.1) corresponding to $Z(t) = Z_1(t)$ (resp. $Z(t) = Z_2(t)$), where

$$Z_1(t) = \psi(t)W(t) \quad \text{and} \quad Z_2(t) = \psi(t)B(t) \quad \text{for} \quad 0 < t < 1.$$

To obtain the KL expansions of $Z_1(\cdot), Z_2(\cdot)$, we will use the following theorems, in the spirit of Kac and Siegert ([17], [18]), and Kac (see, e.g., pp. 199-200 in [15] and Section 2 in [16]).

Theorem 1.1. *Assume $(C.1$–$2)$. Set $Z(t) = Z_1(t) = \psi(t)W(t)$ for $0 < t \leq 1$. Then, the $\{(\lambda_k, e_k(\cdot)) : k \geq 1\}$ in the KL expansion of $Z(\cdot)$ are obtained by setting $\lambda = 1/\gamma$ and $e(t) = y(t)\psi(t)$, where $y(\cdot)$ is a continuous on $[0, 1]$ and twice continuously differentiable on $(0, 1]$ solution of the differential equation*

$$y''(t) + \gamma\psi^2(t)y(t) = 0, \tag{1.15}$$

subject to $\gamma > 0$ and with limit conditions

$$y(0) = 0 \quad and \quad y'(1) = 0. \tag{1.16}$$

Theorem 1.2. *Assume $(C.3)$. Set $Z(t) = Z_1(t) = \psi(t)B(t)$ for $0 < t < 1$. Then, the $\{(\lambda_k, e_k(\cdot)) : k \geq 1\}$ in the KL expansion of $Z(\cdot)$ are obtained by setting $\lambda = 1/\gamma$ and $e(t) = y(t)\psi(t)$, where $y(\cdot)$ is a continuous on $[0, 1]$ and twice continuously differentiable on $(0, 1)$ solution of the differential equation*

$$y''(t) + \gamma\psi^2(t)y(t) = 0, \tag{1.17}$$

subject to $\gamma > 0$ and with limit conditions

$$y(0) = 0 \quad and \quad y(1) = 0. \tag{1.18}$$

The proofs of Theorems 1.1 and 1.2 are postponed until §3. In the sequel, we will concentrate on the particular case where, for some constant $\beta \in \mathbb{R}$,

$$\psi(t) = t^{\beta} \quad \text{for} \quad 0 < t \leq 1. \tag{1.19}$$

We note that $(C.1\text{–}2\text{–}3)$ hold under (1.19) iff $\beta > -1$. In particular,

$$\beta > -1 \quad \Leftrightarrow \quad \int_0^1 t\psi^2(t)dt < \infty \quad \Leftrightarrow \quad \int_0^1 t(1-t)\psi^2(t)dt < \infty.$$

For $\nu > -1$, consider the Bessel function $J_\nu(\cdot)$ of first order and index ν (see §2.1 below for details on the definition and properties of $J_\nu(\cdot)$). For $\nu > -1$, the positive *zeros* of $J_\nu(\cdot)$ (solutions of $J_\nu(z) = 0$) form an infinite sequence, denoted hereafter by $0 < z_{\nu,1} < z_{\nu,2} < \dots$. These zeros are interlaced with the zeros $0 < z_{\nu+1,1} < z_{\nu+1,2} < \dots$ of $J_{\nu+1}(\cdot)$ (see, e.g., [44], p. 479), in such a way that

$$0 < z_{\nu,1} < z_{\nu+1,1} < z_{\nu,2} < z_{\nu+1,2} < z_{\nu,3} < \dots . \tag{1.20}$$

The next two theorems make use of this notation to provide the KL expansions of the processes $\{t^\beta W(t) : 0 \leq t \leq 1\}$ and $\{t^\beta B(t) : 0 \leq t \leq 1\}$ for $\beta > -1$.

Theorem 1.3. *Let $\{W(t) : t \geq 0\}$ denote a standard Wiener process. Then, for each $\beta = \frac{1}{2\nu} - 1 > -1$, or equivalently, for each $\nu = 1/(2(1+\beta)) > 0$, the Karhunen–Loève expansion of $\{t^\beta W(t) : 0 < t \leq 1\}$ is given by*

$$t^\beta W(t) = t^{\frac{1}{2\nu}-1} W(t) = \sum_{k=1}^{\infty} \sqrt{\lambda_k}\, \omega_k e_k(t), \tag{1.21}$$

where $\{\omega_k : k \geq 1\}$ are i.i.d. $N(0,1)$ random variables, and, for $k = 1, 2, \dots$,

$$\lambda_k = \left\{\frac{2\nu}{z_{\nu-1,k}}\right\}^2, \quad e_k(t) = t^{\frac{1}{2\nu}-\frac{1}{2}} \left\{\frac{J_\nu\left(z_{\nu-1,k}\, t^{\frac{1}{2\nu}}\right)}{\sqrt{\nu} J_\nu\left(z_{\nu-1,k}\right)}\right\} \text{ for } 0 < t \leq 1. \tag{1.22}$$

Theorem 1.4. *Let $\{B(t) : 0 \leq t \leq 1\}$ denote a Brownian bridge. Then, for each $\beta = \frac{1}{2\nu} - 1 > -1$, or equivalently, for each $\nu = 1/(2(1+\beta)) > 0$, the Karhunen–Loève expansion of $\{t^\beta B(t) : 0 < t \leq 1\}$ is given by*

$$t^\beta B(t) = t^{\frac{1}{2\nu}-1} B(t) = \sum_{k=1}^{\infty} \sqrt{\lambda_k}\, \omega_k e_k(t), \tag{1.23}$$

where $\{\omega_k : k \geq 1\}$ are i.i.d. $N(0,1)$ random variables, and, for $k = 1, 2, \dots$,

$$\lambda_k = \left\{\frac{2\nu}{z_{\nu,k}}\right\}^2, \quad e_k(t) = t^{\frac{1}{2\nu}-\frac{1}{2}} \left\{\frac{J_\nu\left(z_{\nu,k}\, t^{\frac{1}{2\nu}}\right)}{\sqrt{\nu} J_{\nu-1}\left(z_{\nu,k}\right)}\right\} \text{ for } 0 < t \leq 1. \tag{1.24}$$

Remark 1.3. The KL expansion of $\{t^\beta B(t) : 0 < t \leq 1\}$ in Theorem 1.4 is known for $\beta = 0$ ($\Leftrightarrow \nu = 1/2$) (see (1.37)–(1.38) below and, e.g., [3] and §4.4, pp. 30–32 in [12]). For $\beta = -1/2$ ($\Leftrightarrow \nu = 1$), we refer to Scott ([37]). In the general case where $\beta > -1/2$ ($\Leftrightarrow 0 < \nu < 1$), this KL expansion turns out (see Remark 1.4,4°) below) to be equivalent to a KL expansion given in Li ([24]). We mention further that a version of Theorem 1.4 has been discovered independently by Pycke ([31]).

Remark 1.4. 1°) The simple observation that the $\{\sqrt{\rho}\,t^{\frac{1}{2}(\rho-1)}e_k(t^\rho) : k \geq 1\}$ are orthonormal in $L^2(0,1)$ whenever such is the case for the $\{e_k(t) : k \geq 1\}$ (see, e.g., [27]), allows us to derive, via (1.25)–(1.26)–(1.27) below, through the change of scale $t \to t^\rho$, a series of variants of the KL expansions of Theorems 1.3–1.4. We will make use of the convention of writing these KL expansions under the form

$$Z(t) = \sum_{k=1}^{\infty} \{\lambda_k\}^{1/2} \omega_k e_k(t), \quad 0 < t \leq 1, \tag{1.25}$$

where, as in (1.4), λ_k and $e_k(\cdot)$ for $k \geq 1$, are, respectively, the eigenvalues and eigenfunctions pertaining to $\{Z(t) : 0 < t \leq 1\}$. The arguments above show that, whenever the KL expansion (1.25) holds, then, for each choice of $\rho > 0$, the KL expansion of $t^{\frac{1}{2}(\rho-1)}Z(t^\rho)$ on $(0,1]$ is given by

$$t^{\frac{1}{2}(\rho-1)}Z(t^\rho) = \sum_{k=1}^{\infty} \{\lambda_k/\rho\}^{1/2} \omega_k \Big\{\sqrt{\rho}\,t^{\frac{1}{2}(\rho-1)}e_k(t^\rho)\Big\}, \quad 0 < t \leq 1, \tag{1.26}$$

with eigenvalues λ_k^* and eigenfunctions $e_k^*(\cdot)$ for $k = 1, 2, \ldots$, given by

$$\lambda_k^* = \{\lambda_k/\rho\}^{1/2} \quad \text{and} \quad e_k^*(t) = \Big\{\sqrt{\rho}\,t^{\frac{1}{2}(\rho-1)}e_k(t^\rho)\Big\}. \tag{1.27}$$

2°) By combining (1.25)–(1.26)–(1.27) with Theorems 1.3–1.4, we obtain readily in (1.28)–(1.29) below the KL expansions of $t^\theta W(t^\rho)$ and $t^\theta B(t^\rho)$ (under the form (1.25)). For any $\rho > 0$, $\theta > -\frac{1}{2}(\rho+1)$, and $\nu = \rho/(2\theta+\rho+1) > 0$, we get namely

$$t^\theta W(t^\rho) = \sum_{k=1}^{\infty} \Big\{\frac{2\nu}{z_{\nu-1,k}\sqrt{\rho}}\Big\} \omega_k \Big\{\sqrt{\rho}\,t^{\frac{\rho}{2\nu}-\frac{1}{2}} \Big\{\frac{J_\nu\big(z_{\nu-1,k}\,t^{\frac{\rho}{2\nu}}\big)}{\sqrt{\nu}J_\nu\big(z_{\nu-1,k}\big)}\Big\}\Big\}, \tag{1.28}$$

$$t^\theta B(t^\rho) = \sum_{k=1}^{\infty} \Big\{\frac{2\nu}{z_{\nu,k}\sqrt{\rho}}\Big\} \omega_k \Big\{\sqrt{\rho}\,t^{\frac{\rho}{2\nu}-\frac{1}{2}} \Big\{\frac{J_\nu\big(z_{\nu,k}\,t^{\frac{\rho}{2\nu}}\big)}{\sqrt{\nu}J_{\nu-1}\big(z_{\nu,k}\big)}\Big\}\Big\}. \tag{1.29}$$

In particular, by setting $\rho = 1$, $\theta = \beta$ and $\nu = 1/(2(\beta+1))$ in (1.28)–(1.29), we get (1.21)–(1.22) and (1.23)–(1.24).

3°) Of special interest here is the choice of $\rho = 2\nu$ and $\theta = \frac{1}{2} - \nu$ in (1.28)–(1.29). For these values of the constants ρ, θ, we obtain that, for each $\nu > 0$, the KL expansions of $t^{\frac{1}{2}-\nu}W(t^{2\nu})$ and $t^{\frac{1}{2}-\nu}B(t^{2\nu})$ are given by.

$$t^{\frac{1}{2}-\nu}W(t^{2\nu}) = \sum_{k=1}^{\infty} \Big\{\frac{\sqrt{2\nu}}{z_{\nu-1,k}}\Big\} \omega_k \Big\{\sqrt{2}\,t^{\frac{1}{2}} \Big\{\frac{J_\nu\big(z_{\nu-1,k}\,t\big)}{J_\nu\big(z_{\nu-1,k}\big)}\Big\}\Big\}, \tag{1.30}$$

$$t^{\frac{1}{2}-\nu}B(t^{2\nu}) = \sum_{k=1}^{\infty} \Big\{\frac{\sqrt{2\nu}}{z_{\nu,k}}\Big\} \omega_k \Big\{\sqrt{2}\,t^{\frac{1}{2}} \Big\{\frac{J_\nu\big(z_{\nu,k}\,t\big)}{J_{\nu-1}\big(z_{\nu,k}\big)}\Big\}\Big\}. \tag{1.31}$$

By multiplying both sides of (1.30) by $t^{-\frac{1}{2}}$, we get a *Dini series expansion* of $t^{-\nu}W(t^{2\nu})$ on $(0,1)$. Proceeding likewise with (1.31) one obtains the *Fourier-Bessel series expansion* of $t^{-\nu}B(t^{2\nu})$ on $(0,1)$ (see, e.g., pp 96–103 in [21]). Recall that, under suitable conditions on the functions $f(\cdot)$ and $g(\cdot)$ on $(0,1)$, it is possible to expand $f(\cdot)$ into the Fourier-Bessel expansion $f(t)=\sum_{k=1}^{\infty} a_k J_\nu(z_{\nu,k}t)$, with

$$a_k = \frac{2}{J^2_{\nu-1,k}(z_{\nu,k})}\int_0^1 tf(t)J_\nu(z_{\nu,k}t)dt, \quad k\geq 1, \tag{1.32}$$

and $g(\cdot)$ into a Dini expansion $g(t)=\sum_{k=1}^{\infty} b_k J_\nu(z_{\nu-1,k}t)$, with

$$b_k = \frac{2}{J^2_{\nu,k}(z_{\nu-1,k})}\int_0^1 tg(t)J_\nu(z_{\nu-1,k}t)dt, \quad k\geq 1. \tag{1.33}$$

By setting $f(t)=t^{-\nu}B(t^{2\nu})$ and $g(t)=t^{-\nu}W(t^{2\nu})$ in (1.32)–(1.33), we get

$$a_k = \frac{2\nu\omega_k}{z_{\nu,k}J_{\nu-1}(z_{\nu,k})} \quad \text{and} \quad b_k = \frac{2\nu\omega_k}{z_{\nu-1,k}J_\nu(z_{\nu-1,k})} \quad \text{for} \quad k\geq 1. \tag{1.34}$$

4°) Put $\theta=0$ and $\nu=\rho/(\rho+1)$ in (1.28)–(1.29). Set, for notational simplicity, $z_{\nu,k}=z_{\rho/(\rho+1),k}$ and $z_{\nu-1,k}=z_{\rho/(\rho+1)-1,k}$. We so obtain the KL expansions

$$W(t^\rho)=\sum_{k=1}^{\infty}\Big\{\frac{2\sqrt{\rho}}{z_{\nu-1,k}(\rho+1)}\Big\}\omega_k\Big\{\sqrt{\rho+1}\,t^{\frac{\rho}{2}}\Big\{\frac{J_{\rho/(\rho+1)}\big(z_{\nu-1,k}\,t^{\frac{\rho+1}{2}}\big)}{J_{\rho/(\rho+1)}\big(z_{\nu-1,k}\big)}\Big\}\Big\} \tag{1.35}$$

$$B(t^\rho)=\sum_{k=1}^{\infty}\Big\{\frac{2\sqrt{\rho}}{z_{\nu,k}(\rho+1)}\Big\}\omega_k\Big\{\sqrt{\rho+1}\,t^{\frac{\rho}{2}}\Big\{\frac{J_{\rho/(\rho+1)}\big(z_{\nu,k}\,t^{\frac{\rho+1}{2}}\big)}{J_{\rho/(\rho+1)-1}\big(z_{\nu,k}\big)}\Big\}\Big\}. \tag{1.36}$$

The KL expansion (1.36) has been obtained by Li ([24]) (see the proof of Theorem 1.6, pp. 24–25 in [24]), up to the normalizing factor, for $k=1,2,\ldots$,

$$c_k=\sqrt{\rho+1}/\{J_{\rho/(\rho+1)-1}\big(z_{\nu,k}\big)\},$$

of the eigenfunction in (1.36) (with the notation (1.25))

$$e_k(t)=c_k t^{\rho/2}J_{\rho/(\rho+1)}\big(z_{\nu,k}\,t^{\frac{\rho+1}{2}}\big),$$

left implicit in his work. In spite of the fact that it is possible to revert the previous arguments, starting with (1.36), in order to obtain an alternative proof of Theorem 1.3 based on [24], this does only work for the values of $\nu=\rho/(\rho+1)$ with $0<\nu<1$ (since we must have $\rho>0$). It follows that the KL expansions (1.30) (resp. (1.35)), for $\nu>0$ (resp. $\rho>0$), and (1.29), for $\nu\geq 1$, are new, to our best knowledge, mention being made of the work of Pycke ([31]).

Example 1.1. For $\nu = \frac{1}{2}$, we obtain (see 2.27) in the sequel) that

$$J_{\nu-1}(t) = J_{-\frac{1}{2}}(t) = \sqrt{\tfrac{2}{\pi t}}\,\cos t, \quad J_\nu(t) = J_{\frac{1}{2}}(t) = \sqrt{\tfrac{2}{\pi t}}\,\sin t,$$

$$z_{\nu-1,k} = z_{-\frac{1}{2},k} = \{k - \tfrac{1}{2}\}\pi, \quad z_{\nu,k} = z_{\frac{1}{2},k} = k\pi, \quad k = 1, 2, \ldots,$$

$$J_\nu(z_{\nu-1,k}\,t) = J_{\frac{1}{2}}(z_{-\frac{1}{2},k}t) = \Big\{\frac{2}{\pi\{k-\frac{1}{2}\}\pi t}\Big\}^{1/2} \sin\big(\{k - \tfrac{1}{2}\}\pi t\big),$$

$$J_\nu(z_{\nu,k}\,t) = J_{\frac{1}{2}}(z_{\frac{1}{2},k}t) = \Big\{\frac{2}{\pi \times k\pi t}\Big\}^{1/2} \sin\big(k\pi t\big),$$

$$J_{\nu-1}(z_{\nu,k}\,t) = J_{-\frac{1}{2}}(z_{\frac{1}{2},k}t) = \Big\{\frac{2}{\pi \times k\pi t}\Big\}^{1/2} \cos\big(k\pi t\big).$$

By setting $\nu = \frac{1}{2}$ and $\theta = \rho - 1$ in (1.28)–(1.29), we obtain the KL expansions

$$t^{\frac{\rho-1}{2}} W(t^\rho) = \sum_{k=1}^{\infty} \Big\{\frac{1}{\{k-\frac{1}{2}\}\pi\sqrt{\rho}}\Big\}\omega_k \Big\{\sqrt{2\rho}\,t^{\frac{\rho-1}{2}} \sin\Big(\{k - \tfrac{1}{2}\}\pi t^\rho\Big)\Big\}, \tag{1.37}$$

$$t^{\frac{\rho-1}{2}} B(t^\rho) = \sum_{k=1}^{\infty} \Big\{\frac{1}{k\pi\sqrt{\rho}}\Big\}\omega_k \Big\{\sqrt{2\rho}\,t^{\frac{\rho-1}{2}} \sin\Big(k\pi\, t^\rho\Big)\Big\}, \tag{1.38}$$

where we have dropped the factor $(-1)^k = -\sin(\{k - \frac{1}{2}\}\pi) = \cos(k\pi)$. For $\rho = 1$, (1.37)–(1.38) reduce to the well-known KL expansions of the Wiener process $W(\cdot)$ and Brownian bridge $B(\cdot)$ (see, e.g., [3] and [12]).

Example 1.2. An interesting case is given by $\beta = -2/3$ and $\nu = 3/2$, for which

$$J_{\frac{3}{2}}(x) = \sqrt{\tfrac{2}{\pi x}}\,\Big\{\frac{\sin x}{x} - \cos x\Big\} \quad \text{and} \quad z_{\frac{1}{2},k} = k\pi, \quad k \geq 1.$$

By setting $\theta = -(3+\rho)/6$ and $\nu = 3/2$ in (1.28), we obtain that, for each $\rho > 0$, the following KL expansion holds on $(0,1)$.

$$t^{-\frac{3+\rho}{6}} W(t^\rho) = \sum_{k=1}^{\infty} \Big\{\frac{3}{k\pi\sqrt{\rho}}\Big\}\omega_k\, t^{\frac{\rho}{6}-\frac{1}{2}} \frac{\sqrt{2\rho}}{\pi\sqrt{k}} \Big\{\frac{\sin(k\pi t^{\frac{\rho}{3}})}{k\pi t^{\frac{\rho}{3}}} - \cos(k\pi t^{\frac{\rho}{3}})\Big\}. \tag{1.39}$$

The versions of (1.39) for $\rho = 1$ and $\rho = 3$ are of interest and left to the reader.

1.4. Exact distributions for $L^2(0,1)$ norms of weighted processes

In this subsection, we obtain, for an arbitrary $\nu = 1/(2(1+\beta)) > 0$ (or equivalently, for an arbitrary $\beta = \frac{1}{2\nu} - 1 > -1$), the exact distributions of

$$\begin{aligned} D^2_{W,\nu} &= \int_0^1 t^{1-2\nu} W^2(t^{2\nu})dt = (\beta+1)\int_0^1 t^{2\beta} W^2(t)dt \qquad (1.40)\\ &= \frac{1}{2\nu}\int_0^1 t^{2\beta} W^2(t)dt, \\ D^2_{B,\nu} &= \int_0^1 t^{1-2\nu} B^2(t^{2\nu})dt = (\beta+1)\int_0^1 t^{2\beta} B^2(t)dt \qquad (1.41)\\ &= \frac{1}{2\nu}\int_0^1 t^{2\beta} B^2(t)dt. \end{aligned}$$

Remark 1.5. One of the motivations for the study of the above random variables comes from the theory of nonparametric tests of goodness of fit (Refer to del Barrio, Cuesta-Albertos and Matran ([5]), Durbin ([12]), and the references therein). In particular, it is not too difficult to prove that, if $\alpha_n(\cdot)$ is a uniform empirical process on $[0,1]$ (refer to [40] for the definition of $\alpha_n(\cdot)$), then, as $n \to \infty$

$$C^2_{n;\nu} = \int_0^1 t^{2\beta}\alpha_n^2(t)dt \xrightarrow{d} \int_0^1 t^{2\beta} B^2(t)dt = 2\nu D^2_{B,\nu}, \qquad (1.42)$$

subject to $\beta \geq -1/2$ (see, e.g., Shorack and Wellner [40]). Therefore, the law of $D^2_{B,\nu}$ provides the limiting distribution of the statistic $C^2_{n;\nu}$. The best-known example of the kind is the Cramér-von Mises statistic of goodness of fit, obtained for $\beta = 0$ (see, e.g. [41], [43], [42]). The case $\beta = -1/2$ has been treated by Scott ([37]). An application of (1.42) will be presented in Remark 1.7 below. We will not discuss further the statistical aspects of this problem, nor seek to justify the convergence (1.42), which can be established by the same arguments as for the Anderson-Darling statistic (see, e.g., [40]). This problem will be considered elsewhere.

We keep in mind that $D^2_{W,\nu}$ and $D^2_{B,\nu}$ are the $L^2(0,1)$ norms of the processes with KL expansions given in (1.30)–(1.31), which we recall below for convenience.

$$\begin{aligned} t^{\frac12-\nu}W(t^{2\nu}) &= \sum_{k=1}^{\infty}\Big\{\frac{\sqrt{2\nu}}{z_{\nu-1,k}}\Big\}\omega_k\Big\{\sqrt{2}\,t^{\frac12}\Big\{\frac{J_\nu(z_{\nu-1,k}\,t)}{J_\nu(z_{\nu-1,k})}\Big\}\Big\},\\ t^{\frac12-\nu}B(t^{2\nu}) &= \sum_{k=1}^{\infty}\Big\{\frac{\sqrt{2\nu}}{z_{\nu,k}}\Big\}\omega_k\Big\{\sqrt{2}\,t^{\frac12}\Big\{\frac{J_\nu(z_{\nu,k}\,t)}{J_{\nu-1}(z_{\nu,k})}\Big\}\Big\}. \end{aligned}$$

Theorem 1.5. *For any $\nu > 0$ and $t \geq 0$, we have*

$$\mathbb{P}\Big(D^2_{W,\nu} > t\Big) = \frac{1}{\pi\sqrt{\Gamma(\nu)}}\sum_{k=1}^{\infty}(-1)^{k+1}\int_{z_{\nu-1,2k-1}}^{z_{\nu-1,2k}} \frac{(\frac12 v)^{\frac{\nu-1}{2}-1}e^{-tv^2/(4\nu)}dv}{\sqrt{-J_{\nu-1}(v)}}, \qquad (1.43)$$

$$\mathbb{P}\Big(D^2_{B,\nu} > t\Big) = \frac{1}{\pi\sqrt{\Gamma(\nu+1)}}\sum_{k=1}^{\infty}(-1)^{k+1}\int_{z_{\nu,2k-1}}^{z_{\nu,2k}} \frac{(\frac12 v)^{\frac{\nu}{2}-1}e^{-tv^2/(4\nu)}dv}{\sqrt{-J_{\nu}(v)}}. \qquad (1.44)$$

Proof. By combining (1.9) with Euler's formula (see (2.19) in the sequel), we infer from (1.30) that the Fredholm determinant $\mathbb{F}_{W,\nu}(u)$ for $Z(t) = t^{\frac{1}{2}-\nu}W(t^{2\nu})$ is such that, with $\lambda_k = 1/\gamma_k = 2\nu/z_{\nu-1,k}^2$,

$$\mathbb{F}_{W,\nu}\Big(\frac{v^2}{2\nu}\Big) = \prod_{k=1}^{\infty}\Big\{1-\frac{v^2\lambda_k}{2\nu}\Big\} = \prod_{k=1}^{\infty}\Big\{1-\frac{v^2}{z_{\nu-1,k}^2}\Big\} \tag{1.45}$$

$$= \Gamma(\nu)(\tfrac{1}{2}v)^{-\nu+1}J_{\nu-1}(v).$$

By (1.31), the Fredholm determinant $\mathbb{F}_{B,\nu}(u)$ for $Z(t) = t^{\frac{1}{2}-\nu}B(t^{2\nu})$ is likewise

$$\mathbb{F}_{B,\nu}\Big(\frac{v^2}{2\nu}\Big) = \Gamma(\nu+1)(\tfrac{1}{2}v)^{-\nu}J_{\nu}(v). \tag{1.46}$$

We make the change of variable $u = v^2/(2\nu)$ in (1.45), with $du/u = dv/(\frac{1}{2}v)$, and set $\gamma_k = z_{\nu-1,k}^2/(2\nu)$, in the *Smirnov formula* (1.8) to obtain that, for $t \geq 0$,

$$\mathbb{P}\Big(D_{W,\nu}^2 > t\Big) = \frac{1}{\pi}\sum_{k=1}^{\infty}(-1)^{k+1}\int_{\gamma_{2k-1}}^{\gamma_{2k}}\frac{e^{-tu/2}du}{u\sqrt{|\mathbb{F}_{W,\nu}(u)|}} \tag{1.47}$$

$$= \frac{1}{\pi\sqrt{\Gamma(\nu)}}\sum_{k=1}^{\infty}(-1)^{k+1}\int_{z_{\nu-1,2k-1}}^{z_{\nu-1,2k}}\frac{(\frac{1}{2}v)^{\frac{\nu-1}{2}-1}e^{-tv^2/(4\nu)}dv}{\sqrt{|J_{\nu-1}(v)|}}.$$

The conclusion (1.43) follows from (1.47) and the observation that $J_{\nu-1}(v) < 0$ for $v \in (z_{\nu-1,2k-1}, z_{\nu-1,2k})$. The proof of (1.44) is achieved along the same lines with the formal changes of $\Gamma(\nu)(\frac{1}{2}v)^{-\nu+1}J_{\nu-1}(v)$ into $\Gamma(\nu+1)(\frac{1}{2}v)^{-\nu}J_{\nu}(v)$ in (1.45), and of $z_{\nu-1,2k-1}, z_{\nu-1,2k}$ into $z_{\nu,2k-1}, z_{\nu,2k}$ in (1.47), via (1.46). □

An easy application of (1.10) in combination with the just-given proof of Theorem 1.5 gives the moment-generating functions of $D_{W,\nu}^2$ and $D_{B,\nu}^2$.

Theorem 1.6. *For any $\nu > 0$, we have*

$$\mathbb{E}(\exp(zD_{W,\nu}^2)) = \Big\{\Gamma(\nu)\Big(\sqrt{\nu z}\,\Big)^{-\nu+1}J_{\nu-1}(2\sqrt{\nu z}\,)\Big\}^{-1/2} \tag{1.48}$$

$$\textit{for} \quad |\mathrm{Re}(z)| < \frac{z_{\nu-1,1}^2}{4\nu},$$

$$\mathbb{E}(\exp(zD_{B,\nu}^2)) = \Big\{\Gamma(\nu+1)\Big(\sqrt{\nu z}\,\Big)^{-\nu}J_{\nu}(2\sqrt{\nu z}\,)\Big\}^{-1/2} \tag{1.49}$$

$$\textit{for} \quad |\mathrm{Re}(z)| < \frac{z_{\nu,1}^2}{4\nu}.$$

Proof. As follows from (1.10) and (1.45), we have

$$\mathbb{E}(\exp(zD_{W,\nu}^2)) = \mathrm{F}_{W,\nu}(2z) = \Big[\Gamma(\nu)\Big\{\frac{\sqrt{4\nu z}}{2}\Big\}^{-\nu+1}J_{\nu-1}\Big(\sqrt{4\nu z}\,\Big)\Big]^{-1/2},$$

which is (1.48). The proof of (1.49) is achieved along the same lines with the formal change of $\Gamma(\nu)(\frac{1}{2}v)^{-\nu+1}J_{\nu-1}(v)$ into $\Gamma(\nu+1)(\frac{1}{2}v)^{-\nu}J_{\nu}(v)$. We omit details. □

Remark 1.6. It should be mentioned that, under general assumptions on $\{\psi(t) : 0 < t < 1\}$, it is possible to evaluate directly the moment-generating function

$$\mathbb{E}\Big(\exp\Big\{z\int_0^1 \psi^2(t)B^2(t)dt\Big\}\Big),$$

of the ψ-weighted Brownian bridge (refer to [33]). An application of these methods (which can also apply to the ψ-weighted Wiener process) should enable to obtain alternate proofs of (1.48) and (1.49). This, however, is not completely straightforward, and this problem will be considered elsewhere.

Example 1.3. 1°) For $\nu = \frac{1}{2}$, in view of the fact that (see (2.27) in the sequel)

$$J_{-\frac{1}{2}}(x) = \sqrt{\tfrac{2}{\pi x}}\cos x \quad \text{and} \quad J_{\frac{1}{2}}(x) = \sqrt{\tfrac{2}{\pi x}}\sin x,$$

with $z_{-\frac{1}{2},k} = (k-\frac{1}{2})\pi$ and $z_{\frac{1}{2},k} = k\pi$, $k = 1, 2, \ldots$, we infer from (1.43)–(1.44) the well-known formulas, for $t \geq 0$,

$$\mathbb{P}\Big(\int_0^1 W^2(s)ds > t\Big) = \frac{2}{\pi}\sum_{k=1}^{\infty}(-1)^{k+1}\int_{(2k-\frac{3}{2})\pi}^{(2k-\frac{1}{2})\pi} \frac{e^{-tv^2/2}dv}{v\sqrt{-\cos v}}, \tag{1.50}$$

$$\mathbb{P}\Big(\int_0^1 B^2(s)ds > t\Big) = \frac{2}{\pi}\sum_{k=1}^{\infty}(-1)^{k+1}\int_{(2k-1)\pi}^{2k\pi} \frac{e^{-tv^2/2}dv}{\sqrt{-v\sin v}}. \tag{1.51}$$

Note that (1.51) is equivalent, by setting $v^2 = y$, to (4.4.5), p. 32 in [12].

Likewise, we infer from (1.48)–(1.49) and $\Gamma(\frac{1}{2}) = \sqrt{\pi}$ that

$$\mathbb{E}\left(e^{z\int_0^1 W^2(t)dt}\right) = \left[\cos\left(\sqrt{2z}\,\right)\right]^{-1/2} \quad \text{for} \quad \mathrm{Re}(z) < \frac{\pi^2}{8}, \tag{1.52}$$

$$\mathbb{E}\left(e^{z\int_0^1 B^2(t)dt}\right) = \left[\frac{\sin\left(\sqrt{2z}\,\right)}{\sqrt{2z}}\right]^{-1/2} \quad \text{for} \quad \mathrm{Re}(z) < \frac{\pi^2}{2}. \tag{1.53}$$

In particular, by setting $z = -u$ in (1.53), we get

$$\mathbb{E}\left(e^{-u\int_0^1 B^2(t)dt}\right) = \left[\frac{\sinh\left(\sqrt{2u}\,\right)}{\sqrt{2u}}\right]^{-1/2} \quad \text{for} \quad u \in \mathbb{R}^+. \tag{1.54}$$

The formula (1.54) (see, e.g., p. 200 in [15], and [39]), is a particular case, taken with $\lambda = \sqrt{2u}$ and $a = 0$, of the well-known formula, due to Paul Lévy ([22])

$$\mathbb{E}\Big[\exp\Big(-\frac{\lambda^2}{2}\int_0^1 W^2(t)dt\Big)\Big|W(1) = a\Big] \tag{1.55}$$
$$= \Big[\frac{\sinh\lambda}{\lambda}\Big]^{-1/2}\exp\Big(-\frac{a^2}{2}(\lambda\coth\lambda - 1)\Big).$$

2°) For $\nu = \frac{3}{2}$, making use of the fact that

$$J_{\frac{3}{2}}(x) = \sqrt{\tfrac{2}{\pi x}}\Big\{\frac{\sin x}{x} - \cos x\Big\} \quad \text{and} \quad z_{\frac{1}{2},k} = k\pi, \quad k = 1, 2, \ldots,$$

we infer from (1.43)–(1.44) and (1.83)–(1.84) in the sequel that, for $t \geq 0$,

$$\begin{aligned}\mathbb{P}(D^2_{W,\frac{3}{2}} > t) &= \mathbb{P}\Big(\int_0^1 B^2(s)ds > \frac{t}{3}\Big)\\ &= \mathbb{P}\Big(\int_0^1 s^{-2}W^2(s^3)ds > t\Big) \qquad (1.56)\\ &= \mathbb{P}\Big(\int_0^1 \big\{s^{-2/3}W(s)\big\}^2 ds > 3t\Big)\\ &= \frac{1}{\pi}\sum_{k=1}^{\infty}(-1)^{k+1}\int_{(2k-1)\pi}^{2k\pi}\frac{e^{-tv^2/6}dv}{\sqrt{-v\sin v}};\end{aligned}$$

$$\begin{aligned}\mathbb{P}(D^2_{B,\frac{3}{2}} > t) &= \mathbb{P}\Big(\int_0^1 s^{-2}B^2(s^3)ds > t\Big) \qquad (1.57)\\ &= \mathbb{P}\Big(\int_0^1 \big\{s^{-2/3}B(s)\big\}^2 ds > 3t\Big)\\ &= \frac{1}{\pi}\sum_{k=1}^{\infty}(-1)^{k+1}\int_{z_{\frac{3}{2},2k-1}}^{z_{\frac{3}{2},2k-1}}\frac{e^{-tv^2/6}dv}{\sqrt{\cos v - v^{-1}\sin v}}.\end{aligned}$$

Aside of the case where $\nu = \frac{1}{2}$, to our best knowledge, there is no simple closed form expression for the zeros $z_{\nu,k}$ and $z_{\nu-1,k}$ used in (1.43)–(1.44).

3°) For $\nu = 1$ and $\beta = -1/2$, a version of (1.44)–(1.49) is provided by Scott ([37]).

To tabulate the distributions of $D^2_{W,\nu}$ and $D^2_{B,\nu}$, it is convenient to introduce the following notation. In view of (1.40)–(1.41), for each $\nu = 1/(2(\beta+1)) > 0$, we set

$$D^*_{W,\nu} = 2(\beta+1)\int_0^1 t^{2\beta}W^2(t)dt = 2D^2_{W,\nu}, \qquad (1.58)$$

$$D^*_{B,\nu} = 2(\beta+1)(2\beta+3)\int_0^1 t^{2\beta}B^2(t)dt = 2\Big\{\frac{\nu+1}{\nu}\Big\}D^2_{B,\nu}. \qquad (1.59)$$

As follows from (1.92) and (1.93) in the sequel, we have, independently of $\nu > 0$,

$$\mathbb{E}(D^*_{B,\nu}) = \mathbb{E}(D^*_{W,\nu+1}) = 1, \qquad (1.60)$$

together with the equality in distribution (see (1.84) in the sequel)

$$D^*_{B,\nu} \overset{d}{=} D^*_{W,\nu+1}. \qquad (1.61)$$

In particular, it follows from (1.61) (see Proposition 1.2 in §1.9 below) that the following limit in distribution holds as $\nu \downarrow 0$ (or equivalently, as $\beta \uparrow \infty$). We have

$$\begin{aligned}D^*_{B,\nu} &= 2(\beta+1)(2\beta+3)\int_0^1 t^{2\beta}B^2(t)dt = D^*_{W,\nu+1} \qquad (1.62)\\ &\overset{d}{\to} D^*_{W,1} = \int_0^1 t^{-1}W^2(t)dt \quad \text{as} \quad \nu \downarrow 0.\end{aligned}$$

The following Table 1.1 gives selected quantiles for the distribution of $D^*_{W,\nu}$. A direct consequence of (1.61) is that $p = \mathbb{P}(D^*_{B,\nu-1} \le t) = \mathbb{P}(D^*_{W,\nu} \le t)$, so that Table 1.1 can be used via the formal change of ν into $\nu+1$ to provide quantiles for the distribution of D^*_{B,ν^*} given the corresponding quantiles of the distribution of D^*_{W,ν^*+1} for $\nu^* > 0$. Moreover, in view of (1.103), the quantiles of the distribution of $D^*_{W,0}$ correspond to the quantiles of the limiting distribution of D^*_{B,ν^*} as $\nu^* \downarrow 0$. Further, in view of (1.103) in the sequel, we set $D^*_{W,\infty} = \chi^2_1$ in Table 1.1.

$p = \mathbb{P}(D^*_{W,\nu} \le t)$	$p = 0.900$	0.950	0.990	0.995	0.999
$\nu = 0$ ($\beta = \infty$)	$t = 2.706$	3.841	6.635	7.879	10.828
$\nu = \frac{1}{22}$ ($\beta = 10.00$)	2.668	3.779	6.511	7.728	10.611
$\nu = \frac{1}{12}$ ($\beta = 5.00$)	2.639	3.730	6.413	7.608	10.440
$\nu = \frac{1}{7}$ ($\beta = 2.50$)	2.595	3.657	6.268	7.431	10.187
$\nu = \frac{1}{6}$ ($\beta = 2.00$)	2.579	3.629	6.213	7.364	10.090
$\nu = \frac{1}{5}$ ($\beta = 1.50$)	2.557	3.592	6.138	7.273	9.960
$\nu = \frac{1}{4}$ ($\beta = 1.00$)	2.525	3.539	6.031	7.142	9.773
$\nu = \frac{2}{7}$ ($\beta = 0.75$)	2.504	3.502	5.959	7.053	9.646
$\nu = \frac{1}{3}$ ($\beta = 0.50$)	2.476	3.456	5.866	6.940	9.484
$\nu = \frac{2}{5}$ ($\beta = 0.25$)	2.441	3.395	5.744	6.790	9.270
$\nu = \frac{1}{2}$ ($\beta = 0.00$)	2.392	3.311	5.575	6.584	8.973
$\nu = \frac{2}{3}$ ($\beta = -0.25$)	2.320	3.188	5.326	6.278	8.535
$\nu = 1$ ($\beta = -0.50$)	2.207	2.989	4.918	5.779	7.817
$\nu = \frac{13}{12}$ ($\beta = -\frac{7}{13}$)	2.183	2.947	4.831	5.672	7.663
$\nu = \frac{8}{7}$ ($\beta = -\frac{9}{16}$)	2.167	2.918	4.772	5.599	7.599
$\nu = \frac{7}{6}$ ($\beta = -\frac{8}{14}$)	2.161	2.907	4.749	5.571	7.518
$\nu = \frac{6}{5}$ ($\beta = -\frac{7}{12}$)	2.152	2.892	4.717	5.532	7.462
$\nu = \frac{5}{4}$ ($\beta = -0.60$)	2.140	2.869	4.671	5.475	7.380
$\nu = \frac{9}{7}$ ($\beta = -\frac{11}{18}$)	2.131	2.854	4.639	5.436	7.323
$\nu = \frac{4}{3}$ ($\beta = -0.65$)	2.120	2.834	4.597	5.384	7.250
$\nu = \frac{7}{5}$ ($\beta = -\frac{9}{14}$)	2.105	2.807	4.541	5.315	7.150
$\nu = \frac{3}{2}$ ($\beta = -\frac{2}{3}$)	2.084	2.768	4.461	5.216	7.007
$\nu = \frac{5}{3}$ ($\beta = -0.70$)	2.051	2.709	4.337	5.064	6.787
$\nu = 2$ ($\beta = -0.75$)	1.995	2.606	4.120	4.796	6.400
$\nu = 3$ ($\beta = -\frac{5}{6}$)	1.872	2.381	3.639	4.202	5.539
$\nu = 4$ ($\beta = -0.875$)	1.789	2.229	3.312	3.797	4.950
$\nu = 5$ ($\beta = -0.90$)	1.728	2.118	3.073	3.501	4.518
$\nu = 10$ ($\beta = -0.95$)	1.567	1.821	2.439	2.713	3.365
$\nu = 50$ ($\beta = -0.99$)	1.384	1.455	1.625	1.701	1.882

Table 1.1.

Remark 1.7. Let $X = X_1, X_2, \ldots$ be a sequence of independent and identically distributed random variables with common continuous distribution function $F(x) = \mathbb{P}(X \le x)$. Set $F_n(x) = n^{-1}\#\{X_i \le x : 1 \le i \le n\}$, and denote by $a_n(x) = n^{1/2}(F_n(x) - F(x))$ the empirical process based upon $X_1, \ldots, X_n$ (see, e.g., [40]). In view of Remark 1.5, it is noteworthy that $a_n(\cdot) \stackrel{d}{=} \alpha_n(F(\cdot))$. This allows us to use Table 1.1 for deriving approximate critical points for Cramér-von Mises-type tests of goodness-of-fit, via the limit law, taken with $\nu^* = 1/(2(\beta^* + 1))$ and $\nu = \nu^* + 1 = (2\beta^* + 3)/(2(\beta^* + 1))$,

$$\begin{aligned}\lim_{n\to\infty} \mathbb{P}\Big(n \int_{-\infty}^{\infty} F_n^{2\beta^*}(x) a_n^2(x) dF_n(x) > \frac{t(\nu^*)^2}{\nu^* + 1}\Big) \qquad (1.63)\\ = \lim_{n\to\infty} \mathbb{P}\Big(n \int_{-\infty}^{\infty} F^{2\beta^*}(x) a_n^2(x) dF(x) > \frac{t(\nu^*)^2}{\nu^* + 1}\Big)\\ = \mathbb{P}\Big(D^*_{B,\nu^*} > t\Big) = \mathbb{P}\Big(D^*_{W,\nu} > t\Big) = 1 - p,\end{aligned}$$

which, via (1.51), holds for all $\beta^* \ge -\frac{1}{2}$, or equivalently, for $0 < \nu^* \le 1$ and $1 < \nu \le 2$. Thus, by (1.63), we reject the null hypothesis that $F(x) = \mathbb{P}(X \le x)$ at the (approximate) level $1 - p$ whenever

$$n \int_{-\infty}^{\infty} F_n^{2\beta^*}(x) a_n^2(x) dF_n(x) > \frac{t(\nu^*)^2}{\nu^* + 1}, \qquad (1.64)$$

where t is given, in terms of $\nu = \nu^* + 1$ and p, in Table 1.1. A similar test procedure may be obtained by the formal replacements of $F_n(\cdot)$ and $a_n(\cdot)$ in (1.64) by $F(\cdot)$ and $a_n(\cdot)$.

1.5. Small deviations of the $L^2(0,1)$ norms of the weighted processes

An application of Theorems 1.3–1.4 yields the following identities in distribution.

Corollary 1.1. *For each $\beta > -1$, or equivalently, for each $\nu = 1/(2(1 + \beta)) > 0$,*

$$2\nu D^2_{W,\nu} = \int_0^1 t^{2\beta} W^2(t) dt \stackrel{d}{=} 4\nu^2 \sum_{k=1}^{\infty} \frac{\omega_k^2}{z^2_{\nu-1,k}}, \qquad (1.65)$$

$$2\nu D^2_{B,\nu} = \int_0^1 t^{2\beta} B^2(t) dt \stackrel{d}{=} 4\nu^2 \sum_{k=1}^{\infty} \frac{\omega_k^2}{z^2_{\nu,k}}, \qquad (1.66)$$

where $\{\omega_k : k \ge 1\}$ is a sequence of i.i.d. $N(0,1)$ random variables.

Proof. (1.65) (resp. (1.66)) is the version of (1.5) corresponding to the KL expansion (1.21) (resp. (1.23)) in Theorem 1.3 (resp. Theorem 1.4). □

Remark 1.8. For an arbitrary $\beta > -\frac{1}{2}$, set

$$\alpha = 1/(1 + 2\beta) > 0 \quad \text{and} \quad \nu = 1/(2(\beta + 1)) = \alpha/(\alpha + 1) \in (0, 1).$$

The following equalities are obtained by the change of variable $t = u^{1+2\beta}$ in (1.65) and (1.66). Whenever $\beta > -\frac{1}{2}$ (or equivalently, when $\alpha > 0$, or $0 < \nu < 1$),

$$\int_0^1 t^{2\beta} W^2(t)dt = \alpha \int_0^1 W^2(t^\alpha)dt \stackrel{d}{=} 4\nu^2 \sum_{k=1}^{\infty} \frac{\omega_k^2}{z_{\nu-1,k}^2}, \tag{1.67}$$

$$\int_0^1 t^{2\beta} B^2(t)dt = \alpha \int_0^1 B^2(t^\alpha)dt \stackrel{d}{=} 4\nu^2 \sum_{k=1}^{\infty} \frac{\omega_k^2}{z_{\nu,k}^2}. \tag{1.68}$$

Assuming, as above, that $\beta > -\frac{1}{2}$, Li (see Theorem 6, p. 24 in [24]) showed the existence of a constant $C_\beta > 0$, such that, as $\varepsilon \to 0$,

$$\mathbb{P}\Big(\int_0^1 t^{2\beta} B^2(t)dt \le \varepsilon^2\Big) = (1+o(1))C_\beta\, \varepsilon^{-\frac{\beta}{2(1+\beta)}} \exp\Big(-\frac{1}{8\varepsilon^2(1+\beta)^2}\Big). \tag{1.69}$$

Below, we extend the validity of (1.69) to the case where $-1 < \beta \le -\frac{1}{2}$.

Theorem 1.7. *For each $\beta > -1$, there exist constants $C'_\beta > 0$ and $C''_\beta > 0$ such that, as $\varepsilon \to 0$,*

$$\mathbb{P}\Big(\int_0^1 t^{2\beta} W^2(t)dt \le \varepsilon^2\Big) = (1+o(1))C'_\beta\, \varepsilon^{-\frac{2+3\beta}{2(1+\beta)}} \exp\Big(-\frac{1}{8\varepsilon^2(1+\beta)^2}\Big), \tag{1.70}$$

$$\mathbb{P}\Big(\int_0^1 t^{2\beta} B^2(t)dt \le \varepsilon^2\Big) = (1+o(1))C''_\beta\, \varepsilon^{-\frac{\beta}{2(1+\beta)}} \exp\Big(-\frac{1}{8\varepsilon^2(1+\beta)^2}\Big). \tag{1.71}$$

Proof. We start by proving (1.70). We first recall Theorem 2 of Li [24] and Lemma 1, p. 72 in Li [25]. Given any two sequences $a_n > 0$ and $b_n > 0$ with

$$\sum_{n=1}^{\infty} a_n < \infty, \quad \sum_{n=1}^{\infty} b_n < \infty \quad \text{and} \quad \sum_{n-1}^{\infty} \Big|1 - \frac{a_n}{b_n}\Big| < \infty, \tag{1.72}$$

we have, as $\epsilon \downarrow 0$,

$$\mathbb{P}\Big(\sum_{n=1}^{\infty} a_n \omega_n^2 \le \epsilon\Big) = (1+o(1))\Big\{\prod_{n=1}^{\infty} \frac{a_n}{b_n}\Big\}^{1/2} \mathbb{P}\Big(\sum_{n=1}^{\infty} b_n \omega_n^2 \le \epsilon\Big). \tag{1.73}$$

Since, for each $\nu > -1$, $z_{\nu,k} = \big\{k + \frac{1}{2}(\nu - \frac{1}{2})\big\}\pi + O\big(\frac{1}{k}\big)$, as $k \to \infty$ (see (2.17) below), we see that $a_n = 1/z_{\nu-1,n}^2$ and $b_n = 1/\big\{n + \frac{1}{2}(\nu - \frac{3}{2})\big\}^2\pi^2$ fulfill (1.72). Thus, by (1.65) and (1.73), there exists a constant $c_2 = c_2(\beta) > 0$ such that

$$\mathbb{P}\Big(\int_0^1 t^{2\beta} W^2(t)dt \le \varepsilon^2\Big) = \mathbb{P}\Big(\sum_{k=1}^{\infty} \frac{\omega_k^2}{z_{\nu-1,k}^2} \le \frac{\varepsilon^2}{4\nu^2}\Big) \tag{1.74}$$

$$= \quad (1+o(1))c_2\mathbb{P}\Big(\sum_{k=1}^{\infty} \frac{\omega_k^2}{(k+\frac{1}{2}(\nu-\frac{3}{2}))^2} \le \frac{\varepsilon^2\pi^2}{4\nu^2}\Big) \quad \text{as} \quad \varepsilon \downarrow 0.$$

Also, by (3.4), p. 14 in Li [24] (see also Zolotarev [48]), for an arbitrary $d > -1$, there exists a constant $c_3 = c_3(d) > 0$, such that, as $\epsilon \to 0$,

$$\mathbb{P}\Big(\sum_{k=1}^{\infty} \frac{\omega_k^2}{(k+d)^2} \le \epsilon^2\Big) = (1+o(1))c_3\epsilon^{-2d} \exp\Big(-\frac{\pi^2}{8} \times \frac{1}{\epsilon^2}\Big). \tag{1.75}$$

In view of (1.74), we set $\epsilon^2 = \varepsilon^2\pi^2/(4\nu^2)$ and $d = d(\beta) = \frac{1}{2}(\nu - \frac{3}{2}) = \frac{1}{4}(\frac{1}{1+\beta} - 3)$ in (1.75). Letting $c_1 = c_1(\beta) = c_2(\beta)c_3(d(\beta)) > 0$, we so obtain that, as $\varepsilon \downarrow 0$,

$$\mathbb{P}\Big(\int_0^1 t^{2\beta} W^2(t)dt \le \varepsilon^2\Big) = (1+o(1))c_1\varepsilon^{-(\nu-\frac{3}{2})} \exp\Big(-\frac{\nu^2}{2} \times \frac{1}{\varepsilon^2}\Big), \tag{1.76}$$

which is (1.70). By exactly the same arguments, starting with (1.66) instead of (1.65), we obtain that, for some constant $c_4 = c_4(\beta) > 0$, as $\varepsilon \downarrow 0$,

$$\mathbb{P}\Big(\int_0^1 t^{2\beta} B^2(t)dt \le \varepsilon^2\Big) = (1+o(1))c_4\varepsilon^{-(\nu-\frac{1}{2})} \exp\Big(-\frac{\nu^2}{2} \times \frac{1}{\varepsilon^2}\Big). \tag{1.77}$$

We conclude (1.70)–(1.71) by setting $\nu = 1/(2(\beta+1))$ in (1.76)–(1.77). □

In the following §1.6–§1.9, we present a series of additional results which turn out to be consequences of (1.65)–(1.66).

1.6. Large deviations of the $L^2(0,1)$ norms of the weighted processes

We recall the definitions (1.40)–(1.41) of $D^2_{W,\nu}$ and $D^2_{W,\nu}$, with $\nu = 1/(2(\beta+1))$.

$$D^2_{W,\nu} = \frac{1}{2\nu}\int_0^1 t^{2\beta} W^2(t)dt \quad \text{and} \quad D^2_{B,\nu} = \frac{1}{2\nu}\int_0^1 t^{2\beta} B^2(t)dt.$$

Theorem 1.8. *We have, for any $\nu > 0$, as $t \to \infty$,*

$$\begin{aligned}\mathbb{P}\Big(D^2_{W,\nu} > t\Big) &= (1+o(1))\Big(\frac{z_{\nu-1,1}}{2}\Big)^{\frac{\nu-1}{2}-\frac{3}{2}} \times \frac{\sqrt{\nu}}{\sqrt{\pi\Gamma(\nu)J_\nu(z_{\nu-1,1})}} \\ &\quad \times \frac{\exp(-tz^2_{\nu-1,1}/(4\nu))}{\sqrt{t}},\end{aligned} \tag{1.78}$$

$$\begin{aligned}\mathbb{P}\Big(D^2_{B,\nu} > t\Big) &= (1+o(1))\Big(\frac{z_{\nu,1}}{2}\Big)^{\frac{\nu}{2}-\frac{3}{2}} \times \frac{\sqrt{\nu}}{\sqrt{\pi\Gamma(\nu+1)J_{\nu+1}(z_{\nu,1})}} \\ &\quad \times \frac{\exp(-tz^2_{\nu,1}/(4\nu))}{\sqrt{t}}.\end{aligned} \tag{1.79}$$

Proof. Making use of (1.11), in combination with the definition (1.40) of $D^2_{W,\nu}$, we infer from the KL-expansion (1.30) of $t^{\frac{1}{2}-\nu}W(t^{2\nu})$ that, as $t \to \infty$,

$$\mathbb{P}(D^2_{W,v} > t) = (1+o(1))\sqrt{\frac{2}{\pi}} \frac{1}{\gamma_1\sqrt{-\mathbb{F}'_{W,\nu}(\gamma_1)}} \frac{\exp(-\frac{1}{2}t\gamma_1)}{\sqrt{t}}, \tag{1.80}$$

where $\gamma_1 = z_{\nu-1,1}^2/(2\nu)$ and $\mathbb{F}_{W,\nu}$ is as in (1.45). To evaluate

$$\mathbb{F}'_{W,\nu}(z) = \frac{d}{dz}\Big\{\Gamma(\nu)\Big(\frac{\sqrt{2\nu z}}{2}\Big)^{-\nu+1} J_{\nu-1}\left(\sqrt{2\nu z}\right)\Big\},$$

for $z = \gamma_1$, we observe that $J_{\nu-1}(\gamma_1) = J_{\nu-1}(z_{\nu-1,1}) = 0$, whence

$$\mathbb{F}'_{W,\nu}(\gamma_1) = \Big\{\frac{\nu\Gamma(\nu)}{2}\Big(\frac{z_{\nu-1,1}}{2}\Big)^{-\nu} J'_{\nu-1}\left(z_{\nu-1,1}\right)\Big\}. \tag{1.81}$$

By combining (1.80)–(1.81) with $J'_{\nu-1}\left(z_{\nu-1,1}\right) = -J_\nu\left(z_{\nu-1,1}\right)$, via (2.14), we get

$$\gamma_1\sqrt{-\mathbb{F}'_{W,\nu}(\gamma_1)} = \sqrt{\frac{2}{\nu}}\Big(\frac{z_{\nu-1,1}}{2}\Big)^{2-\frac{\nu}{2}}\Big\{\Gamma(\nu)J_\nu\left(z_{\nu-1,1}\right)\Big\}^{1/2},$$

from where (1.78) is straightforward. The proof of (1.79) is established likewise, with the replacement of (1.45) by (1.46). We omit details. □

1.7. A useful identity relating Wiener processes to Brownian bridges

The corollary of Theorems 1.3–1.4 stated below, gives a simple proof of rather unexpected equalities in distribution relating the Wiener process to the Brownian bridge. The latter equalities have been obtained by Donati-Martin and Yor ([9], [10], [11]) and Yor (see, e.g., [45]), by a completely different methodology making use of a stochastic Fubini theorem for double Wiener integrals

Corollary 1.2. *Let* $\beta \in (-1, -\frac{1}{2})$, $\beta^* \in (-1, \infty)$, $\nu = 1/(2(\beta+1)) \in (0,\infty)$ *and* $\nu^* = 1/(2(\beta^*+1)) \in (-1,\infty)$ *be related through*

$$\nu^* = \nu - 1 \quad \Leftrightarrow \quad \beta = \frac{\beta^*+1}{2\beta^*+3} - 1 \quad \Leftrightarrow \quad \beta^* = -\frac{\beta+1}{2\beta+1} - 1. \tag{1.82}$$

Then, we have the identities in distribution

$$\frac{1}{2\nu}D^2_{W,\nu} = (\beta+1)^2\int_0^1 t^{2\beta}W^2(t)dt \tag{1.83}$$

$$\stackrel{d}{=} \frac{1}{2\nu^*}D^2_{B,\nu^*} = (\beta^*+1)^2\int_0^1 t^{2\beta^*}B^2(t)dt,$$

$$D^2_{W,\nu} = (\beta+1)\int_0^1 t^{2\beta}W^2(t)dt \tag{1.84}$$

$$\stackrel{d}{=} \Big\{\frac{\nu^*+1}{\nu^*}\Big\}D^2_{B,\nu^*} = (\beta^*+1)(2\beta^*+3)\int_0^1 t^{2\beta^*}B^2(t)dt.$$

Proof. Setting $\nu^* = 1/(2(\beta^*+1))$ and $\nu = 1/(2(\beta+1))$, it is readily checked from (1.82) that $\nu^* - 1 = \nu$, so that (1.82) follows readily from (1.65)–(1.66). Since (1.82) implies that $(\beta^*+1)/(\beta+1) = 2\beta^*+3$, (1.84) follows obviously from (1.83). □

Remark 1.9. In a series of papers, Donati-Martin and Yor ([9], [10], [11]), and Yor ([45], [46], [47]) established the following distributional identity (refer to (4.3),

p. 52 in [46]). For all $\delta > 2$,

$$\frac{1}{\delta}\int_0^1 u^{1-\delta}B^2(u^\delta)du \stackrel{d}{=} \frac{1}{\delta-2}\int_0^1 u^{3-\delta}W^2(u^{\delta-2})du. \tag{1.85}$$

The change of variable $u = t^{1/\delta}$ in (1.85) shows readily that

$$\frac{1}{\delta}\int_0^1 u^{1-\delta}B^2(u^\delta)du = (\beta^*+1)^2\int_0^1 t^{2\beta^*}B^2(t)dt, \tag{1.86}$$

$$\text{with } \beta^* = \frac{1}{\delta} - 1 \quad \Leftrightarrow \quad \nu^* = \frac{\delta}{2} = \frac{1}{2(\beta^*+1)}.$$

Likewise, the change of variable $u = t^{1/(\delta-2)}$ in (1.85) shows that

$$\frac{1}{\delta-2}\int_0^1 u^{3-\delta}W^2(u^{\delta-2})du = (\beta+1)^2\int_0^1 t^{2\beta}W^2(t)dt, \tag{1.87}$$

$$\text{with } \beta = \frac{1}{\delta-2} - 1 \quad \Leftrightarrow \quad \nu = \frac{\delta-2}{2} = \frac{1}{2(\beta+1)} = \nu^* - 1.$$

Given (1.86)–(1.87) the equivalence of (1.85) with (1.83) is straightforward. The relation (1.85) turns out to be an indirect consequence of the Ciesielski-Taylor distributional equality (see, e.g., (4.1), p. 50 in [46])

$$\int_0^\infty ds\, 1_{\{R_{\delta+2}(s)\le 1\}} \stackrel{d}{=} T_1(R_\delta). \tag{1.88}$$

where $\{R_\gamma(s) : s \ge 0\}$ denotes a Bessel process of dimension γ starting at 0, and $T_1(R_\delta) = \inf\{t : R_\delta(t) = 1\}$. Since (1.88) holds for all $\delta > 0$, one may also infer from the Ciesielski-Taylor relation the following versions of (1.85), holding respectively for $\delta = 2$ and $0 < \delta < 2$ (refer to (4.3)–(4.5), pp. 52–53 in [46]).

$$(\text{For } \delta = 2), \quad \frac{1}{2}\int_0^1 u^{-1}B^2(u^2)dt \stackrel{d}{=} \int_0^1 uW^2(\log(1/u))du, \tag{1.89}$$

$$(\text{For } \delta < 2), \quad \frac{1}{\delta}\int_0^1 u^{1-\delta}B^2(u^\delta)du \stackrel{d}{=} \frac{1}{\delta-2}\int_0^1 u^{3-\delta}W^2(1-u^{2-\delta})du. \tag{1.90}$$

This hints that the Karhunen–Loève expansions of, respectively,

$$\{u^{-1/2}B(u^2) : 0 < u < 1\} \quad \text{and} \quad \{u^{1/2}W(\log(1/u)) : 0 < u < 1\},$$

and, for $0 < \delta < 2$

$$\{u^{(1-\delta)/2}B(u^\delta) : 0 < u < 1\} \quad \text{and} \quad \{u^{(3-\delta)/2}W(1-u^{2-\delta}) : 0 < u < 1\},$$

are closely related. This problem will be considered elsewhere.

1.8. Applications to Bessel functions

Our next corollary gives a simple proof of the first two Rayleigh extensions of Euler's formula (see, e.g., [44], p. 502), dealing with the positive roots $\{z_{\nu,k} : k \geq 1\}$ of the Bessel function $J_\nu(\cdot)$ (see §2.1 below). Yor (see, e.g. (11.47)–(11.49) in [47]) provides another probabilistic proof of these relations by a different argument.

Corollary 1.3. *Let $\{z_{\nu,k} : k \geq 1\}$ denote the positive roots of the Bessel function $J_\nu(\cdot)$. Then, for each $\nu > -1$, we have*

$$(i)\quad \sum_{k=1}^{\infty} \frac{1}{z_{\nu,k}^2} = \frac{1}{4(\nu+1)} \quad \text{and} \quad (ii)\quad \sum_{k=1}^{\infty} \frac{1}{z_{\nu,k}^4} = \frac{1}{16(\nu+1)^2(\nu+2)}. \tag{1.91}$$

Proof. We start by proving (i) in (1.91). For each $\nu^* = 1/(2(\beta^*+1)) > 0$,

$$\mathbf{E}\Big\{\int_0^1 t^{2\beta^*} W^2(t)dt\Big\} = \int_0^1 t^{2\beta^*+1}dt = \frac{1}{2(\beta^*+1)} = \nu^*, \tag{1.92}$$

and, likewise, for each $\nu = 1/(2(\beta+1)) > 0$,

$$\begin{aligned} \mathbf{E}\Big\{\int_0^1 t^{2\beta} B^2(t)dt\Big\} &= \int_0^1 t^{2\beta}\mathbf{E}(B^2(t))dt \qquad (1.93)\\ &= \int_0^1 (t^{2\beta+1} - t^{2\beta+2})dt = \frac{1}{2(\beta+1)(2\beta+3)} = \frac{\nu^2}{\nu+1}. \end{aligned}$$

Combining (1.66) with (1.93), we obtain readily that, for each $\nu > 0$,

$$\mathbf{E}\Big\{\sum_{k=1}^{\infty} \frac{\omega_k^2}{z_{\nu,k}^2}\Big\} = \frac{1}{4\nu^2}\mathbf{E}\Big\{\int_0^1 t^{2\beta}B^2(t)dt\Big\} = \frac{1}{4(\nu+1)}. \tag{1.94}$$

Repeating the same argument with (1.65) and (1.92), we see that, for each $\nu^* > 0$,

$$\mathbf{E}\Big\{\sum_{k=1}^{\infty} \frac{\omega_k^2}{z_{\nu^*-1,k}^2}\Big\} = \frac{1}{4(\nu^*)^2}\mathbf{E}\Big\{\int_0^1 t^{2\beta}W^2(t)dt\Big\} = \frac{1}{4\nu^*}. \tag{1.95}$$

We conclude (1.91)(i) for $\nu > -1$ by setting $\nu = \nu^* - 1$ in (1.95). This is in agreement with (1.94). However, the validity of (1.94) was established only for $\nu > 0$, whereas (1.91)(i) has just been shown, via (1.95), to hold for all $\nu > -1$.

To establish (1.91)(ii), we observe that $\mathbf{E}(W^2(s)W^2(t)) = 2s^2 + st$ for $0 \leq s \leq t \leq 1$, whence, for an arbitrary $\nu^* = 1/(2(\beta^*+1)) > 0$,

$$\begin{aligned} \mathbf{E}\Big\{\Big(\int_0^1 t^{2\beta^*}W^2(t)dt\Big)^2\Big\} &= \int_0^1 t^{2\beta^*}\Big\{\int_0^1 s^{2\beta^*}\mathbf{E}(W^2(s)W^2(t))ds\Big\}dt \qquad (1.96)\\ &= 2\int_0^t t^{2\beta^*}\Big\{\int_0^t s^{2\beta^*+2}ds + \int_t^1 t^2 s^{2\beta^*}ds\Big\}dt + \Big\{\int_0^1 t^{2\beta^*+1}dt\Big\}^2\\ &= \frac{6\beta^*+7}{(2\beta^*+2)^2(2\beta^*+3)} = \frac{(\nu^*)^2(3+\nu^*)}{\nu^*+1}. \end{aligned}$$

It follows from (1.65), (1.92) and (1.96) that

$$\mathrm{Var}\Big(\int_0^1 t^{2\beta^*} W^2(t)dt\Big) = \frac{(\nu^*)^2(3+\nu^*)}{\nu^*+1} - (\nu^*)^2 = \frac{2(\nu^*)^2}{\nu^*+1} \tag{1.97}$$

$$16(\nu^*)^4 \,\mathrm{Var}\Big(\sum_{k=1}^{\infty} \frac{\omega_k^2}{z_{\nu^*-1,k}^2}\Big) = 32(\nu^*)^4 \sum_{k=1}^{\infty} \frac{1}{z_{\nu^*-1,k}^4}.$$

Finally, we obtain readily (*ii*) in (1.91) by setting $\nu^* = \nu + 1$ in (1.97). □

Remark 1.10. By computing higher moments of $\sum_{k=1}^{\infty} \omega_k^2/z_{\nu^*-1,k}^2$, one obtains likewise explicit expressions of $\sum_{k=1}^{\infty} 1/z_{\nu,k}^{2r}$ for $r = 3, 4, \ldots$ in terms of ν, r, yielding elementary proofs of the higher order Rayleigh formulae (see [44], p. 502).

Some other identities of interest are stated in the corollary below.

Corollary 1.4. *We have, for all* $\nu > 0$

$$\min\{s^\nu, t^\nu\} = 4\nu \sum_{k=1}^{\infty} \frac{1}{z_{\nu-1,k}^2} \Big\{ \frac{J_\nu(z_{\nu-1,k}s) J_\nu(z_{\nu-1,k}t)}{J_\nu^2(z_{\nu-1,k})} \Big\}, \tag{1.98}$$

$$\min\{s^\nu, t^\nu\} - \{st\}^\nu = 4\nu \sum_{k=1}^{\infty} \frac{1}{z_{\nu,k}^2} \Big\{ \frac{J_\nu(z_{\nu,k}s) J_\nu(z_{\nu,k}t)}{J_{\nu-1}^2(z_{\nu,k})} \Big\}, \tag{1.99}$$

where the series are convergent in $L^2\big((0,1)^2\big)$.

Proof. $(st)^{\frac{1}{2}-\nu} \min\{s^{2\nu}, t^{2\nu}\} = (st)^{\frac{1}{2}} \min\{s^\nu, t^\nu\}$ being the covariance function of $Z(t) = t^{\frac{1}{2}-\nu} W(t^{2\nu})$, we infer (1.98) from (1.3), taken in the special case of the KL expansion (1.30). Likewise, $(st)^{\frac{1}{2}-\nu}\big\{\min\{s^{2\nu}, t^{2\nu}\} - (st)^{2\nu}\big\} = (st)^{\frac{1}{2}}\big\{\min\{s^\nu, t^\nu\} - (st)^\nu\big\}$ being the covariance function of $Z(t) = t^{\frac{1}{2}-\nu} B(t^{2\nu})$, we infer (1.99) from (1.3), taken in the special case of the KL expansion (1.31). □

1.9. Limiting distributions

We describe below the limiting behavior of $D_{W,\nu}^2$ and $D_{B,\nu}^2$ as $\nu \uparrow \infty$ and $\nu \downarrow 0$.

Proposition 1.1. *The following limits hold in probability as* $\nu = 1/(2(\beta+1)) \uparrow \infty$, *or equivalently, when* $\beta = \frac{1}{2\nu} - 1 \downarrow -1$.

$$2D_{W,\nu}^2 = 2(\beta+1) \int_0^1 t^{2\beta} W^2(t)dt \overset{\mathbf{P}}{\to} 1, \tag{1.100}$$

$$2D_{B,\nu}^2 = 2(\beta+1) \int_0^1 t^{2\beta} B^2(t)dt \overset{\mathbf{P}}{\to} 1. \tag{1.101}$$

Proof. By (1.92) and (1.97), we have, as $\nu = 1/(2(\beta+1)) \to \infty$,

$$\mathbb{E}\Big(\frac{1}{\nu}\int_0^1 t^{2\beta} W^2(t)dt\Big) = 1 \text{ and } \mathrm{Var}\Big(\frac{1}{\nu}\int_0^1 t^{2\beta} W^2(t)dt\Big) = \frac{2}{\nu+1} \to 0, \tag{1.102}$$

which readily yields (1.100). We obtain likewise (1.101) by combining (1.102) with (1.82) and (1.84). □

Below, we denote by $\mathbb{P}(\chi_1^2 \le x)$ the distribution function of a random variable with a χ_1^2 distribution.

Proposition 1.2. *The following limits hold in probability as* $\nu = 1/(2(\beta+1) \downarrow 0$, *or equivalently, when* $\beta = \frac{1}{2\nu} - 1 \uparrow \infty$.

$$\begin{aligned}\mathbb{P}\Big(2D^2_{W,\nu} \le x\Big) &= \mathbb{P}\Big(2(\beta+1)\int_0^1 t^{2\beta}W^2(t)dt \le x\Big) \qquad (1.103)\\ &\to \mathbb{P}\Big(\chi_1^2 \le x\Big),\end{aligned}$$

$$\begin{aligned}\mathbb{P}\Big(\frac{1}{\nu}D^2_{B,\nu} \le x\Big) &= \mathbb{P}\Big(2(\beta+1)^2\int_0^1 t^{2\beta}B^2(t)dt \le x\Big) \qquad (1.104)\\ &\to \mathbb{P}\Big(D^2_{W,1} \le x\Big).\end{aligned}$$

Proof. To prove (1.103), we observe, via (1.65), that, for $\nu > 0$,

$$\mathbb{P}\Big(2D^2_{W,\nu} \le x\Big) = \mathbb{P}\Big(4\nu\sum_{k=1}^{\infty}\frac{\omega_k^2}{z^2_{\nu-1,k}} \le x\Big).$$

In view of (1.82), setting $\nu^* = \nu - 1$, we first describe the limiting behavior of $z_{\nu^*,1} = z_{\nu-1,1}$ as $\nu^* \to -1 \Leftrightarrow \nu \to 0$. By (2.2) in the sequel, we have

$$\begin{aligned}J_{\nu^*}(z) &= (\tfrac{1}{2}z)^{\nu^*}\sum_{k=0}^{\infty}\frac{(-\frac{1}{4}z^2)^k}{\Gamma(\nu^*+k+1)k!} \qquad (1.105)\\ &= \frac{(\frac{1}{2}z)^{\nu^*}}{\Gamma(\nu^*+1)}\Big\{1 - \frac{(\frac{1}{4}z^2)(1+A(z,\nu)))}{\nu^*+1}\Big\},\end{aligned}$$

where, uniformly over $\nu^* > -1$,

$$|A(z,\nu)| = \sum_{k=0}^{\infty}\frac{(\frac{1}{4}z^2)^{k+1}}{(k+2)!(\nu^*+2)\dots(\nu^*+k+2)} \le \tfrac{1}{8}z^2\exp(\tfrac{1}{4}z^2). \qquad (1.106)$$

It follows readily from (1.105)–(1.106), that, as $\nu^* \downarrow -1$, or equivalently, as $\nu \downarrow 0$,

$$z_{\nu-1,1} = z_{\nu^*,1} = (1+o(1))2\sqrt{\nu^*+1} = (1+o(1))2\sqrt{\nu}. \qquad (1.107)$$

This, when combined with Rayleigh's formula (1.91), entails that, as $\nu \downarrow 0$,

$$1 = 4\nu\sum_{k=1}^{\infty}\frac{1}{z^2_{\nu-1,k}} = 1 + o(1) + 4\nu\sum_{k=2}^{\infty}\frac{1}{z^2_{\nu-1,k}}.$$

Therefore, by Markov's inequality, for each specified $\varepsilon > 0$, we have, as $\nu \downarrow 0$,

$$\begin{aligned}\mathbb{P}\Big(\Big|2D^2_{W,\nu} - \Big\{\frac{4\nu}{z^2_{\nu-1,1}}\Big\}\omega_1^2\Big| \ge \varepsilon\Big) &= \mathbb{P}\Big(4\nu\sum_{k=2}^{\infty}\frac{\omega_k^2}{z^2_{\nu-1,k}} \ge \varepsilon\Big) \qquad (1.108)\\ &\le \Big\{\frac{1}{\varepsilon}\Big\}\Big\{4\nu\sum_{k=2}^{\infty}\frac{1}{z^2_{\nu-1,k}}\Big\} \to 0.\end{aligned}$$

By combining (1.107) with (1.108), we conclude that $2D^2_{W,\nu} \overset{d}{\to} \chi^2_1$, which is (1.103).

In order to establish (1.104), we make use of (1.65), to get

$$\mathbb{P}\Big(D^2_{W,\nu} \le x\Big) \quad = \quad \mathbb{P}\Big(2\nu \sum_{k=1}^{\infty} \frac{\omega_k^2}{z^2_{\nu-1,k}} \le x\Big),$$

and, by another application of (1.65) for $\nu = 1$,

$$\mathbb{P}\Big(D^2_{W,1} \le x\Big) \quad = \quad \mathbb{P}\Big(2 \sum_{k=1}^{\infty} \frac{\omega_k^2}{z^2_{0,k}} \le x\Big),$$

Thus, in view of (1.84), we have, for each $x > 0$, $\nu > 0$, $N \ge 2$ and $\varepsilon > 0$,

$$\mathbb{P}\Big(\Big\{\frac{\nu+1}{\nu}\Big\}D^2_{B,\nu} \le x\Big) = \mathbb{P}\Big(D^2_{W,\nu+1} \le x\Big) = \mathbb{P}\Big(\sum_{k=1}^{\infty} \frac{\omega_k^2}{z^2_{\nu,k}} \le x\Big\{\frac{1}{\nu+1}\Big\}\Big)$$

$$=: \mathbb{P}\Big(\sum_{k=1}^{\infty} \frac{\omega_k^2}{z^2_{\nu,k}} \le x_\nu\Big) \le \mathbb{P}\Big(2\sum_{k=1}^{N-1} \frac{\omega_k^2}{z^2_{\nu,k}} \le x_\nu + \tfrac{1}{4}\varepsilon\Big) + \mathbb{P}\Big(2\sum_{k=N}^{\infty} \frac{\omega_k^2}{z^2_{\nu,k}} \ge \tfrac{1}{4}\varepsilon\Big),$$

where $x_\nu = x\{1/(1+\nu)\}$, and likewise,

$$\mathbb{P}\Big(\Big\{\frac{\nu+1}{\nu}\Big\}D^2_{B,\nu} \le x\Big) = \mathbb{P}\Big(\sum_{k=1}^{\infty} \frac{\omega_k^2}{z^2_{\nu,k}} \le x_\nu\Big)$$

$$\ge \mathbb{P}\Big(2\sum_{k=1}^{N-1} \frac{\omega_k^2}{z^2_{\nu,k}} \le x_\nu - \tfrac{1}{4}\varepsilon\Big) - \mathbb{P}\Big(2\sum_{k=N}^{\infty} \frac{\omega_k^2}{z^2_{\nu,k}} \ge \tfrac{1}{4}\varepsilon\Big).$$

By Fact 2.1 in the sequel, $z_{\nu,k} \downarrow z_{0,k}$ as $\nu \downarrow 0$, and $z_{0,k} = (1 + o(1))k\pi$ as $k \to \infty$. Thus, for any $\varepsilon > 0$, we may select $N = N_\varepsilon$ so large that, uniformly over $\nu \ge 0$,

$$\mathbb{P}\Big(2\sum_{k=N}^{\infty} \frac{\omega_k^2}{z^2_{\nu,k}} \ge \tfrac{1}{4}\varepsilon\Big) \le \frac{4}{\varepsilon}\mathbb{E}\Big(2\sum_{k=N}^{\infty} \frac{\omega_k^2}{z^2_{\nu,k}}\Big) = \frac{8}{\varepsilon}\sum_{k=N}^{\infty} \frac{1}{z^2_{\nu,k}} \le \frac{8}{\varepsilon}\sum_{k=N}^{\infty} \frac{1}{z^2_{0,k}} \le \tfrac{1}{2}\varepsilon.$$

This, in turn, readily implies that

$$\limsup_{\nu\downarrow 0} \mathbb{P}\Big(2\sum_{k=1}^{N-1} \frac{\omega_k^2}{z^2_{\nu,k}} \le x_\nu + \tfrac{1}{4}\varepsilon\Big) \le \mathbb{P}\Big(2\sum_{k=1}^{N-1} \frac{\omega_k^2}{z^2_{0,k}} \le x + \tfrac{1}{2}\varepsilon\Big)$$

$$\le \mathbb{P}\Big(2\sum_{k=1}^{\infty} \frac{\omega_k^2}{z^2_{0,k}} \le x + \varepsilon\Big) + \mathbb{P}\Big(2\sum_{k=N}^{\infty} \frac{\omega_k^2}{z^2_{0,k}} \ge \tfrac{1}{2}\varepsilon\Big) \le \mathbb{P}\Big(D^2_{W,1} \le x + \varepsilon\Big) + \tfrac{1}{2}\varepsilon,$$

So that,

$$\limsup_{\nu\downarrow 0} \mathbb{P}\Big(\Big\{\frac{\nu+1}{\nu}\Big\}D^2_{B,\nu} \le x\Big) \tag{1.109}$$

$$\le \limsup_{\nu\downarrow 0} \mathbb{P}\Big(2\sum_{k=1}^{N-1} \frac{\omega_k^2}{z^2_{0,k}} \le x_\nu + \tfrac{1}{4}\varepsilon\Big) + \tfrac{1}{4}\varepsilon \le \mathbb{P}\Big(D^2_{W,1} \le x + \varepsilon\Big) + \varepsilon.$$

By a similar argument which we omit, we get

$$\liminf_{\nu\downarrow 0}\mathbb{P}\Big(\Big\{\frac{\nu+1}{\nu}\Big\}D^2_{B,\nu}\le x\Big) \;\ge\; \mathbb{P}\Big(D^2_{W,1}\le x-\varepsilon\Big)-\varepsilon. \tag{1.110}$$

By combining (1.109) with (1.110) and the fact that $\varepsilon>0$ may be chosen arbitrarily small, we conclude (1.104). We have used here the fact, obvious by (1.43), that $\mathbb{P}(D^2_{W,1}\le x)$ is a continuous function of x. □

2. KL Expansions via Bessel functions

2.1. Some useful facts about Bessel functions

For each real constant $\nu\in\mathbb{R}$, we denote by $J_\nu(\cdot)$ (resp. $Y_\nu(\cdot)$) the Bessel function of the first kind (resp. of the second kind) of the index ν. For our needs, it will be useful to recall some important properties of these functions (refer to [21] and [44] for details). The second order homogeneous differential equation

$$x^2y''+xy'+(x^2-\nu^2)y=0, \tag{2.1}$$

has a fundamental set of solutions on $(0,\infty)$ of the form $y=Cx^\nu\sum_{k=0}^\infty a_kx^k$, where C is a constant. These solutions are proportional to the *Bessel function of the first kind* (see, e.g., 9.1.69 in [1]), explicitly defined, for an arbitrary $\nu\in\mathbb{R}$, by

$$J_\nu(x)=\frac{(\frac{1}{2}x)^\nu}{\Gamma(\nu+1)}\,{}_0F_1(\nu+1;-\tfrac{1}{4}x^2)=(\tfrac{1}{2}x)^\nu\sum_{k=0}^\infty\frac{(-\frac{1}{4}x^2)^k}{\Gamma(\nu+k+1)\Gamma(k+1)}. \tag{2.2}$$

When $\nu=-n$ is a negative integer, $\Gamma(\nu+k+1)=\Gamma(n+k+1)=\infty$ for $k=0,\dots,n-1$ so that, making use of the convention $a/\infty=0$ when $a\in\mathbb{R}$, the n first terms in the series (2.2) vanish. In this case, we have the relation

$$J_{-n}(x)=(-1)^nJ_n(x). \tag{2.3}$$

In (2.2), we made use of the *generalized hypergeometric function*

$${}_0F_1(b;z)=\sum_{k=0}^\infty\frac{1}{(b)_k}\frac{z^k}{k!}\quad\text{for}\quad z\in\mathbb{C}, \tag{2.4}$$

where the *Pochhammer symbol* $(b)_k$ is defined for $k\in\mathbb{N}$ by $(b)_k=\Gamma(b+k)/\Gamma(b)$ when $b\neq 0,-1,-2,\cdots$, and, for an arbitrary $b\in\mathbb{R}$, by

$$(b)_0=1\quad\text{and}\quad(b)_k=b(b+1)\dots(b+k-1)\quad\text{for}\quad k\ge 1. \tag{2.5}$$

When ν is noninteger, $J_\nu(\cdot)$ and $J_{-\nu}(\cdot)$ are linearly independent solutions of (2.1). On the other hand, when n is integer, (2.3) implies that $J_n(\cdot)$ and $J_{-n}(\cdot)$ are linearly dependent. To obtain, for an arbitrary $\nu\in\mathbb{R}$ a second set of fundamental solutions of (2.1) on $(0,\infty)$, it is convenient to make use of the *Bessel function of the second kind* $Y_\nu(\cdot)$ of the index ν, also called the *Neumann function*, and defined as follows. Whenever $\nu\notin\mathbb{Z}$ is noninteger, we set

$$Y_\nu(x)=\frac{J_\nu(x)\cos\nu\pi-J_{-\nu}(x)}{\sin\nu\pi}, \tag{2.6}$$

and, when $\nu = n \geq 0$ is integer, we set

$$\begin{aligned} Y_\nu(x) &= \lim_{\nu \to n} \frac{J_\nu(x)\cos\nu\pi - J_{-\nu}(x)}{\sin\nu\pi} \\ &= \frac{2}{\pi} J_n(x)\Big(\gamma + \log(\tfrac{x}{2})\Big) - \frac{1}{\pi}(\tfrac{x}{2})^{-n} \sum_{m=0}^{n-1} \frac{(n-m-1)!}{m!}(\tfrac{x^2}{4})^m \qquad (2.7) \\ &\quad - \frac{1}{\pi}(\tfrac{x}{2})^n \sum_{m=0}^{\infty} \frac{(-\frac{x^2}{4})^m}{m!(m+n)!}\left\{\sum_{k=1}^{n+m}\frac{1}{k} + \sum_{k=1}^{m}\frac{1}{k}\right\}. \end{aligned}$$

Here, $\gamma = 0.5772157...$ denotes Euler's constant, and we use the convention that $\sum_\emptyset(\cdot) = 0$. An easy consequence of (2.6)–(2.7) is that, for any $\nu > 0$,

$$Y_\nu(x) = (1+o(1))\frac{(\frac{1}{2}x)^{-\nu}}{\Gamma(1-\nu)\sin\nu\pi} = (1+o(1))\frac{\Gamma(\nu)}{\pi}\,(\tfrac{1}{2}x)^{-\nu} \text{ as } x \downarrow 0 \qquad (2.8)$$

To infer (2.8) from (2.6) when $\nu > 0$ is noninteger, we make use of the well-known relation $\Gamma(\nu)\Gamma(1-\nu) = \pi/\sin\pi\nu$ (see, e.g., Th. 8, p. 21 in [36]). Making use of (2.7), it is readily checked that (2.8) also holds when $\nu = n > 0$ is integer.

For $\nu \geq 0$, $|J_\nu(\cdot)|$ is bounded. We have the inequality (refer to p. 362 in [1])

$$|J_\nu(x)| \leq \min\left\{1, \frac{(\frac{1}{2}x)^\nu}{\Gamma(\nu+1)}\right\} \quad \text{for} \quad \nu \geq 0 \quad \text{and} \quad x \geq 0. \qquad (2.9)$$

For each $\nu \in \mathbb{R}$, the *modified* or *hyperbolic Bessel function* of index ν is defined by (see, e.g. p. 11 in [21])

$$\begin{aligned} I_\nu(x) &= \frac{(\frac{1}{2}x)^\nu}{\Gamma(\nu+1)} \sum_{k=0}^{\infty} \frac{1}{(\nu+1)_k} \frac{(\frac{1}{4}x^2)^k}{k!} \qquad (2.10) \\ &= \frac{(\frac{1}{2}x)^\nu}{\Gamma(\nu+1)} \prod_{k=1}^{\infty}\left\{1 + \frac{x^2}{z_{\nu,k}^2}\right\} = i^{-\nu} J_\nu(ix) \\ &= e^{-\frac{\nu i \pi}{2}} J_\nu(ix) = \left\{\cos\left(\frac{\nu\pi}{2}\right) - i\sin\left(\frac{\nu\pi}{2}\right)\right\} J_\nu(ix). \end{aligned}$$

We now establish some useful recurrence formulas, and assume from now on, and unless otherwise specified, that $\nu > 0$. It follows readily from (2.2) that

$$J_{\nu+1}(x) = \frac{2\nu}{x} J_\nu(x) - J_{\nu-1}(x) \quad \text{for} \quad x \in \mathbb{R}. \qquad (2.11)$$

By combining (2.2) with (2.11) and the obvious relation

$$\frac{d}{dx}\{{}_0F_1(b;x)\} = \sum_{k=1}^{\infty} \frac{1}{(b)_k} \frac{x^{k-1}}{(k-1)!} = \frac{1}{b}\{{}_0F_1(b+1;x)\}, \qquad (2.12)$$

we obtain the formulas

$$\frac{d}{dx}J_\nu(x) = \frac{\nu}{x}J_\nu(x) - J_{\nu+1}(x) = J_{\nu-1}(x) - \frac{\nu}{x}J_\nu(x) \tag{2.13}$$
$$= \tfrac{1}{2}\Big\{J_{\nu-1}(x) - J_{\nu+1}(x)\Big\} \quad \text{for} \quad x \in \mathbb{R},\ x \neq 0.$$

In particular, we see that, for $x > 0$,

$$J_\nu(x) = 0 \quad \Leftrightarrow \quad J_{\nu-1}(x) = -J_{\nu+1}(x) = J'_\nu(x), \tag{2.14}$$

$$J'_\nu(x) = 0 \quad \Rightarrow \quad J_{\nu-1}(x) = J_{\nu+1}(x) = \frac{\nu}{x}J_\nu(x). \tag{2.15}$$

The roots (or zeros) of $J_\nu(\cdot)$ have the following properties, in addition to (1.20) (see, e.g., Ch.XV, pp. 478-521 in [44], p. 96 in [21], and [19], p. 440).

Fact 2.1. *For any $\nu > -1$, $J_\nu(\cdot)$ has only real roots. Moreover, in this case, the positive roots of $J_\nu(\cdot)$ are isolated and form an increasing sequence*

$$0 < z_{\nu,1} < z_{\nu,2} < \dots, \tag{2.16}$$

such that, for any fixed $k \geq 1$, $z_{\nu,k}$ is a continuous and increasing function of $\nu > -1$. In addition, for any specified $\nu > -1$, as $k \to \infty$,

$$z_{\nu,k} = \Big\{k + \tfrac{1}{2}(\nu - \tfrac{1}{2})\Big\}\pi - \frac{4\nu^2 - 1}{8\{k + \frac{1}{2}(\nu - \frac{1}{2})\}\pi} + O\Big(\frac{1}{k^3}\Big). \tag{2.17}$$

Remark 2.1. For $\nu = -\frac{1}{2}$ and $\nu = \frac{1}{2}$, we have (see (2.27) below)

$$z_{-\frac{1}{2},k} = \{k - \tfrac{1}{2}\}\pi \quad \text{and} \quad z_{\frac{1}{2},k} = k\pi \quad \text{for} \quad k = 1, 2, \dots, \tag{2.18}$$

so that, in either of these cases, $z_{\nu,k}$ reduces to the first term in (2.17).

An alternative definition of the Bessel function $J_\nu(\cdot)$ makes use of *Euler's formula* (see, e.g., (2)–(3) p. 498 in [44])

$$J_\nu(z) = \frac{(\frac{1}{2}z)^\nu}{\Gamma(\nu+1)} \prod_{k=1}^{\infty} \Big\{1 - \frac{z^2}{z_{\nu,k}^2}\Big\} \quad \text{for} \quad z > 0. \tag{2.19}$$

The following relation holds between $J_{\nu+1}(\cdot)$ and $J_\nu(\cdot)$ (see, e.g., (1) p. 498 in [44])

$$J_{\nu+1}(z) = -zJ_\nu(z)\Big\{\sum_{k=1}^{\infty} \frac{1}{z_{\nu,k}(z - z_{\nu,k})} - \sum_{k=1}^{\infty} \frac{1}{z_{\nu,k}(z + z_{\nu,k})}\Big\}. \tag{2.20}$$

The next fact will be especially useful (see, e.g., p. 43 in [21] and p. 109 in [6]).

Fact 2.2. *For arbitrary $a > 0$, $b > 0$ with $a \neq b$ and $\nu > -1$, we have*

$$\int_0^1 tJ_\nu(at)J_\nu(bt) = \frac{aJ_\nu(b)J'_\nu(a) - bJ_\nu(a)J'_\nu(b)}{b^2 - a^2} \tag{2.21}$$
$$= \frac{bJ_\nu(a)J_{\nu+1}(b) - aJ_\nu(b)J_{\nu+1}(a)}{b^2 - a^2}$$
$$= \frac{aJ_\nu(b)J_{\nu-1}(a) - bJ_\nu(a)J_{\nu-1}(b)}{b^2 - a^2}$$

$$\int_0^1 tJ_\nu^2(at)dt = \tfrac{1}{2}\Big\{J_\nu^2(a) - J_{\nu-1}(a)J_{\nu+1}(a)\Big\} \tag{2.22}$$
$$= \tfrac{1}{2}\Big\{J_\nu^2(a) + J_{\nu+1}^2(a) - \tfrac{2\nu}{a}J_\nu(a)J_{\nu+1}(a)\Big\}$$
$$= \tfrac{1}{2}\Big\{J_\nu^2(a) + J_{\nu-1}^2(a) + \tfrac{2\nu}{a}J_\nu(a)J_{\nu-1}(a)\Big\}.$$

The next two lemmas will be instrumental in the derivation of our KL expansions.

Lemma 2.1. *For any $\nu > -1$ and $k, \ell \geq 1$, we have*

$$\int_0^1 tJ_\nu(tz_{\nu,k})J_\nu(tz_{\nu,\ell})dt = 0 \quad \text{when} \quad k \neq \ell, \tag{2.23}$$

$$\int_0^1 tJ_\nu^2(tz_{\nu,k})dt = \int_0^1 tJ_{\nu-1}^2(tz_{\nu,k})dt = \int_0^1 tJ_{\nu+1}^2(tz_{\nu,k})dt \tag{2.24}$$
$$= \tfrac{1}{2}J_{\nu+1}^2(z_{\nu,k}) = \tfrac{1}{2}J_{\nu-1}^2(z_{\nu,k}).$$

Proof. To obtain (2.23) , we apply (2.21), taken with $a = z_{\nu,k}$, $b = z_{\nu,\ell}$ and $J_\nu(a) = J_\nu(b) = 0$. Setting $a = z_{\nu,k}$ and $J_\nu(a) = 0$ in (2.22), we get likewise

$$\int_0^1 tJ_\nu^2(tz_{\nu,k})dt = \tfrac{1}{2}J_{\nu+1}^2(z_{\nu,k}) = \tfrac{1}{2}J_{\nu-1}^2(z_{\nu,k}), \quad \text{and}$$

$$\int_0^1 tJ_\nu^2(tz_{\nu\pm1,k})dt = \tfrac{1}{2}J_\nu^2(z_{\nu\pm1,k}),$$

from where (2.24) is straightforward, via (2.14) and the formal changes of ν into $\nu - 1$ and $\nu + 1$ in these last relations. □

Remark 2.2. When $\nu > 0$, an alternate proof of the above well-known result (2.23) (see, e.g., (6.62)-(6.63), p. 108 in [6]), follows from the observation that $y(t) = \sqrt{t}\, J_\nu(z_{\nu,k}t)$ is an eigenfunction, pertaining to the eigenvalue $\sqrt{2\nu}\,/z_{\nu,k}$, of the Fredholm transformation

$$y \in L^2(0,1) \to Ty(t) = \int_0^1 \{st\}^{\frac{1}{2}-\nu}\{\min(s,t) - st\}y(s)ds.$$

For $k \neq \ell$, $\sqrt{t}\, J_\nu(z_{\nu,k}t)$ and $\sqrt{t}\, J_\nu(z_{\nu,\ell}t)$ are eigenfunctions pertaining to different eigenvalues, and are therefore orthogonal in $L^2(0,1)$. The last equality in (2.24) follows from (2.14) and (2.22), when combined with $J_\nu(z_{\nu,k}) = 0$, since

$$\int_0^1 tJ_\nu^2(tz_{\nu,k})dt = -\tfrac{1}{2}J_{\nu-1}(z_{\nu,k})J_{\nu+1}(z_{\nu,k}) = \tfrac{1}{2}J_{\nu-1}^2(z_{\nu,k}) = \tfrac{1}{2}J_{\nu+1}^2(z_{\nu,k}).$$

The next lemma gives a useful variant of (2.23)–(2.24).

Lemma 2.2. *For any $\nu > -1$, $k \geq 1$, $\ell \geq 1$ with $k \neq \ell$, we have*

$$\int_0^1 tJ_\nu(tz_{\nu\pm1,k})J_\nu(tz_{\nu\pm1,\ell})dt = 0, \tag{2.25}$$

$$\int_0^1 tJ_\nu^2(tz_{\nu\pm1,k})dt = \tfrac{1}{2}J_\nu^2(z_{\nu\pm1,k}). \tag{2.26}$$

Proof. By setting either $a = z_{\nu-1,k}$, $b = z_{\nu-1,\ell}$ or $a = z_{\nu+1,k}$, $b = z_{\nu+1,\ell}$ in (2.21), we obtain readily (2.25). The proof of (2.26) follows likewise from (2.22). □

Remark 2.3. As in Remark 2.2, when $\nu > 0$, an alternate proof of (2.25) can be made by observing that $y(t) = \sqrt{t}\ J_\nu(z_{\nu-1,k}t)$ is an eigenfunction, pertaining to the eigenvalue $\sqrt{2\nu}\,/z_{\nu-1,k}$, of the Fredholm transformation

$$y \in L^2(0,1) \to Ty(t) = \int_0^1 \{st\}^{\frac{1}{2}-\nu} \min(s,t)y(s)ds.$$

Since, for $k \neq \ell$, the eigenfunctions $\sqrt{t}\ J_\nu(z_{\nu-1,k}t)$ and $\sqrt{t}\ J_\nu(z_{\nu-1,\ell}t)$ correspond to different eigenvalues, they are orthogonal in $L^2(0,1)$ and we have (2.25).

2.2. Some special cases

The expression (2.2) of the first order Bessel function $J_\nu(\cdot)$ can be simplified when $\nu = m + \frac{1}{2}$ for an integer $m = -1, 0, 1, \ldots$. In particular, for $m = -1$ and $m = 0$,

$$J_{-\frac{1}{2}}(x) = \sqrt{\tfrac{2}{\pi x}}\ \cos x \quad \text{and} \quad J_{\frac{1}{2}}(x) = \sqrt{\tfrac{2}{\pi x}}\ \sin x. \tag{2.27}$$

For $m \geq 0$, we get

$$J_{m+\frac{1}{2}}(x) = (-1)^m\ x\sqrt{\tfrac{2}{\pi x}}\frac{d^m}{dx^m}\Big(\frac{\sin x}{x}\Big). \tag{2.28}$$

In general, for an arbitrary integer $m \geq -1$, $J_{m+\frac{1}{2}}(\cdot)$ is of the form,

$$\sqrt{\tfrac{\pi x}{2}}\ J_{m+\frac{1}{2}}(x) = Q_m\Big(\frac{1}{x}\Big)\sin x - P_m\Big(\frac{1}{x}\Big)\cos x, \tag{2.29}$$

where $P_m(\cdot)$ and $Q_m(\cdot)$ are polynomials. The first terms of the sequence are

$$P_{-1}(u) = -1, \quad Q_{-1}(u) = 0, \quad P_0(u) = 0, \quad Q_0(u) = 1. \tag{2.30}$$

Lemma 2.3. *For an arbitrary $m \geq 0$, we have the recurrence formulas*

$$Q_{m+1}(u) \;=\; (2m+1)wQ_m(w) - Q_{m-1}(w), \tag{2.31}$$

$$P_{m+1}(u) \;=\; (2m+1)wP_m(w) - P_{m-1}(w). \tag{2.32}$$

Proof. By (2.11), we have

$$\sqrt{\tfrac{\pi x}{2}}\, J_{m+\frac{3}{2}}(x) = \frac{2m+1}{x}\sqrt{\tfrac{\pi x}{2}}\, J_{m+\frac{1}{2}}(x) - \sqrt{\tfrac{\pi x}{2}}\, J_{m-\frac{1}{2}}(x),$$

so that (2.31)–(2.32) is straightforward. □

By combining (2.29)–(2.30) with (2.31)–(2.32), we get

$$J_{\frac{3}{2}}(x) = \sqrt{\tfrac{2}{\pi x}}\left\{\frac{\sin x}{x} - \cos x\right\}, \tag{2.33}$$

$$J_{\frac{5}{2}}(x) = \sqrt{\tfrac{2}{\pi x}}\left\{\frac{3\sin x}{x^2} - \frac{3\cos x}{x} - \sin x\right\}. \tag{2.34}$$

The following fact will be important for our needs (see, e.g., [19], p. 666).

Fact 2.3. *Let $\gamma > 0$ and $\beta > 0$ be real constants. Then, the differential equation*

$$y'' + \gamma x^{2\beta} y = 0, \tag{2.35}$$

has fundamental solutions on $(0,\infty)$ given by

$$x^{1/2} J_{\frac{1}{2(\beta+1)}}\left(\frac{\sqrt{\gamma}\, x^{\beta+1}}{\beta+1}\right) \quad \textit{and} \quad x^{1/2} Y_{\frac{1}{2(\beta+1)}}\left(\frac{\sqrt{\gamma}\, x^{\beta+1}}{\beta+1}\right). \tag{2.36}$$

3. Proofs of Theorems 1.1–1.4

3.1. Introduction.

In this section, we prove Theorems 1.1 and 1.3, together with additional results of interest. The proofs of Theorems 1.2 and 1.4 follow along the same lines and will be therefore omitted.

3.2. The general ψ-weighted Wiener process

The general ψ-weighted Wiener process is $Z(t) = \psi(t)W(t)$ on $(0,1)$, where $W(\cdot)$ is a Wiener process, and $\{\psi(t) : 0 < t \le 1\}$ a continuous and positive function on $(0,1]$. We will make, at times, assumptions among the following.

(Ψ.1) $\lim_{t\downarrow 0} t\psi(t) = 0$; ($\Psi$.2) $\int_0^1 t\psi^2(t)dt < \infty$;

(Ψ.3) $\lim_{t\downarrow 0} t\psi^2(t) = C$, for some $C \in [0,\infty)$; (Ψ.4) $\int_0^1 \psi^2(t)dt < \infty$.

Obviously, (Ψ.3) $\Rightarrow$ (Ψ.1). The following useful lemma relates (Ψ.1–4) to (Ψ.2)

Lemma 3.1. *Under (Ψ.1), $t \to t\psi(t)$ is continuous on $[0,1]$, and (Ψ.4) $\Rightarrow$ (Ψ.2).*

Proof. The function $t \to t\psi(t)$ is continuous on $[0,1]$ by (Ψ.1). To establish that (Ψ.4) $\Rightarrow$ (Ψ.2), we combine the Hölder inequality

$$\int_0^1 s\psi^2(s)ds \le \left\{\int_0^1 (s\psi(s))^2 ds\right\}^{1/2}\left\{\int_0^1 \psi^2(s)ds\right\}^{1/2} < \infty, \tag{3.1}$$

with the just-proved continuity of $t\psi(t)$ on $[0,1]$. □

Remark 3.1. Under (Ψ.1), (Ψ.2) is equivalent to the condition (1.1), with

$$R(s,t) = \mathbb{E}(Z(s)Z(t)) = \psi(s)\psi(t)\min(s,t) \quad \text{for} \quad 0 \le s,t \le 1, \tag{3.2}$$

so that, under (Ψ.1–2), the Fredholm equation (1.2) may be rewritten into

$$\lambda e(t) = \psi(t) \int_0^t s\psi(s)e(s)ds + t\psi(t) \int_t^1 \psi(s)e(s)ds \quad \textit{for} \quad t \in [0,1]. \tag{3.3}$$

Below, we will seek the solutions of (3.3) for which $\lambda > 0$. The assumptions (Ψ.1–2) on ψ imply that is no loss of generality in changing variables by setting in (3.3)

$$\gamma = 1/\lambda \quad \text{and} \quad e(t) = y(t)\psi(t) \quad \text{for} \quad t \in (0,1). \tag{3.4}$$

We so obtain the integral equation in $\gamma > 0$ and $y(\cdot)$

$$y(t) = \gamma \int_0^t s\psi^2(s)y(s)ds + \gamma t \int_t^1 \psi^2(s)y(s)ds. \tag{3.5}$$

Lemma 3.2. *Under (Ψ.1), any solution $y(\cdot)$ of (3.5) with $e(\cdot) = y(\cdot)\psi(\cdot) \in L^2[0,1]$ is continuous on $[0,1]$ and fulfills*

$$y(0) = \lim_{t \downarrow 0} y(t) = 0. \tag{3.6}$$

Proof. Under (Ψ.1), the continuity of $y(\cdot)$ on $(0,1]$ is obvious by (3.5) and we need only show (3.6). We recall from (1.2) that any admissible solution $y(\cdot)$ of (3.5) must be such that the function $e(\cdot) = y(\cdot)\psi(\cdot)$ is in $L^2[0,1]$. Whenever this last condition holds, we infer from (Ψ.1), that, as $t \downarrow 0$,

$$\begin{aligned} \left| \int_0^t s\psi^2(s)y(s)ds \right| &\le \Big\{ \int_0^t (s\psi(s))^2 ds \Big\}^{1/2} \Big\{ \int_0^t (\psi(s)y(s))^2 ds \Big\}^{1/2} \\ &\le \Big\{ \sup_{0<s\le t} |s\psi(s)| \Big\} \Big\{ \int_0^1 (\psi(s)y(s))^2 ds \Big\}^{1/2} \to 0. \end{aligned} \tag{3.7}$$

Another application of (Ψ.1) shows that, as $t \downarrow 0$,

$$\left| t \int_t^1 \psi^2(s)y(s)ds \right| \le \Big\{ t \sup_{t \le s \le 1} |\psi(s)| \Big\} \Big\{ \int_0^1 (\psi(s)y(s))^2 ds \Big\}^{1/2} \to 0. \tag{3.8}$$

Recalling (3.5), we conclude (3.6) by combining (3.7) with (3.8). □

Lemma 3.3. *Under (Ψ.1), any solution $y(\cdot)$ of (3.5) with $e(\cdot) = y(\cdot)\psi(\cdot) \in L^2[0,1]$ is continuously differentiable on $(0,1]$, with derivative $y'(\cdot)$ fulfilling*

$$y'(t) = \gamma \int_t^1 \psi^2(s)y(s)ds \quad \textit{for} \quad 0 < t \le 1 \quad \textit{and} \quad y'(1) = 0. \tag{3.9}$$

If (Ψ.1–4) holds, then $y'(\cdot)$ is continuous on $[0,1]$ and (3.9) holds for $0 \le t \le 1$.

Proof. Recall from Lemma 3.2 that, under (Ψ.1), $y(\cdot)$ is continuous on $[0,1]$. Since $\psi(\cdot)$ is continuous on $(0,1]$, it follows readily from (3.5) that $y(\cdot)$ is continuously differentiable on $(0,1]$ with derivative $y'(\cdot)$ fulfilling

$$\begin{aligned} y'(t) &= \gamma t\psi^2(t)y(t) - \gamma t\psi^2(t)y(t) + \gamma\int_t^1 \psi^2(s)y(s)ds \qquad (3.10)\\ &= \gamma\int_t^1 \psi^2(s)y(s)ds \quad \text{for} \quad 0 < t \leq 1. \end{aligned}$$

This, in turn, entails that $y'(1) = 0$, together with

$$\lim_{t\downarrow 0} y'(t) = \gamma\int_0^1 \psi^2(s)y(s)ds. \tag{3.11}$$

Since (Ψ.1) entails, via Lemma 3.2, that $y(\cdot)$ is continuous on $[0,1]$, we see that (Ψ.1–4) jointly imply that the RHS of (3.11) is finite. Making use of (3.6), we see that, for each $t \in (0,1]$, there exists a $\theta(t) \in (0,t)$ such that $t^{-1}(y(t) - y(0)) = y'(\theta(t))$. We infer from this fact, in combination with (3.11), that, as $t \downarrow 0$,

$$t^{-1}y(t) = t^{-1}(y(t) - y(0)) = y'(\theta(t)) \to \gamma\int_0^1 \psi^2(s)y(s)ds.$$

We so obtain that $y(\cdot)$ has derivative at 0 given by

$$y'(0) = \gamma\int_0^1 \psi^2(s)y(s)ds. \tag{3.12}$$

We readily conclude (3.9) by combining (3.10) with (3.12). □

Lemma 3.4. *Under (Ψ.1), any solution $y(\cdot)$ of (3.5) with $e(\cdot) = y(\cdot)\psi(\cdot) \in L^2[0,1]$ is twice continuously differentiable on $(0,1]$, with second derivative $y''(\cdot)$ such that*

$$y''(t) + \gamma\psi^2(t)y(t) = 0 \quad \textit{for} \quad 0 < t \leq 1. \tag{3.13}$$

Under (Ψ.1–3–4), $y(\cdot)$ is twice continuously differentiable on $[0,1]$, and

$$y''(0) = -\gamma\Big\{\lim_{t\downarrow 0} t\psi^2(t)\Big\}y'(0). \tag{3.14}$$

Proof. Obviously, by (3.9), $y''(t)$ is defined and continuous on $(0,1]$, fulfilling

$$y''(t) = -\gamma\psi^2(t)y(t) \quad \text{for} \quad t \in (0,1]. \tag{3.15}$$

An application of (3.6) and (3.12) shows that, as $t \downarrow 0$,

$$y(t) = (1 + o(1))ty'(0) = (1 + o(1))t\gamma\int_0^1 \psi^2(s)y(s)ds. \tag{3.16}$$

When combined with (Ψ.3), (3.16) readily implies that, as $t \downarrow 0$,

$$y''(t) \to y''(0) = -\gamma^2 C\int_0^1 \psi^2(s)y(s)ds = -\gamma C y'(0),$$

which completes the proof of the lemma. □

Proof of Theorem 1.1. Combine Lemmas 3.1, 3.2, 3.3 and 3.4, with Remark 3.1 and the equivalence of (1.2) with (3.3) and (3.5), holding under (Ψ.1–2). □

Remark 3.2. Whereas (Ψ.1–2) allow only, in general, $y'(\cdot)$ and $y''(\cdot)$ to be continuous on $(0,1]$, (Ψ.3–4) imply the continuity of these functions on $[0,1]$.

3.3. The ψ-weighted Wiener process with $\psi(t) = t^\beta$

In this sub-section, we apply the results of §3.2 to the special case where

$$\psi(t) = t^\beta \quad \text{for} \quad t \in (0,1], \tag{3.17}$$

where $\beta \in \mathbb{R}$ is a constant. It is convenient to set $\nu = 1/(2(\beta+1))$. We first observe that the conditions (Ψ.1–2) are fulfilled whenever

$$\beta > -1 \quad \Leftrightarrow \quad 0 < \nu < \infty. \tag{3.18}$$

As follows from Theorem 1.1, the KL expansion of the process $Z(t) = t^\beta W(t)$ is obtained by setting $e(t) = t^\beta y(t)$ and $\lambda = 1/\gamma$, where $y(\cdot)$ ranges through all the solutions on $[0,1]$ of the differential system

$$y'' + \gamma t^{2\beta} y = 0, \quad \text{with} \quad \gamma > 0 \quad \text{and} \quad y(0) = y'(1) = 0. \tag{3.19}$$

Lemma 3.5. *Let $\gamma > 0, \beta > -1$ be fixed. Then, the only solutions on $(0,1]$ of the differential equation $y'' + \gamma t^{2\beta} y = 0$ fulfilling $y(0) = \lim_{t \downarrow 0} y(t) = 0$ are of the form*

$$y(t) = A\sqrt{t}\, J_\nu\left(2\nu\sqrt{\gamma}\, t^{1/(2\nu)}\right) \quad \textit{with} \quad \nu = 1/(2(\beta+1)), \tag{3.20}$$

and where $A \in \mathbb{R}$ is an arbitrary constant.

Proof. By Fact 2.3, the general solution of (3.19) is given by

$$\begin{aligned} y(t) &= A\sqrt{t}\, J_{\frac{1}{2(\beta+1)}}\left(\frac{\sqrt{\gamma}\, t^{\beta+1}}{\beta+1}\right) + B\sqrt{t}\, Y_{\frac{1}{2(\beta+1)}}\left(\frac{\sqrt{\gamma}\, t^{\beta+1}}{\beta+1}\right) \\ &= A\sqrt{t}\, J_\nu\left(2\nu\sqrt{\gamma}\, t^{1/(2\nu)}\right) + B\sqrt{t}\, Y_\nu\left(2\nu\sqrt{\gamma}\, t^{1/(2\nu)}\right), \end{aligned}$$

for suitable choices of the constants A and B. By (2.2) and (2.8), we get, as $t \downarrow 0$,

$$\begin{aligned} y(t) = (1+o(1))&\frac{A\sqrt{t}}{\Gamma(\nu+1)}\left\{\tfrac{1}{2}2\nu\sqrt{\gamma}\, t^{1/(2\nu)}\right\}^\nu \\ +(1+o(1))&\frac{B\sqrt{t}\,\Gamma(\nu)}{\pi}\left\{\tfrac{1}{2}2\nu\sqrt{\gamma}\, t^{1/(2\nu)}\right\}^{-\nu} \to \frac{B\,\Gamma(\nu)}{\pi}\left\{\nu\sqrt{\gamma}\right\}^{-\nu}. \end{aligned} \tag{3.21}$$

Thus, by (3.21), $y(0) = 0$ holds iff $B = 0$. We have therefore (3.20). □

Lemma 3.6. *Fix $\gamma > 0$, $\nu > 0$, and let $y(t) = A\sqrt{t}\, J_\nu\left(2\nu\sqrt{\gamma}\, t^{1/(2\nu)}\right)$. Then,*

$$y'(t) = A\sqrt{\gamma}\, t^{1/(2\nu)-1/2} J_{\nu-1}\left(2\nu\sqrt{\gamma}\, t^{1/(2\nu)}\right). \tag{3.22}$$

Proof. By (2.13), $J'_\nu(x) = J_{\nu-1}(x) - (\nu/x)J_\nu(x)$. Therefore, we get

$$y'(t) = \frac{A}{2\sqrt{t}}\Big\{J_\nu\left(2\nu\sqrt{\gamma}\,t^{1/(2\nu)}\right) + 2\sqrt{\gamma}\,t^{1/(2\nu)}J'_\nu\left(2\nu\sqrt{\gamma}\,t^{1/(2\nu)}\right)\Big\}$$
$$= \frac{A}{2\sqrt{t}}\Big\{2\sqrt{\gamma}\,t^{1/(2\nu)}J_{\nu-1}\left(2\nu\sqrt{\gamma}\,t^{1/(2\nu)}\right)\Big\},$$

which yields (3.22). □

Proof of Theorem 1.3. We have now in hand the necessary ingredients to prove Theorem 1.3. Set $\nu = 1/(2(1+\beta)) > 0$ ($\Leftrightarrow \beta = -1 + \frac{1}{2\nu} > -1$). By Theorem 1.1, applied with $\psi(t) = t^\beta$, we infer from (3.22) that the eigenfunctions $e_k(t) = t^\beta y(t)$ of the KL expansion of $t^\beta W(t)$ fulfill $y(0) = y'(1) = 0$, where $y(\cdot)$ is solution of (3.19). This, when combined with (3.22), implies that the eigenvalues $\lambda_k = 1/\gamma_k$ of the KL expansion are such that, for $k = 1, 2, \ldots$, $J_{\nu-1}\left(2\nu\sqrt{\gamma_k}\,\right) = 0$, whence

$$\lambda_k = \Big\{\frac{2\nu}{z_{\nu-1,k}}\Big\}^2 \quad \text{for} \quad k = 1, 2, \ldots, \tag{3.23}$$

which is (1.22). As follows from (3.20) and (3.23) the eigenfunction $e_k(\cdot)$ pertaining to $\lambda_k = 1/\gamma_k$ is of the form

$$e_k(t) = At^{\frac{1}{2\nu}-1} \times t^{1/2} J_\nu\Big(2\nu\{\frac{z_{\nu-1,k}}{2\nu}\}t^{\frac{1}{2\nu}}\Big) = At^{\frac{1}{2\nu}-\frac{1}{2}} J_\nu\big(z_{\nu-1,k}t^{\frac{1}{2\nu}}\big).$$

In view of (2.26), the value of A is obtained via the equation

$$\begin{aligned} 1 &= \int_0^1 e_k^2(t)dt = A^2 \int_0^1 t^{\frac{1}{\nu}-1} J_\nu^2\big(z_{\nu-1,k}\,t^{\frac{1}{2\nu}}\big)dt \\ &= A^2 \times 2\nu \int_0^1 u J_\nu^2\big(z_{\nu-1,k}\,u\big)du = A^2\nu J_\nu^2(z_{\nu-1,k}). \end{aligned}$$

We so obtain $A = 1/\{\sqrt{\nu}J_\nu(z_{\nu-1,k})\}$, from where (1.22) is straightforward. □

Acknowledgement. We are indebted to Wenbo Li and Marc Yor for a series of remarks and comments. In particular, we thank Marc Yor for pointing out the connection of our results with his work related to Theorem 1.6 and Corollary 1.2. We also thank the referee for a careful reading of our manuscript.

References

[1] Abramowitz, M. and Stegun, I. A. (1965). *Handbook of Mathematical Functions.* Dover, New York.

[2] Adler, R. J. (1990). *An Introduction to Continuity, Extrema and Related Topics for General Gaussian Processes.* IMS Lecture Notes-Monograph Series **12**. Institute of Mathematical Statistics, Hayward, California.

[3] Anderson, T. W. and Darling, D. A. (1952). Asymptotic theory of certain "goodness of fit" criteria based on stochastic processes. *Ann. Math. Statist.* **23** 193–212.

[4] Ash, R. B. and Gardner, M. F. (1975). *Topics in Stochastic Processes.* Academic Press, New York.

[5] del Barrio, E., Cuesta-Albertos, J. A. and Matran, C. (2000). Contributions of empirical and quantile processes to the asymptotic theory of goodness-of-fit tests. *Test* **9** 1–96.

[6] Bowman, F. (1958). *Introduction to Bessel Functions.* Dover, new York.

[7] Ciesielski, Z. and Taylor, S. J. (1962). First passage times and sojourn times for Brownian motion in space and the exact Hausdorff measure of the sample path. *Trans Amer. Math. Soc.* 434–450.

[8] Darling, D. A. (1957). The Kolmogorov-Smirnov, Cramér-von Mises tests. Ann. Math. Statist. **28** 823–838.

[9] Donati-Martin, C. and Yor, M. (1991). Fubini's theorem double Wiener integrals and the variance of the Brownian path. Ann. Inst. Henri Poincaré. **275** 181-200.

[10] Donati-Martin, C. and Yor, M. (1992). Extension d'une formule de Paul Lévy pour la variation quadratique d'un mouvement brownien plan. *Bull. Sci. Math.* Ser. 2 **116** 353–382.

[11] Donati-Martin, C. and Yor, M. (1997). Some Brownian functionals and their laws. Ann. Probab. **25** 1011–1058.

[12] Durbin, J. (1973). *Distribution Theory for Tests Based on the Sample Distribution Function.* Regional Conference Series in Applied Mathematics, **9** S.I.A.M., Philadelphia.

[13] Gikhman, I. I. (1957). On a nonparametric criterion of homogeneity for k samples. *Theor. Probab. Appl.* **2** 369–373.

[14] Hwang, C. (1980). Gaussian measures of large balls in a Hilbert space. *Proc. of the Amer. Math. Soc.* **78** 107–110 [Erratum: (1985) *ibid.* **94** 188].

[15] Kac, M. (1951). On some connections between probability theory and differential and integral equations. *Proc.Second Berkeley Sympos. Math. Statist. Probab.* 180–215.

[16] Kac, M. (1980) *Integration in Function Spaces and Some of its Applications.* Lezioni Ferniane, Academia Nazionale dei Lincei. Pisa.

[17] Kac, M. and Siegert, A. J. F. (1947). On the theory of noise in radio receivers with square law detectors. *J. Appl. Physics.* **18** 383–397.

[18] Kac, M. and Siegert, A. J. F. (1947). An explicit representation of a stationary Gaussian process. *Ann. Math. Statist.* **18** 438–442.

[19] Kamke, E. (1959). *Differentialgleichungen Lösungsmethoden und Lösungen.* Chelsea, New York.

[20] Kiefer, J. (1959). K-sample analogues of the Kolmogorov-Smirnov and Cramér-V. Mises tests. *Ann. Math. Statist.* **30** 420–447

[21] Korenev, B. G. (2002). *Bessel Functions and their Applications.* Taylor & Francis, London.

[22] Lévy, P. (1951). Wiener random functions and other Laplacian random functions. *Proc. 6th Berkeley Sympos. Probab. Theory Math. Statist.* **2** 171–186.

[23] Lévy, P. (1953). La mesure de Hausdorff de la courbe du mouvement brownien. *Giorn. ist. Ital. Attuari.* **16** 1–37.

[24] Li, W. V. (1992a). Comparison results for the lower tail of Gaussian seminorms. *J. Theor. Probab.* **5** 1–31.

[25] Li, W. V. (1992b). Limit theorems for the square integral of Brownian motion and its increments. *Stoch. Processes Appl.* **41** 223–239.

[26] Li, W. V. (1992c). Lim inf results for the Wiener process and its increments under the L_2-norm. *Prob. Th. Rel. Fields.* **92** 69–90.

[27] Maccone, C. (1984). Eigenfunctions and energy for time-rescaled Gaussian processes. *Boll. Un. Mat. ital.* **6** 213–219.

[28] Martynov, G. V. (1975). Computation of the distribution function of quadratic forms in normal random variables. *Theor. Probab. Appl.* **20** 797–809.

[29] Martynov, G. V. (1977). Generalization of the Smirnov formula for the quadratic forms distributions. *Theor. Probab. Appl.* **22** 614–620.

[30] Martynov, G. V. (1992). Statistical tests based on empirical processes and related questions. *J. Soviet. Math.* **61** 2195–2275.

[31] Pycke, J.-R. (2002). Propriétés asymptotiques d'une famille de statistiques de Cramér-von Mises. *Private Communication.*

[32] Pitman, J. W. and Yor, M. (1981). Bessel processes and infinitely divisible laws. In: *Stochastic Integrals*, D. Williams, edit. Lecture Notes in Mathematics **1123** 285–370. Springer, New York.

[33] Pitman, J. W. and Yor, M. (1982). A decomposition of Bessel bridges. *Z. Wahrscheinlichkeit. verw. Gebiete.* **59** 425–457.

[34] Pitman, J. W. and Yor, M. (1986). Some divergent integrals of Brownian motion. *Analytic and Geometric Stochastics.* D. Kendall, edit., Suppl. to *Adv. Appl. Probab.* 109–116.

[35] Pitman, J. and Yor, M. (1999). The law of the maximum of a Bessel bridge. *Electronic J. Probab.* **4** 1–35.

[36] Rainville, E. D. (1960). *Special Functions.* Chelsea, New York.

[37] Scott, W. F. (1999). A wieghted Cramér-von Mises statistic, with some applications to clinical trials. *Commun. Statist. Theor. Methods.* **28** 3001-3008.

[38] Shi, Z. (1996). Small ball probabilities for a Wiener process under weighted sup-norms, with an application to the supremum of Bessel local times. *J. Theor. Probab.* **9** 915–929.

[39] Shi, Z. and Yor, M. (1997). On an identity in law for the variance of the Brownian bridge. *Bull. London Math. Soc.* **29** 103–108.

[40] Shorack, G. R. and Wellner, J. A. (1986). *Empirical Processes with Applications to Statistics.* Wiley, New York.

[41] Smirnov, N. V. (1936). Sur la distribution de ω^2. *C. R. Acad. Sci. Paris.* **202** 449–452.

[42] Smirnov, N. V. (1937). On the distribution of the ω^2 criterion. *Rec. Math. (Mat. Sbornik).* **6** 3–26.

[43] Smirnov, N. V. (1948). Table for estimating the goodness of fit of empirical distributions. *Ann. Math. Statist.* **19** 279–281.

[44] Watson, G. N. (1952). *A Treatise on the Theory of Bessel Functions.* Cambridge University Press, Cambridge.

[45] Yor, M. (1991). Une explication du théorème de Ciesielski-Taylor. *Ann. Inst. H. Poincaré.* **27** 201-213.

[46] Yor, M. (1992). *Some Aspects of Brownian Motion, Part I: Some Special Functionals.* Lectures in Math., ETH Zürich, Birkhäuser, Basel.

[47] Yor, M. (1997). *Some Aspects of Brownian Motion, Part II: Some Recent Martingale Problems.* Lectures in Math., ETH Zürich, Birkhäuser, Basel.

[48] Zolotarev, V. M. (1961). Concerning a certain probability problem. *Theor. Probab. Appl.* **6** 201–204;

Paul Deheuvels, 7 avenue du Château, 92340 Bourg-la-Reine, France
E-mail address: pd@ccr.jussieu.

Guennady Martynov, Information Transmission Institute, Russian Academy of Sciences, Moscow, Russia
E-mail address: martynov@iitp.ru

Progress in Probability, Vol. 55, 95–102

Extension du théorème de Cameron–Martin aux translations aléatoires. II. Intégrabilité des densités.

Xavier Fernique

Abstract. Let G be a Gaussian vector on a probability space $(\Omega, \mathcal{F}, \mathbf{P})$ taking its values in a separable Fréchet space E. We denote by γ its law and by $(H, \|.\|_H)$ its reproducing kernel Hilbert space. Let moreover X be an E-valued random vector of law μ.

We know that if μ is absolutely continuous relatively to γ, then there exist necessarly a Gaussian vector G' of the law γ and an H-valued random vector Z such that $G' + Z$ has the law μ of X.

In this work, we will compare the γ-integrability of the density $D = d\mu/d\gamma$ and the $\mathbf{P}$-integrability of $\|Z\|_H$.

1. Introduction, notations

On note G un vecteur gaussien à valeurs dans un espace de Fréchet localement convexe séparable E ; on note γ la loi de G et $(H, \|.\|_H)$ l'espace auto-reproduisant associé à γ. Soit X un vecteur aléatoire à valeurs dans E ; on se propose d'évaluer la continuité absolue de la loi μ de X relativement à γ.
Dans un travail précédent ([3], [4]) , on a montré que si μ est absolument continue relativement à γ, alors il existe nécessairement deux vecteurs aléatoires G' de loi γ et Z à valeurs dans H tels que $X' = G' + Z$ ait aussi la loi μ. On compare maintenant la γ-intégrabilité de la densité $D = d\mu/d\gamma$ et l'intégrabilité de la v.a. $\|Z\|_H$.
On notera que la variable aléatoire Z est loin d'être unique : elle peut être construite à partir de différentes contraintes et a alors des propriétés d'intégrabilité différentes ; on préfèrera parler d'intégrabilité éventuelle.

2000 *Mathematics Subject Classification.* Primary : 60 G 15, 60 G 30 ; secundary : 28 D 05.
Key words and phrases. Cameron-Martin Theorem, Integrability of densities.

2. L'intégrabilité éventuelle de $\|Z\|_H$

Théorème 2.1. *On suppose que μ est absolument continue relativement à γ et que la densité $D = d\mu/d\gamma$ appartient à $L_p(\gamma), p \in]1, \infty]$. Dans ces conditions et pour tout $a \in]0, (p-1)/(4p-2)[$, il existe deux vecteurs aléatoires G' de loi γ et X' de loi μ tels que $Z = G' - X'$ prenne ses valeurs dans H et que*

$$\mathbf{E}\{\exp[\, a\, \|Z\|_H^2] - 1\} \leq \|D\|_{L_p} \times [C(a,p) + 1] \tag{2.1}$$

$$\text{où } C(a,p) = (1-2aq)^{-1/2q}\left[2 + \frac{2a}{1-2a(1+q)}\right] \text{ et } q = \frac{p}{p-1}. \tag{2.2}$$

La démonstration du théorème utilisera, comme dans [3], les inégalités isopérimétriques gaussiennes et le théorème de Kantorovitch-Rubinstein, mais sous une forme différente.

2.1. Le lemme fondamental

Lemme 2.2. *On suppose que H est de dimension finie et donc que $\gamma(H) = \mu(H) = 1$. On fixe un nombre $p \in]1, \infty[$ et un nombre $a \in \left]0, (p-1)/(4p-2)\right[$. On note $F(a)$ l'ensemble des couples (f_1, f_2) de fonctions mesurables bornées sur H telles que*

$$\forall (x,y) \subset H, f_1(x) + f_2(y) \leq \exp(a\|x-y\|_H^2) - 1. \tag{2.3}$$

Pour tout $(f_1, f_2) \in F(a)$, on a alors :

$$\int f_1 d\gamma + \int f_2 d\mu \leq \|D\|_{L_p} \times C(a,p). \tag{2.4}$$

Un lemme auxiliaire. La preuve du lemme 2.2 utilisera un résultat technique classique, mais un peu délicat que le rapporteur a suggéré de détailler :

Lemme 2.3. *Soient f une fonction réelle mesurable et m une probabilité sur $\mathbb{R}$. Pour tout t réel, on pose*

$$A(t) = \{x : m\{f \geq f(x)\} \leq t\};$$

de plus, pour tout $\alpha \in]0,1[$, on pose

$$I(\alpha, f) = \int \frac{dm(x)}{m^\alpha\{f \geq f(x)\}}.$$

Dans ces conditions, on a :

$$(a) \quad \forall t \in [0,1], \qquad m(A(t)) \leq t,$$

$$(b) \quad \forall \alpha \in]0,1[, \qquad I(\alpha, f) \leq \frac{1}{1-\alpha}.$$

Démonstration du lemme auxiliaire. La preuve de (a) est simple et on la laisse au lecteur. On donne ici la preuve de (b) :
On introduit sur $(\mathbb{R}, m)$ les deux v.a. U et V définies par

$$U(x) = m\{f \geq f(x)\}, V = U^{-\alpha}\mathbf{I}_{U>0}.$$

Dans ces conditions, la propriété (a) implique que la loi de V est telle que :

$$\forall t > 1, m\{V \geq t\} = m\{Z \leq t^{-1/\alpha}\} \leq t^{-1/\alpha}.$$

On en déduit par le calcul usuel des moments d'une v.a. supérieure ou égale à 1 :

$$I(\alpha, f) = \mathbf{E}V = \int_0^\infty m\{V \geq t\}\, dt \leq 1 + \int_1^\infty \frac{dt}{t^{1/\alpha}} = 1 + \frac{\alpha}{1-\alpha},$$

c'est le résultat (b). □

Démonstration du lemme 2.2. Pour tout couple (x, y) d'éléments de E, on pose

$$d_H(x, y) = \|x - y\|_H \text{ si } x - y \in H, d_H(x, y) = \infty \text{ sinon};$$

Pour toute partie mesurable A de E et tout $h \geq 0$, on pose aussi

$$V_h(A) = \{x \in E : \exists y \in A, d_H(x, y) \leq h\}.$$

On fixe un couple (f_1, f_2) de $F(a)$. On fixe aussi deux nombres $t \in \mathbb{R}$, $\alpha \in \overline{\mathbb{R}}$ tels que

$$\gamma\{f_1 \geq t\} = \Phi(\alpha) = (1/\sqrt{2\pi}) \int_{-\infty}^{\alpha} \exp[-u^2/2] du\,; \tag{2.5}$$

alors les inégalités isopérimétriques gaussiennes ([2],théorème 1.5.2, [5], théorème 1.2) indiquent que

$$\forall h \geq 0, \gamma[V_h\{f_1 \geq t\}] \geq \Phi(\alpha + h);$$

or si y appartient à $V_h\{f_1 \geq t\}$, alors il existe un élément x de H tel que

$$d_H(x, y) \leq h, f_1(x) \geq t\,;$$

la définition de $F(a)$ implique alors que

$$f_2(y) \leq \exp[ah^2] - 1 - t.$$

On en déduit que pour tout $s > 0$, on a :

$$\begin{aligned}\gamma\{y : f_2(y) + t > s\} &\leq 1 - \Phi(\alpha + \sqrt{1/a \log(1+s)}\,) \\ &\leq (1/\sqrt{2\pi}) \int_{\alpha+\sqrt{a^{-1}\log(1+s)}}^{\infty} \exp[-u^2/2] du.\end{aligned} \tag{2.6}$$

On utilise maintenant cette majoration (2.6) pour évaluer $\int [[f_2(y) + t]^+]^q d\gamma(y)$: c'est en effet le moment d'ordre q de la v.a.r. $[f_2 + t]^+$ positive sur (E, γ) de sorte que

$$\begin{aligned}\int [[f_2(y) + t]^+]^q d\gamma(y) &= q \int_0^\infty s^{q-1}\, \gamma\{y : f_2(y) + t \geq s\} ds \\ &\leq q \int_0^\infty s^{q-1} \Big\{ \frac{1}{\sqrt{2\pi}} \int_{\alpha+\sqrt{a^{-1}\log(1+s)}} \exp[-u^2/2]\, du \Big\} ds,\end{aligned}$$

et donc en permutant les intégrations,

$$\int [[f_2(y)+t]^+]^q d\gamma(y) \le \int_\alpha^\infty \left[\exp[a(u-\alpha)^2]-1\right]^q \exp[-u^2/2]du/\sqrt{2\pi}$$
$$\le \int_\alpha^\infty \exp[aq(u-\alpha)^2 - u^2/2]du/\sqrt{2\pi}. \tag{2.7}$$

On notera dans la suite $J(a,\alpha,q)$ cette dernière intégrale ; on remarquera que a est inférieur à $(p-1)/(4p-2) = 1/(2q+1)$ et donc à $1/(2q)$ de sorte que cette intégrale est convergente ; on remarquera aussi que c'est une fonction décroissante de α.

En utilisant alors la formule de Hölder et l'hypothèse d'intégrabilité de la densité, on déduit de (2.7) :

$$\int [[f_2(y)+t]^+]d\mu(y) = \int [[f_2(y)+t]^+]\, D(y)d\gamma(y) \le J(a,\alpha,q)^{1/q}\|D\|_{L_p}. \tag{2.8}$$

On utilisera deux fois cette majoration (2.8).

La première fois, on substitue à t la valeur τ d'une γ-médiane de f_1 ; ce choix de τ assure que $\gamma\{f_1 \ge \tau\}$ est supérieur ou égal à 1/2 ; la valeur α_0 de α associée à τ par la formule (2.5) est donc positive ou nulle de sorte que $J(a,\alpha_0,q)$ est inférieur ou égal à $J(a,0,q) = (1-2aq)^{-1/2}$; on obtient alors par (2.8) :

$$\int [f_2(y)+\tau]^+ d\mu(y) \le \|D\|_{L_p} \times J(a,0,q)^{1/q}$$
$$\le \|D\|_{L_p} \times (1-2aq)^{-1/2q}. \tag{2.9}$$

On utilise une seconde fois (2.8) pour tout $t = f_1(x) > \tau$; $\alpha = \alpha(x)$ est alors négatif ou nul ; en intégrant le premier membre de (2.8) relativement à $d\gamma(x)$ dans les deux domaines $\{f_1(x) \le \tau\}$ et $\{f_1(x) > \tau\}$, on obtient :

$$\iint [f_2(y)+f_1(x)]^+ d\mu(y)d\gamma(x) \le \|D\|_{L_p} \times \left[(1-2aq)^{-1/2q} + J'\right],$$
$$J' = \int_{f_1(x)>\tau} \left\{\int_\alpha^\infty \exp[aq(u-\alpha)^2 - u^2/2]\frac{du}{\sqrt{2\pi}}\right\}^{1/q} d\gamma(x).$$

Pour réduire la forme quadratique, on opère le changement de variables: $v = u(1-2aq)^{1/2} + 2aq\alpha(1-2aq)^{-1/2}$ qui fournit:

$$J' \le \int_{f_1(x)>\tau} \left\{\int_{-\infty}^\infty \frac{\exp[-u^2/2]\, du}{\sqrt{1-2aq}\sqrt{2\pi}}\right\}^{1/q} \exp[\frac{a\alpha^2}{1-2aq}]d\gamma(x),$$
$$\alpha = \alpha(x) = \Phi^{-1} \circ \gamma\{f_1 \ge f_1(x)\} \le 0,$$

de sorte que dans le domaine d'intégration,

$$\gamma\{f_1 \geq f_1(x)\} \leq \exp[-\alpha^2/2],$$

$$\text{et donc } J' \leq (1-2aq)^{-1/2q} \int \frac{1}{\left[\gamma\{f_1 \geq f_1(x)\}\right]^{\frac{2a}{1-2aq}}} d\gamma(x).$$

Le lemme auxiliaire 2.3 majore alors cette dernière intégrale indépendamment de la γ-loi de f_1. On obtient finalement puisque $1-2a(1+q)$ est positif:

$$\int f_1 d\gamma + \int f_2 d\mu \leq \iint [f_2(y) + f_1(x)]^+ d\mu(y) d\gamma(x) \tag{2.10}$$

$$\leq \|D\|_{L_p} \times (1-2aq)^{-1/2q} \times \left[1 + \frac{(1-2aq)}{1-2a(1+q)}\right],$$

c'est le résultat annoncé. □

Remarque: La démonstration du lemme présente successivement des majorations en (2.9) et (2.10) ; la première qui suppose $2aq < 1$ est réaliste, mais elle ne semble pas suffisante pour conclure ; la seconde qui exige $2a(1+q) < 1$ est moins satisfaisante (cf. exemple 3.2 ci-dessous).

2.2. Démonstration du théorème

Nous opérons en deux étapes.

2.2.1. Dans la première étape, nous supposons que H est de dimension finie et donc que $\gamma(H) = \mu(H) = 1$; dans ces conditions une forme du théorème de Kantorovich-Rubinstein adaptée à la fonction de coût c définie par $c(x,y) = \exp[a\, d_H^2(x,y)] - 1, x, y \in H$ ([6], théorème 2.3.1) fournit le résultat à partir du lemme 1 ci-dessus.

2.2.2. Pour conclure dans la situation générale, on utilise le même procédé d'approximation que dans la preuve du théorème 1 de la rédaction [3]:
Nous notons G et X des vecteurs aléatoires de lois respectives γ et μ ; nous notons aussi θ l'application canonique ([2], 2.1) de H sur l'adhérence du dual E' de E dans $L_2(\gamma)$; nous choisissons une base orthonormale $(h_n, n \in \mathbb{N}^*)$ de H de sorte que son image par θ soit contenue dans E' ; pour tout $N \in \mathbb{N}^*$, nous notons F_N l'application de E dans E définie par :

$$\forall x \in E, F_N(x) = \sum_{n=1}^{N} \langle x, \theta h_n \rangle h_n,$$

et nous posons :

$$G_N = F_N(G) = \sum_{n=1}^{N} \langle G, \theta h_n \rangle h_n,$$

$$X_N = F_N(X) = \sum_{n=1}^{N} \langle X, \theta h_n \rangle h_n.$$

Nous notons H_N le sous-espace de H engendré par $(h_n, n \leq N)$. Alors G_N et X_N sont des vecteurs aléatoires à valeurs dans $H_N \subset E$; G_N est Gaussien et prend ses valeurs dans H_N qui est son espace auto-reproduisant ; G_N et X_N sont les images de G et X par la même application F_N de sorte que la loi de X_N est absolument continue par rapport à celle de G_N et sa densité, obtenue par conditionnement appartient à L_p ; on peut donc leur appliquer les conclusions de la première étape et construire deux vecteurs aléatoires G'_N de même loi que G_N et X'_N de même loi que X_N tels que

$$\mathbf{E}\{\exp[\, a\, \|Z_n\|_H^2] - 1\} \leq \|D\|_{L_p} \times [C(a,p)+1].$$

Remarquons maintenant que la suite $\{G_N, N \in \mathbb{N}^*\}$ définit un développement de Karhunen-Loeve de G et converge donc p.s. vers G ([2], théorème 2.4.2, [5] théorème 6.1) dans E, cette propriété s'écrit en fonction de la loi γ de G :

$$\gamma\Big\{x \in E : \sum_n \langle x, \theta h_n\rangle h_n \text{ converge vers } x\Big\} = 1,$$

l'absolue continuité de μ relativement à γ implique alors que :

$$\mu\Big\{x \in E : \sum_n \langle x, \theta h_n\rangle h_n \text{ converge vers } x\Big\} = 1;$$

ceci signifie que la suite $\{X_N, N \in \mathbb{N}^*\}$ converge p.s. vers X et dans ces conditions, l'isonomie (l'égalité des lois) implique que les suites $\{G'_N, N \in \mathbb{N}^*\}$ et $\{X'_N, N \in \mathbb{N}^*\}$ convergent en loi dans E ; la suite double $\{(G'_N, X'_N), N \in \mathbb{N}^*\}$ dont les composantes sont tendues dans E est alors elle-même tendue dans $E \times E$ et on peut en extraire une suite partielle convergeant en loi qu'on peut même réaliser de sorte qu'elle converge p.s. dans E vers une limite (G', X') ; G' a la même loi que G, X' a la même loi que X; le lemme de Fatou assure que

$$\mathbf{E}\{\exp[\, a\, \|Z\|_H^2] - 1\} \leq \|D\|_{L_p} \times [C(a,p)+1].$$

La preuve est complète. □

3. Des propriétés réciproques

Dans ce paragraphe, on suppose que X est de la forme $X = G + Y$ où G et Y sont indépendants et que X a même loi μ que $G + Z$ où Z est à valeurs dans H.
On sait ([3], Exemple 2.3) que dans cette situation, Y n'est pas nécessairement à valeurs dans H et que Z ne peut pas en général être choisi pour être indépendant de G. On ne peut donc pas confondre Y et Z.
Le théorème 2 de [3] affirme d'ailleurs que pour tout $\lambda \in]0,1[$, la loi μ_λ de $X_\lambda = G + \lambda Y$ est absolument continue relativement à γ ; on note D_λ sa densité.

On suppose qu'il existe un nombre $\alpha > 0$ tel que $\mathbf{E}\exp\left[\alpha\, \|Z\|_H^2\right] = C$ soit fini et sous cette hypothèse, on étudie l'intégrabilité de D_λ.

Théorème 3.1. *Sous les hypothèses ci-dessus, pour tout couple (λ, p) tel que*

$$\lambda \in]0,1[, \quad p \in \Big]1, [1+\sqrt{1+\frac{8\alpha(1-\lambda^2)}{\lambda^2}}]/2\Big],$$

la densité D_λ appartient à $L_p(\gamma)$.
En particulier, D_λ appartient à $L_2(\gamma)$ pour tout $\lambda \in \big]0, \sqrt{\frac{\alpha}{1+\alpha}}\,\big]$.

3.1. Démonstration du théorème

La démonstration du théorème utilisera une variante du lemme 1 de [3] :

3.1.1. Pour tout nombre $\lambda \in]0,1[$ et toute fonction f mesurable et bornée sur E, on note $M_\lambda f$ la fonction définie sur E par

$$\forall z \in E, \quad M_\lambda f(z) = \int f[\lambda(G+z) + \sqrt{1-\lambda^2}\, g']\, d\gamma(g') \tag{3.1}$$

Lemme 3.2. *(a) On suppose que z appartient à l'espace autoreproduisant H de G; dans ces conditions et pour tout $p > 1$, on a :*

$$|M_\lambda f(z)| \le \left[\int |f(\lambda G + \sqrt{1-\lambda^2}\, g')|^q d\gamma(g')\right]^{1/q} \times \exp\left[\frac{(p-1)\,\lambda^2\,\|z\|_H^2}{2(1-\lambda^2)}\right],\ q = \frac{p}{p-1}. \tag{3.2}$$

(b) Soit de plus Z un vecteur aléatoire à valeurs dans H ; on suppose que

$$\mathbf{E}\exp\left[\alpha\,\|Z\|_H^2\right] = C \tag{3.3}$$

est fini. Pour tout p tel que $\frac{p(p-1)\lambda^2}{2(1-\lambda^2)} \le \alpha$, on a alors

$$\mathbf{E}|M_\lambda f(Z)| \le C^{1/p}\left[\mathbf{E}|f(G)|^q\right]^{1/q}. \tag{3.4}$$

(c) Soit Y un vecteur aléatoire à valeurs dans E indépendant de G. On suppose que $G+Y$ a même loi que $G+Z$; dans ces conditions et pour tout $\lambda \in]0,1[$, $M_\lambda f(Y)$ a même loi que $M_\lambda f(Z)$.
De plus pour toute fonction f mesurable et bornée sur E, on a :

$$\mathbf{E}f(G+\lambda Y) = \mathbf{E}M_\lambda f(Y). \tag{3.5}$$

Démonstration. Les propriétés (a) et (c) ont été démontrées dans [3]. Pour démontrer (b), on utilise la propriété 3.2 et l'inégalité de Hölder qui fournissent :

$$\mathbf{E}|M_\lambda f(Z)| \le \mathbf{E}|f(G)|^q\big]^{1/q} \times \left[\mathbf{E}\exp\left\{\frac{p(p-1)\lambda^2\|Z\|_H^2}{2(1-\lambda^2)}\right\}\right]^{1/p},$$

d'où la conclusion (b). □

3.1.2. Pour démontrer le théorème 3.1, on utilise les conclusions (b) et (c) du lemme : elles impliquent que sous les hypothèses de ce théorème, pour tout $\lambda \in]0,1[$ et toute fonction mesurable et bornée f sur E, on a :

$$\int f D_\lambda d\gamma \leq C^{1/p} \Big[\int |f|^q d\gamma \Big]^{1/q} \text{ pourvu que } \frac{p(p-1)\lambda^2}{2(1-\lambda^2)} \leq \alpha,$$

et cette inégalité suffit pour conclure. □

3.2. L'efficacité des évaluations

Exemple. Supposons que Y soit indépendant de G, que $X = G + Y$ ait une loi μ absolument continue relativement à la loi γ de G et que la densité D soit de carré intégrable.

Dans ces conditions, le théorème 2.1 construit deux vecteurs aléatoires G' de loi γ et X' de loi μ tels que $Z = G' - X'$ prenne ses valeurs dans H et que

$$\mathbf{E} \exp[a\|Z\|_H^2] < \infty \text{ pour } a \in]0, 6^{-1/2}[.$$

Si on appliquait aux données ainsi obtenues le théorème 3.1, il ne fournirait une densité D_λ de carré intégrable que pour $\lambda < 7^{-1/2}$.

Cet exemple montre donc que la chaine des évaluations devrait être améliorée.

References

[1] X. Fernique. Sur l'équivalence de certaines mesures produit. Probab. Theory Relat. Fields, 98, 77-90, 1994.

[2] X. Fernique. Fonctions aléatoires gaussiennes, vecteurs aléatoires gaussiens, Les Publications C.R.M., 1997.

[3] X. Fernique. Extension du théorème de Cameron-Martin aux translations aléatoires I, Ann.Probab., *à paraître.*

[4] X. Fernique. Extension du théorème de Cameron-Martin aux translations aléatoires. C. R. Acad. Sci. Paris, Ser I 335 (2002) 65-68.

[5] M. Ledoux, M. Talagrand. Probability in Banach spaces. Ergebnisse der Mathematik, série 3, vol. 23, Berlin-Heidelberg, New-York, Springer, 1991.

[6] S.T. Rachev, L. Ruschendorf. Mass Transportation Problems. Probability and it Applications, Berlin-Heidelberg, New-York, Springer, 1998.

Institut de Recherche Mathématique Avancée,
Université Louis Pasteur et C.N.R.S.
7 rue René-Descartes,
67084 Strasbourg Cedex, France.
http://perso.wanadoo.fr/xavier.fernique/
E-mail address: `xavier.fernique@wanadoo.fr, fernique@math.u-strasbg.fr`

Part 3

Limit Theorems

Progress in Probability, Vol. 55, 105–109

Rates of Convergence for Lévy's Modulus of Continuity and Hinchin's Law of the Iterated Logarithm

Vladimir Dobric and Lisa Marano

Abstract. For a standard Wiener process $\{W_t : t\varepsilon[0,1]\}$, denoting $X(\rho) = \sup_{|s-t|\le\rho} |W_s - W_t|$, $g(x) = \sqrt{2x\ln(\frac{1}{x})}$ and $h(x) = \sqrt{2x\ln\ln\frac{1}{x}}$, we have proved the following two results: $E\sup_{\rho\le\delta}\frac{X(\rho)}{g(\rho)}$ converges down to 1 at least as fast as $3.27/\sqrt{\ln\frac{1}{\delta}}$ and $\sup_{t\le\delta}\frac{W_t}{h(t)}$ converges down to 1 at least as fast as $3\ln\ln\ln\frac{1}{\delta}/\ln\ln\frac{1}{\delta}$.

1. Rates of convergence in Lévy's modulus of continuity

Let $\{W_t : t\varepsilon[0,1]\}$ denote a standard Wiener process,

$$X(\rho) = \sup_{|s-t|\le\rho} |W_s - W_t|\,, \text{ and } \quad g(x) = \sqrt{2x\ln(\frac{1}{x})}.$$

Proposition 1.1. *For each $\delta, 0 < \delta \le 2^{-4}$, the following bounds hold:*

$$1 \le \mathbb{E}\left(\sup_{\rho\le\delta}\frac{X(\rho)}{g(\rho)}\right) \le 1 + \frac{3.27}{\sqrt{\ln\frac{1}{\delta}}} + 9.07\frac{\ln\ln\frac{1}{\delta}}{\ln\frac{1}{\delta}} + O\left(\delta\ln^{\frac{1}{2}}\frac{1}{\delta}\right).$$

Proof. The proof proceeds in three steps. In the first step we use [1]: for each δ, $0 < \delta \le 2^{-4}$, and each ε, $\varepsilon > 0$,

$$\mathbb{P}\left(\sup_{\rho\le\delta}\frac{X(\rho)}{r(\rho)\,g(\rho)} \ge \sqrt{1+\varepsilon}\right) \le K(\varepsilon)\,\delta^{\varepsilon}\left(\ln\frac{1}{\delta}\right)^{\frac{3}{2}}, \tag{1.1}$$

where

$$K(\varepsilon) = 2^{\varepsilon}\left(253\cdot\mathbf{1}_{[1,\infty)}(\varepsilon) + 298\frac{\mathbf{1}_{(0,1)}(\varepsilon)}{\varepsilon^{3+\varepsilon}}\right), \text{ and } r(x) = \left(1 + \frac{3.27}{\sqrt{\ln\frac{1}{|x|}}}\right),$$

2000 *Mathematics Subject Classification.* Primary 60G17, 60G15; Secondary 62EA17.

Key words and phrases. Brownian motion, Modulus of continuity, rates of convergence.

to obtain an upper bound for the expected value of

$$Z(\delta) = \sup_{\rho \le \delta} \frac{X(\rho)}{r(\rho)\, g(\rho)}.$$

The upper bound is calculated as follows: for any $\varepsilon > 0$, and $\delta \le 2^{-4}$, we have

$$\begin{aligned}
\mathbb{E}Z(\delta) &= \int_0^{\sqrt{1+\varepsilon}} \mathbb{P}(Z(\delta) \ge t)\, dt + \int_{\sqrt{1+\varepsilon}}^{\infty} \mathbb{P}(Z(\delta) \ge t)\, dt \\
&\le \sqrt{1+\varepsilon} + \int_{\varepsilon}^{\infty} \frac{1}{2\sqrt{1+\alpha}} \mathbb{P}\left(Z(\delta) \ge \sqrt{1+\alpha}\right) d\alpha \\
&\le \sqrt{1+\varepsilon} + \left(\ln \frac{1}{\delta}\right)^{\frac{3}{2}} \int_{\varepsilon}^{\infty} (2\delta)^{\alpha} \left[\frac{2981_{(0,1)}(\alpha)}{2\alpha^{3+\alpha}\sqrt{1+\alpha}} + \frac{2531_{[1,\infty)}(\alpha)}{2\sqrt{1+\alpha}}\right] d\alpha \\
&\le \sqrt{1+\varepsilon} + \left(\ln \frac{1}{\delta}\right)^{\frac{3}{2}} \left[\frac{149\,(2\delta)^{\varepsilon}\, \mathbf{1}_{(0,1)}(\varepsilon)}{(2+\varepsilon)\,\varepsilon^{2+\varepsilon}} + \frac{179\delta}{\ln \frac{1}{2\delta}}\right] \\
&\le 1 + \frac{\varepsilon}{2} + \left(\ln \frac{1}{\delta}\right)^{\frac{3}{2}} \left[\frac{74.5\,(2\delta)^{\varepsilon}}{\varepsilon^{2+\varepsilon}} + \frac{179\delta}{\ln \frac{1}{2\delta}}\right].
\end{aligned}$$

In the second step we seek to find an ε that minimizes the obtained upper bound. Using the substitution $\varepsilon = k \ln\ln \frac{1}{\delta} / \ln \frac{1}{\delta}$ the problem boils down to finding an appropriate value of k. Since $\delta^{\varepsilon} = \ln^{-k}(\frac{1}{\delta})$ and $1/\ln \frac{1}{2\delta} \le 1.34/\ln \frac{1}{\delta}$, it follows:

$$\mathbb{E}Z(\delta) \le 1 + \frac{\ln\ln \frac{1}{\delta}}{\ln \frac{1}{\delta}} \left(\frac{k}{2} + \frac{74.5\left(\ln \frac{1}{\delta}\right)^{\frac{7}{2}-k}}{\left(k \ln\ln \frac{1}{\delta}\right)^2} \left(\frac{2\ln \frac{1}{\delta}}{k \ln\ln \frac{1}{\delta}}\right)^{\frac{k \ln\ln \frac{1}{\delta}}{\ln \frac{1}{\delta}}}\right) + 240\delta \left(\ln \frac{1}{\delta}\right)^{\frac{1}{2}}.$$

The function

$$k \to \frac{k}{2} + 74.5 \left(\ln \frac{1}{\delta}\right)^{\frac{7}{2}-k} \left(\frac{1}{k \ln\ln \frac{1}{\delta}}\right)^2 \left(\frac{2\ln \frac{1}{\delta}}{k \ln\ln \frac{1}{\delta}}\right)^{\frac{k \ln\ln \frac{1}{\delta}}{\ln \frac{1}{\delta}}}$$

is minimized if k is approximately 5.5. The minimum is less than 3.06. (Those estimates are obtained using Maple). As a consequence,

$$\mathbb{E} \sup_{\rho \le \delta} \frac{X(\rho)}{r(\rho)\, g(\rho)} \le 1 + 3.06 \frac{\ln\ln \frac{1}{\delta}}{\ln \frac{1}{\delta}} + 240\delta \left(\ln \frac{1}{\delta}\right)^{\frac{1}{2}}.$$

In the third and final step, we first observe that the function $\delta \to \ln\ln \frac{1}{\delta} / \ln \frac{1}{\delta}$ achieves its maximum $1/e$ at $\delta = e^{-e}$, and, since $\delta \le 2^{-4} < e^{-e}$, it is an increasing function. It follows that

$$\begin{aligned}
\mathbb{E}\sup_{\rho\leq\delta}\frac{X(\rho)}{g(\rho)} &\leq r(\delta)\,\mathbb{E}\sup_{\rho\leq\delta}\left(\frac{X(\rho)}{r(\rho)\,g(\rho)}\right)\\
&\leq \left(1+\frac{3.27}{\sqrt{\ln\frac{1}{\delta}}}\right)\left(1+3.06\frac{\ln\ln\frac{1}{\delta}}{\ln\frac{1}{\delta}}+240\delta\left(\ln\frac{1}{\delta}\right)^{\frac{1}{2}}\right)\\
&\leq 1+\frac{3.27}{\sqrt{\ln\frac{1}{\delta}}}+9.07\frac{\ln\ln\frac{1}{\delta}}{\ln\frac{1}{\delta}}+O\left(\delta\left(\ln\frac{1}{\delta}\right)^{\frac{1}{2}}\right).
\end{aligned}$$

The lower bound in the statement of the proposition is a straightforward consequence of Lévy's modulus of continuity. □

2. Rates of Convergence in Hinchin's Law of the Iterated Logarithm

In what follows below we will use the following notations

$$Y(\delta)=\sup_{t\leq\delta}\frac{W_t}{h(t)}\quad\text{and}\quad h(x)=\sqrt{2x\ln\ln\frac{1}{x}}.$$

Proposition 2.1. *For each δ, $0<\delta<2^{-4}$, the following bounds hold*

$$1\leq\mathbb{E}Y(\delta)\leq 1+\frac{3\ln\ln\ln\frac{1}{\delta}}{\ln\ln\frac{1}{\delta}}+\frac{5.4}{\ln\ln\frac{1}{\delta}\sqrt{\ln\ln\ln\frac{1}{\delta}}}.$$

Proof. The proof proceeds in three steps. We first recall [1]: for each δ, $0<\delta<2^{-4}$, and each ε, $\varepsilon>0$,

$$\mathbb{P}\left(Y(\delta)>r(\delta,\varepsilon)\sqrt{1+\varepsilon}\right)\leq\frac{1}{\left(\ln\frac{1}{\delta}\right)^{\frac{\varepsilon}{2}}\sqrt{\ln\ln\frac{1}{\delta}}}l(\varepsilon),\tag{2.1}$$

where

$$r(\delta,\varepsilon)=1+\frac{3.7(0.87)^{\frac{\varepsilon}{4}}}{\left(\ln\frac{1}{\delta}\right)^{\frac{\varepsilon}{4}}}\left(\frac{\mathbf{1}_{(0,1)}(\varepsilon)}{\sqrt{\varepsilon}}+\mathbf{1}_{[1,\infty)}(\varepsilon)\right)$$

and

$$l(\varepsilon)=\frac{1.6}{\varepsilon}\mathbf{1}_{(0,1)}(\varepsilon)+0.85\,(0.87)^{\frac{\varepsilon}{2}}\,\mathbf{1}_{[1,\infty)}(\varepsilon).$$

Using the fact that $Y(\delta)\geq 0$ for every $\varepsilon>0$ we have

$$\mathbb{E}Y(\delta)\leq r(\delta,\varepsilon)\sqrt{1+\varepsilon}+\int_{r(\delta,\varepsilon)\sqrt{1+\varepsilon}}^{\infty}\mathbb{P}\left(Y(\delta)>s\right)ds,$$

and the estimation boils down to computing the right-hand side integral. In that regard this proof is similar to the proof given in the previous proposition. However, now a minor complication arises: the function r in ont only δ but also ε

dependent. Since $x \to r(\delta, x)$ is decreasing, the function $f : (0, \infty) \to \mathbb{R}$ defined by $f(x) = r(\delta, x)\sqrt{1+x}$ is not monotone increasing. This statement becomes clear if we observe that $f(x) \to \infty$ when $x \to 0$ and when $x \to \infty$. Choose an $\eta(\delta)$ such that $\eta(\delta) = \varepsilon$ and $f(x) \geq f(\varepsilon)$ for all $x\epsilon[\eta(\delta), \infty)$. Using the substitution $s = f(x)$ and applying (2.1) we derive:

$$\begin{aligned}
\int_{f(\varepsilon)}^{\infty} \mathbb{P}\left(Y(\delta) > s\right) ds &= \int_{\eta(\delta)}^{\infty} \mathbb{P}\left(Y(\delta) > f(x)\right)\left(\frac{r(\delta,x)}{2\sqrt{1+x}} + \sqrt{1+x}\frac{\partial r(\delta,x)}{\partial x}\right) dx \\
&\leq \int_{\eta(\delta)}^{\infty} \mathbb{P}\left(Y(\delta) > f(x)\right) \frac{r(\delta,x)}{2\sqrt{1+x}} dx \\
&\leq \frac{r(\delta,\varepsilon)}{2\sqrt{\ln\ln\frac{1}{\delta}}}\left(\frac{1.6\ln\frac{1}{\varepsilon}}{\left(\ln\frac{1}{\delta}\right)^{\frac{\varepsilon}{2}}}\mathbf{1}_{(0,1)}(\varepsilon) + 0.61\left(\frac{0.87}{\ln\frac{1}{\delta}}\right)^{\frac{1}{2}}\frac{2}{\ln\left(\frac{\ln\frac{1}{\delta}}{0.87}\right)}\right) \\
&\leq \frac{r(\delta,\varepsilon)}{\sqrt{\ln\ln\frac{1}{\delta}}}\left(\frac{0.8\ln\frac{1}{\varepsilon}}{\left(\ln\frac{1}{\delta}\right)^{\frac{\varepsilon}{2}}}\mathbf{1}_{(0,1)}(\varepsilon) + \frac{0.57}{\left(\ln\frac{1}{\delta}\right)^{\frac{1}{2}}\ln\ln\frac{1}{\delta}}\right).
\end{aligned}$$

The next step of our estimation starts by setting $\varepsilon = 6\ln\ln\ln\frac{1}{\delta}/\ln\ln\frac{1}{\delta}$. The restriction on δ assures

$$\frac{\ln\frac{1}{\varepsilon}}{\ln\ln\frac{1}{\delta}} < 2.12 \text{ and } \frac{1}{\left(\ln\frac{1}{\delta}\right)^{\frac{1}{2}}} \leq \frac{1.21}{\ln\ln\frac{1}{\delta}},$$

leading to

$$\frac{r(\delta,\varepsilon)}{\sqrt{\ln\ln\frac{1}{\delta}}}\left(\frac{0.8\ln\frac{1}{\varepsilon}}{\left(\ln\frac{1}{\delta}\right)^{\frac{\varepsilon}{2}}}\mathbf{1}_{(0,1)}(\varepsilon) + \frac{0.57}{\left(\ln\frac{1}{\delta}\right)^{\frac{1}{2}}\ln\ln\frac{1}{\delta}}\right) \leq \frac{2.39 r(\delta,\varepsilon)}{\left(\ln\ln\frac{1}{\delta}\right)^{\frac{5}{2}}}$$

and

$$r(\delta,\varepsilon) \leq 1 + \frac{3.7}{\left(\ln\ln\frac{1}{\delta}\right)^{\frac{3}{2}}}\sqrt{\frac{\ln\ln\frac{1}{\delta}}{6\ln\ln\ln\frac{1}{\delta}}} \leq 1 + \frac{1.52}{\ln\ln\frac{1}{\delta}\sqrt{\ln\ln\ln\frac{1}{\delta}}}.$$

Finally,

$$\begin{aligned}
\mathbb{E}Y(\delta) &\leq r(\delta,\varepsilon)\left(1 + \frac{3\ln\ln\ln\frac{1}{\delta}}{\ln\ln\frac{1}{\delta}} + \frac{2.39}{\left(\ln\ln\frac{1}{\delta}\right)^{\frac{5}{2}}}\right) \\
&\leq 1 + \frac{3\ln\ln\ln\frac{1}{\delta}}{\ln\ln\frac{1}{\delta}} + \frac{5.4}{\ln\ln\frac{1}{\delta}\sqrt{\ln\ln\ln\frac{1}{\delta}}}.
\end{aligned}$$

The lower bound in the statement of the theorem is a straightforward consequence of Hinchin's Law of the Iterated Logarithm. □

Remark 2.2. Goran Peskir of Aarhus University has informed us that he was able to obtain results for $Y(\delta)$ that are similar to our (2.1) estimates[3].

Remark 2.3. Methods developed in the theory of small ball problems may be also used to obtain the rates for $\mathbb{E}Y(\delta)$. W. Li and Q-M. Shao have written an excellent overview of the subject [2]; they have pointed to us that possibility.

References

[1] Dobric, V. and L. Marano, Lower bounds for the distribution of suprema of Brownian increments and Brownian motion normalized by the corresponding modulus functions (2002) (Submitted)

[2] Li, W and Q-M. Shao, Gaussian processes: Inequalities, Small Ball Probabilities and Applications, Stochastic Processes: Theory and Methods. Handbook of statistics, Vol. 19.(to appear)

[3] Peskir, G., On nonlinear equations arising in problems of optimal stopping. Research Report No. 426 (2002). Dept. Theoret. Stat. Aarhus (with J.L. Pedersen) (Submitted)

Department of Mathematics, Lehigh University,
14 Packer Av., Bethlehem, PA 18015
E-mail address: `vd00@lehigh.edu`
URL: `http://www.lehigh.edu/~vd00/vd00.html`

Department of Mathematics, West Chester University of PA,
Anderson Hall, West, Chester, PA 19383
E-mail address: `lmarano@wcupa.edu`
URL: `http://math.wcupa.edu/~marano.`

Progress in Probability, Vol. 55, 111–126

On the Limit Set in the Law of the Iterated Logarithm for U-statistics of Order Two

Stanislaw Kwapień[1], Rafał Latała[1], Krzysztof Oleszkiewicz[1] and Joel Zinn[2]

Abstract. We find the cluster set in the Law of the Iterated Logarithm for U-statistics of order 2 in some interesting special cases. The lim sup is an unusual function of the quantities that determine the Bounded LIL.

1. Introduction and Notation

In [GKLZ] necessary and sufficient conditions were obtained for the law of the iterated logarithm for canonical U-statistics of order 2 to hold. Here we continue the investigation of the LIL for U-statistics of order 2 by describing the cluster (or limit) set for the examples in [GZ], which helped motivate [GKLZ]. Namely, let $X_1, X_2, \ldots$ denote a sequence of iid r.v.'s with values in some measurable space $(S, \mathcal{S})$. In the general case for a measurable kernel h on S^2 we define symmetrized U-statistics by the formula

$$U_n = \sum_{1 \le i < j \le n} \varepsilon_i \varepsilon_j h(X_i, X_j),$$

where (ε_i) is a Rademacher sequence (i.e. a sequence of independent symmetric ± 1 valued r.v's) independent of (X_i). In our case we will assume that each X_i has a uniform distribution on $[0, 1]$ and

$$h(x, y) = \sum_{k=1}^{\infty} a_k h_k(x) h_k(y), \tag{1.1}$$

where

$$h_k(x) = I_{A_k}(x) \text{ and } A_k = (2^{-k}, 2^{-k+1}], k = 1, 2, \ldots.$$

We also assume that $0 \le a_k \le k^{-1/2} 2^k$ (this assumption seems not to be necessary but makes the calculations easier).

2000 *Mathematics Subject Classification.* Primary 60F15.
Key words and phrases. cluster set, limit set, LIL.
[1]Research partially supported by KBN grant #2 PO3A 027 22,
[2]Research partially supported by NSA Grant No. MDA904-01-1-0027.

The (nas) conditions for the (bounded) LIL for $\{U_n\}$ that were obtained in [GKLZ] imply that the conditions for the LIL in our case be in terms of

$$A = \sup\{\mathbf{E}h(X_1,X_2)f(X_1)f(X_2) : \mathbf{E}f^2(X_i) \le 1\} = \sup_k |a_k 2^{-k}| \tag{1.2}$$

and

$$B = \limsup_{u\to\infty} \frac{\mathbf{E}(h^2(X_1,X_2)\wedge u)}{L_2 u}. \tag{1.3}$$

However, what is not so clear is the form of the function of A and B that determines the lim sup. It turns out that the lim sup is

$$C = \phi(A,B) = \begin{cases} A + \frac{B^2}{4A} & \text{if } B \le 2A \\ B & \text{if } B \ge 2A. \end{cases}$$

In the sequel letters like K, K_1, etc., will denote universal constants that may change from line to line, but do not depend on any parameters. To simplify the notation we define $Lx = \log(x \vee e)$ and $L_2x = LLx$. We also write $\log_2$ for the logarithm to the base 2.

Now, a few comments about the organization of the paper. After presenting in Section 2 some known results, we present in Section 3 a few results for general U-statistics. Finally, in Section 4 we concentrate on the types of kernels of the form (1.1) that are the main focus of this paper.

2. Preliminary results

In this section we gather a few inequalities proven elsewhere that we will use in the sequel.

Lemma 1. *([KW], Theorem 6.2.1) There exists a universal constant K such that for any $t > 0$ and any sequence of real numbers $(a_{ij})_{1\le i<j\le n}$ we have*

$$\mathbf{P}\Big(\max_{1\le k\le n}\Big|\sum_{1\le i<j\le k} a_{ij}\varepsilon_i\varepsilon_j\Big| > t\Big) \le K\mathbf{P}\Big(\Big|\sum_{1\le i<j\le n} a_{ij}\varepsilon_i\varepsilon_j\Big| > t\Big).$$

Lemma 2 (Bernstein inequality).
([de la P,G] Lemma 4.1.9 and Remark 4.1.10, [D] Th. 1.3.2) If Z_i are independent r.v's with $\mathbf{E}Z_i = 0$, $\|Z_i\|_\infty \le a$ and $b^2 = \sum \mathbf{E}Z_i^2$, then for all $t \ge 0$

$$\mathbf{P}\Big(\Big|\sum_i Z_i\Big| \ge t\Big) \le 2\exp\left(-\frac{t^2}{2b^2 + \frac{2}{3}at}\right).$$

Lemma 3 (Kolmogorov's converse exponential inequality).
([S] Th. 5.2.2) For any $\gamma > 0$ there exist numbers $\tilde{K}(\gamma) < \infty$ and $\tilde{\varepsilon}(\gamma) > 0$ such that if Z_i are independent r.v's with $\mathbf{E}Z_i = 0$, $\|Z_i\|_\infty \le a$, $b^2 = \sum \mathbf{E}Z_i^2$ satisfying $t \ge \tilde{K}(\gamma)b$ and $ta \le \tilde{\varepsilon}(\gamma)b^2$ for some $t > 0$, then

$$\mathbf{P}\Big(\sum_i Z_i \ge t\Big) \ge \exp\left(-\frac{(1+\gamma)t^2}{2b^2}\right).$$

We will, however, prefer to use the following simple corollary of Kolmogorov's converse exponential inequality (one may take below $\varepsilon(\gamma) = (1+\gamma)\tilde{\varepsilon}(\gamma)^2/2$ and $K(\gamma) = \exp((1+\gamma)\tilde{K}(\gamma)^2/2)$.)

Corollary 1. *For any $\gamma > 0$ there exist numbers $K(\gamma) < \infty$ and $\varepsilon(\gamma) > 0$ such that if Z_i are independent r.v's with $\mathbf{E}Z_i = 0$, $\|Z_i\|_\infty \le a$, $b^2 = \sum \mathbf{E}Z_i^2$, then for all $t > 0$*

$$\mathbf{P}\Big(\sum_i Z_i \ge t\Big) \ge \frac{1}{K(\gamma)} \exp\left(-\frac{(1+\gamma)t^2}{2b^2}\right) - \exp\left(-\varepsilon(\gamma)\frac{b^2}{a^2}\right).$$

Lemma 4. *([GLZ], Corollary 3.4) There exists a universal constant $K < \infty$ such that for all $t > 0$*

$$\mathbf{P}(|U_n| \ge t) \le K \exp\Big[-\frac{1}{K}\min\Big(\frac{t^2}{n^2 Eh^2},$$
$$\frac{t}{n\|h\|_{L_2\to L_2}}, \frac{t^{2/3}}{[n(\|E_Y h^2\|_\infty + \|E_X h^2\|_\infty)]^{1/3}}, \frac{t^{1/2}}{\|h\|_\infty^{1/2}}\Big)\Big].$$

3. Technical Lemmas. General Kernels

In this section we present few technical lemmas that do not require additional assumptions on the form of the kernel h.

Lemma 5. *We have*

$$\mathbf{E}\exp\Big(\lambda\big(\sum_{i=1}^n \varepsilon_i\big)^2\Big) \le \frac{1}{\sqrt{1-2\lambda n}} \quad \textit{for all } 0 \le \lambda < \frac{1}{2n}. \tag{3.1}$$

Moreover, for each $\gamma > 0$, there exist positive numbers $K(\gamma)$ and $\delta(\gamma)$ such that for any n

$$\mathbf{P}\Big(\sum_{i=1}^n \varepsilon_i \ge t\sqrt{n}\Big) \ge \frac{1}{K(\gamma)}\exp\left(-\frac{(1+\gamma)t^2}{2}\right) - \exp(-\delta(\gamma)n). \tag{3.2}$$

Proof. Notice that for any t

$$\mathbf{E}\exp\Big(t\sum_{i=1}^n \varepsilon_i\Big) = \Big(\frac{1}{2}e^t + \frac{1}{2}e^{-t}\Big)^n \le e^{\frac{nt^2}{2}}.$$

So if g is $\mathcal{N}(0,1)$ r.v. independent of ε_i, then

$$\mathbf{E}\exp\Big(\lambda\big(\sum_{i=1}^n \varepsilon_i\big)^2\Big) = \mathbf{E}_\varepsilon \mathbf{E}_g e^{\sqrt{2\lambda}(\sum_{i=1}^n \varepsilon_i)g}$$
$$= \mathbf{E}_g E_\varepsilon e^{\sqrt{2\lambda}g\sum_{i=1}^n \varepsilon_i} \le \mathbf{E}e^{n\lambda g^2} = \frac{1}{\sqrt{1-2\lambda n}}.$$

Inequality (3.2) is an immediate consequence of Kolmogorov's converse exponential inequality (Corollary 1). □

Lemma 6. *Suppose that $a_{ij}^{(n)}$ is a triply indexed sequence of numbers such that*

$$\limsup_{n\to\infty}\Big|\sum_{i,j=1}^{n} a_{ij}^{(n)}\varepsilon_i\varepsilon_j\Big| \le C \ a.s.$$

Then,

$$\limsup_{n\to\infty}\Big|\sum_{i=1}^{n} a_{ii}^{(n)}\Big| \le C.$$

Proof. Let $t > C$, then $I(\sum_{i,j=1}^{n} a_{ij}^{(n)}\varepsilon_i\varepsilon_j \ge t) \to 0$ a.s. so in particular

$$\mathbf{P}\Big(\sum_{i,j=1}^{n} a_{ij}^{(n)}\varepsilon_i\varepsilon_j \ge t\Big) \to 0.$$

However

$$\mathbf{P}\Big(\sum_{i,j=1}^{n} a_{ij}^{(n)}\varepsilon_i\varepsilon_j \ge \sum_{i=1}^{n} a_{ii}^{(n)}\Big) = \mathbf{P}\Big(\sum_{1\le i\ne j\le n} a_{ij}^{(n)}\varepsilon_i\varepsilon_j \ge 0\Big) \ge \frac{1}{K}$$

for some universal K ([de la P,G] Proposition 3.3.7 combined with Theorem 3.2.2). This implies $\sum_{i=1}^{n} a_{ii}^{(n)} \le t$ for large enough n, so $\limsup_{n\to\infty}\sum_{i=1}^{n} a_{ii}^{(n)} \le C$. In a similar way we prove that $\limsup_{n\to\infty}(-\sum_{i=1}^{n} a_{ii}^{(n)}) \le C$. □

Lemma 7. *a)If $C < \infty$ is a number such that*

$$\forall_{\varepsilon>0}\exists_{K,N}\forall_{n\ge N}\ \mathbf{P}(|U_n| \ge C(1+\varepsilon)nL_2n) \le \frac{K}{\log n(L_2n)^{1+\varepsilon}}, \tag{3.3}$$

then

$$\limsup_{n\to\infty}\frac{|U_n|}{nL_2n} \le C \ a.s.$$

b)If $C < \infty$ is a number such that

$$\forall_{\varepsilon>0,n_0}\exists_{K,N>n_0}\forall_{N\le n\le N^2}\ \mathbf{P}(|U_n| \ge C(1+\varepsilon)nL_2n) \ge \frac{1}{K\log n}, \tag{3.4}$$

then

$$\limsup_{n\to\infty}\frac{|U_n|}{nL_2n} \ge C \ a.s.$$

Proof. We start with the proof of part a). Let $\alpha > 1$, in this part of the proof we will denote $U_a = U_{\lfloor a\rfloor}$ for all $a \ge 0$. Let $\varepsilon > 0$ and K, N be given by formula (3.3). Let us choose k_0 such that $\alpha^{k_0} \ge N$. Then, we have for all $t > 0$

$$\mathbf{P}\Big(\max_{n\ge\alpha^{k_0}}\frac{|U_n|}{nL_2n} \ge t\Big) \le \sum_{k=k_0}^{\infty}\mathbf{P}\Big(\max_{\alpha^k\le n\le\alpha^{k+1}}\frac{|U_n|}{nL_2n} \ge t\Big)$$

$$\le \sum_{k=k_0}^{\infty}\mathbf{P}\Big(\max_{1\le n\le\alpha^{k+1}}|U_n| \ge t\alpha^kL_2(\alpha^k)\Big) \le \sum_{k=k_0}^{\infty}K\mathbf{P}(|U_{\alpha^{k+1}}| \ge t\alpha^kL_2(\alpha^k)),$$

where in the last line we used the maximal inequality (Lemma 1). Since for large enough k we have $L_2(\alpha^k) \geq \alpha^{-1} L_2(\alpha^{k+1})$ we get that for sufficiently large k_0

$$\mathbf{P}\left(\max_{n\geq\alpha^{k_0}} \frac{|U_n|}{nL_2 n} \geq C\alpha^2(1+\varepsilon)\right) \leq \sum_{k=k_0}^{\infty} K\mathbf{P}(|U_{\alpha^{k+1}}| \geq C(1+\varepsilon)\alpha^{k+1}L_2(\alpha^{k+1}))$$
$$\leq \sum_{k=k_0}^{\infty} \frac{K}{\log\lfloor\alpha^{k+1}\rfloor (L_2\lfloor\alpha^{k+1}\rfloor)^{1+\varepsilon}}.$$

This implies that

$$\lim_{k\to\infty} \mathbf{P}\Big(\max_{n\geq\alpha^k} \frac{|U_n|}{nL_2 n} \geq C\alpha^2(1+\varepsilon)\Big) = 0,$$

so $\limsup_{n\to\infty} \frac{|U_n|}{nL_2 n} \leq C\alpha^2(1+\varepsilon)$ a.s. and part a) follows, when $\alpha \to 1^+$ and $\varepsilon \to 0^+$.

To prove part b) suppose that

$$\limsup_{n\to\infty} \frac{|U_n|}{nL_2 n} \leq C_1 < C \text{ a.s.}$$

(By the 0-1 Law we know that the lim sup is constant a.s.). Let $m > 1$ be an integer (to be chosen later) and $\tilde{\varepsilon}_i$ be another Rademacher sequence independent of ε_i and X_i. Since for any choice of signs $\eta_i = \pm 1$ the sequence $\eta_i\varepsilon_i$ has the same distribution as ε_i we get that

$$\limsup_{n\to\infty} \frac{|\sum_{k,l=1}^{n} \tilde{\varepsilon}_k\tilde{\varepsilon}_l \sum_{m^{k-1}\leq i<m^k, m^{l-1}\leq j<m^l, i<j} \varepsilon_i\varepsilon_j h(X_i,X_j)|}{m^n L_2(m^n)} \leq C_1 \text{ a.s..}$$

So

$$\mathbf{P}_{\varepsilon,X}\Big(\limsup_{n\to\infty} \frac{|\sum_{k,l=1}^{n} \tilde{\varepsilon}_k\tilde{\varepsilon}_l \sum_{m^{k-1}\leq i<m^k, m^{l-1}\leq j<m^l, i<j} \varepsilon_i\varepsilon_j h(X_i,X_j)|}{m^n L_2(m^n)} \leq C_1 \ \tilde{\varepsilon}\text{-a.s.}\Big) = 1.$$

However by Lemma 6 it implies

$$\mathbf{P}\Big(\limsup_{n\to\infty} \frac{|\sum_{k=1}^{n} \sum_{m^{k-1}\leq i<j<m^k} \varepsilon_i\varepsilon_j h(X_i,X_j)|}{m^n L_2(m^n)} \leq C_1\Big) = 1.$$

Let $1/2 > \delta > 0$ to be chosen later and $C_1 < C_2 < C$, then

$$\mathbf{P}\Big(\max_{n\geq n_0} \frac{|\sum_{k=1}^{n} \sum_{m^{k-1}\leq i<j<m^k} \varepsilon_i\varepsilon_j h(X_i,X_j)|}{m^n L_2(m^n)} > C_2\Big) < \delta$$

for sufficiently large n_0. Notice that if $|s_n| \leq C_2 m^n L_2(m^n)$ for $n \geq n_0$, then $|s_n - s_{n-1}| \leq C_2(m^n + m^{n-1})L_2(m^n)$ for $n > n_0$. Therefore

$$\mathbf{P}\Big(\max_{n>n_0} \frac{|\sum_{m^{n-1}\leq i<j<m^n} \varepsilon_i\varepsilon_j h(X_i,X_j)|}{m^n L_2(m^n)} > C_2(1+\frac{1}{m})\Big) < \delta.$$

Thus by the independence (since $\mathbf{P}(\bigcup A_i) \geq 1/2 \sum \mathbf{P}(A_i)$ if A_i are independent and $\mathbf{P}(\bigcup A_i) \leq 1/2$)

$$\sum_{n>n_0} \mathbf{P}\Big(|\sum_{1\leq i<j\leq m^n - m^{n-1}} \varepsilon_i\varepsilon_j h(X_i, X_j)| \geq C_2 m^n (1+\frac{1}{m}) L_2(m^n)\Big)$$

$$= \sum_{n>n_0} \mathbf{P}\Big(|\sum_{m^{n-1}\leq i<j<m^n} \varepsilon_i\varepsilon_j h(X_i, X_j)| \geq C_2 m^n (1+\frac{1}{m}) L_2(m^n)\Big) < 2\delta.$$

Now choose m and increase n_0, if necessary, in such a way that

$$C_2 m^n \Big(1+\frac{1}{m}\Big) L_2(m^n) \leq C(1+\varepsilon)(m^n - m^{n-1}) L_2(m^n - m^{n-1})$$

for $n > n_0$. By our assumption (3.4) we can find $N > m^{n_0}$ such that

$$\mathbf{P}\big(|U_{m^n - m^{n-1}}| \geq C(1+\varepsilon)(m^n - m^{n-1}) L_2(m^n - m^{n-1})\big)$$
$$\geq \frac{1}{K \log(m^n - m^{n-1})} \geq \frac{1}{Kn \log m}$$

for all n such that $N \leq m^n - m^{n-1} \leq N^2$. However

$$\sum_{n: N\leq m^n - m^{n-1}\leq N^2} \frac{1}{Kn\log m} \gtrsim \frac{\log 2}{K \log m} > 2\delta$$

if we choose δ small enough. □

The next Lemma shows why the LIL-limit depends on two quantities in a very non-obvious way.

Lemma 8. *Suppose that S_1, S_2 are independent r.v's, $A, B > 0$ and*

$$C = \begin{cases} A + \frac{B^2}{4A} & \text{if } B \leq 2A \\ B & \text{if } B \geq 2A \end{cases}$$

a) If for some $K \geq 1$ and $\varepsilon > 0$

$$\mathbf{P}(S_1 \geq sAn) \geq \frac{1}{K} e^{-s(1+\varepsilon)} - \frac{1}{(\log n)^{1+\varepsilon}} \text{ for all } s \geq 0$$

and

$$\mathbf{P}(S_2 \geq sBn\sqrt{L_2 n}) \geq \frac{1}{K} e^{-s^2(1+\varepsilon)^2} - \frac{1}{(\log n)^{1+\varepsilon}} \text{ for all } s \geq 0,$$

then for sufficiently large n

$$\mathbf{P}(S_1 + S_2 \geq (1+\varepsilon)^{-1} CnL_2 n) \geq \frac{1}{K^2}\frac{1}{\log n} - \frac{2}{(\log n)^{1+\varepsilon}}.$$

b) On the other hand if for some $K, \varepsilon > 0$

$$\mathbf{P}(S_1 \geq sAn) \leq Ke^{-\frac{s}{1+\varepsilon}} + \frac{1}{(\log n)^{1+\varepsilon}} \text{ for all } s \geq 0$$

and

$$\mathbf{P}(S_2 \geq sBn\sqrt{L_2 n}) \leq Ke^{-\frac{s^2}{(1+\varepsilon)^2}} + \frac{1}{(\log n)^{1+\varepsilon}} \text{ for all } s \geq 0,$$

then

$$\mathbf{P}(S_1 + S_2 \geq (1+\varepsilon)^3 CnL_2n) \leq \Big(\frac{1}{\varepsilon}+1\Big)\frac{(K+2)^2}{(\log n)^{1+\varepsilon}}.$$

Proof. For the first part of the statement it is enough to notice that in the case when $B \geq 2A$ we get for sufficiently large n

$$\mathbf{P}(S_1 + S_2 \geq (1+\varepsilon)^{-1} CnL_2n) \geq \mathbf{P}(S_1 \geq 0)\mathbf{P}(S_2 \geq (1+\varepsilon)^{-1} BnL_2n)$$
$$\geq \left(\frac{1}{K} - \frac{1}{(\log n)^{1+\varepsilon}}\right)\left(\frac{1}{K}e^{-L_2n} - \frac{1}{(\log n)^{1+\varepsilon}}\right) \geq \frac{1}{K^2 \log n} - \frac{2}{(\log n)^{1+\varepsilon}}.$$

In the case when $B \leq 2A$ we have for large enough n

$$\begin{aligned}
&\mathbf{P}\left(S_1 + S_2 \geq (1+\varepsilon)^{-1} CnL_2n\right)\\
&\geq \mathbf{P}\left(S_1 \geq (1+\varepsilon)^{-1}\left(A - \frac{B^2}{4A}\right) nL_2n\right)\mathbf{P}\left(S_2 \geq (1+\varepsilon)^{-1}\frac{B^2}{2A} nL_2n\right)\\
&\geq \left(\frac{1}{K}\exp\left(-\Big(1 - \frac{B^2}{4A^2}\Big)L_2n\right) - \frac{1}{(\log n)^{1+\varepsilon}}\right)\\
&\qquad \times \left(\frac{1}{K}\exp\left(-\frac{B^2}{4A^2}L_2n\right) - \frac{1}{(\log n)^{1+\varepsilon}}\right)\\
&\geq \frac{1}{K^2 \log n} - \frac{2}{(\log n)^{1+\varepsilon}}.
\end{aligned}$$

To prove part b) first notice that for all $x \in [0, C]$

$$\frac{x}{A} + \frac{(C-x)^2}{B^2} \geq 1.$$

Hence, for such x

$$\begin{aligned}
&\mathbf{P}\left(S_1 \geq (1+\varepsilon)^2 xnL_2n, S_2 \geq (1+\varepsilon)^2 (C-x) nL_2n\right)\\
&\leq \left(K\exp\left(-(1+\varepsilon)\frac{x}{A}L_2n\right) + \frac{1}{(\log n)^{1+\varepsilon}}\right)\\
&\qquad \times \left(K\exp\left(-(1+\varepsilon)\frac{(C-x)^2}{B^2}L_2n\right) + \frac{1}{(\log n)^{1+\varepsilon}}\right) \leq \frac{(K+1)^2}{(\log n)^{1+\varepsilon}}.
\end{aligned}$$

Moreover,

$$\mathbf{P}(S_1 \leq 0, S_1 + S_2 \geq (1+\varepsilon)^2 CnL_2n) \leq \mathbf{P}(S_2 \geq (1+\varepsilon)^2 CnL_2n)$$

$$\leq K\exp\left(-(1+\varepsilon)\frac{C^2}{B^2}L_2n\right) + \frac{1}{(\log n)^{1+\varepsilon}} \leq \frac{K+1}{(\log n)^{1+\varepsilon}}$$

and

$$\mathbf{P}(S_1 \geq (1+\varepsilon)^2 CnL_2n) \leq \frac{K+1}{(\log n)^{1+\varepsilon}}.$$

Let $k_0 = \lfloor \varepsilon^{-1} \rfloor$. Then,

$$\begin{aligned}
&\mathbf{P}(S_1 + S_2 \geq (1+\varepsilon)^3 C n L_2 n) \\
&\leq \mathbf{P}\left(S_1 \leq 0, \frac{S_1 + S_2}{(1+\varepsilon)^2 n L_2 n} \geq C\right) + \mathbf{P}\left(\frac{S_1}{(1+\varepsilon)^2 n L_2 n} \geq C\right) \\
&\quad + \sum_{k=0}^{k_0} \mathbf{P}\left(\frac{S_1}{(1+\varepsilon)^2 n L_2 n} \in [k\varepsilon C, (k+1)\varepsilon C], \frac{S_2}{(1+\varepsilon)^2 n L_2 n} \geq C - k\varepsilon C\right) \\
&\leq \frac{2K+2}{(\log n)^{1+\varepsilon}} + (k_0+1)\frac{(K+1)^2}{(\log n)^{1+\varepsilon}} \leq \left(\frac{1}{\varepsilon}+1\right)\frac{(K+2)^2}{(\log n)^{1+\varepsilon}}.
\end{aligned}$$

□

4. Special Kernels

From this point on we will assume that our kernel is of the form (1.1). We consider the following (undecoupled) U-statistics Let

$$\tilde{U}_n = \sum_{k=1}^{\infty} a_k \sum_{1 \leq i < j \leq N_k} \varepsilon_i^k \varepsilon_j^k = \sum_{k=1}^{\infty} \frac{a_k}{2}\Big(\big(\sum_{i=1}^{N_k} \varepsilon_i^k\big)^2 - N_k\Big),$$

where

$$N_k = \#\{1 \leq i \leq n : X_i \in (2^{-k}, 2^{-k+1}]\}, k = 1, 2 \ldots.$$

Notice that

$$\mathcal{L}(U_n | \sigma(X_1, X_2, \ldots)) = \mathcal{L}(\tilde{U}_n | \sigma(X_1, X_2, \ldots)),$$

so U_n and $\tilde{U}_n$ have the same distribution.

Lemma 9. *We have for all $\delta > 0$*

$$\mathbf{P}\big(\exists_{k \leq m} |N_k - n2^{-k}| \geq \delta n 2^{-k}\big) \leq \frac{2^{m+1}}{\delta^2 n}.$$

Proof. Notice that

$$\begin{aligned}
\mathbf{P}(\exists_{k \leq m} |N_k - n2^{-k}| \geq \delta n 2^{-k}) &\leq \sum_{k=1}^{m} \mathbf{P}(|N_k - \mathbf{E}N_k| \geq \delta n 2^{-k}) \\
&\leq \sum_{k=1}^{m} \frac{2^{2k}}{\delta^2 n^2} \mathrm{Var}(N_k) \leq \frac{1}{\delta^2 n} \sum_{k=1}^{m} 2^k \leq \frac{2^{m+1}}{\delta^2 n}.
\end{aligned}$$

□

Lemma 10. *Suppose that $s > 0$ and $|n_k - n2^{-k}| \leq \varepsilon n 2^{-k-1}$ for $k = 1, \ldots, m$. Let $\alpha = \max\{2^{-k}|a_k| : 1 \leq k \leq m\}$, then*

$$\mathbf{P}\Big(\Big|\sum_{k=1}^{m} \frac{a_k}{2}\big(\big(\sum_{i=1}^{n_k} \varepsilon_i^k\big)^2 - n_k\big)\Big| \geq \alpha s n\Big) \leq \Big(\frac{2e(1+\varepsilon)}{\varepsilon}\Big)^{m/2} e^{-\frac{s}{1+\varepsilon}} \tag{4.1}$$

On the other hand, if $\alpha_1 = \max\{2^{-k}a_k : 1 \le k \le m\} > 0$, *then*

$$\mathbf{P}\left(\sum_{k=1}^{m} \frac{a_k}{2}\Big(\big(\sum_{i=1}^{n_k} \varepsilon_i^k\big)^2 - n_k\Big) \ge \alpha_1 sn\right) \ge \frac{1}{K(\varepsilon)} e^{-(1+\varepsilon)s} - \exp(-\delta(\varepsilon)2^{-m}n), \quad (4.2)$$

and if $\alpha_2 = \max\{-2^{-k}a_k : 1 \le k \le m\}$, *then*

$$\mathbf{P}\left(-\sum_{k=1}^{m} \frac{a_k}{2}\Big(\big(\sum_{i=1}^{n_k} \varepsilon_i^k\big)^2 - n_k\Big) \ge \alpha_2 sn\right) \ge \frac{1}{K(\varepsilon)} e^{-(1+\varepsilon)s} - \exp(-\delta(\varepsilon)2^{-m}n), \quad (4.3)$$

where $K(\varepsilon)$ *and* $\delta(\varepsilon)$ *depend only on* ε.

Proof. Let $S = \sum_{i=1}^{m} \frac{|a_k|}{2}(\sum_{i=1}^{n_k} \varepsilon_i^k)^2$, then by (3.1) we have

$$\mathbf{E}e^{\lambda S} \le \prod_{i=1}^{m} \frac{1}{\sqrt{1 - \lambda |a_k| n_k}}.$$

But by our assumptions $|a_k|n_k \le (1 + \frac{\varepsilon}{2})\alpha n$, so

$$\mathbf{E}\exp\left(\frac{1}{\alpha n(1+\varepsilon)} S\right) \le \left(1 - \frac{1+\frac{\varepsilon}{2}}{1+\varepsilon}\right)^{-m/2} = \left(\frac{2(1+\varepsilon)}{\varepsilon}\right)^{m/2}.$$

Notice that

$$\Big|\sum_{k=1}^{m} \frac{a_k}{2}\Big(\big(\sum_{i=1}^{n_k} \varepsilon_i^k\big)^2 - n_k\Big)\Big| \le S + \frac{1}{2}\sum_{k=1}^{m} |a_k| n_k \le S + \frac{1}{2}(1+\varepsilon)\alpha n m,$$

so (4.1) immediately follows, since

$$\mathbf{P}\left(\Big|\sum_{k=1}^{m} \frac{a_k}{2}\Big(\big(\sum_{i=1}^{n_k} \varepsilon_i^k\big)^2 - n_k\Big)\Big| \ge \alpha sn\right) \le \mathbf{P}\left(S \ge \alpha n\Big(s - \frac{1}{2}(1+\varepsilon)m\Big)\right).$$

To get (4.2) let k_0 be such that $a_{k_0} = \alpha_1 2^{k_0}$, then

$$\begin{aligned}
&\mathbf{P}\left(\sum_{k=1}^{m} \frac{a_k}{2}\Big(\big(\sum_{i=1}^{n_k} \varepsilon_i^k\big)^2 - n_k\Big) \ge \alpha_1 sn\right) \\
&\quad \ge \mathbf{P}\Big(\frac{a_{k_0}}{2}\big(\sum_{i=1}^{n_{k_0}} \varepsilon_i\big)^2 \ge \alpha_1 sn\Big)\mathbf{P}\Big(\sum_{k \ne k_0} a_k \sum_{1 \le i < j \le n_k} \varepsilon_i^k \varepsilon_j^k \ge 0\Big) \\
&\quad \ge \frac{1}{K}\mathbf{P}\Big(\big(\sum_{i=1}^{n_{k_0}} \varepsilon_i\big)^2 \ge 2^{-k_0+1} sn\Big),
\end{aligned}$$

where in the last inequality we used the same properties of Rademacher chaoses as in the proof of Lemma 6 (see [de la P,G], Proposition 3.3.7). Thus (4.2) follows by (3.2). The proof of (4.3) is similar. □

Lemma 11. *Suppose that $0 < \delta < 1, k_1 \geq 1, |n_k - n2^{-k}| \leq \delta n 2^{-k}$ for $k_1 \leq k \leq k_2$ and*

$$a = \sup\{|a_k|2^{-k} : k_1 \leq k \leq k_2\}, b^2 = \sum_{k=k_1}^{k_2} a_k^2 2^{-2k}.$$

Then, for any $s > 0$ and $t > 0$, we have

$$\mathbf{P}\Big(\frac{1}{2}\sum_{k=k_1}^{k_2} a_k\Big(\big(\sum_{i=1}^{n_k}\varepsilon_i^k\big)^2 - n_k\Big) \geq t + 4k_2 nae^{-s/8}\Big) \leq \exp\Big(-\frac{t^2}{(1+\delta)^2 n^2 b^2 (1+50e^{-s/8}) + 2tsan}\Big) + 2k_2 e^{-s/4} \tag{4.4}$$

and

$$\mathbf{P}\Big(\frac{1}{2}\sum_{k=k_1}^{k_2} a_k\Big(\big(\sum_{i=1}^{n_k}\varepsilon_i^k\big)^2 - n_k\Big) \geq t - 4k_2 nae^{-s/8}\Big) \geq \frac{1}{K(\delta)}\exp\Big(-\frac{(1+\delta)t^2}{(1-\delta)^2 n^2 b^2(1-50e^{-s/8}) - 2^{k_2+1}nb^2}\Big) - \exp\Big(-\frac{\varepsilon(\delta)b^2[(1-\delta)^2(1-50e^{-s/8}) - n^{-1}2^{k_2+1}]}{s^2a^2}\Big) - 2k_2e^{-s/4}, \tag{4.5}$$

where positive constants $K(\delta)$ and $\varepsilon(\delta)$ depend only on δ.

Proof. Let

$$S_k = \Big(\sum_{i=1}^{n_k}\varepsilon_i^k\Big)^2 I_{(\sum_{i=1}^{n_k}\varepsilon_i^k)^2 \leq sn_k},$$

then

$$\|a_k(S_k - ES_k)\|_\infty \leq sa_kn_k \leq 2san.$$

Notice that by (3.1) we have

$$\mathbf{P}\Big(\Big|\sum_{i=1}^{n_k}\varepsilon_i^k\Big| \geq \sqrt{sn_k}\Big) \leq 2e^{-s/4},$$

so

$$|n_k - \mathbf{E}S_k| = \mathbf{E}\Big(\sum_{i=1}^{n_k}\varepsilon_i^k\Big)^2 I_{(\sum_{i=1}^{n_k}\varepsilon_i^k)^2 > sn_k} \leq \sqrt{\mathbf{E}\Big(\sum_{i=1}^{n_k}\varepsilon_i^k\Big)^4}\sqrt{\mathbf{P}\Big(\Big|\sum_{i=1}^{n_k}\varepsilon_i^k\Big| \geq \sqrt{sn_k}\Big)} \leq 4n_ke^{-s/8}.$$

Therefore

$$\sum_{k=k_1}^{k_2}|a_k(\mathbf{E}S_k - n_k)| \leq 8n\sum_{k=k_1}^{k_2}|a_k|2^{-k}e^{-s/8} \leq 8k_2nae^{-s/8} \tag{4.6}$$

and

$$\mathbf{P}\Big(\sum_{k=k_1}^{k_2} a_k\big(\sum_{i=1}^{n_k}\varepsilon_i^k\big)^2 \neq \sum_{k=k_1}^{k_2} a_k S_k\Big) \le \sum_{k=k_1}^{k_2} \mathbf{P}\Big(S_k \neq \big(\sum_{i=1}^{n_k}\varepsilon_i^k\big)^2\Big) \le 2k_2 e^{-s/4}. \quad (4.7)$$

We have

$$\Big|\mathbf{E}\Big(\sum_{i=1}^{n_k}\varepsilon_i^k\Big)^4 - \mathbf{E}S_k^2\Big| = \mathbf{E}\big(\sum_{i=1}^{n_k}\varepsilon_i^k\big)^4 I_{(\sum_{i=1}^{n_k}\varepsilon_i^k)^2 > sn_k}$$

$$\le \sqrt{\mathbf{E}\big(\sum_{i=1}^{n_k}\varepsilon_i^k\big)^8}\sqrt{\mathbf{P}\Big(\Big|\sum_{i=1}^{n_k}\varepsilon_i^k\Big| \ge \sqrt{sn_k}\Big)} \le 80n_k^2 e^{-s/8}$$

by the Khinchine inequality. Moreover,

$$\Big|(\mathbf{E}S_k)^2 - \Big(\mathbf{E}\big(\sum_{i=1}^{n_k}\varepsilon_i^k\big)^2\Big)^2\Big| = |(\mathbf{E}S_k)^2 - n_k^2| = |\mathbf{E}S_k + n_k|\cdot|\mathbf{E}S_k - n_k|$$

$$\le 2n_k \cdot 4n_k e^{-s/8} = 8n_k^2 e^{-s/8},$$

so

$$|\mathrm{Var}(S_k) - \mathrm{Var}((\sum_{i=1}^{n_k}\varepsilon_i^k)^2)| \le 100n_k^2 e^{-s/8}.$$

Therefore

$$\mathrm{Var}\left(\frac{1}{2}\sum_{k=k_1}^{k_2} a_k S_k\right) \le \sum_{k=k_1}^{k_2} a_k^2\Big(\frac{1}{2}n_k(n_k-1) + 25n_k^2 e^{-s/8}\Big)$$

$$\le \frac{1}{2}(1+\delta)^2 n^2 b^2 (1+50e^{-s/8})$$

and by the Bernstein inequality (Lemma 2) we have

$$\mathbf{P}\Big(\frac{1}{2}\sum_{k=k_1}^{k_2} a_k(S_k - ES_k) \ge t\Big) \le \exp\left(-\frac{t^2}{(1+\delta)^2 n^2 b^2 (1+50e^{-s/8}) + 2stan}\right). \tag{4.8}$$

Inequality (4.4) follows by (4.6), (4.7) and (4.8). To get the other estimate notice that

$$2\mathrm{Var}\left(\frac{1}{2}\sum_{k=k_1}^{k_2} a_k S_k\right) \ge \sum_{k=k_1}^{k_2} a_k^2(n_k(n_k-1) - 50n_k^2 e^{-s/8})$$

$$\ge (1-\delta)^2 n^2 b^2 (1-50e^{-s/8}) - \sum_{k=k_1}^{k_2} a_k^2 n_k \ge (1-\delta)^2 n^2 b^2 (1-50e^{-s/8}) - 2^{k_2+1} n b^2.$$

So by Kolmogorov's converse exponential inequality (Corollary 1) we get

$$\mathbf{P}\Big(\frac{1}{2}\sum_{k=k_1}^{k_2} a_k(S_k - ES_k) \geq t\Big)$$
$$\geq \frac{1}{K(\delta)}\exp\left(-\frac{(1+\delta)t^2}{(1-\delta)^2 n^2 b^2 (1-50e^{-s/8}) - 2^{k_2+1}nb^2}\right)$$
$$-\exp\left(-\frac{\varepsilon(\delta)b^2[(1-\delta)^2(1-50e^{-s/8}) - n^{-1}2^{k_2+1}]}{s^2a^2}\right). \tag{4.9}$$

Inequality (4.5) follows by (4.6), (4.7) and (4.9). □

Lemma 12. *Suppose that* $|n_k - n2^{-k}| \leq \delta n 2^{-k}$, $|a_k| \leq k^{-1/2}2^k$ *for* $k \leq k_2$ *and*

$$k_0 = \sqrt{L_2 n}, k_1 = (L_2 n)^{10}, k_2 = \log_2 n - 10L_2 n.$$

Let, moreover,

$$A_n = \sup\{|a_k|2^{-k} : k \leq k_0\}, B_n^2 = \frac{1}{L_2 n}\sum_{k=k_1}^{k_2} a_k^2 2^{-2k}$$

and

$$C_n = \begin{cases} A_n + \frac{B_n^2}{4A_n} & \text{if } B_n \leq 2A_n \\ B_n & \text{if } B_n \geq 2A_n \end{cases}.$$

Then, for any $\varepsilon > 0$, *there exists* $K(\varepsilon)$ *such that for sufficiently large* n *and sufficiently small* δ *we have*

$$\mathbf{P}\left(\Big|\Big(\sum_{k\leq k_0} + \sum_{k=k_1}^{k_2}\Big)\frac{a_k}{2}\Big(\big(\sum_{i=1}^{n_k}\varepsilon_i^k\big)^2 - n_k\Big)\Big| \geq (1+\varepsilon)C_n n L_2 n\right) \leq \frac{1}{(\log n)^{1+\varepsilon}}$$

and

$$\mathbf{P}\left(\Big|\Big(\sum_{k\leq k_0} + \sum_{k=k_1}^{k_2}\Big)\frac{a_k}{2}\Big(\big(\sum_{i=1}^{n_k}\varepsilon_i^k\big)^2 - n_k\Big)\Big| \geq (1-\varepsilon)C_n n L_2 n\right) \geq \frac{1}{K(\varepsilon)\log n}.$$

Proof. Let

$$S_1 = \sum_{k\leq k_0}\frac{a_k}{2}\Big(\big(\sum_{i=1}^{n_k}\varepsilon_i^k\big)^2 - n_k\Big) \text{ and } S_2 = \sum_{k=k_1}^{k_2}\frac{a_k}{2}\Big(\big(\sum_{i=1}^{n_k}\varepsilon_i^k\big)^2 - n_k\Big).$$

We will show that for sufficiently small δ and sufficiently large n

$$\mathbf{P}\big(|S_2| \geq un(B_n\sqrt{L_2 n} + 1)\big) \leq 2\exp\left(-\frac{u^2}{(1+\varepsilon/10)^2}\right) + \frac{1}{(\log n)^2}. \tag{4.10}$$

Obviously we may assume $0 < \varepsilon < 1$. It is enough to show that

$$\mathbf{P}\big(\pm S_2 \geq un(B_n\sqrt{L_2 n} + 1)\big) \leq \exp\left(-\frac{u^2}{(1+\varepsilon/10)^2}\right) + \frac{1}{4(\log n)^2} \tag{4.11}$$

for $u \in [1/2, 4\sqrt{L_2 n}]$. Indeed, for $u < 1/2$ the right hand side of (4.10) is greater than 1 and for $u = 4\sqrt{L_2 n}$ the right hand side of (4.11) is less than $(2\log n)^{-2}$. Now apply Lemma 11 with $s = 20L_2 n$, $t = un(B_n\sqrt{L_2 n} + 1/2)$ and $b^2 = \max(B_n^2 L_2 n, 1/4)$ (notice that then $t^2/(n^2 b^2) \geq u^2$ and that part (4.4) of Lemma 11 holds also under the assumption $b^2 \geq \sum_{k=k_1}^{k_2} a_k^2 2^{-2k}$ - the estimates are monotone in b^2). Since

$$a = \sup\{|a_k|2^{-k} : k_1 \leq k \leq k_2\} \leq k_1^{-1/2} \leq (L_2 n)^{-5}$$

we have

$$\begin{aligned} 2tsan &\leq 2 \cdot 4\sqrt{L_2 n} \cdot n(B_n\sqrt{L_2 n} + 1) \cdot 20L_2 n \cdot (L_2 n)^{-5} \cdot n \\ &\leq 160(L_2 n)^{-3} n^2 (B_n\sqrt{L_2 n} + 1) \leq \delta n^2 b^2 \end{aligned}$$

for sufficiently large n. Also

$$2k_2 e^{-s/4} \leq 2(\log_2 n)(\log n)^{-5} < (4\log n)^{-1},$$

$$4k_2 nae^{-s/8} \leq 4(\log_2 n)ne^{-s/8} \leq n/4 \leq un/2$$

and $50e^{-s/8} < \delta$ for sufficiently large n. Now it is enough to choose sufficiently small δ (which will depend on ε). Lemma 12 easily follows by Lemmas 8 and 10. □

Lemma 13. *If $\varepsilon > 0$, $|a_k| \leq k^{-1/2}2^k$ for all k and*

$$k_0 = \sqrt{L_2 n}, k_1 = (L_2 n)^{10}, k_2 = \log_2 n - 10L_2 n,$$

then for sufficiently large n

$$\mathbf{P}\left(\Big|\Big(\sum_{k=k_0}^{k_1} + \sum_{k=k_2}^{\infty}\Big)\frac{a_k}{2}\Big(\Big(\sum_{i=1}^{N_k}\varepsilon_i^k\Big)^2 - N_k\Big)\Big| \geq \varepsilon n L_2 n\right) \leq \frac{5}{\log n (L_2 n)^{3/2}}$$

Proof. In this proof K denotes a universal constant that may change from line to line. Let us additionally define

$$k_3 = \log_2 n, k_4 = \log_2 n + \frac{1}{4}\log_2\log n \text{ and } k_5 = \log_2 n + \frac{1}{2}\log_2\log n + \frac{3}{4}\log_2(L_2 n)$$

Notice that

$$\begin{aligned} &\mathbf{P}\left(\sum_{k=k_5}^{\infty}\frac{a_k}{2}\Big(\Big(\sum_{i=1}^{N_k}\varepsilon_i^k\Big)^2 - N_k\Big) \neq 0\right) \leq \mathbf{P}(\exists_{k \geq k_5} N_k > 1) \\ &\leq \mathbf{P}(\exists_{i,j \leq n}|X_i|, |X_j| \leq 2^{-k_5+1}) \leq n^2 2^{-2k_5+1} \leq \frac{2}{\log n (L_2 n)^{3/2}}. \end{aligned} \tag{4.12}$$

For $k \leq k_5$ we have $|a_k| \leq k_5^{-1/2}2^{k_5} \leq Kn(L_2 n)^{3/4}$, therefore

$$\begin{aligned} \left|\sum_{k=k_4}^{k_5-1}\frac{a_k}{2}\Big(\Big(\sum_{i=1}^{N_k}\varepsilon_i^k\Big)^2 - N_k\Big)\right| &\leq Kn(L_2 n)^{3/4}\left(\sum_{i=k_4}^{k_5-1} N_k\right)^2 \\ &\leq Kn(L_2 n)^{3/4}(\#\{i \leq n : |X_i| \leq 2^{-k_4+1}\})^2. \end{aligned}$$

Thus for fixed ε and sufficiently large n

$$\mathbf{P}\Big(\Big|\sum_{k=k_4}^{k_5-1}\frac{a_k}{2}\Big(\big(\sum_{i=1}^{N_k}\varepsilon_i^k\big)^2-N_k\Big)\Big|\geq\frac{\varepsilon}{2}nL_2n\Big)$$

$$\leq\mathbf{P}\big(\#\{i\leq n:|X_i|\geq 2^{-k_4+1}\}\geq(L_2n)^{1/8}\big)$$

$$\leq\left(\frac{en2^{-k_4+1}}{(L_2n)^{1/8}}\right)^{(L_2n)^{1/8}}\leq\frac{1}{\log n(L_2n)^{3/2}}. \tag{4.13}$$

Here we used the fact that

$$\mathbf{P}(X\geq k)\leq\binom{n}{k}p^k\leq(\frac{enp}{k})^k \text{ if } X\sim\mathrm{Bin}(n,p).$$

Similarly, for $k\leq k_4$, $|a_k|\leq k_4^{-1/2}2^{k_4}\leq Kn(\log n)^{-1/4}$, so

$$\left|\sum_{k=k_3}^{k_4-1}\frac{a_k}{2}\Big(\big(\sum_{i=1}^{N_k}\varepsilon_i^k\big)^2-N_k\Big)\right|\leq Kn(\log n)^{-1/4}\left(\sum_{i=k_3}^{k_4-1}N_k\right)^2$$

$$\leq Kn(\log n)^{-1/4}(\#\{i\leq n:|X_i|\leq 2^{-k_3+1}\})^2.$$

Therefore, for sufficiently large n

$$\mathbf{P}\Big(\Big|\sum_{k=k_3}^{k_4-1}\frac{a_k}{2}\Big((\sum_{i=1}^{N_k}\varepsilon_i^k)^2-N_k\Big)\Big|\geq\frac{\varepsilon}{2}nL_2n\Big)$$

$$\leq\mathbf{P}\big(\#\{i\leq n:|X_i|\geq 2^{-k_3+1}\}\geq(\log n)^{1/8}\big) \tag{4.14}$$

$$\leq\left(\frac{en2^{-k_3+1}}{(\log n)^{1/8}}\right)^{(\log n)^{1/8}}\leq\frac{1}{\log n(L_2n)^{3/2}}. \tag{4.15}$$

Finally

$$\mathcal{L}\Big(\sum_{k=k_0}^{k_1}+\sum_{k=k_2}^{k_3-1}\Big)\frac{a_k}{2}\Big(\big(\sum_{i=1}^{N_k}\varepsilon_i^k\big)^2-N_k\Big)\Big)=\mathcal{L}\Big(\sum_{i,j=1}^{n}\varepsilon_i\varepsilon_j\tilde{h}(X_i,X_j)\Big),$$

where

$$\tilde{h}(x,y)=\Big(\sum_{k=k_0}^{k_1}+\sum_{k=k_2}^{k_3-1}\Big)a_kh_k(x)h_k(y).$$

Let $A=[k_0,k_1]\cup[k_2,k_3-1]$, notice that

$$\|\tilde{h}\|_{L^2\to L^2}=\max_{k\in A}|a_k2^k|\leq\frac{1}{\sqrt{k_0}}\leq\frac{1}{(L_2n)^{1/4}},$$

$$E\tilde{h}^2=\sum_{k\in A}a_k^22^{2k}\leq\sum_{k\in A}\frac{1}{k}\leq CL_3n,$$

$$\|E_X\tilde{h}^2\|_\infty=\|E_Yh^2\|_\infty=\max_{k\in A}a_k^22^{-k}\leq\max_{k\in A}\frac{2^k}{k}\leq\frac{2^{k_3}}{k_3}\leq\frac{n}{\log_2 n}$$

and

$$\|\tilde{h}\|_\infty = \max_{k\in A}|a_k| \le \frac{2^{k_3}}{\sqrt{k_3}} \le \frac{n}{\sqrt{\log_2 n}}.$$

So by Lemma 4 it easily follows that

$$\mathbf{P}\left(\Big|\Big(\sum_{k=k_0}^{k_1}+\sum_{k=k_2}^{k_3}\Big)\frac{a_k}{2}\Big(\Big(\sum_{i=1}^{N_k}\varepsilon_i^k\Big)^2 - N_k\Big)\Big| \ge \frac{\varepsilon}{2}nL_2n\right) \le \frac{1}{\log n(L_2n)^{3/2}}. \quad (4.16)$$

The lemma follows by (4.12)–(4.16). □

Theorem 1. *If* $|a_k| \le \frac{2^k}{\sqrt{k}}$ *and* A *and* B *are given by (1.2) and (1.3), then*

$$\limsup_{n\to\infty}\frac{|U_n|}{nL_2n} = \begin{cases} A + \frac{B^2}{4A} & \textit{if } B \le 2A \\ B & \textit{if } B \ge 2A \end{cases} \quad a.s.$$

Proof. Let A_n, B_n be as in Lemma 12 notice that $\lim_{n\to\infty} A_n = A$ and

$$(L_2n)B_n^2 \le \mathbf{E}(h^2 \wedge n) \le (L_2n)B_n^2 + \Big(\sum_{k\le k_1} + \sum_{k=k_2}^{\log_2 n}\Big)a_k^2 2^{-2k} + n\sum_{k\ge\log_2 n} 2^{-2k}$$

$$\le (L_2n)B_n^2 + CL_3n.$$

Since $L_2n/L_2(n^2) \to 1$ as $n\to\infty$ we get that $\limsup_{n\to\infty} B_n \le B$ and

$$\forall_{\varepsilon>0}\forall_{n_0}\exists_{n\ge n_0}\forall_{N\le n\le N^2} B_n \ge B - \varepsilon.$$

So the theorem follows by Lemmas 7, 9, 12 and 13. □

References

[de la P,G] V. de la Peña and E. Giné, *Decoupling. From dependence to independence. Randomly stopped processes. U-statistics and processes. Martingales and beyond*, Probability and its Applications. Springer-Verlag, New York, 1999.

[D] R. M. Dudley, Uniform central limit theorems. *Cambridge Studies in Advanced Mathematics*, 63. Cambridge University Press, Cambridge, 1999.

[GKLZ] E. Giné, S. Kwapień, R. Latała, J. Zinn The LIL for canonical U-statistics of order 2 *Ann. Probab.* **29** (2001), 502–557.

[GLZ] E. Giné, R. Latała, J. Zinn Exponential and moment inequalities for U-statistics, *High Dimensional Probability II*, 13–38, Progress in Probability 47, Birkhauser, Boston 2000.

[GZ] E. Giné, C.-H. Zhang On the integrability in the LIL for degenerate U-statistics *J. Theoret. Probab.* **9** (1996), 385–412.

[KW] S. Kwapień, W. Woyczyński *Random series and stochastic integrals: single and multiple*, Probability and Its Applications, Birkhauser, Boston 1992.

[S] W. F. Stout, *Almost sure convergence*, Probability and Mathematical Statistics, Vol. 24. Academic Press, New York-London, 1974.

Institute of Mathematics, Warsaw University,
Banacha 2, 02-097 Warszawa, Poland
E-mail address: kwapstan@mimuw.edu.pl, rlatala@mimuw.edu.pl,
koles@mimuw.edu.pl

Department of Mathematics, Texas A&M University,
College Station, Texas 77843
E-mail address: jzinn@math.tamu.edu

Progress in Probability, Vol. 55, 127–134

Perturbation Approach Applied to the Asymptotic Study of Random Operators

André Mas and Ludovic Menneteau

Abstract. We prove that, for the main kind of limit theorems (laws of large numbers, central limit theorems, large deviations principles, laws of the iterated logarithm) asymptotic results for selfadjoint random operators yield equivalent results for their eigenvalues and associated projections.

1. Introduction and statement of the main results

The goal of this note is to present a simple derivation of several facts about asymptotics of random operators that have been frequently used in statistical applications (see e.g. [1], [4], [7], [8], [9], [10], [11]) and follow from known results of perturbation theory. Up to our best knowledge, these facts have not been published before. Consider a separable Hilbert Space H (with norm $\|\cdot\|$ and scalar product $\langle .,. \rangle$), denote by $\mathcal{L}(H)$ the separable Banach space of bounded linear operators from H to H endowed with the norm

$$\|\cdot\|_{\mathcal{L}} : x \in \mathcal{L}(H) \mapsto \sup_{\|h\| \leq 1} \|x(h)\|,$$

and define the subspace of $\mathcal{L}(H)$ of Hilbert-Schmidt operators by

$$\mathcal{S} = \left\{ s \in \mathcal{L}(H) : \sum_{p \geq 1} \|s(e_p)\|^2 < \infty \right\},$$

where $(e_p)_{p \geq 1}$ is any complete orthonormal system in H. It is well known (see [3] p. 1013 or [6] p. 262) that if we define the scalar product

$$\langle s, t \rangle_{\mathcal{S}} = \sum_{p \geq 1} \langle s(e_p), t(e_p) \rangle, \tag{1.1}$$

$\mathcal{S}$ becomes a separable Hilbert space. Let C be a self-adjoint Hilbert-Schmidt operator and consider a sequence (C_n) of random self-adjoint elements of $\mathcal{S}$ defined on

Received by the editors December 1, 2002.
2000 *Mathematics Subject Classification.* Primary: 60B12, 60F05, 60F10, 60F15.
Key words and phrases. Random operators, Limit Theorems, transfer Principle, Functional Principal Components Analysis .

a common probability space $(\Omega, \mathcal{A}, \mathbb{P})$. Since C (resp. C_n) is Hilbert-Schmidt and self-adjoint, its repeated non null eigen-values $(\mu_j)_{j\geq 1}$ (resp. $(\mu_{j,n})_{j\geq 1}$) are uniformly bounded real numbers. Without loss of generality, we assume that $(\mu_j)_{j\geq 1}$ (resp. $(\mu_{j,n})_{j\geq 1}$) is a decreasing sequence and we denote by $(\lambda_k)_{k\geq 1}$ the non-increasing sequence of non repeated elements of $(\mu_j)_{j\geq 1}$. For all $k \geq 1$, m_k denotes the multiplicity degree of λ_k, we set $\lambda_{k,n} = \{\mu_{j,n} : j \text{ such that } \mu_j = \lambda_k\}$, and Π_k (resp. $\Pi_{k,n}$) is the orthogonal projection onto $\ker(C - \lambda_k \mathrm{Id}_H)$ (resp. $\oplus_{\mu\in\lambda_{k,n}} \ker(C_n - \mu \mathrm{Id}_H)$) (where Id_H is the identity operator on H).
Here, $(C_n)_{n\geq 1}$ will be considered as a sequence of estimators of C and our aim is to study how several limit theorems, namely law of large numbers (L.L.N.), central limit theorem (C.L.T.), large deviations principle (L.D.P.) and compact law of the iterated logarithm (L.I.L.), characterizing the convergence of C_n to C can be used to infer informations about the convergence of $(\lambda_{k,n})_{n\geq 1}$ to λ_k and of $(\Pi_{k,n})_{n\geq 1}$ to Π_k.

For every $l \geq 1$, we consider the product spaces $\mathbb{R}^l$ and $\mathcal{S}^l$ endowed with the respective norms

$$|(\alpha_1, ..., \alpha_l)|_l = \max_{j\leq l} |\alpha_j|$$

and

$$\|(s_1, ..., s_l)\|_{\mathcal{S},l} = \max_{j\leq l} \|s_j\|_{\mathcal{S}}.$$

For all $k \geq 1$, we set

$$S_k : h \in H \mapsto \sum_{p\neq k} (\lambda_k - \lambda_p)^{-1} \Pi_p(h), \tag{1.2}$$

$$\varphi_k : s \in \mathcal{S} \mapsto S_k s \Pi_k + \Pi_k s S_k \in \mathcal{S}, \tag{1.3}$$

and

$$p_k : s \in \mathcal{S} \mapsto \langle \Pi_k, s \rangle_{\mathcal{S}} \in \mathbb{R}. \tag{1.4}$$

Moreover, for all finite subset $L \subset \mathbb{N}^*$ of cardinality $l \geq 1$, we define

$$\varphi_L : s \in \mathcal{S} \mapsto \{\varphi_k(s) : k \in L\} \in \mathcal{S}^l$$

and

$$p_L : s \in \mathcal{S} \mapsto \{p_k(s) : k \in L\} \in \mathbb{R}^l.$$

Note for further references that φ_L and p_L are continuous and linear.
Here are our main results.

Theorem 1.1. *If $(C_n - C)_{n\geq 1}$ converges almost surely to 0 in $\mathcal{S}$, then, for all finite subset $L \subset \mathbb{N}^*$ of cardinality l,*
i) $(\{\Pi_{k,n} - \Pi_k : k \in L\})_{n\geq 1}$ converges almost surely to 0 in $\mathcal{S}^l$.
ii) $\left(\sup_{k\geq 1} \max_{\mu\in\lambda_{k,n}} |\mu - \lambda_k|\right)_{n\geq 1}$ converges almost surely to 0 in $\mathbb{R}$.

Theorem 1.2. *If for some $b_n \uparrow \infty$, $(b_n (C_n - C))_{n\geq 1}$ converges in law in $\mathcal{S}$ to the limit law* G_C, *then, for all finite subset $L \subset \mathbb{N}^*$ of cardinality l,*
i) $(b_n \{\Pi_{k,n} - \Pi_k : k \in L\})_{n\geq 1}$ *converges in law in $\mathcal{S}^{\,l}$ to the limit law*

$$\mathrm{G}_{\Pi_L} : A \in \mathcal{B}(\mathcal{S}^l) \mapsto \mathrm{G}_C\left(\varphi_L^{-1}(A)\right).$$

ii) $\left(b_n \left\{\sum_{\mu\in\lambda_{k,n}} \mu - m_k\lambda_k : k \in L\right\}\right)_{n\geq 1}$ *converges in law in $\mathbb{R}^l$ to the limit law*

$$\mathrm{G}_{\lambda_L} : B \in \mathcal{B}(\mathbb{R}^l) \mapsto \mathrm{G}_C\left(p_L^{-1}(B)\right).$$

Theorem 1.3. *If, for some $b_n \uparrow \infty$, $(b_n (C_n - C))_{n\geq 1}$ satisfies the large deviations principle in $\mathcal{S}$ with speed (v_n) and rate function* J_C, *then, for all finite subset $L \subset \mathbb{N}^*$ of cardinality l,*
i) $(b_n \{\Pi_{k,n} - \Pi_k : k \in L\})_{n\geq 1}$ *satisfies the large deviation principle in $\mathcal{S}^l$ with speed (v_n) and rate function*

$$\mathrm{J}_{\Pi_L} : t \in \mathcal{S}^l \mapsto \inf\{\mathrm{J}_C(s) : \varphi_L(s) = t\}.$$

ii) $\left(b_n \left\{\sum_{\mu\in\lambda_{k,n}} \mu - m_k\lambda_k : k \in L\right\}\right)_{n\geq 1}$ *satisfies the large deviations principle in $\mathbb{R}^l$ with speed (v_n) and rate function*

$$\mathrm{J}_{\lambda_L} : \alpha \in \mathbb{R}^l \mapsto \inf\{\mathrm{J}_C(s) : p_L(s) = \alpha\}.$$

Theorem 1.4. *If, for some $b_n \uparrow \infty$, $(b_n (C_n - C))_{n\geq 1}$ is almost surely compact in $\mathcal{S}$ with limit set K_C then, for all subset $L \subset \mathbb{N}^*$ of cardinality l,*
i) $(b_n \{\Pi_{k,n} - \Pi_k : k \in L\})_{n\geq 1}$ *is almost surely relatively compact in $\mathcal{S}^l$ with limit set*

$$K_{\Pi_L} = \varphi_L(K_C).$$

ii) $\left(b_n \left\{\sum_{\mu\in\lambda_{k,n}} \mu - m_k\lambda_k : k \in L\right\}\right)_{n\geq 1}$ *is almost surely relatively compact in $\mathbb{R}^l$ with limit set*

$$K_{\lambda_L} = p_L(K_C).$$

2. Proofs

In the next lemmas we give some results related to perturbation theory for linear operators useful for our needs.

Lemma 2.1. *Let Δ be a self-adjoint element of $\mathcal{L}(H)$ and let σ be a finite set of isolated points of the spectrum of Δ. We call Γ an admissible contour for σ and Δ whenever Γ is a contour around σ which contains no other eigenvalues of Δ. Then, for every contour Γ, admissible for σ and Δ, the mapping*

$$\Pi = \frac{1}{2i\pi}\int_\Gamma (z\mathrm{Id}_H - \Delta)^{-1}\, dz, \tag{2.1}$$

where $i^2 = -1$, is the orthogonal projection onto $\oplus_{\mu\in\sigma}\ker(\Delta - \mu\mathrm{Id}_H)$.

Proof. See e.g. Proposition 6.3 of [5]. □

Set

$$\delta_k = \inf_{p \neq k} |\lambda_k - \lambda_p| = \min(\lambda_k - \lambda_{k+1}, \lambda_{k-1} - \lambda_k). \tag{2.2}$$

Let Γ_k be the oriented circle with center λ_k and radius $\rho_k = \delta_k/2$. Note that Γ_k is an admissible contour for λ_k and C.
Moreover, define the event

$$\mathcal{O}_{k,n} = \{\|C_n - C\|_{\mathcal{S}} < \delta_k/4\}. \tag{2.3}$$

Since

$$\sup_{j\geq 1} |\mu_{j,n} - \mu_j| \leq \|C_n - C\|_{\mathcal{S}}, \tag{2.4}$$

(see e.g. [3] p. 1091), we can prove :

Lemma 2.2. *i) For all $\omega \in \mathcal{O}_{k,n}$, Γ_k is an admissible contour for $\lambda_{k,n}(\omega)$ and $C_n(\omega)$.*
ii)

$$\sup_{z\in\Gamma_k} \left\{ \left\| (z\mathrm{Id}_H - C_n)^{-1} \right\|_{\mathcal{L}} \right\} \mathbf{1}_{\mathcal{O}_{k,n}} \leq 4\delta_k^{-1}. \tag{2.5}$$

iii)

$$\sup_{z\in\Gamma_k} \left\{ \left\| (z\mathrm{Id}_H - C)^{-1} \right\|_{\mathcal{L}} \right\} \leq 2\delta_k^{-1}. \tag{2.6}$$

Proof. i) Set $\omega \in \mathcal{O}_{k,n}$. By (2.3) and (2.4),

$$\sup_{\mu\in\lambda_{k,n}(\omega)} |\mu - \lambda_k| \leq \delta_k/4 < \rho_k \tag{2.7}$$

and

$$\begin{aligned} \inf_{p\neq k} \inf_{\mu\in\lambda_{p,n}(\omega)} |\mu - \lambda_k| &\geq \inf_{p\neq k} |\lambda_p - \lambda_k| - \sup_{p\neq k} \sup_{\mu\in\lambda_{p,n}(\omega)} |\mu - \lambda_p| \\ &= 3\delta_k/4 > \rho_k. \end{aligned} \tag{2.8}$$

Hence, the result holds by (2.7) and (2.8).
ii) Set $\omega \in \mathcal{O}_{k,n}$. By Theorem 5.8 of [5],

$$\sup_{z\in\Gamma_k} \left\{ \left\| (z\mathrm{Id}_H - C_n(\omega))^{-1} \right\|_{\mathcal{L}} \right\} \leq \sup_{z\in\Gamma_k} \sup_{p\geq 1} \sup \left\{ |z - \mu|^{-1} : \mu \in \lambda_{p,n}(\omega) \right\}. \tag{2.9}$$

Moreover, for all $z \in \Gamma_k$,

$$\begin{aligned} \inf_{\mu\in\lambda_{k,n}(\omega)} |z - \mu| &\geq |z - \lambda_k| - \sup_{\mu\in\lambda_{k,n}(\omega)} |\lambda_k - \mu| \\ &\geq \rho_k - \delta_k/4 = \delta_k/4, \end{aligned}$$

and

$$\inf_{p\neq k}\inf_{\mu\in\lambda_{p,n}(\omega)}|\mu-\lambda_k| = \inf_{p\neq k}\inf_{\mu\in\lambda_{p,n}(\omega)}|(\lambda_p-\lambda_k)+(\lambda_k-z)+(\lambda_{p,n}(\omega)-\lambda_p)|$$
$$\geq \inf_{p\neq k}|\lambda_p-\lambda_k| - |\lambda_k - z| - \sup_{p\neq k}\sup_{\mu\in\lambda_{p,n}(\omega)}|\mu-\lambda_p|$$
$$\geq \delta_k - \rho_k - \delta_k/4 = \delta_k/4.$$

Therefore,

$$\inf_{z\in\Gamma_k}\inf_{p\geq 1}\inf_{\mu\in\lambda_{p,n}(\omega)}|z-\mu| \geq \delta_k/4,$$

which, combined with (2.9), give the result.
iii) (2.2) and the definition of Γ_k entails, for all $z\in\Gamma_k$,

$$\inf_{p\neq k}|\lambda_p - z| \geq \inf_{p\neq k}|\lambda_p-\lambda_k| - |\lambda_k - z|$$
$$\geq \delta_k - \delta_k/2 = \delta_k/2.$$

Therefore, using Theorem 5.8 of [5], we get,

$$\sup_{z\in\Gamma_k}\left\{\left\|(z\mathrm{Id}_H - C)^{-1}\right\|_{\mathcal{L}}\right\} \leq \sup\left\{|z-\lambda_p|^{-1} : p\geq 1, z\in\Gamma_k\right\}$$
$$\leq 2\delta_k^{-1}.$$

□

For all finite subset $L\subset\mathbb{N}^*$ of cardinality l, set

$$\delta_L = \min_{k\in L}\delta_k \text{ and } \mathcal{O}_{L,n} = \{\|C_n - C\|_{\mathcal{S}} < \delta_L/4\} = \bigcap_{k\in L}\mathcal{O}_{k,n}.$$

Now, we can state the main tools used in the proof of our theorems.

Proposition 2.3. *For all finite subset $L\subset\mathbb{N}^*$ of cardinality l and all $n\geq 1$,*
i) There exists a $\mathcal{S}^l$-valued random variable $R_{L,n}$ such that

$$\{\Pi_{k,n} - \Pi_k : k\in L\} = \varphi_L(C_n - C) + R_{L,n}, \tag{2.10}$$

and

$$\|R_{L,n}\|_{\mathcal{S},l}\mathbf{1}_{\mathcal{O}_{L,n}} \leq 8\delta_L^{-2}\|C_n - C\|_{\mathcal{S}}^2. \tag{2.11}$$

ii) There exists a $\mathbb{R}^l$-valued random variable $r_{L,n}$ such that

$$\left\{\sum_{\mu\in\lambda_{k,n}}\mu - m_k\lambda_k : k\in L\right\} = p_L(C_n - C) + r_{L,n}, \tag{2.12}$$

and,

$$|r_{L,n}|_l\mathbf{1}_{\mathcal{O}_{L,n}} \leq \tau_L\|C_n - C\|_{\mathcal{S}}^2 + \gamma_L\|C_n - C\|_{\mathcal{S}}^3, \tag{2.13}$$

where

$$\tau_L = 8\delta_L^{-2}\|C\|_{\mathcal{S}} + 2\delta_L^{-1} \text{ and } \gamma_L = 8\delta_L^{-2}.$$

Proof. i) Set $\omega \in \mathcal{O}_{L,n}$ and take any $k \in L$. Since, by the first part of Lemma 2.2, Γ_k is an admissible contour for $\lambda_{k,n}(\omega)$ and $C_n(\omega)$ (and also for λ_k and C), Lemma 2.1. implies that

$$\Pi_{k,n}(\omega) - \Pi_k = \frac{1}{2i\pi}\int_{\Gamma_k} (z\mathrm{Id}_H - C_n(\omega))^{-1} - (z\mathrm{Id}_H - C)^{-1}\, dz. \qquad (2.14)$$

For convenience, set

$$a = z\mathrm{Id}_H - C_n(\omega) \text{ and } b = z\mathrm{Id}_H - C.$$

Note that

$$a^{-1} - b^{-1} = b^{-1}(b-a)b^{-1} + b^{-1}(b-a)a^{-1}(b-a)b^{-1}. \qquad (2.15)$$

Therefore, if we set

$$U_{k,n} = \frac{\mathbf{1}_{\mathcal{O}_{k,n}}}{2i\pi}\int_{\Gamma_k} (z\mathrm{Id}_H - C)^{-1}(C_n - C)(z\mathrm{Id}_H - C_n)^{-1}(C_n - C)(z\mathrm{Id}_H - C)^{-1}\, dz,$$

we get, by (2.14) and (2.15),

$$\Pi_{k,n}(\omega) - \Pi_k = \frac{1}{2i\pi}\left(\int_{\Gamma_k} (z\mathrm{Id}_H - C)^{-1}(C_n(\omega) - C)(z\mathrm{Id}_H - C)^{-1}\, dz\right) + U_{k,n}(\omega).$$

Now, in [2] p. 145, it is shown that

$$\varphi_k : s \in \mathcal{S} \mapsto \frac{1}{2i\pi}\left(\int_{\Gamma_k} (z\mathrm{Id}_H - C)^{-1} s (z\mathrm{Id}_H - C)^{-1}\, dz\right).$$

Hence, if we define

$$R_{k,n} = U_{k,n} + (\Pi_{k,n} - \Pi_k - \varphi_k(C_n - C))\mathbf{1}_{\mathcal{O}_{k,n}^c},$$

and

$$R_{L,n} = \{R_{k,n} : k \in L\},$$

(2.10) holds. Moreover, following [2] p. 142 (line 2 and 3), we obtain, using (2.5) and (2.6), that

$$\begin{aligned}
&\|U_{k,n}\|_{\mathcal{S}} \\
&\le \rho_k \sup_{z\in\Gamma_k}\left(\left\|(z\mathrm{Id}_H - C)^{-1}(C_n - C)(z\mathrm{Id}_H - C_n)^{-1}(C_n - C)(z\mathrm{Id}_H - C)^{-1}\right\|_{\mathcal{S}}\right)\mathbf{1}_{\mathcal{O}_{k,n}} \\
&\le \frac{\delta_k}{2}\|C_n - C\|_{\mathcal{S}}^2 \sup_{z\in\Gamma_k}\left\{\left\|(z\mathrm{Id}_H - C)^{-1}\right\|_{\mathcal{L}}^2 \left\|(z\mathrm{Id}_H - C_n)^{-1}\right\|_{\mathcal{L}}\right\}\mathbf{1}_{\mathcal{O}_{k,n}} \\
&\le 8\delta_k^{-2}\|C_n - C\|_{\mathcal{S}}^2.
\end{aligned}$$

Therefore,

$$\begin{aligned}
\|R_{L,n}\|_{\mathcal{S},l}\mathbf{1}_{\mathcal{O}_{L,n}} &= \max_{k\in L}\|U_{k,n}\|_{\mathcal{S}} \\
&\le 8\delta_L^{-2}\|C_n - C\|_{\mathcal{S}}^2.
\end{aligned}$$

ii) First, observe that, if Δ is a self-adjoint element of $\mathcal{L}(H)$ with (real) eigenvalues $(l_k)_{k\geq 1}$ of respective multiplicity degrees $(d_k)_{k\geq 1}$ and associated projectors $(P_k)_{k\geq 1}$, for all $k \geq 1$,

$$\langle P_k, \Delta \rangle_{\mathcal{S}} = d_k l_k.$$

Hence,

$$\begin{aligned}\sum_{\mu \in \lambda_{k,n}} \mu - m_k \lambda_k &= \langle \Pi_{k,n}, C_n \rangle_{\mathcal{S}} - \langle \Pi_k, C \rangle_{\mathcal{S}} \\ &= p_k (C_n - C) + \langle \varphi_k (C_n - C), C \rangle_{\mathcal{S}} \\ &\quad + \langle R_{k,n}, C \rangle_{\mathcal{S}} + \langle \Pi_{k,n} - \Pi_k, C_n - C \rangle_{\mathcal{S}}.\end{aligned} \tag{2.16}$$

Furthermore, let β_k be an orthonormal basis of the eigen-subspace of C associated with λ_k and let β be any orthonormal basis of H such that $\beta_k \subset \beta$. Then, by (1.1), (1.3) and (1.2), for all $s \in \mathcal{S}$, easy calculations give

$$\langle \varphi_k (s), C \rangle_{\mathcal{S}} = 0. \tag{2.17}$$

Hence, if we combine (2.16) and (2.17), we get

$$\left\{ \sum_{\mu \in \lambda_{k,n}} \mu - m_k \lambda_k : k \in L \right\} = p_L (C_n - C) + r_{L,n},$$

where

$$r_{L,n} = \left\{ \langle R_{k,n}, C \rangle_{\mathcal{S}} + \langle \Pi_{k,n} - \Pi_k, C_n - C \rangle_{\mathcal{S}} : k \in L \right\}.$$

Now $r_{L,n}$ satisfies

$$\begin{aligned}|r_{L,n}|_l \mathbf{1}_{O_{L,n}} &\leq \|C\|_{\mathcal{S}} \|R_{L,n}\|_{\mathcal{S},l} \mathbf{1}_{O_{L,n}} + \|C_n - C\|_{\mathcal{S}} \|\Pi_{L,n} - \Pi_L\|_{\mathcal{S},l} \mathbf{1}_{O_{L,n}} \\ &\leq \left(8\delta_L^{-2} \|C\|_{\mathcal{S}} + \max_{k \in L} \|\varphi_k\|_{\mathcal{L}(\mathcal{S})} \right) \|C_n - C\|_{\mathcal{S}}^2 + 8\delta_L^{-2} \|C_n - C\|_{\mathcal{S}}^3,\end{aligned} \tag{2.18}$$

where $\|\cdot\|_{\mathcal{L}(\mathcal{S})}$ is the usual norm on the space $\mathcal{L}(\mathcal{S})$ of bounded linear $\mathcal{S}$-valued operators. Finally, since for all $k \geq 1$, as easily seen, $\|\Pi_k\|_{\mathcal{L}} = 1$ and $\|S_k\|_{\mathcal{L}} = \delta_k^{-1}$, we get using (1.3) that

$$\|\varphi_k\|_{\mathcal{L}(\mathcal{S})} \leq 2\delta_k^{-1},$$

which, combined with (2.18), lead to (2.11) and completes our proof. □

Proof of Theorems 1.1–1.4 : Using Proposition 2.3, the proofs of our theorems are now simple exercises. Hence, we just give a sketch in the case of the projectors. By (2.10), and the linearity of φ_L,

$$b_n \{\Pi_{k,n} - \Pi_k : k \in L\} = \varphi_L (b_n (C_n - C)) + b_n R_{L,n}.$$

Now, since the L.L.N. (resp. C.L.T., L.D.P., L.I.L.) is preserved by continuous functions, the continuity of φ_L entails that the L.L.N. (resp. C.L.T., L.D.P., L.I.L.) for $(b_n (C_n - C))_{n\geq 1}$ implies the corresponding limit theorem for $(\varphi_L(b_n (C_n - C)))_{n\geq 1}$.

Besides, by (2.11), it is easily seen that $(b_n R_{L,n})_{n\geq 1}$ vanishes for all the limit theorems considered. □

References

[1] Bosq D. (2000), *Linear processes in function spaces* , Lecture Notes in Statistics, 149, Springer-Verlag.

[2] Dauxois J., Pousse A. and Romain Y. (1982), Asymptotic theory for the principal component analysis of a vector random function : some applications to statistical inference. *J. Multivar. Anal.* **12** 136-154.

[3] Dunford N. and Schwartz J.T. (1988), *Linear Operators* Vol II, Wiley Classics Library.

[4] Fine J. (1987), On the validity of the perturbation method in asymptotic theory. *Statistics* **18**, 401-414.

[5] Hislop P.D. and Sigal I.M. (1996), *Introduction to Spectral Theory*, Applied Mathematical Sciences, 113, Springer.

[6] Kato T. (1976), *Perturbation theory for linear operators*, Grundlehren der mathematischen Wissenschaften. 132, Springer-Verlag.

[7] Mas A. (2002), Weak convergence for covariance operators of a linear Hilbertian process, *Stochastic Processes Appl.,* **99** (1), 117-135.

[8] Mas A. and Menneteau L. (2001), Large and moderate deviations for infinite dimensional autoregressive processes, submitted.

[9] Menneteau L. (2002), Some laws of the iterated logarithm in hilbertian autoregressive models, submitted.

[10] Ruymgaart F.H. and Yang S. (1997), Some applications of Watson's perturbation approach to random matrices. *J. Multivar. Anal.,* **60**, 48-60 doi: 10.1006/jmva.1996.1640.

[11] Tyler D.E. (1981), Asymptotic inference for eigenvectors, *Ann. Stat.*, **9**, 725-736.

Université Toulouse III and CREST, 118, Route de Narbonne 31062 TOULOUSE, Cedex 4 FRANCE
E-mail address: mas@cict.fr

Université Montpellier II and CREST, Département des Sciences Mathématiques, Place Eugène Bataillon, 34095 MONTPELLIER Cedex 5 FRANCE
E-mail address: mennet@math.univ-montp2.fr

Progress in Probability, Vol. 55, 135–151

A Uniform Functional Law of the Logarithm for a Local Gaussian Process

David M. Mason

Abstract. We establish a uniform functional law of the logarithm for a Gaussian process closely related to the local empirical process. We then discuss the necessity of the polynomial covering assumption on the indexing class of functions.

1. Introduction

Let $\{W_n\}_{n\geq 1}$ be a sequence of independent standard Wiener processes on $\mathbb{R}$ and define the process

$$s_n(t) = n^{-1/2}\sum_{i=1}^{n} W_i(t), \ t \in [0,1],$$

and for any $h \in (0,1)$ and $n \geq 1$ consider the oscillation modulus of s_n,

$$\omega_{s_n}(h) = \sup\{|s_n(t+s) - s_n(t)| : 0 \leq t, \ t+s \leq 1, 0 \leq s \leq h\}.$$

A special case of the Csörgő and Révész strong law [4] implies that whenever $\{h_n\}_{n\geq 1}$ is a sequence of positive constants converging to zero at the rate:

(H.i) $h_n \searrow 0$, $nh_n \nearrow \infty$; (H.ii) $\log(1/h_n)/\log\log n \to \infty$,

then the following uniform law of the logarithm [ULL] holds:

$$\lim_{n\to\infty} \omega_{s_n}(h_n)/\sqrt{2h_n \log(1/h_n)} = 1, \text{ a.s.} \tag{1.1}$$

Now let $\{U_n\}_{n\geq 1}$ be a sequence of independent Uniform $[0,1]$ random variables. Consider for each integer $n \geq 1$ the empirical process based on $U_1, \ldots,$ U_n,

$$\alpha_n(t) = \sqrt{n}\{G_n(t) - t\}, \ t \in [0,1],$$

2000 *Mathematics Subject Classification.* Primary 60F15, 60F17; Secondary 62G07, 62G30.
Key words and phrases. Gaussian process, functional law of the logarithm, kernel density estimators, empirical processes.
[1]This work was completed with the support of NSA Grant MDA904–02–1–0034 and NSF Grant DMS–0203865.

where G_n is the uniform empirical distribution function

$$G_n(t) = n^{-1} \sum_{i=1}^{n} 1\{U_i \le t\}, \ t \in \mathbb{R}.$$

Guided by the limit result (1.1), Stute [31] showed that whenever $\{h_n\}_{n\ge 1}$ is a sequence of positive constants satisfying, in addition, to (H.i–ii), the condition

(H.iii) $\quad nh_n / \log(1/h_n) \to \infty,$

then $\omega_{\alpha_n}(h_n)$ also obeys the ULL,

$$\lim_{n\to\infty} \omega_{\alpha_n}(h_n)/\sqrt{2h_n \log(1/h_n)} = 1, \text{ a.s.,} \tag{1.2}$$

where for any $h \in (0,1)$,

$$\omega_{\alpha_n}(h) = \sup \{|\alpha_n(t+s) - \alpha_n(t)| : 0 \le t, \ t+s \le 1, 0 \le s \le h\}.$$

Conditions (H.ii) and (H.iii) are sharp in the sense that if either of them is not satisfied, then (1.2) no longer holds. Refer, for instance, to Mason, Shorack and Wellner [22] for details.

A suboptimal version of the Stute [31] result (1.2) can be easily inferred from the Komlós, Major and Tusnády [KMT] [18] *Kiefer process strong approximation* to the uniform empirical process, which says that on the same probability space there exist a sequence of i.i.d. Uniform $[0,1]$ random variables $\{U_n\}_{n\ge 1}$ and a sequence of independent standard Wiener processes $\{W_n\}_{n\ge 1}$ such that

$$\sup_{0\le t\le 1} \left| \alpha_n(t) - n^{-1/2} \sum_{i=1}^{n} B_i(t) \right| = O\left((\log n)^2 / \sqrt{n} \right), \text{ a.s.,} \tag{1.3}$$

where for each $n \ge 1$, B_n is the Brownian bridge

$$B_n(t) = W_n(t) - tW_n(1), \ t \in [0,1].$$

Since by the law of the iterated logarithm

$$n^{-1/2} \sum_{i=1}^{n} W_i(1) = O\left(\sqrt{\log\log n}\right), \text{ a.s.}$$

one readily infers (1.2) from (1.1) and (1.3) as long as (H.i–ii) hold and (H.iii) is replaced by the more restrictive condition that $nh_n/(\log n)^2 \to \infty$. (For more about this approach, along with closely related results, refer to Chapter 14 of Shorack and Wellner [29].) This meant that Stute had to devise a new set of techniques, independent of the strong approximation (1.3), in order to establish (1.2) under the optimal conditions (H.i–iii).

Now let $\{Z_n\}_{n\ge 1}$ be i.i.d. random variables taking values in $\mathbb{R}$, with common Lebesgue density function f. Stute [32] obtained from his result (1.2) and the probability integral transformation a ULL for the kernel density estimator f_n over a compact interval J, namely he showed that under certain regularity conditions on f, among them being continuity on J,

$$\lim_{n\to\infty} \sqrt{nh_n} \sup_{z\in J} \left\{ |f_n(z) - Ef_n(z)| / \sqrt{2||K||_2^2 f(z) \log(1/h_n)} \right\} = 1, \text{ a.s.,} \tag{1.4}$$

where f_n is defined for $z \in \mathbb{R}$ to be

$$f_n(z) = (nh_n)^{-1} \sum_{i=1}^{n} K\left(h_n^{-1}(z - Z_i)\right), \tag{1.5}$$

with K being a kernel with compact support and of bounded variation satisfying

$$0 < \int_{\mathbb{R}} K^2(x)dx = ||K||_2^2 < \infty. \tag{1.6}$$

Earlier work of Csörgő and Révész [5] had pointed the way to the Stute [32] result (1.2). They had previously derived a version of (1.4), via the KMT [18] strong approximation (1.3), by first establishing, under many smoothness conditions, the ULL

$$\lim_{n\to\infty} \sqrt{nh_n} \sup_{z\in J} \left\{ |\zeta_n(z, K)| / \sqrt{2||K||_2^2 f(z) \log(1/h_n)} \right\} = 1, \text{ a.s.,} \tag{1.7}$$

where $\zeta_n(z, K)$ is the local Gaussian process

$$\zeta_n(z, K) := (nh_n)^{-1/2} \sum_{i=1}^{n} \int_{\mathbb{R}} K\left(h_n^{-1}(z - t)\right) dW_i\left(F(t)\right), \; z \in J, \tag{1.8}$$

with F denoting the cumulative distribution function determined by f. In Section 3 we shall prove a uniform functional law of the logarithm [UFLL] for a generalized version of this process, which yields (1.7) as a special case. We shall shortly describe through an example what we mean by a UFLL.

Later, Stute [33] established a version of his ULL (1.2) for certain oscillations of the empirical process based upon $\{Z_n\}_{n\geq 1}$ i.i.d. d–dimensional random vectors with common Lebesgue density function f. He used it to derive precise results on the uniform consistency of the d–variate kernel density estimator, which is defined as in (1.5), but with the h_n inside K replaced by $h_n^{1/d}$.

Deheuvels and Mason [7] extended the Stute [31] ULL (1.2) to a UFLL for the cluster of random increment functions on $[0, 1]$,

$$\{\xi_n(t, \cdot) : 0 \leq t \leq 1 - h_n\}, \tag{1.9}$$

where for each $0 \leq t \leq 1 - h_n$, $\xi_n(t, \cdot)$ is the local empirical process:

$$\xi_n(t, s) = h_n^{-1/2} \{\alpha_n(t + h_n s) - \alpha_n(t) - h_n s\} \; s \in [0, 1]. \tag{1.10}$$

To state their result let $\mathcal{S}_0$ denote the set of Strassen functions on $[0, 1]$. These are the absolutely continuous functions φ on $[0, 1]$ with Lebesgue derivative φ' satisfying

$$\varphi(0) = 0 \text{ and } \int_{[0,1]} (\varphi'(s))^2 ds \leq 1.$$

Let $l_\infty[0,1]$ denote the space of bounded functions on $[0,1]$ and let $||\cdot||$ be the supremum norm on $l_\infty[0,1]$. For any $\varepsilon > 0$ and $A \subset l_\infty[0,1]$ let

$$A^\varepsilon = \left\{ g \in l_\infty[0,1] : \inf_{\varphi \in A} ||g - \varphi|| < \varepsilon \right\}.$$

Deheuvels and Mason [7] proved the following UFLL for (1.9), whenever (H.i-iii) hold, namely, with probability 1, for every $\varepsilon > 0$ there exists an $n_\varepsilon \geq 1$ such for all $n \geq n_\varepsilon$

$$\left\{ \xi_n(t,\cdot)/\sqrt{2\log(1/h_n)} : 0 \leq t \leq 1 - h_n \right\}^\varepsilon \subset \mathcal{S}_0^\varepsilon \tag{1.11}$$

and for every $\varepsilon > 0$ and $\varphi \in \mathcal{S}_0$ there exists an $n_{\varphi,\varepsilon} \geq 1$ such that for every $n \geq n_{\varphi,\varepsilon}$ there is a $t_{\varphi,n} \in [0, 1 - h_n]$ satisfying

$$||\xi_n(t_{\varphi,n},\cdot)/\sqrt{2\log(1/h_n)} - \varphi|| < \varepsilon. \tag{1.12}$$

For the corresponding results when (H.i) or (H.ii) are not satisfied refer to Deheuvels and Mason [7] and Deheuvels [6].

Révész [27] had earlier obtained a similar UFLL for the Wiener process (for closely related work consult Mueller [23]). A version of the Révész result says that whenever (H.i–ii) are satisfied then (1.11) and (1.12) hold with $\xi_n(t,\cdot)$ replaced for each $t \in [0, 1 - h_n]$ by the local Wiener process,

$$\zeta_n(t,s) = (nh_n)^{-1/2} \sum_{i=1}^{n} \{W_i(t + h_n s) - W_i(t)\}, \ s \in [0,1]. \tag{1.13}$$

The proof of the Deheuvels and Mason UFLL for (1.9) relies strongly on the KMT [17] *Wiener process approximation* to partial sums of i.i.d. Poisson random variables, coupled with a functional large deviation result for the Wiener process. Presently such a precise and powerful approximation as is found in [17] does not exist in the general empirical process setting.

The KMT approximations have frequently helped workers in large sample theory formulate the 'right' limit result to prove. Furthermore, the limitations of the KMT approximations, especially in multi-dimensional problems, have also been a great impetus to the development of more general methods.

Mason [21] has recently been able to meld the techniques from classical and abstract empirical process theory, Gaussian distributional approximation and probability on Banach spaces to prove a UFLL for a general indexed by class of functions version of (1.9) formed by a sequence of i.i.d. random vectors $\{Z_n\}_{n\geq 1}$ taking values in $\mathbb{R}^d$ with common Lebesgue density function f. Among other results, his methods are powerful enough to obtain the Deheuvels and Mason UFLL (without the use of KMT [17]) as a corollary of his main result. In the next section we shall describe the UFLL for the local empirical process proved in [21] and then in Section 3 we shall prove a UFLL for a local Gaussian process indexed by a class of functions, motivated by (1.8) and closely related to the local empirical process. Finally in Section 4 we shall provide some heuristics on the necessity of the polynomial covering number assumption in the proof of UFLL proved in Section 3.

2. A UFLL for the local empirical process

Let $\{Z_n\}_{n\geq 1}$ be an i.i.d. sequence of $\mathbb{R}^d$ valued random vectors with common Lebesgue density function f. Let $\{h_n\}_{n\geq 1}$ be a sequence of positive constants less than 1 converging to zero. Using the classic kernel density estimator (1.5) as our initial motivation, we define the local empirical process at $z \in \mathbb{R}^d$ indexed by a class of functions $\mathcal{G}$ to be for $g \in \mathcal{G}$,

$$E_n(z,g) := (nh_n)^{-1/2} \sum_{i=1}^{n} \left\{ g\left(h_n^{-1/d}(z - Z_i)\right) - Eg\left(h_n^{-1/d}(z - Z)\right) \right\}. \tag{2.1}$$

Observe that by choosing $\mathcal{G} = \{K\}$, we get

$$E_n(z,K) = \sqrt{nh_n}\left\{f_n(z) - Ef_n(z)\right\}. \tag{2.2}$$

Einmahl and Mason [11] and [12] obtained central limit theorems, strong approximations and functional laws of the iterated logarithm for the local empirical process at a fixed z. (Mason [20] had treated a special case of this process, which he called the tail uniform empirical process.) They showed how to apply their results to obtain the exact rate of point–wise consistency for a number of well–known non-parametric kernel–type function estimators. The definition of the local empirical process given by Einmahl and Mason [11] and [12] is a bit more general in that the $h_n^{1/d}$ is replaced by a sequence of bi–measurable functions. It extends an earlier notion introduced by Deheuvels and Mason [8]. Recently Einmahl and Mason [13] established a ULL for a local process closely related to (2.1) that yielded as a special case the exact rate of uniform consistency of certain kernel–type regression function estimators.

In this section we shall describe via a UFLL the uniform limiting behavior of the cluster of processes (2.1) indexed by $g \in \mathcal{G}$ as z moves over a compact set J. Towards this end we shall introduce the following normed version of E_n : For any $z \in \mathbb{R}^d$ such that $f(z) > 0$ and $g \in \mathcal{G}$, set

$$L_n(z,g) := E_n(z,g)/\sqrt{2\log(1/h_n)f(z)}. \tag{2.3}$$

We shall now fix notation and assumptions. Throughout the remainder of this paper $\mathcal{G}$ will denote a class of measurable real valued functions defined on $\mathbb{R}^d$, which have support contained in $I^d := [-1/2, 1/2]^d$ and are bounded by some $\kappa > 0$. Let $|\cdot|_2$ denote the usual Euclidean norm on $\mathbb{R}^d$. Assume that the class $\mathcal{G}$ satisfies

(G.i) $\lim_{|w|_2\to 0} \sup_{g\in\mathcal{G}} \int_{\mathbb{R}^d} [g(x) - g(x+w)]^2\, dx = 0$;

(G.ii) $\lim_{\lambda\to 1} \sup_{g\in\mathcal{G}} \int_{\mathbb{R}^d} [g(x) - g(\lambda x)]^2\, dx = 0$.

In addition, let $\mathcal{F}$ denote the class of functions formed from $\mathcal{G}$ satisfying

(F.i) for each $\lambda \geq 1$, $z \in \mathbb{R}^d$ and $g \in \mathcal{G}$, $g(z - \cdot\lambda) \in \mathcal{F}$.

To avoid using outer probability measures in all of our statements, we impose the measurability assumption:

(F.ii) $\mathcal{F}$ is a pointwise measurable class, i.e. there exists a countable subclass $\mathcal{F}_0$ of $\mathcal{F}$ such that we can find for any function $g \in \mathcal{F}$ a sequence of functions $\{g_m\}$ in $\mathcal{F}_0$ for which $g_m(z) \to g(z)$, $z \in \mathbb{R}^d$. (See Example 2.3.4 in van der Vaart and Wellner [35].)

For $\epsilon > 0$, let

$$N(\epsilon, \mathcal{F}) = \sup_Q N(\kappa\epsilon, \mathcal{F}, d_Q),$$

where the supremum is taken over all probability measures Q on $(\mathbb{R}^d, \mathcal{B})$, d_Q is the $L_2(Q)$-metric, and, as usual, $N(\epsilon, \mathcal{F}, d_Q)$ is the minimal number of balls $\{g : d_Q(g, g') < \epsilon\}$ of d_Q-radius ϵ needed to cover $\mathcal{F}$. We shall assume that $\mathcal{F}$ satisfies the following uniform polynomial covering number condition:

(F.iii) for some $C_0 > 0$ and $\nu_0 > 0$, $N(\epsilon, \mathcal{F}) \leq C_0 \epsilon^{-\nu_0}$, $\epsilon \in (0, 1)$.

Consider the inner product defined for $g_1, g_2 \in \mathcal{G}$ by

$$(g_1, g_2) := \int_{I^d} g_1(u) g_2(u) du. \tag{2.4}$$

Let $G_2(I^d)$ be the Hilbert subspace of $L_2(I^d)$ spanned by $\mathcal{G}$. Now let $\mathcal{S}$ denote its reproducing kernel Hilbert space generated by the inner product $(\cdot, \cdot)$. Applying Theorem 4D of Parzen [26], the space $\mathcal{S}$ can be represented as follows: Let $l_\infty(\mathcal{G})$ denote the class of bounded functions on $\mathcal{G}$. For any $\xi \in G_2(I^d)$ denote $\varphi_\xi \in l_\infty(\mathcal{G})$ by $\varphi_\xi(g) := (g, \xi)$, $g \in \mathcal{G}$. Each φ_ξ is uniquely defined by ξ in the sense that $\varphi_{\xi_1} = \varphi_{\xi_2}$ if and only if $\xi_1 = \xi_2$, in $L_2(I^d)$. The space $\mathcal{S} = \{\varphi_\xi : \xi \in G_2(I^d)\}$ has the inner product

$$\langle \varphi_{\xi_1}, \varphi_{\xi_2} \rangle := (\xi_1, \xi_2). \tag{2.5}$$

Let $\mathcal{S}_0$ denote the unit ball in $\mathcal{S}$ and for any $\vartheta \in \mathcal{S}_0$ and $\varepsilon > 0$ set

$$B_\varepsilon(\vartheta) = \{\psi \in l_\infty(\mathcal{G}) : ||\psi - \vartheta||_{\mathcal{G}} < \varepsilon\}, \tag{2.6}$$

where for any class of functions $\mathcal{C}$ and $\psi \in l_\infty(\mathcal{C})$ (the class of bounded functions on $\mathcal{C}$)

$$||\psi||_{\mathcal{C}} = \sup_{g \in \mathcal{C}} |\psi(g)|. \tag{2.7}$$

Write for any $\varepsilon > 0$,

$$\mathcal{S}_0^\varepsilon = \left\{ \psi \in l_\infty(\mathcal{G}) : \inf_{\vartheta \in \mathcal{S}_0} ||\psi - \vartheta||_{\mathcal{G}} < \varepsilon \right\}. \tag{2.8}$$

Throughout this section J will denote a compact subset of $\mathbb{R}^d$ with *non-empty interior*. For any $\delta > 0$, we set

$$J_\delta = \left\{ x : \inf_{z \in J} |x - z|_2 \leq \delta \right\}. \tag{2.9}$$

The following UFLL for the local empirical process is the main result in Mason [21].

Theorem 1. *Assume that $\mathcal{G}$ satisfies (G.i–ii), $\mathcal{F}$ fulfills (F.i–iii), $\{h_n\}_{n\geq 1}$ converges to zero at the rate (H.i–iii), and the density f is continuous and positive on J_δ for some $\delta > 0$ Then, with probability 1,*

(I) *for all $\varepsilon > 0$, there exists an $n(\varepsilon) \geq 1$ such that for each $n \geq n(\varepsilon)$, $\{L_n(z,\cdot) : z \in J\} \subset \mathcal{S}_0^{\varepsilon}$;*

(II) *for any $\vartheta \in \mathcal{S}_0$ and $\varepsilon > 0$ there is an $n(\vartheta, \varepsilon) \geq 1$ such that for every $n \geq n(\vartheta, \varepsilon)$ there is a $z_n \in J$ such that $L_n(z_n, \cdot) \in B_\varepsilon(\vartheta)$.*

For numerous examples of the application of Theorem 1 refer to Mason [21]. In particular, consult [21] to see how Theorem 1 implies the Deheuvels and Mason [7] UFLL described in (1.11) and (1.12), along with UFLL versions of the consistency results of Stute [32] and [33] and Giné and Guillou [14] for kernel density estimators on $\mathbb{R}^d$. Of course, (1.4) follows from Theorem 1.

Remark 1. It has long been recognized that the uniform polynomial covering number assumption (F.iii) is the natural condition to impose upon the indexing class, when studying the local behavior of the empirical process. For instance, when Alexander [1] made the first steps towards the investigation of the increments of the empirical process in a general indexed by a class of sets framework, he considered classes of index sets, which satisfy (F.iii). Nolan and Pollard [25] and Nolan and Marron [24] pointed out how the assumption (F.iii) on the class $\mathcal{F}$ arises naturally when investigating the large sample behavior of the kernel density estimator via empirical process indexed by a class of functions theory. Later, Rio [28] found that (F.iii) was the right assumption to impose on $\mathcal{F}$ when he derived his local invariance principle for the Uniform $\left([0,1]^d\right)$ empirical process indexed by a class of functions, and applied it to kernel density estimation; as did Einmahl and Mason [11] and [13] in their treatment of local empirical processes and, more recently, Giné and Guillou [14] in their derivation of rates of strong consistency for multivariate kernel density estimators. Classes of functions satisfying (F.iii) play an important role in Devroye and Lugosi's [9] derivation of bounds in the L_1 error for certain kinds of density estimators. This assumption also plays a critical role in the work of Giné, Koltchinskii and Wellner [15] on ratio limit theorems for empirical processes. For a discussion closely related to the uniform polynomial covering number assumption (F.iii) refer to Section 4.

3. UFLL for a local Gaussian process

With the goal in mind of defining a generalized version of the Csörgő and Révész [5] local Gaussian process (1.8), we introduce a sequence of $\{W_n^F\}_{n\geq 1}$ i.i.d. d–parameter F-Wiener processes on $\mathbb{R}^d$. By F–Wiener process, we mean that for each $n \geq 1$, W_n^F is a Gaussan process on $\mathbb{R}^d$ satisfying $W_n^F(0) = 0$, $EW_n^F(s) = 0$

for all $s \in \mathbb{R}^d$ and for any choice of rectangles A and B in $\mathbb{R}^d$,

$$E\left(W_n^F(A)\, W_n^F(B)\right) = F(A \cap B) := \int_{A\cap B} dF(t),$$

where

$$W_n^F(A) = \int_A dW_n^F(t),\ W_n^F(B) = \int_B dW_n^F(t),$$

with F being a cumulative distribution function on $\mathbb{R}^d$. From now on we will assume that F has a Lebesgue density function f on $\mathbb{R}^d$.

Let $\{h_n\}_{n\geq 1}$ be a sequence of positive constants less than 1 converging to zero at the rate (H.i–ii). Motivated by the version of the Révész [27] UFLL for the local Wiener process defined in (1.13) and the Csörgő and Révész [5] ULL (1.7) for the local Gaussian process (1.8), and in light of Theorem 1, we shall now formulate a UFLL for the local Gaussian process $\zeta_n(z, g)$ defined for each $z \in \mathbb{R}^d$ and indexed by $g \in \mathcal{G}$ to be

$$\zeta_n(z,g) := (nh_n)^{-1/2} \sum_{i=1}^{n} \int_{\mathbb{R}^d} g\left(h_n^{-1/d}(z-t)\right) dW_i^F(t), \tag{3.1}$$

where the multiple Wiener integrals are defined as in Itô [16]. (A somewhat more general version of this local Gaussian process at a fixed point is considered in [11].)

As in Section 2, $\mathcal{G}$ will denote a class of measurable real valued functions defined on $\mathbb{R}^d$, which have support contained in $I^d := [-1/2, 1/2]^d$ and are bounded by some $\kappa > 0$. Let $\mathcal{F}$ be formed from $\mathcal{G}$ as in (F.i) and consider now the sequence of Gaussian processes indexed by $f \in \mathcal{F}$,

$$Y_n(f) = n^{-1/2} \sum_{i=1}^{n} \int_{\mathbb{R}^d} f(t)\, dW_i^F(t). \tag{3.2}$$

Notice that for f_1 and $f_2 \in \mathcal{F}$,

$$\varrho(f_1, f_2) := \sqrt{E\left(Y_n(f_1) - Y_n(f_2)\right)^2} = \sqrt{\int_{\mathbb{R}^d} \left(f_1(t) - f_2(t)\right)^2 dF(t)}. \tag{3.3}$$

For each $\epsilon > 0$, let $N(\epsilon, \varrho, \mathcal{F})$ denote the minimal number of ϱ balls of radius ϵ needed to cover $\mathcal{F}$ and introduce the polynomial covering number assumption on $\mathcal{F}$,

(F$'$.iii) for some $C_0 > 0$ and $\nu_0 > 0$, $N(\epsilon, \varrho, \mathcal{F}) \leq C_0\, \epsilon^{-\nu_0}$, $\epsilon \in (0, 1)$.

Set

$$b_n = \sqrt{2nh_n \log(1/h_n)} \tag{3.4}$$

and consider the normed local Gaussian process defined for $z \in \mathbb{R}^d$ such that $f(z) > 0$ and indexed by $g \in \mathcal{G}$,

$$\mathrm{L}_n(z,g) := \frac{\zeta_n(z,g)}{\sqrt{2f(z)\log(1/h_n)}} = \frac{1}{\sqrt{f(z)}\, b_n} \sum_{i=1}^{n} \int_{\mathbb{R}^d} g\left(h_n^{-1/d}(z-t)\right) dW_i^F(t).$$

The following UFLL for $\mathrm{L}_n(z,g)$ can be proved along the same lines as that of Theorem 1 of Mason [21]. As in Theorem 1, J denotes a compact subset of $\mathbb{R}^d$ with non–empty interior.

Theorem 2. *Assume that $\mathcal{G}$ satisfies (G.i-ii), $\mathcal{F}$ fulfills (F.i–ii) and (F'.iii), $\{h_n\}_{n\geq 1}$ converges to zero at the rate (H.i–ii), and the density f is continuous and positive on J_δ for some $\delta > 0$. Then, with probability 1,*

(I) *for all $\varepsilon > 0$, there exists an $n(\varepsilon) \geq 1$ such that for each $n \geq n(\varepsilon)$,*

$$\{\mathrm{L}_n(z,\cdot) : z \in J\} \subset \mathcal{S}_0^{\varepsilon};$$

(II) *for any $\vartheta \in \mathcal{S}_0$ and $\varepsilon > 0$ there is an $n(\vartheta,\varepsilon) \geq 1$ such that for every $n \geq n(\vartheta,\varepsilon)$ there is a $z_n \in J$ such that*

$$\mathrm{L}_n(z_n,\cdot) \in B_\varepsilon(\vartheta).$$

Remark 2. Note that just as both the Csörgő–Révész strong law [4] and the version of Révész [27] UFLL stated in the Introduction do not require assumption (H.iii), neither does Theorem 2. Also the Révész [27] UFLL for the process in (1.13) can be derived as a special case of Theorem 2.

Remark 3. Consult Einmahl and Mason [11] for a *functional law of the iterated logarithm* for $\zeta_n(z,)$ at a fixed point z.

Sketch of the Proof of Theorem 2: The proof of Theorem 2 closely parallels that of Theorem 1, however with numerous simplifications. Therefore we shall only provide here a sketch of its proof. First, we shall gather together the basic tools for proving Theorem 2. These are the following two Gaussian process facts, taken from Appendix A.2 of van der Vaart and Wellner [35], and a functional large deviation result stated in the proposition below, which is a special case of a result of Arcones [3].

Some basic facts about Gaussian processes

Let X be a separable mean zero Gaussian process indexed by a set T. Define the natural metric ρ on T by

$$\rho(s,t) = \sqrt{E(X_t - X_s)^2}. \tag{3.5}$$

For each $\varepsilon > 0$ let $N(\varepsilon,\rho,T)$ denote the minimal number of ρ balls of radius ε needed to cover T. Let

$$\|X\| = \sup_{t\in T} |X_t|.$$

Gaussian Fact 1 (Borell). For all $t > 0$,

$$P\{|\|X\| - E\|X\|| > t\} \leq 2\exp\left(-\frac{t^2}{2\sigma^2(X)}\right),$$

where

$$\sigma^2(X) = \sup_{t\in T} E\left(X_t^2\right).$$

Gaussian Fact 2. For some universal constant $K > 0$, for all $\varepsilon > 0$,

$$E \sup_{\rho(s,t)<\varepsilon} |X_t - X_s| \le K \int_0^\varepsilon \sqrt{\log N(s, \rho, T)}ds.$$

A large deviation result

Define the rate function $I(\cdot)$ on $l_\infty(\mathcal{G})$ as follows. For any $\psi \in l_\infty(\mathcal{G})$,

$$I(\psi) = \begin{cases} \frac{1}{2}\int_{I^d} \xi^2(u)du \text{ if } \psi = \varphi_\xi \text{ for some } \xi \in G_2(I^d), \\ \infty, \text{ otherwise.} \end{cases} \tag{3.6}$$

Recall the definitions of φ_ξ and $G_2(I^d)$ between (2.4) and (2.5) in Section 2. Also denote for any subset $B \subset l_\infty(\mathcal{G})$,

$$I(B) = \inf \{I(\psi) : \psi \in B\}. \tag{3.7}$$

We endow $l_\infty(\mathcal{G})$ with the topology generated by the norm $||\cdot||_\mathcal{G}$, defined as in (2.7). Applying Theorem 5.2 of Arcones [3], coupled with a special case of a convolution limit result to be found on pp. 62-63 of Stein [30], (using here our assumption that f is positive and continuous on J_δ, for some $\delta > 0$, with J compact) and arguing as in the proof of Proposition 1 of [21], we readily obtain the following result:

Proposition. *Under the assumptions of Theorem 2, for any sequence* $\{m_n\}_{n\ge1}$ *of positive integers and any triangular array of points* $z_{i,n}$, $i = 1, \ldots, m_n$, $n \ge 1$, *in* J, *we have*

(i) for all closed subsets F *of* $l_\infty(\mathcal{G})$,

$$\limsup_{n\to\infty} \max_{1\le i\le m_n} \varepsilon_n \log P\{\mathrm{L}_{i,n}(\cdot) \in F\} \le -I(F);$$

(ii) for all open subsets G *of* $l_\infty(\mathcal{G})$,

$$\liminf_{n\to\infty} \min_{1\le i\le m_n} \varepsilon_n \log P\{\mathrm{L}_{i,n}(\cdot) \in G\} \ge -I(G),$$

where $\mathrm{L}_{i,n}(\cdot) = \mathrm{L}_n(z_{i,n}, \cdot)$, $i = 1, \ldots, m_n$, $n \ge 1$ *and*

$$\varepsilon_n = (2\log(1/h_n))^{-1}. \tag{3.8}$$

Much of the proof of Theorem 1 of Mason [21] is devoted to the establishment of a large deviation result like the above proposition for a Poissonized version of $L_n(z, \cdot)$. This is accomplished through the Talagrand [34] exponential inequality for the empirical process indexed by functions, a tight bound due to Einmahl and Mason [13] for the absolute moment of the supremum of the empirical process indexed by a class of functions satisfying (F.i–iii), a Gaussian distributional approximation of multivariate sums by Zaitsev [36] and [37] and functional large deviation results for stochastic processes by Arcones [2] and [3]. It was at this step in the proof that assumption (H.iii) played a crucial role to insure that the

Poissonized version of $L_n(z,\cdot)$ has the same large deviation behavior as the Gaussian proccss $\mathrm{L}_n(z,\cdot)$. This step is unnecessary in the proof of Theorem 2. Also a Poissonization inequality was needed to derive an independence property essential in the proof of part (II) of Theorem 1. This property is already possessed by $\mathrm{L}_n(z,\cdot)$, so the Poissonization inequality is not required in the proof of Theorem 2.

The proof of part (II) of Theorem 2, is almost a copy of the proof of Lemma 2 in Mason [21], applying the above Proposition, ignoring the Poissonization bound, and using instead the independence over disjoint rectangles property of W^F. We shall include the details in the following lemma.

Lemma. *With probability 1, for any $\vartheta \in \mathcal{S}_0$ and $\varepsilon > 0$ there is an $n(\vartheta,\varepsilon) \geq 1$ such that for all $n \geq n(\vartheta,\varepsilon)$ there is a $z_n \in J$ such that*

$$\mathrm{L}_n(z_n,\cdot) \in B_\varepsilon(\vartheta). \tag{3.9}$$

Proof. Choose any $\vartheta \in \mathcal{S}_0$ with $0 < 0 < \langle\vartheta,\vartheta\rangle = 2I(\vartheta) \leq 1$ and $\varepsilon > 0$ small enough so that

$$0 < 2I(B_\varepsilon(\vartheta)) < 1. \tag{3.10}$$

Select $z_{1,n},\ldots,z_{m_n,n} \in J$ such that the components of $z_{i,n}$ and $z_{j,n}$, $i \neq j$, differ in absolute value by more than $h_n^{1/d}$ and

$$\log m_n / \log(1/h_n) \to 1 \text{ as } n \to \infty. \tag{3.11}$$

The existence of such a sequence $\{m_n\}_{n\geq 1}$ is guaranteed by the assumptions that the density f is continuous and positive on J_δ for some $\delta > 0$ and that J has non–empty interior. Note that these two assumptions imply that for some $0 < \beta_1 \leq \beta_2 < \infty$ and all $z \in J_\delta$,

$$0 < \beta_1 \leq f(z) \leq \beta_2 < \infty \tag{3.12}$$

and $[a_1,b_1] \times \cdots \times [a_d,b_d] \subset J$, for some $-\infty < a_i < b_i < \infty$, $i = 1,\ldots,d$, from which the existence of such a sequence $\{m_n\}_{n\geq 1}$ readily follows.

Now by using the independence property of the Gaussian processes $\mathrm{L}_{i,n}(\cdot) = \mathrm{L}_n(z_{i,n},\cdot)$, $i = 1,\ldots,m_n$, $n \geq 1$, following from the choice of the $z_{i,n}$, $i = 1,\ldots,m_n$, and the assumption that the functions g have support in I^d, we get that

$$P_n := P\{\mathrm{L}_{i,n} \notin B_\varepsilon(\vartheta),\ i = 1,\ldots,m_n\} = \prod_{i=1}^{m_n} P\{\mathrm{L}_{i,n} \notin B_\varepsilon(\vartheta)\}.$$

Applying part (ii) of the above Proposition , we see that this last expression is for any $\rho > 0$ and all n sufficiently large

$$P_n \leq [1 - \exp(-2(1+\rho)\log(1/h_n) I(B_\varepsilon(\vartheta)))]^{m_n}.$$

which, in turn, by (3.10) and an appropriate choice of $0 < \rho < 1$ is for some $0 < \tau < 1$ and for all n sufficiently large

$$\leq 2\,[1 - \exp(-\tau \log(1/h_n))]^{m_n} = 2\,(1 - h_n^\tau)^{m_n} \leq 2\exp(-m_n h_n^\tau).$$

Since we assume (3.11) and (H.ii), we see that for all $\gamma > 1$ and n large $P_n \le \exp(-(\log n)^\gamma)$, from which we readily conclude (3.9) by the Borel–Cantelli lemma. The case $I(\vartheta) = 0$ is readily inferred from the $0 < 2I(\vartheta) \le 1$ case. □

The proof of part (I) of Theorem 2 is also nearly a rewrite, with obvious notational changes, of the proof of part (I) of Theorem 1 in Mason [21]. One first establishes, using the above large deviation Proposition, that

$$P\{\mathrm{L}_{\nu_r}(z_{i,\nu_r},\cdot) \notin \mathcal{S}_0^\varepsilon, \text{ for some } i = 1,\dots,m_{\nu_r}, \text{i.o.}\} = 0 \tag{3.13}$$

on an appropriately chosen grid of $z_{i,\nu_r} \in J$, $i = 1,\dots,m_{\nu_r}$, and subsequence $\{\nu_r\}_{r\ge1}$ of $\{n\}_{n\ge1}$ defined for some $\gamma > 0$ to be

$$\nu_r = [(1+\gamma)^r], \text{ for } r = 1,2,\dots \tag{3.14}$$

Then instead of applying the maximal form of the Talagrand inequality, c.f. [13] or [14], coupled with the moment inequality of Einmahl and Mason [13] to take care of the oscillations between the grid points, one uses the Lévy inequality for sums of i.i.d. symmetric random vectors in a Banach space (refer for instance to Appendix A.1 of van der Vaart and Wellner [35]) in combination with Gaussian Facts 1 and 2. (For a few more details refer to the next section.) With these indications of the necessary changes, it should be clear that the remainder of the proof of Theorem 2 follows along exactly the same lines as that of Theorem 1. □

4. Some heuristics on the necessity of the polynomial covering number assumption

In this section we provide some heuristics to indicate how the proof of Theorem 2 breaks down when the class $\mathcal{F}$ is significantly enlarged. Suppose, for instance, that the polynomial covering assumption (F$'$.iii) were replaced by the entropy condition

$$c_1 l(\epsilon)\,\epsilon^{-2\nu} \le \log\ N(\epsilon,\varrho,\mathcal{F}) \le c_2 l(\epsilon)\,\epsilon^{-2\nu},\ 0<\epsilon<1, \tag{4.1}$$

where $0 < c_1 < c_2 < \infty$, $0 < \nu < 1$ and $l(\epsilon)$ is a positive slowly varying function at 0. (Chapter 8 of Dudley [10] provides examples of classes of functions that satisfy (4.1).)

In addition to the two Gaussian facts stated in Section 3, we shall also require the following Fact 3, which like Facts 1 and 2, is also to be found in Appendix A.2 of van der Vaart and Wellner [35].

Gaussian Fact 3 (Sudakov). For all $\delta > 0$,

$$\delta\sqrt{\log N(\delta,\rho,T)} \le 3E\,\|X\|\,.$$

The major step in the proof of Theorem 2 is the establishment of part (I), namely, to show that for all $\varepsilon > 0$,

$$P\{\mathrm{L}_n(z,\cdot) \notin \mathcal{S}_0^\varepsilon, \text{ for some } z \in J, \text{ i.o.}\} = 0.$$

As pointed out in the sketch of the proof of Theorem 2, to do this one first verifies that (3.13) holds on an appropriately chosen grid of $z_{i,\nu_r} \in J$, $i = 1,\dots,m_{\nu_r}$, and subsequence $\{\nu_r\}_{r\geq 1}$ of $\{n\}_{n\geq 1}$. Afterwards, one must control the oscillations between the grid points. To accomplish this, keeping in mind that we assume f is bounded away from 0 on J, one is forced to prove a limit result like:

$$\lim_{\delta\searrow 0}\limsup_{n\to\infty}\frac{\Delta_n(\delta h_n)}{\sqrt{h_n\log(1/h_n)}}=0,\ \text{a.s.} \tag{4.2}$$

where for any $h>0$,

$$\Delta_n(h)=\sup\left\{|Y_n(f_1)-Y_n(f_2)|:\varrho(f_1,f_2)\leq\sqrt{h}\right\}$$

and Y_n is defined as in (3.2). Moreover, to do this it is necessary to block in n using the Lévy inequality, coupled with (H.i), and then apply a good exponential inequality for $\Delta_n(h)$. In fact, here we shall make use of the Borell inequality, stated as Gaussian Fact 1, to the process

$$X(f_1,f_2)=Y(f_1)-Y(f_2)$$

indexed by (f_1,f_2) satisfying $\varrho(f_1,f_2)\leq\sqrt{\delta h}$. We get

$$\sigma^2(X)=\sup\left\{E(Y(f_1)-Y(f_2))^2:\varrho(f_1,f_2)\leq\sqrt{\delta h}\right\}\leq\delta h$$

and thus for all δ, $\eta>0$ and $h\in(0,1)$,

$$P\left\{|\Delta_n(\delta h)-E\Delta_n(\delta h)|>\eta\sqrt{h\log(1/h)}\right\}\leq 2\exp\left(-\frac{\eta^2\log(1/h)}{2\delta}\right). \tag{4.3}$$

One then obtains after a routine argument based on the Borel-Cantelli lemma, using an appropriate blocking in n and inequality (4.3), combined with assumptions (H.i–ii), that

$$\Delta_n(\delta h_n)=E\Delta_n(\delta h_n)+o\left(\sqrt{h_n\log(1/h_n)}\right),\ \text{a.s.}$$

It is the nature of concentration inequalities like that of Borell to transfer the difficulty of bounding a probability to bounding a moment. At this step we see that we are far from done.

Notice that for any $h>0$, $E\Delta_n(h)=E\Delta(h)$, where

$$\Delta(h)=\sup\left\{|Y(f_1)-Y(f_2)|:f_1,f_2\in\mathcal{F},\ \varrho(f_1,f_2)\leq\sqrt{h}\right\},$$

and $Y(f)=\int_{\mathbb{R}^d}f(t)\,dW^F(t)$, with $W^F=_d W_1^F$. It is clear that to finish the proof of (4.2) we must now verify that

$$\lim_{\delta\searrow 0}\limsup_{n\to\infty}\frac{E\Delta(\delta h_n)}{\sqrt{h_n\log(1/h_n)}}=0. \tag{4.4}$$

Assuming (F′.iii), an application of Gaussian Fact 2, above, readily shows this. (At an analogous step in Mason [21] a moment inequality due to Einmahl and Mason

[13] is applied.) However, if one replaces (F$'$.iii) by (4.1), one gets using Gaussian Fact 2 that for some constants K and K_1 and all $\delta \in (0,1)$,

$$E\Delta(\delta h_n) \le K \int_0^{\sqrt{\delta h_n}} \sqrt{\log N(s,\rho,\mathcal{F})}ds \le K_1 l\left(\sqrt{\delta h_n}\right)(\delta h_n)^{(1-\nu)/2}.$$

This bound obviously does not suffice to prove (4.4), since by noting that $\nu \in (0,1)$, we get for any $\delta > 0$,

$$\lim_{n\to\infty} \frac{l\left(\sqrt{\delta h_n}\right)(\delta h_n)^{(1-\nu)/2}}{\sqrt{h_n \log(1/h_n)}} = \infty,$$

which implies that we are in serious difficulty to prove (4.4) under the entropy assumption (4.1).

In fact, we shall show that (4.4) cannot be true under assumption (4.1). Clearly to establish (4.4) it suffices to prove that for some $\lambda > 0$ and $h_0 \in (0,1]$,

$$\sup_{0<h\le h_0} \frac{E\Delta(h)}{\sqrt{h\log(1/h)}} \le \lambda. \tag{4.5}$$

An application of the following inequality will show that it is impossible for (4.5) to be true when (4.1) holds.

Inequality. *Whenever Y is bounded and continuous on $\mathcal{F}$, $\mathcal{F}$ is totally bounded in the ϱ metric as defined in (3.3), and condition (4.5) holds then for all* $k > 1$ and $h \in (0,h_0]$,

$$\sqrt{\log N\left(\sqrt{h}/k,\varrho,\mathcal{F}\right)} \le \sqrt{\log N\left(\sqrt{h},\varrho,\mathcal{F}\right)} + \lambda 3k\sqrt{\log(1/h)}. \tag{4.6}$$

Proof. Here we borrow ideas from the proof of Corollary 3.19 of Ledoux and Talagrand [19]. Let A be a finite set of functions in $\mathcal{F}$ of cardinality $N\left(\sqrt{h},\varrho,\mathcal{F}\right)$ such that balls $\mathcal{B}_{\sqrt{h}}(f)$ of radius $\sqrt{h}$ centered at functions $f \in A$ cover $\mathcal{F}$. For every $f \in A$ there exists a set $A_f \subset \mathcal{F}$ of cardinality $Card\ A_f = N\left(\sqrt{h}/k,\varrho,\mathcal{B}_{\sqrt{h}}(f)\right)$ such that if $g \in \mathcal{B}_{\sqrt{h}}(f)$ then there is a function $r \in A_f$ with $\varrho(g,r) < \sqrt{h}/k$. By the Sudakov minorization result given in Gaussian Fact 3 we have

$$\begin{aligned} k^{-1}\sqrt{h}\sqrt{\log Card\ A_f} &= k^{-1}\sqrt{h}\sqrt{\log N\left(\sqrt{h}/k,\varrho,\mathcal{B}_{\sqrt{h}}(f)\right)} \\ &\le 3E\left[\sup\left\{|Y(f)-Y(g)| : g \in \mathcal{B}_{\sqrt{h}}(f)\right\}\right], \end{aligned}$$

which by (4.5) is for all $h \in (0,h_0]$,

$$\le \lambda 3\sqrt{h}\sqrt{\log(1/h)}.$$

Define $B = \cup_{f\in A} A_f$. We see that each function in $\mathcal{F}$ is within $\sqrt{h}/k$ of a function in B. Thus

$$N\left(\sqrt{h}/k,\varrho,\mathcal{F}\right) \le Card\ B \le Card\ A \max_{f\in A} Card\ A_f.$$

Therefore

$$\sqrt{\log N\left(\sqrt{h}/k, \varrho, \mathcal{F}\right)} \le \sqrt{\log N\left(\sqrt{h}, \varrho, \mathcal{F}\right)} + \lambda 3k\sqrt{\log(1/h)}.$$

□

Inequality (4.6) and condition (4.1) imply that for all $h \in (0, h_0]$ and $k > 1$,

$$c_1^{1/2} k^{\nu} h^{-v/2}\sqrt{l\left(\sqrt{h}/k\right)} \le c_2^{1/2} h^{-v/2}\sqrt{l\left(\sqrt{h}\right)} + \lambda 3k\sqrt{\log(1/h)},$$

which for $c_1^{1/2}k^{\nu} > c_2^{1/2}$ is clearly impossible. Therefore (4.5) cannot hold under the entropy condition (4.1). This means that if one assumes (4.1), in order to stabilize $\{\zeta_n(z,\cdot) : z \in J\}$ to obtain a UFLL based on an oscillation result like (4.2) one must replace the norming $\sqrt{2\log(1/h_n)}$ in

$$\left\{\frac{\zeta_n(z,\cdot)}{\sqrt{2f(z)\log(1/h_n)}} : z \in J\right\} = \{\mathrm{L}_n(z,\cdot) : z \in J\},$$

appearing in part (I) of Theorem 2, by a norming sequence γ_n satisfying

$$\sqrt{2\log(1/h_n)}/\gamma_n \to 0. \tag{4.7}$$

This would imply that whenever $\{h_n\}_{n\ge 1}$ converges to zero at the rate specified by (H.i-ii), then for each fixed $g \in \mathcal{G}$,

$$\lim_{n\to\infty} \sup_{z\in J} \gamma_n^{-1} |\zeta_n(z,g)| = 0, \text{ a.s. },$$

which provides no exact information about $\sup_{z\in J} |\zeta_n(z,g)|$. Recall we already know by our present Theorem 2 that for any fixed $g \in \mathcal{G}$ such that $0 < ||g||_2 < \infty$,

$$\lim_{n\to\infty} \sup_{z\in J} \frac{|\zeta_n(z,g)|}{\sqrt{2||g||_2^2 \log(1/h_n)}} = 1, \text{ a.s.}$$

Moreover, if we were to fill in completely the details of the proof that we just sketched, we would obtain the stability result

$$\lim_{n\to\infty} \sup_{g\in\mathcal{G}} \sup_{z\in J} \gamma_n^{-1} |\zeta_n(z,g)| = 0, \quad \text{a.s.}$$

The implication of these heuristics is there is strong evidence that, under assumption (4.1), a UFLL of the type described in parts (I) and (II) of our Theorem 2 either simply does not exist or one has to develop an approach that does not rely on a result like (4.2). These same remarks are applicable to the necessity of assumption (F.iii) in the proof of Theorem 1.

Acknowledgements. The author would like to thank Uwe Einmahl, Evarist Giné and Paul Deheuvels for useful comments on various drafts of this paper, as well as the referee, who cleared up a serious ambiguity in the original definition of the F–Wiener process.

References

[1] K.S. Alexander, *Rates of growth and sample moduli for weighted empirical processes.* Probab. Th. Rel. Fields **75** (1987), 379–423.

[2] M. Arcones, *The large deviation principle of stochastic processes. I.* Theor. Probab. Appl. **47** (2002a). In press.

[3] M. Arcones, *The large deviation principle of stochastic processes. II.* Theor. Probab. Appl. **47** (2002b). In press.

[4] M. Csörgő and P. Révész, *How big are the increments of the Wiener process?* Ann. Probab. **7** (1979), 731–737.

[5] M. Csörgő and P. Révész, *Strong Approximations in Probability and Statistics.* Academic Press, New York, 1981.

[6] P. Deheuvels, *Functional laws of the iterated logarithm for large increments of empirical and quantile processes.* Stoch. Proc. Appl .**43** (1992), 133-163

[7] P. Deheuvels and D.M. Mason, *Functional laws of the iterated logarithm for increments of empirical and quantile processes.* Ann. Probab. **20** (1992), 1248–1287.

[8] P. Deheuvels and D.M. Mason, *Functional laws of the iterated logarithm for local empirical processes indexed by sets.* Ann. Probab. **22** (1994), 1619-1661.

[9] L. Devroye and G. Lugosi, *Combinatorial Methods in Density Estimation.* Springer, New York, 2000.

[10] R. M. Dudley, *Uniform Central Limit Theorems,* Cambridge University Press, New York, 1999.

[11] U. Einmahl and D.M. Mason, *Gaussian approximation of local empirical processes indexed by functions.* Probab. Th. Rel. Fields **107** (1997), 283–311.

[12] U. Einmahl and D.M. Mason, *Strong approximations for local empirical processes.* In: *Progress in Probability 43, Proceedings of High Dimensional Probability, Oberwolfach 1996, (*E. Eberlein, M. Hahn and J. Kuelbs, eds.), pp. 75–92, Birkhäuser, Basel, 1998.

[13] U. Einmahl and D.M. Mason, *An empirical process approach to the uniform consistency of kernel–type function estimators.* J. Theoretical Prob. **13** (2000), 1–37.

[14] E. Giné and A. Guillou, *Rates of strong consistency for multivariate kernel density estimators.* Ann. Inst. H. Poincaré. **38** (2002), 907–921.

[15] E. Giné, V. Koltchinskii and J. Wellner, *Ratio limit theorems for empirical processes.* Preprint.

[16] K. Itô, *Multiple Wiener integral.* J. Math. Soc. Japan **3** (1951),157–169.

[17] J. Komlós, P. Major and G. Tusnády, *An approximation of partial sums of independent rv's and the sample df I.* Z. Wahrsch. verw. Gebiete. **32** (1975), 111–131.

[18] J. Komlós, P. Major and G. Tusnády, *An approximation of partial sums of independent rv's and the sample df II.* Z. Wahrsch. verw. Gebiete. **34** (1976), 33–58.

[19] M. Ledoux and M. Talagrand, *Probability in Banach Spaces,* Springer, New York, 1991.

[20] D.M. Mason, *A strong invariance theorem for the tail empirical process.* Ann. Inst. H. Poincaré **24** (1988), 491–506.

[21] D.M. Mason, *A Uniform functional law of the logarithm for the local empirical process.* Preprint.

[22] D.M. Mason, G.R. Shorack and J.A. Wellner, *Strong limit theorems for the oscillation moduli of the uniform empirical process.* Z. Wahrsch. verw. Gebiete **65** (1983), 83–97.

[23] C. Mueller, *A unification of Strassen's law and Levy's modulus of continuity.* Z.Wahrsch. **56** (1981), 163-179.

[24] D. Nolan and J.S. Marron, *Uniform consistency of automatic and location–adaptive delta–sequence estimators. Probab.* Th. Rel. Fields **80** (1989), 619–632.

[25] D. Nolan and D. Pollard, *U–processes: rates of convergence.* Ann. Statist. **15** (1987), 780–799.

[26] E. Parzen, *An approach to time series analysis.* Ann. Math. Statist. **32** (1961), 951–989.

[27] P. Révész, *A generalization of Strassen's functional law of the iterated logarithm.* Wahrsch. verw. Gebiete. **50** (1979), 257–264.

[28] E. Rio, *Local invariance principles and their applications to density estimation.* Probab. Th. Rel. Fields **98** (1994), 21–45.

[29] G.R. Shorack and J.A. Wellner, *Empirical Processes with Applications to Statistics.* Wiley, 1986.

[30] E.M. Stein, *Singular Integrals and Differentiability Properties of Functions.* Princeton University Press, Princeton, New Jersey, 1970.

[31] W. Stute, *The oscillation behavior of empirical processes.* Ann. Probab. **10** (1982a), 86–107.

[32] W. Stute, *The law of the iterated logarithm for kernel density estimators.* Ann. Probab. **10** (1982b), 414–422.

[33] W. Stute, *The oscillation behavior of empirical processes: the multivariate case.* Ann. Probab. **12** (1984), 361–379.

[34] M. Talagrand, *Sharper bounds for Gaussian and empirical processes.* Ann. Probab. (1994), 28–76.

[35] A.W. van der Vaart and J.A. Wellner, *Weak Convergence and Empirical Processes.* Springer, New York, 1996.

[36] A.Yu Zaitsev, *Estimates of the Lévy–Prokhorov distance in the multivariate central limit theorem for random variables with finite exponential moments.* Theory Probab. Appl. **31** (1987a), 203–220.

[37] A.Yu Zaitsev, *On the Gaussian approximation of convolutions under multidimensional analogues of S. N. Bernstein's inequality conditions.* Probab. Th. Rel. Fields **74** (1987b), 534–566.

Department of Food & Resource Economics, University of Delaware, 206 Townsend Hall, Newark, DE 19716, USA

E-mail address: davidm@udel.edu

Progress in Probability, Vol. 55, 153–174

Strong Limit Theorems for Mixing Random Variables with Values in Hilbert Space and their Applications

Olimjon Sh. Sharipov

Abstract. We consider strong limit theorems such as bounded and compact laws of the iterated logarithm and an almost sure invariance principle for mixing random variables with values in Hilbert space. Applying these results we prove an almost sure invariance principle for U-statistics as well.

1. Introduction and main results

Let $\{X_n, n \geq 1\}$ be a sequence of the centered random variables (r.v.s) with values in separable Banach space B (with a norm $\|\cdot\|$ and topological dual B^*).

Set

$$S_n = X_1 + \cdots + X_n, \quad a_n = \sqrt{2nLLn}, \quad n \geq 3, \quad Lx = \max(1, lnx).$$

We say that $\{X_n, n \geq 1\}$ satisfies the bounded law of the iterated logarithm (BLIL), if

$$\limsup_{n\to\infty} \frac{\|S_n\|}{a_n} < \infty \quad \text{a.s.}$$

We say that $\{X_n, n \geq 1\}$ satisfies the compact law of the iterated logarithm (CLIL), if there exists a compact set $K \subset H$ such that

$$C\left(\left\{\frac{S_n}{a_n}\right\}\right) = K \quad \text{a.s.}$$

$$\lim_{n\to\infty} \left\|\frac{S_n}{a_n} - K\right\| = 0 \quad \text{a.s.}$$

where $C(\{b_n\})$ means a cluster set of $\{b_n\}$ and

$$\left\|\frac{S_n}{a_n} - K\right\| = \inf_{x\in K} \left\|\frac{S_n}{a_n} - x\right\|.$$

Received by the editors December 15, 2002.
2000 *Mathematics Subject Classification.* Primary 60B12; Secondary 60F15.
Key words and phrases. Hilbert space, mixing coefficients, law of the iterated logarithm, invariance principle, U–statistics.

The BLIL and CLIL for sequences of r.v.s with values in infinite dimensional spaces were investigated by many authors. We refer to Ledoux, Talagrand (1988), (1991), Chen (1993), (1994) for the results in this direction. Necessary and sufficient conditions for the BLIL and CLIL for independent identically distributed (i.i.d.) r.v.s with values in B were obtained by Ledoux, Talagrand (1988). In particulars if B is a type 2 Banach space (see the definition of type 2 Banach space, for example, in Ledoux, Talagrand (1991)) then $\{X_n, n \geq 1\}$ satisfies the BLIL iff

$$EX_1 = 0, \quad Ef^2(X_1) < \infty, \text{ for all } f \in B^*, \tag{1.1}$$

$$E\left(\frac{\|X_1\|^2}{LL\|X_1\|}\right) < \infty, \tag{1.2}$$

and $\{X_n, n \geq 1\}$ satisfies CLIL iff (1.1), (1.2) and

$$C(f,g) = Ef(X_1)g(X_1), \quad f, g \in B^*, \quad \text{is weak-star sequentially continuous} \tag{1.3}$$

hold. (1.3) is equivalent to

$$\{f^2(X_1), f \in B^*\} \text{ is uniformly integrable.}$$

We call $C(f,g)$ a covariance function.

The BLIL and CLIL for weakly dependent r.v.s with values in Hilbert and Banach spaces were obtained by Kuelbs, Philipp (1980), Dehling, Philipp (1982). Moreover one can obtain BLIL and CLIL as consequences from an almost sure invariance principle (ASIP). We refer to Philipp (1986) for results on the ASIP. But all above mentioned results on the BLIL and CLIL were proved assuming $E\|X_k\|^{2+\delta} < \infty$, $k = 1, 2, \ldots$ for some $\delta > 0$.

Sharipov (1991) proved that a strictly stationary sequence of φ - mixing r.v.s $\{X_n, n \geq 1\}$ for which (1.1) holds satisfies the BLIL iff (1.2) holds. Under the additional condition (1.3) $\{X_n, n \geq 1\}$ satisfies the CLIL iff (1.2) holds

The results obtained by Ledoux, Talagrand (1988) were extended to the case of m - dependent r.v.s by Chen (1997).

One of our goals is to obtain the BLIL and the CLIL for non-stationary sequences of r.v.s with values in a separable Hilbert space H (with the norm $\|\cdot\|$, the inner product $<\cdot,\cdot>$ and topological dual H^*) under minimal moment conditions. We assume that r.v.s satisfy mixing conditions.

Introduce the mixing coefficients

$$\varphi(n) = \sup\{\mid P(A/B) - P(A) \mid : k \in N, B \in F_1^k, A \in F_{n+k}^\infty, P(B) > 0\},$$

where F_a^b denotes the σ–field generated by $X_a, \ldots, X_b$.

In our case the following will play the role of the covariance function $C(f,g)$ from (1.3)

$$T(f,g) = \lim_{n\to\infty} \frac{Ef(S_n)g(S_n)}{n}, \quad f, g \in H^*.$$

We note that throughout this paper we will prefer using the topological dual H^* (with a norm $\|\cdot\|_*$) in formulations and proofs of theorems.

Now we can formulate the laws of the iterated logarithm.

Theorem 1.1. *Let $\{X_n, n \geq 1\}$ be a sequence of identically distributed r.v.s with values in H for which the following conditions hold:*

$$Ef(X_1) = 0, \ \ Ef^2(X_1) < \infty \text{ for all } f \in H^* \tag{1.4}$$

$$\sum_{k=1}^{\infty} \varphi^{\frac{1}{\theta}}(2^k) < \infty \text{ for some } \theta > 3 \tag{1.5}$$

$$\lim_{n\to\infty} \frac{Ef^2(S_n)}{n} = \sigma_f < \infty \text{ for all } f \in H^* \tag{1.6}$$

Then $\{X_n, n \geq 1\}$ satisfies the BLIL iff

$$E\left(\frac{\|X_1\|^2}{LL\|X_1\|}\right) < \infty. \tag{1.7}$$

Theorem 1.2. *Let $\{X_n, n \geq 1\}$ satisfy the conditions of Theorem 1.1 and assume*

$$T(f,g) \text{ is weak-star sequentially continuous.} \tag{1.8}$$

Then $\{X_n, n \geq 1\}$ satisfies the CLIL iff (1.7) holds.

Our next goal is to obtain an ASIP for U - statistics. But first we give an ASIP for strictly stationary sequences of H–valued r.v.s.

Theorem 1.3. *Let $\{X_n, n \geq 1\}$ be a strictly stationary sequence of r.v.s with values in H for which the following conditions hold*

$$EX_1 = 0, \quad E\|X_1\|^2 < \infty$$

$$Ef^2(S_n) \to \infty \text{ under } n \to \infty \text{ for all } f \in H^* \tag{1.9}$$

$$\sum_{k=1}^{\infty} k^{-1/2}\varphi^{1/2}(k) < \infty \tag{1.10}$$

Then without changing its distribution we can redefine $\{X_n, n \geq 1\}$ on a new probability space on which there exists a Brownian motion $W(t)$, $t \geq 0$ (with covariance function $T(f,g)$) such that

$$\|\sum_{i\leq t} X_i - W(t)\| = o((tLLt)^{1/2}) \ a.s.$$

We note that (1.9) and (1.10) imply (1.6). It is well-known that $\varphi(1) < 1$ implies (1.9). Theorem 1.3 was proved by Sharipov (1991) but here we will give the proof of this theorem for completeness.

Now we can consider U-statistics. Let $\xi_1, \xi_2, ...$ be a strictly stationary sequence of centered r.v.s with common distribution μ.

We consider U-statistics with real valued, symmetric and square-integrable (respectively $\mu \times \mu$) kernel $h(x,y)$. Assume that the kernel is degenerate i.e.

$$Eh(\xi_1, x) = 0 \text{ for all } x \in R$$

It is well-known that in this case the following representation takes place (see Hoeffding (1948))

$$h(\xi_i, \xi_j) = \sum_{k=1}^{\infty} \lambda_k g_k(\xi_i) g_k(\xi_j)$$

where $\{\lambda_k,\ k \geq 1\}$ and $\{g_k(x),\ k \geq 1\}$ respectively eigenvalues and eigenfunctions of the operator $T_h : L_2 \to L_2$ defined by

$$T_h f(x) = E(h(\xi_1, x) f(\xi_1)), \qquad f \in L_2.$$

Introduce the U-statistics $S(t)$ by the following equation

$$S(t) = 2 \sum_{1 \leq i \leq j \leq [t]} h(\xi_i, \xi_j) \qquad t \geq 0$$

Denote:

$$U(t) = 2 \sum_{j=1}^{\infty} \lambda_j \sigma_j^2 (W_j^2(t) - \frac{t}{\sigma_j^2}), \qquad t \geq 0$$

$$\sigma_j^2 = \lim_{n \to \infty} \frac{1}{n} E(g_j(\xi_1) + ... + g_j(\xi_n))^2, \quad j = 1, 2, ...,$$

$\{W_j(t)\}_{j \geq 1}$ is a sequence of Wiener processes such that

$$EW_i(t) W_j(s) = \lim_{n \to \infty} \frac{1}{n \sigma_i \sigma_j} \sum_{k=1}^{[nt]} \sum_{l=1}^{[ns]} E g_i(\xi_k) g_j(\xi_l),$$

Now we state our result

Theorem 1.4. *Let $\{\xi_i\}_{i \geq 1}$ additionally satisfies the following conditions*

$$\sup_{j \geq 1} E|g_j(\xi_1)|^4 < \infty,$$

$$\sum_{j=1}^{\infty} |\lambda_j| < \infty, \tag{1.11}$$

$$\sum_{k=1}^{\infty} k^{-1/2} \varphi^{1/2}(k) < \infty, \ \varphi(1) < 1.$$

Then without changing the distributions of $\{S(t), t \geq 0\}$ and $\{U(t), t \geq 0\}$, we can redefine them on a new probability space such that the following takes place

$$|S(t) - U(t)| = o(tLLt) \text{ a.s. under } t \to \infty$$

Theorem 1.4 is partial improvement of the results by Kanagawa, Yoshihara (1994). In Kanagawa, Yoshihara (1994) (1.11) was replaced by

$$\sum_{j=N+1}^{\infty} |\lambda_j| = O(N^{-\varepsilon}) \text{ under } N \to \infty \text{ for some } \varepsilon > 0.$$

We note that Kanagawa, Yoshihara (1994) proved their results using the ASIP by Dehling (1983).

In Section 2 we will prove Theorem 1.1. The prove is based on Bernstein's blocking method and the method of the paper by Acosta, Kuelbs (1983). In Section 3 we will prove Theorem 1.2 using methods by Acosta, Kuelbs (1983) and Theorem 1.1. In Section 4 we will prove Theorems 1.3 and 1.4. The proof of Theorem 1.3 is based on the results by Berger (1990) and we will prove Theorem 1.4 using method by Kanagawa, Yoshihara (1994) and Theorem 1.3.

2. Proof of Theorem 1.1

First we will prove the sufficiency of condition (1.7). In what follows C denotes absolute and non-absolute constants possibly different even in one chain of inequalities. C may depend on various parameters but it does not depend on summation indices or on the number of summands (i.e. $n, p, q, m, r, k, i, j, \ldots$).

Let $\beta > 1$. Introduce the notation

$$n_k = [\beta^k], \quad I(k) = \{n_k + 1, ..., n_{k+1}\}, \quad \tau_k = 2n_{k+1}LLn_{k+1} = a^2_{n_{k+1}},$$

$$\Gamma = \sup_{x \in K} \|x\| = \sup_{\|f\| \leq 1} (T(f,f))^{1/2},$$

$[k]$ – integer part of k. $a_n \sim b_n$ means the existence of absolute constants $c_1 > 0$, $c_2 > 0$ such that $c_1 < \frac{a_n}{b_n} < c_2$ $n = 1, 2, \ldots$.

K is the unit ball of the reproducing kernel Hilbert space H_μ which is defined in the following lemma by Goodman, Kuelbs, Zinn (1981).

Lemma 2.1. *Let μ denote a Borel probability measure on H with covariance function $T(f,g)$. Define the mapping $S : H^* \to H$ by*

$$Sf = \int_H x f(x) d\mu(x)$$

where the integral is Pettis integral. Denote by H_μ the completion of the range of S with respect to the norm obtained from the inner product

$$\langle Sf, Sg \rangle = \int_H f(x) g(x) d\mu(x) = T(f,g).$$

If K is the unit ball of H_μ, then K is a closed, symmetric, convex subset of H and for each $f \in H^*$

$$\sup_{x \in K} f(x) = \left(\int_H f^2(y) d\mu(y) \right)^{1/2}.$$

Further, K is a compact subset of H iff the covariance function $T(f,g)$ is weak-star sequentially continuous.

The existence of

$$T(f,g) = \lim_{n \to \infty} \frac{Ef(S_n)g(S_n)}{n}, \quad f, g \in H^*.$$

follows from (1.6).

Denote

$$u_j = X_j I(\|X_j\|^2 \le \tau_k) - EX_j I(\|X_j\|^2 \le \tau_k),\ j \in I(k),$$
$$w_j = X_j I(\|X_j\|^2 > \tau_k) - EX_j I(\|X_j\|^2 > \tau_k),\ j \in I(k),$$
$$U_n = \sum_{j=1}^{n} u_j,\quad W_n = \sum_{j=1}^{n} w_j.$$

Here $I(A)$ means the indicator function of .

We will prove that for an arbitrary $\varepsilon > 0$

$$\limsup_{n\to\infty} \frac{\|U_n\|}{a_n} \le \Gamma + \varepsilon \text{ a.s.}, \tag{2.1}$$

$$\limsup_{n\to\infty} \frac{\|W_n\|}{a_n} = 0 \text{ a.s.} \tag{2.2}$$

which imply

$$\limsup_{n\to\infty} \frac{\|S_n\|}{a_n} \le \Gamma \text{ a.s.} \tag{2.3}$$

It is enough to prove (2.3) assuming $\Gamma > 0$. If $\Gamma = 0$, one can easily prove using the same argument what was used in Chen (1997), p.712, that

$$\limsup_{n\to\infty} \frac{\|S_n\|}{a_n} = 0 \text{ a.s.}$$

Lemma 2.2. *Under the condition (1.7), the relation (2.2) holds.*

This lemma was proved by Acosta, Kuelbs (1983) for the case of independent r.v.s but without using the independence and therefore their proof remains valid in our case.

Now we will prove (2.1). We will show that there exists an arbitrary close to 1, $\beta > 1$ such that

$$\limsup_{r\to\infty} \max_{n\in I(r)} \left\| \frac{U_n}{a_n} \right\| \le \Gamma + \varepsilon \text{ a.s.} \tag{2.4}$$

Since a_n is increasing we have

$$\max_{n\in I(r)} \left\| \frac{U_n}{a_n} \right\| \le \frac{a_{n_{r+1}}}{a_{n_r}} \max_{n\in I(r)} \left\| \frac{U_n}{a_{n_{r+1}}} \right\| \le \sqrt{\beta} \limsup_{r\to\infty} \max_{n\in I(r)} \left\| \frac{U_n}{a_{n_{r+1}}} \right\| \text{ a.s.}$$

Since we can choose β an arbitrary close to 1 (2.4) follows from the following

$$\limsup_{r\to\infty} \max_{n\in I(r)} \left\| \frac{U_n}{a_{n_{r+1}}} \right\| \le \Gamma + \varepsilon \text{ a.s.} \tag{2.5}$$

Now we define the blocks. For fixed $n \in I(r)$ set

$$\xi_i = \sum_{j=(p+q)(i-1)+1}^{(p+q)(i-1)+p} u_j,\quad \eta_i = \sum_{j=(p+q)i-q+1}^{(p+q)i} u_j.$$

So for $n \in I(r)$ we have

$$U_n = \sum_{i=1}^{m} \xi_i + \sum_{i=1}^{k} \eta_i + U_o,$$

where U_o is the "rest" which appears after blocking $\xi_1, \eta_1, \xi_2, \eta_2, \dots$ till for the next ξ or η we will not have enough u_i. Note

$$m = m(n), \ \ k = k(n), \ \ | m - k | \le 1.$$

Choose

$$p = \left[\frac{\beta^r}{r^\gamma}\right], \ \ 2 < \gamma < \theta - 1, \ \ q = [\beta^{\alpha r}], \ \ 0 < \alpha < 1. \tag{2.6}$$

Then $m \sim k \sim r^\gamma$.

We have

$$\begin{aligned}\limsup_{r\to\infty} \max_{n\in I(r)} \left\| \frac{U_n}{a_{n_{r+1}}} \right\| &\le \limsup_{r\to\infty} \max_{n\in I(r)} \left\| \frac{\sum_{i=1}^{m} \xi_i}{a_{n_{r+1}}} \right\| \\ &+ \limsup_{r\to\infty} \max_{n\in I(r)} \left\| \frac{\sum_{i=1}^{k} \eta_i}{a_{n_{r+1}}} \right\| \\ &+ \limsup_{r\to\infty} \max_{n\in I(r)} \left\| \frac{U_o}{a_{n_{r+1}}} \right\| \ \text{a.s.}\end{aligned} \tag{2.7}$$

Using the following lemma by Utev (1989) we will show that the last two summands in the right hand side of (2.7) a. s. equal to zero.

Lemma 2.3. *Let $\{X_n, n \ge 1\}$ be a sequence of the centered r.v.s with values in H for which (1.5) holds. Then for $t \ge 2$ the following inequality takes place:*

$$E \max_{1\le k\le n} \|S_k\|^t \le C\left(\sum_{k=1}^{n} E\|X_k\|^t + \left(\sum_{k=1}^{n} E\|X_k\|^2\right)^{t/2} \right).$$

Using lemma 2.3 and the following inequality

$$E\left\| X_1 I(\|X_1\|^2 \le \tau_r) - EX_1 I(\|X_1\|^2 \le \tau_r) \right\|^2 \le 4E\|X_1\|^2 I(\|X_1\|^2 \le \tau_r),$$

we obtain

$$\begin{aligned}\sum_{r=1}^{\infty} P\left(\max_{n\in I(r)} \left\| \frac{U_o}{a_{n_{r+1}}} \right\| > \varepsilon \right) &\le \sum_{r=1}^{\infty} \frac{E \max_{n\in I(r)} \|U_o\|^2}{\varepsilon^2 a^2_{n_{r+1}}} \\ &\le \sum_{r=1}^{\infty} \frac{CpE(\|X_1\|^2 I(\|X_1\|^2 \le \tau_r))}{\varepsilon^2 a^2_{n_{r+1}}} \\ &\le \sum_{r=1}^{\infty} \frac{Cp}{\varepsilon^2 \beta^r} E\left(\frac{\|X_1\|^2}{LL\|X_1\|} I(\|X_1\|^2 \le \tau_r) \frac{LL\|X_1\|}{LL\tau_r} \right).\end{aligned}$$

(2.6) implies the convergence of the last series.

By the same way we obtain

$$\sum_{r=1}^{\infty} P\Big(\max_{n\in I(r)} \Big\|\frac{\sum_{i=1}^{k}\eta_i}{a_{n_{r+1}}}\Big\| > \varepsilon\Big) \le \sum_{r=1}^{\infty} \frac{E \max_{n\in I(r)} \|\sum_{i=1}^{k}\eta_i\|^2}{\varepsilon^2 a_{n_{r+1}}^2}$$

$$\le \sum_{r=1}^{\infty} \frac{Ckq}{\varepsilon^2 a_{n_{r+1}}^2} E(\|X_1\|^2 I(\|X_1\|^2 \le \tau_r))$$

$$\le \sum_{r=1}^{\infty} \frac{Ckq}{\varepsilon^2 \beta^r} E\Big(\frac{\|X_1\|^2}{LL\|X_1\|} I(\|X_1\|^2 \le \tau_r) \frac{LL\|X_1\|}{LL\tau_r}\Big).$$

Again (2.6) implies the convergence of the last series. Hence using Borel-Cantelli lemma we conclude that the last two summands in the right hand side of (2.7) a.s. equal to zero.

It remains to estimate the first summand in the right hand side of (2.7). At the first we obtain the following

$$\sup_{n\in I(r)} P\Big(\Big\|\sum_{i=1}^{m(n)} \xi_i - \sum_{i=1}^{m([\beta^{r+1}])} \xi_i\Big\| > \frac{\varepsilon}{2} a_{n_{r+1}}\Big)$$

$$\le \sup_{n\in I(r)} P\Big(\Big\|\sum_{i=m(n)+1}^{m([\beta^{r+1}])} \xi_i\Big\| > \frac{\varepsilon}{2} a_{n_{r+1}}\Big) \le C\frac{1}{a_{n_{r+1}}} \sup_{n\in I(r)} \Big\|\sum_{i=m(n)+1}^{m([\beta^{r+1}])} \xi_i\Big\|$$

$$\le C\frac{\Big(\Big[\frac{\beta^{r+1}}{p}\Big] - \Big[\frac{\beta^r}{p}\Big]\Big)^{1/2}}{a_{n_{r+1}}} \sup_{n\in I(r)} \max_{m(n)\le i\le m([\beta^{r+1}])} (E\|\xi_i\|^2)^{1/2}$$

$$\le C\frac{r^{\gamma/2}p^{1/2}}{a_{n_{r+1}}}\Big(E\|X_1 I(\|X_1\|^2 \le \tau_r) - EX_1 I(\|X_1\|^2 \le \tau_r)\|^2\Big)^{1/2}$$

$$\le C\frac{\beta^{r/2} LL\tau_r}{a_{n_{r+1}}}\Big\{E\Big(\frac{\|X_1\|^2}{LL\|X_1\|} I(\|X_1\|^2 \le \tau_r)\frac{LL\|X_1\|}{LL\tau_r}\Big)\Big\}^{1/2}$$

$$\le C\Big\{E\Big(\frac{\|X_1\|^2}{LL\|X_1\|} I(\|X_1\|^2 \le \tau_r)\frac{LL\|X_1\|}{LL\tau_r}\Big)\Big\}^{1/2} \tag{2.8}$$

Latter tends to zero under $r \to \infty$.

Now we use the following lemma.

Lemma 2.4. *Let $\{X_n, n \ge 1\}$ be φ–mixing sequence of r.v.s with values in H. If for some $a > 0$*

$$\min_{1\le j\le n} P\Big(\|S_n - S_j\| < a\Big) - \varphi(1) > 0$$

then for any x

$$P\Big(\max_{1\le k\le n}\|S_k\|>x\Big)\le\Big(\min_{1\le j\le n}P(\|S_n-S_j\|<a)-\varphi(1)\Big)^{-1}P(\|S_n\|>x-a).$$

This lemma was proved in Iosifescu, Theodorescu (1969) for real valued r.v.s but their proof remains valid also for Hilbert valued r.v.s.

Applying (2.8), lemma 2.4 and choosing r such that $\varphi(q)<1$ we have

$$P\Big(\max_{n\in I(r)}\Big\|\sum_{i=1}^{m(n)}\xi_i\Big\|>(\Gamma+\varepsilon)a_{n_{r+1}}\Big)\le CP\Big(\Big\|\sum_{i=1}^{m([\beta^{r+1}])}\xi_i\Big\|>\Big(\Gamma+\frac{\varepsilon}{2}\Big)a_{n_{r+1}}\Big).$$

Now we have to prove

$$\sum_{r=1}^{\infty}P\Big(\Big\|\sum_{i=1}^{m([\beta^{r+1}])}\xi_i\Big\|>\Big(\Gamma+\frac{\varepsilon}{2}\Big)a_{n_{r+1}}\Big)<\infty.$$

From Berkes–Philipp (1979) it follows that there exists the sequence of independent r.v.s $\{y_i\}$ in H such that y_i and ξ_i have the same distribution and

$$P\Big(\|\xi_i-y_i\|>6\varphi(q)\Big)\le 6\varphi(q),\ \ i=1,2,\dots$$

Since (1.5) and (2.6) imply

$$\sum_{r=1}^{\infty}P\Big(\Big\|\sum_{i=1}^{m([\beta^{r+1}])}(\xi_i-\eta_i)\Big\|>\frac{\varepsilon}{4}a_{n_{r+1}}\Big)\le\sum_{r=1}^{\infty}Cm([\beta^{r+1}])\varphi(q)<\infty,$$

it remains to prove

$$\sum_{r=1}^{\infty}P\Big(\Big\|\sum_{i=1}^{m([\beta^{r+1}])}y_i\Big\|>\Big(\Gamma+\frac{\varepsilon}{4}\Big)a_{n_{r+1}}\Big)<\infty. \tag{2.9}$$

Now we need the following lemma (see Acosta, Kuelbs (1983)).

Lemma 2.5. *Let* $\{Z_i, i\ge1\}$ *be a sequence of independent pregaussian r.v.s with values in* B *where* B *is type 2 Banach space the norm of which satisfies the following conditions*

a) $\sup_{\|x\|=1}\|D_x^2\|<\infty$,

b) D_x^2 *is* $Lip(\alpha)$ *away from zero for some* $\alpha>0$. *Here* D_x^2 *is the second derivative of the norm.*

Then

$$\Big|E\Big(g\Big(\sum_{j=1}^{n}Z_j\Big)\Big)-E\Big(g\Big(\sum_{j=1}^{n}G_j\Big)\Big)\Big|<C(\delta,\Lambda,\alpha)\sum_{j=1}^{n}\|Z_j\|^{2+\alpha},$$

where $\{G_i, i \geq 1\}$ *is the sequence of independent Gaussian r.v.s in* B *such that* G_i *has the same covariance operator as* Z_i,

$$g(x) = \Phi(\|x\|),$$

$$\Phi(t) = \begin{cases} 0, & 0 \leq t \leq \Lambda \\ \textit{increasing}, & \Lambda \leq t \leq \Lambda + \delta, \ \ \Lambda > 0, \quad \delta > 0, \\ 1, & t \geq \Lambda + \delta \end{cases}$$

$C(\delta, \Lambda, \alpha)$ *is the constant which depend on* δ, Λ, α *and doesn't depend on* $\{Z_i, i \geq 1\}$ *and* $\{G_i, i \geq 1\}$.

Since the sequence $\{y_i, i \geq 1\}$ and Hilbert space satisfy the conditions of Lemma 2.5 (with $\alpha = 1$) choosing $\Lambda = \Gamma + \frac{\varepsilon}{8}$, $\delta = \frac{\varepsilon}{8}$, we obtain

$$P\left(\left\| \sum_{i=1}^{m([\beta^{r+1}])} y_i \right\| > \left(\Gamma + \frac{\varepsilon}{4}\right) a_{n_{r+1}} \right) \leq E\left(g\left(\sum_{i=1}^{m([\beta^{r+1}])} y_i / a_{n_{r+1}} \right) \right)$$

$$\leq E\left(g\left(\sum_{i=1}^{m([\beta^{r+1}])} G_i / a_{n_{r+1}} \right) \right) + C \sum_{i=1}^{m([\beta^{r+1}])} \frac{E\|y_i\|^3}{a^3_{n_{r+1}}}. \tag{2.10}$$

Setting (2.10) into (2.9) we estimate the second summand (using lemma 2.3)

$$\sum_{r=1}^{\infty} \sum_{i=1}^{m([\beta^{r+1}])} \frac{E\|y_i\|^3}{a^3_{n_{r+1}}} = \sum_{r=1}^{\infty} \sum_{i=1}^{m([\beta^{r+1})]} \frac{E\|\xi_i\|^3}{a^3_{n_{r+1}}}$$

$$\leq \sum_{r=1}^{\infty} \sum_{i=1}^{m([\beta^{r+1}])} \frac{C}{a^3_{n_{r+1}}}$$

$$\times \left[\sum_{j=(p+q)(i-1)+1}^{(p+q)(i-1)+p} E\|X_j I(\|X_j\|^2 \leq \tau_r) - EX_j I(\|X_j\|^2 \leq \tau_r)\|^3 \right.$$

$$\left. + \left(\sum_{j=(p+q)(i-1)+1}^{(p+q)(i-1)+p} E\|X_j I(\|X_j\|^2 \leq \tau_r) - EX_j I(\|X_j\|^2 \leq \tau_r)\|^2 \right)^{\frac{3}{2}} \right]$$

$$= I + II.$$

Now we estimate each summand separately.

$$II = \sum_{r=1}^{\infty}\left(\sum_{i=1}^{m([\beta^{r+1}])} \frac{C}{a^3_{n_{r+1}}} \times \sum_{j=(p+q)(i-1)+1}^{(p+q)(i-1)+p} E\|X_j I(\|X_j\|^2 \le \tau_r) - EX_j I(\|X_j\|^2 \le \tau_r)\|^2\right)^{\frac{3}{2}}$$

$$\le \sum_{r=1}^{\infty}\sum_{i=1}^{m([\beta^{r+1}])} \frac{C}{a^3_{n_{r+1}}} p^{\frac{3}{2}}\Big(E\Big(\|X_1\|^2 I(\|X_1\|^2 \le \tau_r)\Big)\Big)^{\frac{3}{2}}$$

$$\le \sum_{r=1}^{\infty} \frac{C\beta^{r+1}p^{\frac{3}{2}}}{pa^3_{n_{r+1}}}\Big(E\Big(\|X_1 I(\|X_1\|^2 \le \tau_r)\Big)\Big)^{\frac{3}{2}}$$

$$\le \sum_{r=1}^{\infty} \frac{C\beta^{r+1}p^{\frac{1}{2}}(LL\tau_r)^{\frac{3}{2}}}{\beta^{\frac{3(r+1)}{2}}(LL\beta^{r+1})^{\frac{3}{2}}}\left(E\left(\frac{\|X_1\|^2}{LL\|X_1\|} I(\|X_1\|^2 \le \tau_r)\frac{LL\|X_1\|}{LL\tau_r}\right)\right)^{\frac{3}{2}}$$

$$\le \sum_{r=1}^{\infty} \frac{C}{r^{\frac{\gamma}{2}}}\left(E\left(\frac{\|X_1\|^2}{LL\|X_1\|} I(\|X_1\|^2 \le \tau_r)\frac{LL\|X_1\|}{LL\tau_r}\right)\right)^{\frac{3}{2}}.$$

(2.6) implies the convergence of the last series.

Estimate I. (Denote $\psi(x) = log_\beta x$)

$$I = \sum_{r=1}^{\infty}\sum_{i=1}^{m([\beta^{r+1}])} \frac{C}{a^3_{n_{r+1}}}\left(\sum_{j=(p+q)(i-1)+1}^{(p+q)(i-1)+p} E\|X_1 I(\|X_1\|^2 \le \tau_r) - EX_1 I(\|X_1\|^2 \le \tau_r)\|^3\right)$$

$$\le \sum_{r=1}^{\infty} \frac{C\beta^{r+1}p}{pa^3_{n_{r+1}}} E(\|X_1\|^3 I(\|X_1\|^2 \le \tau_r))$$

$$\le \sum_{r=1}^{\infty}\sum_{k=1}^{a^2_{n_{r+1}}} \frac{k^{\frac{3}{2}}P(k-1 \le \|X_1\|^2 \le k)}{\beta^{\frac{r\alpha}{2}}(LL\beta^r)^{\frac{3}{2}}}$$

$$\le C\sum_{k=1}^{\infty}\sum_{r \ge \psi(\frac{k}{4LLk})-1} \frac{k^{\frac{3}{2}}P(k-1 \le \|X_1\|^2 \le k)}{\beta^{\frac{r\alpha}{2}}(LL\beta^r)^{\frac{3}{2}}}$$

$$\le C\sum_{k=1}^{\infty} \frac{k^{\frac{3}{2}}P(k-1 \le \|X_1\|^2 \le k)}{(\frac{k}{LLk})^{1/2}\left(LL\left(\frac{k}{LLk}\right)\right)^{\frac{3}{2}}}$$

$$\le C\sum_{k=1}^{\infty} \frac{k}{LLk}P(k-1 \le \|X_1\|^2 \le k) \le CE\left(\frac{\|X_1\|^2}{LL\|X_1\|}\right) < \infty$$

In order to prove (2.9) it remains to prove

$$\sum_{k=1}^{\infty} E\left(g\left(\sum_{i=1}^{m([\beta^{r+1}])} G_i/a_{n_{r+1}} \right)\right) < \infty \tag{2.11}$$

We have

$$E\, g\left(\sum_{i=1}^{m([\beta^{r+1}])} G_i/a_{n_{r+1}} \right) \le E\left(I\left(\left\| \sum_{i=1}^{m([\beta^{r+1}])} G_i/a_{n_{r+1}} \right\| > \Gamma + \frac{\varepsilon}{8}\right)\right)$$
$$= P\left(\left\| \sum_{i=1}^{m([\beta^{r+1}])} G_i/a_{n_{r+1}} \right\| > \Gamma + \frac{\varepsilon}{8}\right).$$

We will show that

$$\sum_{k=1}^{\infty} P\left(\left\| \sum_{i=1}^{m([\beta^{r+1}])} G_i/a_{n_{r+1}} \right\| > \Gamma + \frac{\varepsilon}{8}\right) < \infty. \tag{2.12}$$

For Gaussian r.v. G in H the following inequality takes place (for the proof see Acosta, Kuelbs (1983), p.116)

$$P(\|G\| > t) = \exp\left\{ \frac{(1-\delta)}{\delta\Lambda} E\|G\|^2 - \frac{(1-\delta)}{\Lambda} t^2 \right\},$$

where $\delta > 0$, $\Lambda = 2 \sup_{\|f\|_* \le 1} Ef^2(G)$.

Set

$$G = \sum_{i=1}^{m([\beta^{r+1}])} \frac{G_i}{\sqrt{n_{r+1}}}, \quad t = \left(\Gamma + \frac{\varepsilon}{8}\right)\sqrt{2LLn_{r+1}}.$$

Then $\Gamma_r = (\Lambda_r/2)^{1/2}$ and $E\|G\|^2 = o(LLn_{r+1})$ since

$$\lim_{r\to\infty} E\left\| \sum_{i=1}^{m([\beta^{r+1}])} G_i/a_{n_{r+1}} \right\|^2 \le \lim_{r\to\infty} \frac{1}{a^2_{n_{r+1}}} \sum_{i=1}^{m([\beta^{r+1}])} E\|G_i\|^2$$

$$\le \lim_{r\to\infty} \frac{1}{a^2_{n_{r+1}}} \sum_{i=1}^{m([\beta^{r+1}])} E\|\xi_i\|^2 \le \lim_{r\to\infty} \frac{C\beta^{r+1}p}{pa^2_{n_{r+1}}} E(\|X_1\|^2 I(\|X_1\|^2 \le \tau_r))$$
$$\le C \lim_{r\to\infty} \frac{E(\|X_1\|^2 I(\|X_1\|^2 \le \tau_r))}{LLn_{r+1}}$$
$$= C \lim_{r\to\infty} E\left(\frac{\|X_1\|^2}{LL\|X_1\|} I(\|X_1\|^2 \le \tau_r) \frac{LL\|X_1\|}{LLn_{r+1}} \right) = 0.$$

Above we used the following result by Kruglov (1973) concerning the convergence of the second moments in central limit theorem

$$E\|G_i\|^2 = \lim_{n\to\infty} E\|(y_{i1} + \cdots + y_{in})/\sqrt{n}\|^2 = E\|y_{i1}\|^2 = E\|\xi_i\|^2,$$

where $\{y_{ik} : k \ge 1\}$ are independent copies of y_i.

Now we prove that $\Gamma_r \to \Gamma$ under $r \to \infty$. Recall

$$\Gamma_r = \left(\sup_{\|f\|_* \le 1} Ef^2(G)\right)^{1/2} = \sup_{\|f\|_* \le 1} \left(Ef^2\left(\sum_{i=1}^{m([\beta^{r+1}])} G_i/\sqrt{n_{r+1}}\right)\right)^{1/2},$$

$$\Gamma = \sup_{\|f\|_* \le 1} (T(f,f))^{1/2} = \sup_{\|f\|_* \le 1} \left(\lim_{n\to\infty} \frac{Ef^2(S_n)}{n}\right)^{1/2}.$$

First we prove the following relation

$$\left|Ef^2\left(\sum_{i=1}^{m([\beta^{r+1}])} G_i/\sqrt{n_{r+1}}\right) - \frac{Ef^2(S_n)}{n_{r+1}}\right| \to 0 \text{ under } r \to \infty. \tag{2.13}$$

Using independence of $\{G_i, i \ge 1\}$ and the fact that G_i and y_i have the same covariance operator we obtain

$$Ef^2\left(\sum_{i=1}^{m([\beta^{r+1}])} G_i/\sqrt{n_{r+1}}\right) = \frac{1}{n_{r+1}} \sum_{i=1}^{m([\beta^{r+1}])} Ef^2(G_i)$$

$$= \frac{1}{n_{r+1}} \sum_{i=1}^{m([\beta^{r+1}])} Ef^2(y_i) = \frac{1}{n_{r+1}} \sum_{i=1}^{m([\beta^{r+1}])} Ef^2(\xi_i) \tag{2.14}$$

Now we prove that under $r \to \infty$

$$\left|\frac{1}{n_{r+1}} \sum_{i=1}^{m([\beta^{r+1}])} Ef^2(\xi_i) - \frac{1}{n_{r+1}} Ef^2\left(\sum_{i=1}^{m([\beta^{r+1}])} \xi_i\right)\right| \to 0. \tag{2.15}$$

We need the following lemma which is the partial case of the results by Utev (1989)

Lemma 2.6. *For the sequence of centered r.v.s $\{X_n, n \ge 1\}$ with values in H the following inequality takes place*

$$\left|\sum_{i,j=1,\ |i-j|>1}^{n} E\langle X_i, X_j\rangle\right|$$

$$\le C \exp\left(\sum_{j=0}^{[log_2 n]} \varphi^{1/2}(2^{i/3})\right) \sum_{j=0}^{[log_2 n]} \varphi^{1/2}(2^{i/3}) \sum_{j=1}^{n} E\|X_j\|^2$$

here C is an absolute constant.

Using lemma 2.6 for real valued r.v.s $f(\xi_i)$ and taking into account lemma 2.3 and (1.5) we have

$$\left|\frac{1}{n_{r+1}}\left(\sum_{i=1}^{m([\beta^{r+1}])} Ef^2(\xi_i) - Ef^2\left(\sum_{i=1}^{m([\beta^{r+1}])} \xi_i\right)\right)\right|$$
$$= \left|\frac{1}{n_{r+1}}\left(\sum_{i,j=1,\ |i-j|>1}^{n} Ef(\xi_i)f(\xi_j)\right|$$
$$\leq \frac{C}{n_{r+1}} \exp\left(2C\frac{log_2 r}{r^2}\right)\frac{log_2 r}{r^2}\sum_{i=1}^{m(\beta^{r+1})} Ef^2(\xi_i)$$
$$\leq \frac{Cexp\left(2C\frac{log_2 r}{r^2}\right) log_2 r}{\beta^r r^2}\sum_{i=1}^{m([\beta^{r+1}])} Ef^2\left(\sum_{j=(p+q)(i-1)+1}^{(p+q)(i-1)+p} u_j\right)$$
$$\leq \frac{Cexp\left(2C\frac{log_2 r}{r^2}\right) log_2 r}{\beta^r r^2} r^{\gamma} p Ef^2\left(u_{(p+q)(m([\beta^{r+1}])-1)+p}\right)$$
$$\leq \frac{Clog_2 r}{r^2} Ef^2\left(u_{(p+q)(m([\beta^{r+1}])-1)+p}\right) \to 0 \text{ under } r \to \infty.$$

Now we will prove that under $r \to \infty$

$$\frac{1}{n_{r+1}}\left|Ef^2\left(\sum_{i=1}^{m([\beta^{r+1}])} \xi_i\right) - Ef^2\left(\sum_{i=1}^{n_{r+1})} u_i\right)\right| \to 0. \tag{2.16}$$

We have

$$\left|\frac{1}{n_{r+1}}\left(Ef^2\left(\sum_{i=1}^{m([\beta^{r+1}])} \xi_i\right) - Ef^2\left(\sum_{i=1}^{m([\beta^{r+1}])} u_i\right)\right)\right|$$
$$= \left|\frac{1}{n_{r+1}}\left(Ef^2\left(\sum_{i=1}^{m([\beta^{r+1}])} \xi_i\right) - Ef^2\left(\sum_{i=1}^{m([\beta^{r+1}])} \xi_i + \sum_{i=1}^{k}\eta_i + U_o\right)\right)\right|$$
$$= \left|\frac{1}{n_{r+1}}\left(Ef^2\left(\sum_{i=1}^{k}\eta_i\right) + Ef^2(U_o) + 2Ef\left(\sum_{i=1}^{m([\beta^{r+1}])} \xi_i\right) f\left(\sum_{i=1}^{k}\eta_i\right)\right.\right.$$
$$\left.\left. + 2Ef\left(\sum_{i=1}^{m([\beta^{r+1}])} \xi_i\right) f(U_o) + 2Ef\left(\sum_{i=1}^{k}\eta_i\right) f(U_o)\right)\right|.$$

Latter tends to zero because of the following 5 relations:

1)

$$\frac{1}{n_{r+1}} Ef^2\Big(\sum_{i=1}^{k} \eta_i\Big) \le \frac{C}{n_{r+1}} \sum_{i=1}^{n} Ef^2(\eta_i) \le \frac{Ck \cdot q}{n_{r+1}} Ef^2(u_{(p+q)k})$$
$$\le \frac{Cr^\gamma \beta^{\alpha r}}{\beta^{r+1}} Ef^2(u_{(p+q)k}) \to 0 \text{ under } r \to \infty;$$

2)

$$\frac{1}{n_{r+1}} Ef^2(U_o) \le \frac{C \cdot p}{n_{r+1}} Ef^2(u_{n_{r+1}}) \le \frac{C\frac{\beta^r}{r^\gamma}}{\beta^{r+1}} Ef^2(u_{n_{r+1}}) \to 0 \text{ as } r \to \infty;$$

3)

$$\left| \frac{2}{n_{r+1}} Ef\Big(\sum_{i=1}^{m([\beta^{r+1}])} \xi_i \Big) f\Big(\sum_{i=1}^{k} \eta_i \Big) \right|$$
$$\le \frac{2}{n_{r+1}} \Big(Ef^2\Big(\sum_{i=1}^{m([\beta^{r+1}])} \xi_i \Big)\Big)^{1/2} \Big(Ef^2\Big(\sum_{i=1}^{k} \eta_i \Big)\Big)^{1/2}$$
$$\le \frac{C\sqrt{m([\beta^{r+1}])p}\sqrt{kq}}{n_{r+1}} \le \frac{C\sqrt{r^\gamma \frac{\beta^r}{r^\gamma} r^\gamma \beta^{\alpha r}}}{\beta^{r+1}} \to 0 \text{ under } r \to \infty.$$

4)

$$\left| \frac{2}{n_{r+1}} Ef\Big(\sum_{i=1}^{m([\beta^{r+1}])} \xi_i \Big) f(U_o) \right|$$
$$\le \frac{2}{n_{r+1}} \Big(Ef^2\Big(\sum_{i=1}^{m([\beta^{r+1}])} \xi_i \Big)\Big)^{1/2} \Big(Ef^2(U_o) \Big)^{1/2}$$
$$\le \frac{C\sqrt{r^\gamma \frac{\beta^r}{r^\gamma}}\sqrt{\frac{\beta^r}{r^\gamma}}}{\beta^r} \to 0 \text{ under } r \to \infty.$$

5)

$$\left| \frac{2}{n_{r+1}} Ef\Big(\sum_{i=1}^{k} \eta_i \Big) f(U_o) \right| \le \frac{2}{n_{r+1}} \Big(Ef^2\Big(\sum_{i=1}^{k} \eta_i \Big)\Big)^{1/2} \Big(Ef^2(U_o) \Big)^{1/2}$$
$$\le \frac{C\sqrt{r^\gamma \beta^{\alpha r}}\sqrt{\frac{\beta^r}{r^\gamma}}}{\beta^r} \to 0 \text{ under } r \to \infty.$$

It remains to prove that under $r \to \infty$

$$\left| \frac{1}{n_{r+1}} \Big(Ef^2\Big(\sum_{i=1}^{n_{r+1}} u_i \Big) - Ef^2(S_{n_{r+1}}) \Big) \right| \to 0 \tag{2.17}$$

We have

$$\left|\frac{1}{n_{r+1}}\left(Ef^2\left(\sum_{i=1}^{n_{r+1}} u_i\right) - Ef^2(S_{n_{r+1}})\right)\right|$$
$$= \left|\frac{1}{n_{r+1}}\left(Ef^2\left(\sum_{i=1}^{n_{r+1}} u_i\right) - Ef^2\left(U_{n_{r+1}} + W_{n_{r+1}}\right)\right)\right|$$
$$= \frac{1}{n_{r+1}}\left|Ef^2(W_{n_{r+1}}) + Ef(U_{n_{r+1}})f(W_{n_{r+1}})\right|.$$

Lemma 2.3 implies

$$\frac{1}{n_{r+1}} Ef^2(W_{n_{r+1}}) \leq C\frac{\sum_{i=1}^{n_{r+1}} Ef^2(w_i)}{n_{r+1}}. \tag{2.18}$$

Since $Ef^2(w_{n_{r+1}}) \to 0$ under $r \to \infty$ using Teplitz lemma we conclude that under $r \to \infty$

$$\frac{\sum_{i=1}^{n_{r+1}} Ef^2(w_i)}{n_{r+1}} \to 0. \tag{2.19}$$

We have

$$\frac{1}{n_{r+1}}\left|Ef(U_{n_{r+1}})f(W_{n_{r+1}})\right| \frac{1}{n_{r+1}}\left(Ef^2(U_{n_{r+1}})\right)^{1/2} \left(Ef^2(W_{n_{r+1}})\right)^{1/2}$$

$$\leq \frac{C\sqrt{n_{r+1}}\left(Ef^2(W_{n_{r+1}})\right)^{1/2}}{n_{r+1}} \leq C\left(\frac{Ef^2(W_{n_{r+1}})}{n_{r+1}}\right)^{1/2} \to 0 \text{ as } r \to \infty.$$

The last relation follows from (2.18), (2.19).

So (2.14)–(2.17) imply (2.13) which in turn implies $\Gamma_r \to \Gamma$ under $r \to \infty$. Taking into account this fact and choosing $\delta > 0$ in such a way that for sufficiently large r

$$\frac{(1-\delta)\left(\Gamma + \frac{\varepsilon}{8}\right)^2}{2\Gamma_r^2} > \frac{1}{2}$$

holds, we conclude that there exists $\eta > 0$ such that for sufficiently large r

$$P\left(\left\|\sum_{j=1}^{m([\beta^{r+1}])} G_j/\sqrt{n_{r+1}}\right\| > \left(\Gamma + \frac{\varepsilon}{8}\right)\right) \leq \exp\{-(1+\eta)LLn_{r+1}\},$$

from which (2.12) follows. So we have proved (2.9) and the sufficiency of the condition (1.7) is proved.

In order to prove the necessity of the condition (1.7) we note that $\limsup_{n\to\infty} \frac{\|S_n\|}{a_n} < \infty$ a.s. implies the existence of the constant $A > 0$ such that

$$P\Big(\ \|X_n/a_n\| > A \text{ infinitely often }\Big) = 0 \tag{2.20}$$

We will use the following lemma which is an immediate consequence of the results by Cohn (1965).

Lemma 2.7. *For the sequence $\{X_n, n \geq 1\}$ of φ–mixing r.v.s with values in H (2.20) implies*

$$\sum_{n=1}^{\infty} P\Big(\ \|X_n/a_n\| > A\Big) < \infty.$$

Using lemma 2.7 we have

$$\sum_{n=1}^{\infty} P\Big(\ \|X_n/a_n\| > A\Big) = \sum_{n=1}^{\infty} P\Big(\ \|X_1\| > Aa_n\Big) < \infty$$

The convergence of the last series is equivalent to (1.7).

Theorem 1.1 is proved.

3. Proof of Theorem 1.2

Necessity follows from Theorem 1.1.

Sufficiency. We must prove the existence of the compact set $K \subset H$ such that

$$C\Big(\Big\{\frac{S_n}{a_n}\Big\}\Big) = K \text{ a.s.} \tag{3.1}$$

$$\lim_{n\to\infty} \Big\|\frac{S_n}{a_n} - K\Big\| = 0 \text{ a.s.} \tag{3.2}$$

We take as K the unit ball of the Hilbert space H_μ from lemma 1.1. (1.8) implies the compactness of K. First we prove (3.2). We denote by T the covariance operator corresponding to covariance function $T(f, g)$.

We will use the next lemma by Acosta, Kuelbs (1983).

Lemma 3.1. *T is a bounded, symmetric, non-negative defined operator and*

$$T^{1/2}(V) = K$$

where $V = \{x \in H : \|x\| = (\langle x, x\rangle)^{1/2} \leq 1\}$.

Let $\varepsilon > 0$ and I is identity map in H ($I(x) = x$). Then there exists operator $(T^{1/2} + \varepsilon I)^{-1}$ which is bounded. Set

$$q(x) = \Big\|(T^{1/2} + \varepsilon I)^{-1}(x)\Big\| = \Big(\langle (T^{1/2} + \varepsilon I)^{-1}(x), (T^{1/2} + \varepsilon I)^{-1}(x)\rangle\Big)^{1/2}. \tag{3.3}$$

Consider Hilbert space H with the norm $q(\cdot)$. In the space (H,q) Theorem 1.1 takes place and from (2.3) we have

$$\limsup_{n\to\infty} q(S_n/a_n) \le \sup_{x\in K} q(x) \quad \text{a.s.} \tag{3.4}$$

From lemma 3.1 we obtain

$$\sup_{x\in K} q(x) = \sup_{x\in K} \left\|(T^{1/2}+\varepsilon I)^{-1}(x)\right\| = \sup_{y\in V} \left\|(T^{1/2}+\varepsilon I)^{-1}T^{1/2}(y)\right\|$$
$$= |||(T^{1/2}+\varepsilon I)^{-1}T^{1/2}||| \tag{3.5}$$

where $|||\cdot|||$ is the norm of operator.

Denote $U = T^{1/2}+\varepsilon I$. The following was proved in Acosta, Kuelbs (1983), p. 119,

$$\langle x,x\rangle \ge \langle U^{-1}T^{1/2}(x), U^{-1}T^{1/2}(x)\rangle \tag{3.6}$$

Relations (3.4)–(3.6) imply

$$\limsup_{n\to\infty} q\left(\frac{S_n}{a_n}\right) \le 1 \text{ a.s.} \tag{3.7}$$

Since $q(x) \le 1$ iff $(T^{1/2}+\varepsilon I)^{-1}(X) = y$ for some $y \in V$ then lemma 3.1 implies

$$\{x:\ g(x) \le 1\} \le K + \varepsilon V. \tag{3.8}$$

Using (3.7) and (3.8) we have

$$\limsup_{n\to\infty} \left\|\frac{S_n}{a_n} - K\right\| \le \varepsilon \text{ a.s.}$$

Since ε is an arbitrary we have (3.2). It remains to prove (3.1). We will use the following lemmas.

Lemma 3.2. *Let H_μ be a Hilbert space which is obtained using lemma 1.1 and $T(f,g)$ satisfies (1.8). Moreover, assume that the sequence $\{X_n, n\ge 1\}$ of r.v.s with values in H satisfies the following condition*

$$\limsup_{n\to\infty} \frac{f(S_n)}{a_n} = \sup_{x\in K} f(x) \ a.s. \text{ for all } f \in H^*$$

where K is an unit ball of H_μ.
Then
1) if $\left\{\frac{S_n}{a_n}\right\}$ is a.s. relatively compact then

$$C\left(\left\{\frac{S_n}{a_n}\right\}\right) = K \ a.s.$$

2) $\left\{\frac{S_n}{a_n}\right\}$ is a.s. relatively compact iff

$$\lim_{n\to\infty} \left\|\frac{S_n}{a_n} - K\right\| = 0 \ a.s.$$

Lemma 3.3. *Let $\{\xi_i, i > 1\}$ be a sequence of identically distributed r.v.s which satisfies the following conditions*

$$E\xi_i = 0, \quad E\xi_i^2 < \infty, \; \lim_{n\to\infty} \frac{1}{n} E(\xi_1 + \cdots + \xi_n)^2 = \sigma < \infty, \; \sum_{k=1}^{\infty} \varphi^{1/2}(2^k) < \infty.$$

Then

$$\limsup_{n\to\infty} \frac{\xi_1 + \cdots + \xi_n}{a_n} = \sigma.$$

Lemma 3.2 is a partial case of theorem 3.1 by Kuelbs (1976). But it was proved under condition $\int_H \|x\|^2 d\mu(x) < \infty$. This hypothesis was assumed in order to have compactness of K. But compactness of K follows from (1.8). Hence the proof of theorem 3.1 by Kuelbs (1976) remains valid under condition (1.8). Moreover in Kuelbs (1976) part 1) of the lemma 3.2 was proved assuming that H_μ is infinite dimensional. But one can prove the lemma 3.2 without this condition by the same way as it was done in Chen (1997) (proposition 1) for m-dependent case (we can use lemma 3.3 instead the relation (2.2) in Chen (1997).

Lemma 3.3 is a corollary of the results by Utev (1989). But it was additionally assumed that $\sigma > 0$. One can easily prove lemma 3.3 in the case $\sigma = 0$ using the same argument that was used in Chen (1997) p. 712.

Lemma 3.3 implies

$$\limsup_{n\to\infty} \frac{f(S_n)}{a_n} = \left(\int_H f^2(y) d\mu(y) \right)^{\frac{1}{2}} \quad \text{a.s.}$$

Using lemma 1.1 we have

$$\sup_{x\in K} f(x) = \left(\int_H f^2(y) d\mu(y) \right)^{\frac{1}{2}}.$$

So the conditions of lemma 3.2 hold and taking into account (3.2) we obtain (3.1). Theorem 1.2 is proved. □

4. Proofs of Theorems 1.3–1.4

Proof of Theorem 1.3. We will use the following partial cases of theorem 3.2 by Berger (1990) and a central limit theorem by Zhurbenko, Zuparov (1986).

Lemma 4.1. *(Berger (1990)). Let $\{X_n, n \geq 1\}$ be a strictly stationary sequence of r.v.s with values in H for which the following conditions hold*

$$Ef(X_1) = 0, \quad Ef^2(X_1) < \infty \text{ for all } f \in H^*$$

$$\sum_{k=1}^{\infty} k^{-1/2} \varphi^{1/2}(k) < \infty$$

Then there exists

$$T(f,g) = \lim_{n\to\infty} \frac{Ef(S_n)g(S_n)}{n} \text{ for all} f, g \in H^*$$

and the following two statements are equivalent:

1) there exists Gaussian distribution μ with mean zero and covariance function $T(f,g)$ and $\{\frac{S_n}{a_n}\}$ is a.s. conditionally compact in H.

2) without changing its distribution we can redefine $\{X_n, n \geq 1\}$ on a new probability space on which there exists a Brownian motion $W(t)$, $t \geq 0$ (with covariance function $T(f,g)$) such that

$$\|\sum_{i\leq[t]} X_i - W(t)\| = o((tLLt)^{1/2}) \text{ a.s.}$$

Lemma 4.2. *(Zhurbenko, Zuparov (1986)) Let $\{X_n, n \geq 1\}$ satisfies the conditions of Theorem 1.3. Then under $n \to \infty$*

$$\frac{1}{\sqrt{n}} S_n \to \mu(0,T) \text{ in distribution}$$

where $\mu(0,T)$ is Gaussian r.v. with mean zero and covariance function $T(f,g)$.

Now it remains to use lemmas 4.1–4.2 and Theorem 1.2. Theorem 1.3 is proved. □

Proof of Theorem 1.4. We will use the method developed by Kanagawa, Yoshihara (1994). Introduce a separable Hilbert space H (with the inner product $<\cdot,\cdot>$ and the norm $\|\cdot\|$) as follows:

$$H = \{x = (x_1, x_2, ...) : \sum_{k=1}^{\infty} |\lambda_k| x_k^2 < \infty\}$$

$$< x, y > = \sum_{k=1}^{\infty} |\lambda_k| x_k y_k, \quad \|x\| = \left(\sum_{k=1}^{\infty} |\lambda_k| x_k^2\right)^{1/2}$$

As in Kanagawa, Yoshihara (1994) we can define $G_i = (g_1(\xi_i), g_2(\xi_i), ...)$ for each $i \geq 1$ and under conditions of Theorem 1.4

$$E\|G_i\|^4 < \infty$$

Kanagawa, Yoshihara (1994) p. 353-354 proved that

$$|S(t) - U(t)| \leq J_1 + J_2 \tag{4.1}$$

and

$$J_1 \leq \left(\left\|\sum_{i=1}^{[t]} G_i - W(t)\right\|\right)\left(\left\|\sum_{i=1}^{[t]} G_i - W(t)\right\| + 2\|W(t)\|\right)$$

$$J_2 = \sum_{k=1}^{\infty} |\lambda_k| \left|\sum_{k=1}^{[t]} g_k^2(\xi_i) - t\right|$$

where $W(t)$ is H - valued Brownian motion
$W(t) = (\sigma_1 W_1(t), \sigma_2 W_2(t), ...)$.

Since all conditions of Theorem 1.3 hold we have

$$\left\| \sum_{i=1}^{[t]} G_i - W(t) \right\| = o((tLLt)^{1/2}) \text{ a.s.}$$

which together with the law of the iterated logarithm for the Brownian motion implies

$$J_1 = o(tLLt) \text{ a.s.} \tag{4.2}$$

As in Kanagawa, Yoshihara (1994) we have

$$J_2 \leq \left(\sum_{k=1}^{\infty} |\lambda_k| \right)^{1/2} \left\| \sum_{k=1}^{[t]} X_i \right\|$$

where $X_i = (g_1^2(\xi_i) - 1, g_2^2(\xi_i) - 1, ...)$, $i \geq 1$, are H - valued r.v.s. Using Theorem 1.1 (instead the Gaal - Koksma strong law of large numbers which was used in Kanagawa, Yoshihara (1994)) we obtain

$$J_2 = O((tLLt)^{1/2}) \text{ a.s.} \tag{4.3}$$

Theorem 1.4 follows now from (4.1)–(4.3). □

References

[1] Acosta A.de, Kuelbs J.(1983) *Some results on the cluster sets $C(\{S_n/a_n\})$ and the LIL.* Ann. Probab. **11**, (1), 102-105

[2] Berkes I., Philipp W.(1979)*Approximation theorems for independent and weakly dependent random vectors.* Ann. Probab.**7**,(1), 29-54

[3] Berger E. (1990) *An almost sure invariance principles for stationary ergodic sequences of Banach space valued random variables.* Probab. Theor. and rel. fields **84**, 161-201.

[4] Chen X. (1993) *On the law of the iterated logarithm for independent Banach space valued random variables.* Ann. Probab. **21**, (4), 1991-2011

[5] Chen X. (1994) *On Strassen's law of the iterated logarithm in Banach space.* Ann. Probab. **V.22**, (2), 1026-1043

[6] Chen X. (1997) *The law of the iterated logarithm for m–dependent Banach space valued random variables.* J. Theor. Probab. **10**, (3), 695–732.

[7] Cohn H. (1965) *On a class of dependent random variables.* Rev. Roum. Math. Pures et Appl. **10**, 1593-1606

[8] Dehling H., Phillipp W. (1982) *Almost sure invariance principles for weakly dependent vector-valued random variables.* Ann. Probab. **10**, (3), 689-701

[9] Dehling H. (1983) *Limit theorems for sums of weakly dependent Banach space valued random variables.* Z. Wahrsch. verw. Gebiete **63**, 393-432

[10] Goodman V., Kuelbs J., Zinn J. (1981)*Some results on the LIL in Banach space with apllications to weighted empirical processes.* Ann. Probab. **9**, (5),713-752.

[11] Hoeffding W. (1948) *A non - parametric test of independence.* Ann. Math. Statist. **19**, 546-557

[12] Iosifescu M., Theodorescu R. (1969)*Stochastic Processes and Learning.* Springer.

[13] Kanagawa S., Yoshihara K. (1994) *The almost sure invariance principles of degenerate U - statistics of degree two for stationary random variables.* Stoch. Processes and their Appl. **49**, 347-356

[14] Kruglov V.M. (1973)*Convergence of numerical characteristic of the sum of independent random variables with values in Hibert space.* Probab. Theor. Appl. (bf 18), (4), 734-752 (in Russian)

[15] Kuelbs J. (1976)*A strong convergence theorem for Banach space valued random variables.* Ann. Probab. **4**, (5), 744-771

[16] Kuelbs J., Philipp W.(1980) *Almost sure invariance principles for partial sums of mixing B-valued random variables.* Ann. Probab. **8**, (6), 1003-1036

[17] Ledoux M., Talagrand M.(1988)*Characterization of the law of the iterated logarithm in Banach spaces.* Ann. Probab. **16**, (3), 1241-1264

[18] Ledoux M., Talagrand M.(1991)*Probability in Banach spaces.* Springer, Berlin.

[19] Philipp W. (1986)*Invariance principles for independent and weakly dependent random variables.* In: Dependence in Probability and Statistics (Proc. Conf. Oberwolfach. 1985). Boston–Basel–Stutgart: Birkhauser. 225–268.

[20] Sharipov O.Sh. (1991)*Laws of the iterated logarithm and an almost sure invariance principle for weakly dependent random variables with values in Hilbert space.* Dep. VINITI (USSR). Dep. No.3493-B91.(in Russian)

[21] Utev S. (1991) *Sums of random variables with φ-mixing.* Siberian Adv. Math. 1, 124-155.

[22] Zhurbenko I.G., Zuparov T.M. (1986)*A central limit theorem for mixing stationary processes in Hilbert space.* Dokl. Acad. Sciences USSR, **286**, (2), 272-276 (in Russian)

Acknowledgment

The author thanks professor M.Marcus and the referee for carefully reading the paper and for their help in preparing the revised form of the paper.

Department of Probability Theory, Institute of Mathematics, Uzbek Academy of Sciences, 29, F. Khodjaev str., Tashkent, 700143, Uzbekistan
E-mail address: `root@im.tashkent.su`

Part 4

Local Times

Progress in Probability, Vol. 55, 177–192

Local Time-Space Calculus and Extensions of Itô's Formula

R. Ghomrasni and G. Peskir

Abstract. Let $X = (X_t)_{t\geq 0}$ be a continuous semimartingale and let $F : I\!R_+ \times I\!R \to I\!R$ be a C^1 function. Then the change-of-variable formula is valid:

$$F(t, X_t) = F(0, X_0) + \int_0^t F_t(s, X_s)\, ds + \int_0^t F_x(s, X_s)\, dX_s - \frac{1}{2}\int_0^t \int_{I\!R} F_x(s, x)\, d\ell_s^x$$

where ℓ_s^x is the local time of X at x defined by:

$$\ell_s^x = \mathbb{P}\text{-}\lim_{\varepsilon \downarrow 0} \frac{1}{\varepsilon} \int_0^s I(x \leq X_r < x + \varepsilon)\, d\langle X, X\rangle_r$$

and $d\ell_s^x$ refers to an area integration with respect to $(s, x) \mapsto \ell_s^x$. Further extensions of this formula for non-smooth functions F are also briefly examined. The approach leads to a formal $d\ell_t^x$ calculus which appears useful in guessing a candidate formula for $F(t, X_t)$ before a rigorous proof is known or given.

1. Introduction

The fundamental result of stochastic calculus is Itô's formula (2.1) firstly established by Itô [5] for a standard Brownian motion and then later extended to continuous semimartingales by Kunita and Watanabe [7]. [For simplicity in this article we will not consider semimartingales with jumps.] The function F appearing in Itô's formula is C^2 in the space variable, and the correction to the classic Leibnitz-Newton formula (the final term in (2.1) below) is expressed by means of the quadratic variation (2.2).

Various extensions of the Itô formula have been established for functions F which are not C^2 in the space variable. The best know of these extensions is the Itô-Tanaka formula (2.3) firstly derived by Tanaka [13] for $F(x) = |x|$ and then extended to absolutely continuous F with F' of bounded variation by Meyer [9] and Wang [14]. The correction term appearing in this formula is expressed by means of the local time (2.4) which goes back to Lévy [8] (see e.g. [6] or [12]).

2000 *Mathematics Subject Classification.* Primary 60H05, 60J55; Secondary 60G44, 60J65.
Key words and phrases. Itô's formula, Tanaka's formula, local time, Brownian motion, continuous semimartingale, stochastic integral, local time-space integral, local time-space calculus.

A different extension to absolutely continuous F with locally bounded F' due to Bouleau and Yor [1] is given in (2.5) below. The correction term appearing in this formula is also expressed by means of the local time (2.4), however, in a different manner which suggests a formal integration by parts. Both formulas (2.3) and (2.5) are derived only in dimension one.

Motivated by applications in free-boundary problems of optimal stopping [10] we have recently derived an extension of Itô's formula [11] stated in (2.11) below. The most interesting in this formula, in comparison with the extensions above, is its final term where the possible jump of $F_x(s,\cdot)$ along the curve $s \mapsto b(s)$ is integrated with respect to the *time variable* s of the local time ℓ_s^b from (2.12) below.

In our attempts to understand a more general rule unifying the various correction terms reviewed above, we have noticed that Eisenbaum [3] made a fundamental contribution in the case of standard Brownian motion by deriving the extension of Itô's formula stated in (2.8) below. The correction term in this formula is expressed as an area integral with respect to both the time variable s and the space variable x of the local time ℓ_s^x from (2.4) below. The arguments of Eisenbaum rely on combining the Bouleau-Yor extension (2.5) with the Föllmer-Protter-Shiryaev extension (2.6) and thus strongly depend on the time-reversal property of standard Brownian motion.

The main aim of the present article is to make use of the formula (2.11) below and show that the representation of the correction term in the Eisenbaum's extension (2.11) as an area integral with respect to the local time-space (as we call it in short) is not only restricted to a time-invariant Brownian motion process but extends quite generally to all continuous semimartingales. This is firstly done for C^1 functions F in Section 3, and then extended to absolutely continuous functions F with F_t and F_x of bounded variation in Section 4.

We make no attempt in the present article to specify the most general class of functions H for which the double integral $\int_0^t \int_{I\!R} H(s,x)\, d\ell_s^x$ makes sense. Instead in Section 5 we will show how a number of known extensions of the Itô formula can be obtained by formal manipulations of the $d\ell_s^x$ integral. This formalism (or formal $d\ell_t^x$ calculus as we call it) appears to be useful when one needs to guess a candidate formula for $F(t, X_t)$ before a rigorous proof is known or given. A typical example of this guessing mechanism is given in the end of Section 5. It remains a challenging task, however, to carry out this programme on firm mathematical grounds to a more satisfactory completion.

2. Itô formula and extensions

In this section we will review various extensions of the Itô formula for the purpose of comparison and for further reference.

1. **Itô formula** ([5], [7]). Let $X = (X_t)_{t>0}$ be a continuous semimartingale (see e.g. [12]) and let $F : I\!R_+ \times I\!R \to I\!R$ be a $C^{1,2}$ function. Then we have:

$$F(t, X_t) = F(0, X_0) + \int_0^t F_t(s, X_s)\, ds + \int_0^t F_x(s, X_s)\, dX_s \tag{2.1}$$
$$+ \frac{1}{2}\int_0^t F_{xx}(s, X_s)\, d\langle X, X\rangle_s$$

where $\langle X, X\rangle_s$ is the quadratic variation of X given by:

$$\langle X, X\rangle_s = I\!P - \lim_{n\to\infty} \sum_{s_i \in D_s^n} (X_{s_i} - X_{s_{i-1}})^2 \tag{2.2}$$

and the set D_s^n consists of arbitrary points $0 = s_0 < s_1 < \ldots < s_{n-1} < s_n = s$ satisfying $\max_{1\le i\le n}(s_i - s_{i-1}) \to 0$ as $n \to \infty$.

2. **Itô-Tanaka formula** ([13], [9], [14]). Let $X = (X_t)_{t\ge0}$ be a continuous semimartingale and let $F : I\!R \to I\!R$ be an AC function with F' of BV . [Throughout AC stands for absolutely continuous, and BV for bounded variation. It is equivalent to $F = F_1 - F_2$ where F_1 and F_2 are convex functions.] Then we have:

$$F(X_t) = F(X_0) + \int_0^t F'_-(X_s)\, dX_s + \frac{1}{2}\int_{I\!R} \ell_t^x\, dF'(x) \tag{2.3}$$

where ℓ_t^x is the local time of X at the point x defined by:

$$\ell_t^x = I\!P\ -\ \lim_{\varepsilon\downarrow 0} \frac{1}{\varepsilon}\int_0^t I(x \le X_s < x+\varepsilon)\, d\langle X, X\rangle_s. \tag{2.4}$$

The formula (2.3) remains valid if the left-derivative F'_- is replaced by the right-derivative F'_+ provided that $I(x \le X_s < x + \varepsilon)$ in (2.4) is replaced by $I(x - \varepsilon < X_s \le x)$.

3. **Bouleau-Yor extension** [1]. Let $X = (X_t)_{t\ge0}$ be a continuous semimartingale and let $F : I\!R \to I\!R$ be an AC function with F' locally bounded. Then:

$$F(X_t) = F(X_0) + \int_0^t F'(X_s)\, dX_s - \frac{1}{2}\int_{I\!R} F'(x)\, d_x\ell_t^x \tag{2.5}$$

where $d_x\ell_t^x$ refers to an integration with respect to $x \mapsto \ell_t^x$. The latter function is known to be of unbounded variation generally, and so is the final term in (2.5) as a function of t .

4. **Föllmer-Protter-Shiryaev extension** [4]. Let $B = (B_t)_{t\ge0}$ be a standard Brownian motion and let $F : I\!R_+ \times I\!R \to I\!R$ be a measurable function satisfying the following conditions: (i) $t \mapsto F(t,x)$ is AC ; (ii) $x \mapsto F(t,x)$ is AC; (iii) $x \mapsto F_t(t,x)$ belongs to L^1_{loc} ; (iv) $x \mapsto F_x(t,x)$ belongs to L^2_{loc} ; (v) $t \mapsto F_t(t,\cdot)$ is continuous from $I\!R_+$ to L^1_{loc} ; (vi) $t \mapsto F_t(t,\cdot)$ is continuous from $I\!R_+$ to L^2_{loc} .

Then we have:

$$F(t,B_t) = F(0,B_0) + \int_0^t F_t(s,B_s)\,ds + \int_0^t F_x(s,B_s)\,dB_s + \frac{1}{2}[F_x(\,\cdot\,,B),B]_t \quad (2.6)$$

where $[F_x(\,\cdot\,,B),B]_t$ is the quadratic covariation given by:

$$[F_x(\,\cdot\,,B),B]_t = \mathbf{P}-\lim_{n\to\infty} \sum_{t_i\in D_t^n} \Big(F_x(t_i,B_{t_i})-F_x(t_{i-1},B_{t_{i-1}})\Big)\big(B_{t_i}-B_{t_{i-1}}\big) \quad (2.7)$$

and the set D_t^n consists of arbitrary points $0 = t_0 < t_1 < \ldots < t_{n-1} < t_n = t$ satisfying $\max_{1\le i\le n}(t_i - t_{i-1}) \to 0$ as $n \to \infty$.

5. **Eisenbaum's extension** [3]. Let $B = (B_t)_{t\ge 0}$ be a standard Brownian motion and let $F : I\!R_+{\times}I\!R \to I\!R$ be a measurable function satisfying the following conditions: (i) $t \mapsto F(t,x)$ is AC ; (ii) $x \mapsto F(t,x)$ is AC; (iii) $\int_0^t \int_{I\!R} |F_t(s,x)|\,(1/\sqrt{s})\,ds\,dx < \infty$; (iv) $\int_0^t \int_{I\!R} |F_x(s,x)|^2\,(1/\sqrt{s})\,\varphi(x/\sqrt{s})\,ds\,dx < \infty$; (v) $\int_0^t \int_{I\!R} |F_x(s,x)|\,(|x|/\sqrt{s})\,\varphi(x/\sqrt{s})\,ds\,dx < \infty$. [We recall that $\varphi(x) = (1/\sqrt{2\pi})\exp(-x^2/2)$ for $x \in I\!R$.] Then we have:

$$\begin{aligned} F(t,B_t) = F(0,B_0) &+ \int_0^t F_t(s,B_s)\,ds + \int_0^t F_x(s,B_s)\,dB_s \\ &- \frac{1}{2}\int_0^t \int_{I\!R} F_x(s,x)\,d\ell_s^x \end{aligned} \quad (2.8)$$

where ℓ_s^x is the local time of B at the point x given by (2.4) above, and $d\ell_s^x$ refers to an area integration with respect to $(s,x) \mapsto \ell_s^x$.

6. **Cherny's extension** [2]. Let $B = (B_t)_{t\ge 0}$ be a standard Brownian motion and let $F : I\!R \to I\!R$ be an AC function with F' being AC on $I\!R \setminus \{0\}$. Then:

$$\begin{aligned} F(B_t) = F(B_0) &+ \int_0^t F'(B_s)\,dB_s + \frac{1}{2}\lim_{\varepsilon\downarrow 0}\,(F'(\varepsilon){-}F'(-\varepsilon))\,\ell_t^0 \\ &+ \frac{1}{2}\Big(v.p. \int_0^t F''(B_s)\,ds\Big) \end{aligned} \quad (2.9)$$

whenever the principal value integral:

$$v.p. \int_0^t F''(B_s)\,ds = \lim_{\varepsilon\downarrow 0} \int_0^t F''(B_s)\,I(|B_s|{>}\varepsilon)\,ds \quad (2.10)$$

exists as a limit in probability.

7. **Extension with local time on curves** [11]. Let $X = (X_t)_{t\ge 0}$ be a continuous semimartingale and let $b : I\!R_+ \to I\!R$ be a continuous function of bounded variation. Setting $C = \{\,(s,x) \in I\!R_+ \times I\!R \mid x < b(s)\,\}$ and $D = \{\,(s,x) \in I\!R_+ \times I\!R \mid x > b(s)\,\}$ suppose that a continuous function $F : I\!R_+ \times I\!R \to I\!R$ is given such that F is $C^{1,2}$ on $\overline{C}$ and F is $C^{1,2}$ on $\overline{D}$. Then we have:

$$F(t,X_t) = F(0,X_0) + \int_0^t F_t(s,X_s-)\,ds + \int_0^t F_x(s,X_s-)\,dX_s \tag{2.11}$$
$$+ \frac{1}{2}\int_0^t F_{xx}(s,X_s)\,I(X_s \neq b(s))\,d\langle X,X\rangle_s$$
$$+ \frac{1}{2}\int_0^t \Big(F_x(s,X_s+) - F_x(s,X_s-)\Big)\,I(X_s = b(s))\,d\ell_s^b$$

where ℓ_s^b is the local time of X at the curve b defined by:

$$\ell_s^b = \mathbb{P}\text{-}\lim_{\varepsilon\downarrow 0} \frac{1}{\varepsilon}\int_0^s I(b(r) \le X_r < b(r)+\varepsilon)\,d\langle X,X\rangle_r \tag{2.12}$$

and $d\ell_s^b$ refers to the integration with respect to the continuous increasing function $s \mapsto \ell_s^b$. The formula (2.11) remains valid if the left-limits $F_t(s,X_s-)$ and $F_x(s,X_s-)$ are replaced by the right-limits $F_t(s,X_s+)$ and $F_x(s,X_s+)$ provided that $I(b(r) \le X_r < b(r)+\varepsilon)$ in (2.12) is replaced by $I(b(r)-\varepsilon \le X_r < b(r))$.

8. **Formal $d\ell_t^x$ extension** [Sections 3–5]. Let $X = (X_t)_{t\ge 0}$ be a continuous semimartingale and let $F : \mathbb{R}_+ \times \mathbb{R} \to \mathbb{R}$ be a measurable function satisfying certain regularity conditions. Then we have:

$$F(t,X_t) = F(0,X_0) + \int_0^t D_t F(s,X_s)\,ds + \int_0^t D_x F(s,X_s)\,dX_s \tag{2.13}$$
$$- \frac{1}{2}\int_0^t\int_{\mathbb{R}} D_x F(s,x)\,d\ell_s^x$$

where ℓ_s^x is the local time of X at the point x given by (2.4) above, and $d\ell_s^x$ refers to an area integration with respect to $(s,x) \mapsto \ell_s^x$. The operators D_t and D_x refer to a differentiation with respect to t and x, respectively.

9. **The occupation times formula** (cf. [12] or [6]). Let $X = (X_t)_{t\ge 0}$ be a continuous semimartingale. Comparing the final terms in (2.1) and (2.3) when F is C^2 and using the monotone class theorem (see e.g. [12] page 3) it follows:

$$\int_0^t G(X_s)\,d\langle X,X\rangle_s = \int_{\mathbb{R}} G(x)\,\ell_t^x\,dx \tag{2.14}$$

for every bounded measurable function $G : \mathbb{R} \to \mathbb{R}$. Similarly, inspecting first the case of $G = 1_{(t_1,t_2]\times(x_1,x_2]}$ and then using the monotone class theorem, one gets the following extension of (2.14) to the time-dependent case:

$$\int_0^t G(s,X_s)\,d\langle X,X\rangle_s = \int_{\mathbb{R}}\left(\int_0^t G(s,x)\,d_s\ell_s^x\right)dx \tag{2.15}$$

for every bounded measurable function $G : \mathbb{R}_+ \times \mathbb{R} \to \mathbb{R}$.

3. Local time-space formula

This section consists of two parts. In the first part we will define a local time-space integral of simple functions. In the second part we will extend it to more general functions.

1. Let $X = (X_t)_{t \geq 0}$ be a continuous semimaringale, and let ℓ_t^x be the local time of X at the point x given by (2.4) above.

With $t > 0$ given and fixed, let us introduce the following families of functions:

$$\mathcal{P}_1 = \{1_{(t_1,t_2]}\, 1_{(x_1,x_2]} \mid 0 \leq t_1 < t_2 \leq t, -\infty < x_1 < x_2 < \infty\} \tag{3.1}$$

$$\mathcal{P}_2 = \{C\, 1_{(x_1,x_2]} \mid C \in C^1([0,t]), -\infty < x_1 < x_2 < \infty\} \tag{3.2}$$

$$\mathcal{P}_3 = \{C\, d \mid C \in C^1([0,t]), d \in C(I\!R)\} \tag{3.3}$$

$$\mathcal{H}_1 = \{H \mid H = F_x \text{ for some } F \in C^1([0,t] \times I\!R)\} \tag{3.4}$$

$$\mathcal{H}_2 = \{H \mid H = F_x \text{ for some } F \in C^2([0,t] \times I\!R)\} \tag{3.5}$$

$$\mathcal{H}_{1,2} = \{H \mid H = F_x \text{ for some } F \in C^{1,2}([0,t] \times I\!R)\} \tag{3.6}$$

$$\mathcal{M} = \{H \mid H : [0,t] \times I\!R \to I\!R \text{ is measurable}\}. \tag{3.7}$$

For $\mathcal{C} \subset \mathcal{M}$ let $sp(\mathcal{C})$ denote the smallest subspace of $\mathcal{M}$ containing $\mathcal{C}$. For example, we have:

$$sp(\mathcal{P}_1) = \{\textstyle\sum_{i=1}^n \alpha_i\, 1_{(t_{i-1},t_i]}\, 1_{(x_{i-1},x_i]} \mid \alpha_1, \ldots, \alpha_n \in I\!R,\, n \geq 1\} \tag{3.8}$$

where there is no restriction to assume that $0 \leq t_1 < t_2 < \ldots < t_n$ and $x_1 < x_2 < \ldots < x_n$.

Let L^0 denote the space of all random variables (defined on $(\Omega, \mathcal{F}, I\!P)$ where X is defined) equipped with the metric of convergence in $I\!P$-probability.

Define an operator $\Lambda : \mathcal{P}_1 \to L^0$ by setting:

$$\Lambda(1_{(t_1,t_2]}\, 1_{(x_1,x_2]}) = \ell_{t_2}^{x_2} - \ell_{t_1}^{x_2} + \ell_{t_1}^{x_1} - \ell_{t_2}^{x_1} \tag{3.9}$$

and extend it by linearity to $sp(\mathcal{P}_1)$ as follows:

$$\Lambda(\textstyle\sum_{i=1}^n \alpha_i\, 1_{(t_{i-1},t_i]}\, 1_{(x_{i-1},x_i]}) = \sum_{i=1}^n \alpha_i\, (\ell_{t_2}^{x_2} - \ell_{t_1}^{x_2} + \ell_{t_1}^{x_1} - \ell_{t_2}^{x_1}) \tag{3.10}$$

both in reminiscence to the classic area integral in plane.

Let Λ also denote the linear extension of Λ from $sp(\mathcal{P}_1)$ to $\mathcal{D}(\Lambda)$ satisfying the following continuity condition:

$$H_n \xrightarrow{\ell} H_0 \Rightarrow \Lambda(H_n) \to \Lambda(H_0) \text{ in } I\!P\text{-probability} \tag{3.11}$$

where $H_n \xrightarrow{\ell} H_0$ if and only if the following two conditions hold:

$$H_n \to H_0 \text{ with } |H_n| \leq G \text{ for all } n \geq 1 \text{ where } G \text{ is locally bounded} \tag{3.12}$$

$$\partial_s \textstyle\int H_n\, dy \xrightarrow{\sim} \partial_s \int H_0\, dy \tag{3.13}$$

the meaning of (3.13) being that:

$$\textstyle\int_0^t (\partial_s \int_0^x H_n(s,y)\, dy)\big|_{x=h(s)} \to \int_0^t (\partial_s \int_0^x H_0(s,y)\, dy)\big|_{x=h(s)} \tag{3.14}$$

for every continuous function $h : [0,t] \to I\!R$ as $n \to \infty$, where $\partial_s \int_0^x H_n(s,y)\,dy$ denotes the signed measure associated with the BV mapping $s \mapsto \int_0^x H_n(s,y)\,dy$ for $n \geq 0$.

The question then arises to determine if there exists such an extension of Λ and if it is unique. More precisely, it is of interest to determine (characterize) the maximal subspace of $\mathcal{M}$ to which Λ can be extended uniquely. This subspace is then denoted by $\mathcal{D}(\Lambda)$ and is called the (maximal) domain of Λ .

We will not deal with the latter question in full generality. Instead we will display subspaces of $\mathcal{M}$ to which Λ can be extended uniquely and its action determined explicitly. In this section we will show that $\mathcal{H}_1$ from (3.4) above is one such subspace (with a further subspace $\mathcal{H}_{1,2}$ of particular interest). In the next section we will make attempts to go beyond the subspace $\mathcal{H}_1$ at some moderate depth.

2. It is naturally guessed using (3.9) that $\mathcal{P}_2 \subset \mathcal{D}(\Lambda)$ and that we have:

$$\Lambda(C\,1_{(x_1,x_2]}) = \int_0^t C(s)\,(d_s\ell_s^{x_2} - d_s\ell_s^{x_1}) \tag{3.15}$$

for $C\,1_{(x_1,x_2]}$ from $\mathcal{P}_2$. A formal verification can be carried out using that $sp(\mathcal{P}_1)$ is dense in $\mathcal{P}_2$ relative to the ℓ-convergence in (3.11) (recalling also that $s \mapsto \ell_s^{x_i}$ is increasing and continuous). We take (3.11) to be the initial point in the proof of the following theorem.

Theorem 3.1. *There exists a unique linear extension of Λ from $sp(\mathcal{P}_1)$ to $sp(\mathcal{P}_1 \cup \mathcal{P}_2 \cup \mathcal{P}_3 \cup \mathcal{H}_1)$ satisfying (3.10) and (3.11). In view of (3.9) this extension will be denoted by either of the area integrals:*

$$\Lambda(H) = \int_0^t \int_{I\!R} H(s,x)\,d\ell_s^x = \int_{I\!R} \int_0^t H(s,x)\,d\ell_s^x \tag{3.16}$$

where the order of integration is interchanged formally.

Proof. The key argument in the proof will be provided using the following extension of the formula (2.11) to the case when instead of one function b we are given finitely many functions $b_1, \ldots, b_n$ which do not intersect (cf. Remarks 3.2 and 3.3 in [11]).

More precisely, let us assume that the following conditions are satisfied:

$$b_i : I\!R_+ \to I\!R \text{ is continuous and of bounded variation for } 1 \leq i \leq n \tag{3.17}$$

$$F_i : I\!R_+ \times I\!R \to I\!R \text{ is } C^{1,2} \text{ for } 1 \leq i \leq n+1 \tag{3.18}$$

$$\begin{aligned} F(t,x) &= F_1(t,x) \text{ if } x < b_1(t) \\ &= F_i(t,x) \text{ if } b_{i-1}(t) < x < b_i(t) \text{ for } 2 \leq i \leq n \\ &= F_{n+1}(t,x) \text{ if } x > b_n(t) \end{aligned} \tag{3.19}$$

where $F : I\!R_+ \times I\!R \to I\!R$ is continuous. Then (2.11) extends as follows:

$$F(t,X_t) = F(0,X_0) + \int_0^t F_t(s,X_s-)\,ds + \int_0^t F_x(s,X_s-)\,dX_s \tag{3.20}$$
$$+ \frac{1}{2}\int_0^t F_{xx}(s,X_s)\,I\big(X_s\notin\{b_1(s),\dots,b_n(s)\}\big)\,d\langle X,X\rangle_s$$
$$+ \frac{1}{2}\sum_{i=1}^n \int_0^t \Big(F_x(s,X_s+)-F_x(s,X_s-)\Big)\,I(X_s=b_i(s))\,d\ell_s^{b_i}(X)$$

where $\ell_s^{b_i}$ is the local time of X at the curve b_i given by (2.12) above, and $d\ell_s^{b_i}(X)$ refers to the integration with respect to $s\mapsto \ell_s^{b_i}(X)$ for $i=1,\dots,n$. [The formula (3.20) remains valid if the left-limits $F_t(s,X_s-)$ and $F_x(s,X_s-)$ are replaced by the right-limits $F_t(s,X_s+)$ and $F_x(s,X_s+)$ provided that $I(b_i(r)\le X_r<b_i(r)+\varepsilon)$ in the definition (2.12) is replaced by $I(b_i(r)-\varepsilon\le X_r<b_i(r))$ for $i=1,\dots,n$.]

1. Take $C\,1_{(x_1,x_2]}$ from $\mathcal{P}_2$, set $d(x)=1_{(x_1,x_2]}(x)$, and let $D(x)=\int_0^x d(y)\,dy$. Setting $b_1(s)\equiv x_1$, $b_2(s)\equiv x_2$, $F_1(s,x)\equiv 0$, $F_2(s,x)=C(s)(x-x_1)$ and $F_3(s,x)=C(s)(x_2-x_1)$ we see that (3.17)-(3.19) are satisfied so that (3.20) gives:

$$C(t)D(X_t) = C(0)D(X_0) + \int_0^t C'(s)D(X_s)\,ds + \int_0^t C(s)d(X_s)\,dX_s \tag{3.21}$$
$$+\frac{1}{2}\bigg(\int_0^t C(s)\,d_s\ell_s^{x_1} - \int_0^t C(s)\,d_s\ell_s^{x_2}\bigg)$$

using that $F_x(s,b_1(s)+)-F_x(s,b_1(s)+) = C(s)\,(d(x_1+)-d(x_1-)) = C(s)$ and $F_x(s,b_2(s)+)-F_x(s,b_2(s)+) = C(s)\,(d(x_2+)-d(x_2-)) = -C(s)$.

In view of (3.15) above, the final term in (3.21) may be recognized as a negative action of the operator Λ on $C\,1_{(x_1,x_2]}$, or in other words:

$$C(t)D(X_t) = C(0)D(X_0) + \int_0^t C'(s)D(X_s)\,ds \tag{3.22}$$
$$+\int_0^t C(s)\,d(X_s)\,dX_s - \frac{1}{2}\Lambda(C\,d)$$

yielding a stochastic-integral representation for the action of Λ on $\mathcal{P}_2$.

2. Given $c=1_{(t_1,t_2]}$ with $0\le t_1<t_2\le t$ introduce the convolution approximation by defining a sequence of functions C_n with $n\ge 1$ as follows:

$$C_n(s) = \int_{I\!R_+} c(s-r/n)\,\Omega(r)\,dr \tag{3.23}$$

where $\Omega : I\!R\to I\!R$ is C^∞ with $\mathrm{supp}(\Omega)=[0,1]$ and $\int_{I\!R}\Omega(r)\,dr=1$. Then $C_n : [0,t]\to I\!R$ is C^∞ and $C_n\to c$ as $n\to\infty$ with $|C_n|\le 1$ for all $n\ge 1$. Moreover $C_n' \stackrel{\sim}{\to} \delta_{t_1}-\delta_{t_2}$ i.e. $\int_0^t h(s)\,dC_n(s)\to h(t_1)-h(t_2)$ for every continuous function $h:[0,t]\to I\!R$ as $n\to\infty$.

Inserting C_n in place of C in (3.22) and letting $n\to\infty$, using the convergence relations just exhibited and the stochastic dominated convergence theorem (see e.g. [12] page 142), as well as (3.11) above, we get:

$$c(t)D(X_t) = c(0)D(X_0) + (D(X_{t_1}) - D(X_{t_2})) \tag{3.24}$$
$$+ \int_0^t c(s)d(X_s)\, dX_s - \frac{1}{2}\Lambda(c\,d)$$

yielding a stochastic-integral representation for the action of Λ on $\mathcal{P}_1$.

3. Take $C\,d$ from $\mathcal{P}_3$ and choose a sequence of simple functions:

$$d_n = \sum_{i=1}^{k_n} \beta_i^n \, 1_{(x_{i-1}^n, x_i^n]} \tag{3.25}$$

so that $d_n \to d$ as $n \to \infty$ with $|d_n| \le g$ for all $n \ge 1$ where g is locally bounded. Setting $D_n(x) = \int_0^x d_n(y)\, dy$ we also have $D_n \to D$ as $n \to \infty$ with $|D_n| \le \Gamma$ for all $n \ge 1$ where Γ is locally bounded.

Inserting D_n in place of D in (3.22) above, and likewise d_n in place of d , letting $n \to \infty$ and using the dominated convergence theorem (both deterministic and stochastic) and localization, as well as (3.11) above, it follows that (3.22) extends to all $C \in C^1(I\!R_+)$ and $D \in C^1(I\!R)$. In particular, it shows that (3.22) holds for all polynomials in two variables s and x .

4. Take H from $\mathcal{H}_2$ and let F be from $C^2([0,t] \times I\!R)$ such that $H = F_x$. For this F there exists a sequence of polynomials $P^n(s,x) = \sum_{i=1}^{k_n} C^{i,n}(s) D^{i,n}(x)$ such that $P^n \to F$, $P_t^n \to F_t$ and $P_x^n \to F_x$ uniformly on a compact set in $I\!R_+ \times I\!R$ as $n \to \infty$ [for this use that F_{tx} is continuous and find a sequence of polynomials $\widetilde{P}^n$ such that $\widetilde{P}^n \to F_{tx}$ on a given compact set by means of the Weierstrass theorem; the sequence $P^n(s,x) = \int_0^s \int_0^x \widetilde{P}^n\, dr dy$ then has the desired properties]. Extending then (3.22) by linearity from $C(s)D(x)$ to $P^n(s,x)$, letting $n \to \infty$ in the resulting formula, and using the dominated convergence theorem (both deterministic and stochastic) and localization, as well as (3.11) above, we obtain:

$$\int_0^{X_t} H(t,y)\, dy = \int_0^{X_0} H(0,y)\, dy + \int_0^t \Big(\frac{\partial}{\partial s} \int_0^x H(s,y)\, dy \Big) \Big|_{x=X_s} ds \tag{3.26}$$
$$+ \int_0^t H(s,X_s)\, dX_s - \frac{1}{2}\Lambda(H)$$

yielding a stochastic-integral representation for the action of Λ on $\mathcal{H}_2$.

5. Take H from $\mathcal{H}_1$ and let F be from $C^1([0,t] \times I\!R)$ such that $H = F_x$. Associate with this F the convolution approximation:

$$F^n(s,x) = \int_{I\!R_+} \int_{I\!R} F(s - r/n, x - y/n)\, \Omega(r)\, \Omega(y)\, dr\, dy \tag{3.27}$$

where Ω is the same as in (3.23) above. Then F^n is C^∞ and we have $F^n \to F$, $F_t^n \to F_t$, $F_x^n \to F_x$ uniformly on a compact set in $I\!R_+ \times I\!R$ as $n \to \infty$.

Setting $H^n = F_x^n$ and inserting this H^n in place of H in (3.26) above, letting $n \to \infty$ and using the dominated convergence theorem (both deterministic and stochastic) and localization, as well as (3.11) above, it follows that (3.26) extends to all $H \in \mathcal{H}_1$.

6. Combining this fact with (3.22) and (3.24) we see that the stochastic-integral representation (3.26) (with $(\partial/\partial s)ds$ replaced by ∂_s when needed) for the action of Λ on H holds for all $H \in sp(\mathcal{P}_1 \cup \mathcal{P}_2 \cup \mathcal{P}_3 \cup \mathcal{H}_1)$ from where by (3.12)-(3.14) using the dominated convergence theorem (both deterministic and stochastic) it follows that (3.11) holds on the entire $sp(\mathcal{P}_1 \cup \mathcal{P}_2 \cup \mathcal{P}_3 \cup \mathcal{H}_1)$. This shows that the definition of Λ is not dependent on the particular choice of the approximating sequence used above, and the proof of the theorem is complete. □

Corollary 3.2 (Local time-space formula). *Let $X = (X_t)_{t \geq 0}$ be a continuous semimartingale and let $F : \mathbb{R}_+ \times \mathbb{R} \to \mathbb{R}$ be a C^1 function. Then the following change-of-variable formula holds:*

$$F(t, X_t) = F(0, X_0) + \int_0^t F_t(s, X_s)\, ds + \int_0^t F_x(s, X_s)\, dX_s - \frac{1}{2} \int_0^t \int_{\mathbb{R}} F_x(s, x)\, d\ell_s^x \tag{3.28}$$

where ℓ_s^x is the local time of X at the point x given by (2.4) above, and $d\ell_s^x$ refers to the area integration with respect to $(s, x) \mapsto \ell_s^x$ established in Theorem 3.1.

Proof. It follows by (3.16) and (3.26) above. □

Corollary 3.3 (Integration by parts). *If $H \in \mathcal{H}_{1,2}$ then we have:*

$$\int_{\mathbb{R}} \int_0^t H(s, x)\, d\ell_s^x = -\int_{\mathbb{R}} \left(\int_0^t H_x(s, x)\, d_s \ell_s^x \right) dx \tag{3.29}$$

and if $H \in \mathcal{H}_2$ then we have:

$$\int_0^t \int_{\mathbb{R}} H(s, x)\, d\ell_s^x = \int_{\mathbb{R}} H(t, x)\, d_x \ell_t^x - \int_0^t \left(\int_{\mathbb{R}} H_t(s, x)\, d_x \ell_s^x \right) ds \tag{3.30}$$

where $d\ell_s^x$ refers to the area integration with respect to $(s, x) \mapsto \ell_s^x$ established in Theorem 3.1.

Proof. The first identity follows by comparing (3.28) with (2.1) and using (2.15). [It also follows by a formal partial integration upon setting $u(x) = H(s, x)$ and $dv(x) = d_x(\int_0^t d_s \ell_s^x)$ so that $du(x) = H_x(s, x)\, dx$ and $v(x) = \int_0^t d_s \ell_s^x$ with $\ell_s^{\pm\infty} \equiv 0$.] The second identity follows by a formal partial integration upon setting $u(s) = H(s, x)$ and $dv(s) = d_s(\int_{\mathbb{R}} d_x \ell_s^x)$ so that $du(s) = H_t(s, x)\, ds$ and $v(s) = \int_{\mathbb{R}} d_x \ell_s^x$ with $\int_{\mathbb{R}} d_x \ell_0^x \equiv 0$. [It can be justified using (2.5) above but we will omit the details.] □

In contrast to the semimartingale decomposition of $F(t, X_t)$ in Itô's formula (2.1) when F is $C^{1,2}$, it is important to realize that the final term in (3.28) is not necessarily of bounded variation in t when F is C^1 only. [In dimension one the same remark holds for the formulas (2.3) and (2.5), respectively.]

A comparison of (3.28) with (2.6)+(2.7) suggests that Λ admits a representation as the difference of a forward and backward stochastic integral at least when X is time-reversal invariant such as a Brownian motion process (cf. [3]). We will not examine this interesting point any further in the present article.

4. Extended local time-space formula

The level of generality reached in the preceding section by extending the operator Λ to $\mathcal{H}_1$ is somewhere between (2.1) and (2.3). It is thus natural to pursue further extensions so to recover the Itô-Tanaka formula (2.3) as a special case in dimension one.

Convexity in two dimensions is more complicated and there seem to be no simple $AC+BV$ characterization as in dimension one according to which $F = F_1 - F_2$ where F_1 and F_2 are convex if and only if F is AC with F' of BV. Moreover, considering the setting of (2.11) above and letting b to oscillate wildly (being also a more general curve in $I\!R_+ \times I\!R$ rather than just a function of time) we see that the assumption of the existence of the one-sided limits $F_t(s-,x-)$ and $F_x(s-,x-)$ may generally be rather restrictive. It is however exceedingly difficult to incorporate all special cases under a general condition which is technically not too demanding. We therefore state the theorem below more as an indication of what can be done in special cases than to give a final word on all possible extensions.

In the setting of Section 3 above, with $t > 0$ given and fixed, let us denote by $C^1_-([0,t] \times I\!R)$ the family of functions $F : [0,t] \times I\!R \to I\!R$ satisfying the following conditions:

$$\text{limits } F_t(s-,x-) \text{ and } F_x(s-,x-) \text{ exist at all } (s,x) \in [0,t] \times I\!R \tag{4.1}$$

$$(s,t) \mapsto F_t(s-,x-) \text{ and } (s,t) \mapsto F_x(s-,x-) \text{ are locally bounded} \tag{4.2}$$
$$\text{on } [0,t] \times I\!R.$$

Extend the family $\mathcal{H}_1$ from (3.4) as follows:

$$\mathcal{H}_1^- = \{\, H \mid H(s,x) = F_x(s-,x-) \text{ for some } F \in C^1_-([0,t] \times I\!R) \,\}. \tag{4.3}$$

Denoting $\mathcal{C} = \mathcal{P}_1 \cup \mathcal{P}_2 \cup \mathcal{P}_3 \cup \mathcal{H}_1$ we then have the following sequel to Theorem 3.1 above.

Theorem 4.1. *There exists a unique linear extension of Λ from $sp(\mathcal{C})$ to $sp(\mathcal{C} \cup \mathcal{H}_1^-)$ satisfying (3.10) and (3.11). In view of (3.9) this extension will also be denoted by (3.16) above.*

Proof. Take $H \in \mathcal{H}_1^-$ and let F be from $C^1_-([0,t] \times I\!R)$ such that $H(s,x) = F_x(s-,x-)$ for all $0 \le s \le t$ and all $x \in I\!R$. Associate with this F the convolution approximation F^n from (3.27) above. Then F^n is C^∞ and $F^n(s,x) \to F(s,x)$, $F^n_t(s,x) \to F_t(s-,x-)$ and $F^n_x(s,x) \to F_x(s-,x-)$ for all $(s,x) \in [0,t] \times I\!R$. Moreover, by means of (4.2) we can achieve that $|F^n_t| \le \gamma$ and $|F^n_x| \le G$ for all $n \ge 1$ where γ and G are locally bounded on $[0,t] \times I\!R$.

Setting $H^n = F^n_x$ and inserting this H^n in place of H in (3.26) above, letting $n \to \infty$ and using the dominated convergence theorem (both deterministic and stochastic) and localization, as well as (3.11) above, it follows that (3.26) extends to all $H \in \mathcal{H}_1^-$. □

Corollary 4.2 (Extended local time-space formula). *Let $X = (X_t)_{t \geq 0}$ be a continuous semimartingale and let $F : I\!R_+ \times I\!R \to I\!R$ be a function from the class $C^1_-([0,t] \times I\!R)$ with $t > 0$. Then the following change-of-variable formula holds:*

$$F(t, X_t) = F(0, X_0) + \int_0^t F_t(s-, X_s-)\, ds + \int_0^t F_x(s-, X_s-)\, dX_s \qquad (4.4)$$
$$- \frac{1}{2} \int_0^t \int_{I\!R} F_x(s-, x-)\, d\ell_s^x$$

where ℓ_s^x is the local time of X at the point x given by (2.4) above, and $d\ell_s^x$ refers to the area integration with respect to $(s,x) \mapsto \ell_s^x$ established in Theorem 4.1.

Proof. It follows from the proof of Theorem 4.1 above. □

Corollary 4.3 (Integration by parts for products). *If $C : I\!R_+ \to I\!R$ and $D : I\!R \to I\!R$ are of BV then we have:*

$$\int_{I\!R} \int_0^t C(s) D(x)\, d\ell_s^x = - \int_{I\!R} \Big(\int_0^t C(s)\, d_s \ell_s^x \Big)\, dD(x) \qquad (4.5)$$
$$= C(t) \int_{I\!R} D(x)\, d_x \ell_t^x - \int_0^t \Big(\int_{I\!R} D(x)\, d_x \ell_s^x \Big)\, dC(s)$$
$$= -C(t) \int_{I\!R} \ell_t^x\, dD(x) + \int_0^t \Big(\int_{I\!R} \ell_s^x\, dD(x) \Big)\, dC(s)$$

where $d\ell_s^x$ refers to the area integration with respect to $(s,x) \mapsto \ell_s^x$ established in Theorem 4.1.

Proof. It follows by combining the results of Corollary 4.2 and Corollary 3.3. □

In particular when $C(s) \equiv 1$ and $F_x(s-, x-) = D(x)$ then formula (4.4) above together with the final identity in (4.5) reduces to the Itô-Tanaka formula (2.3).

In a general case of (3.29) the first identity in (4.5) can be written as follows:

$$\int_{I\!R} \int_0^t H(s,x)\, d\ell_s^x = - \int_{I\!R} \int_0^t d_x H(s,x)\, d_s \ell_s^x \qquad (4.6)$$

but it may be not so obvious to determine its sense.

Clearly, replacing the left-limits in (4.1)-(4.3) above with the right-limits it follows that Theorem 4.1, Corollary 4.2 and Corollary 4.3 extend to the case where $\mathcal{H}_1^-$ is replaced by $\mathcal{H}_1^+$ provided that $I(x \leq X_s < x+\varepsilon)$ in the definition (2.4) is replaced by $I(x-\varepsilon < X_s \leq x)$.

5. Formal $d\ell_t^x$ calculus

In this section we will show how extensions of Itô's formula reviewed in Section 2 above can be obtained by formal manipulations of the $d\ell_s^x$ integral in the formula (2.13) which we recall here:

$$F(t,X_t) = F(0,X_0) + \int_0^t D_tF(s,X_s)\,ds + \int_0^t D_xF(s,X_s)\,dX_s \tag{5.1}$$
$$- \frac{1}{2}\int_0^t \int_{\mathbb{R}} D_xF(s,x)\,d\ell_s^x.$$

Formulas (3.28) and (4.4) derived above are special cases of this general formula.

This formalism appears to be useful for at least two reasons. Firstly, it helps to develop intuition needed to understand and compare known formulas. Secondly, if a new function F is given and one needs to write down a change-of-variable formula for $F(t,X_t)$, then such a formalism can be helpful in guessing a candidate formula before a rigorous proof is found. This will be illustrated in the final example below.

Let us now apply the formal $d\ell_t^x$ calculus and show how formulas (2.3), (2.5), (2.9) and (2.11) follow from (5.1). We assume throughout that $X = (X_t)_{t\geq 0}$ is a continuous semimartingale and F is a given function defined on $\mathbb{R}_+$ or $\mathbb{R}_+ \times \mathbb{R}$ with values in $\mathbb{R}$. Relevant properties of F will be inherited from Section 2 without further mentioning.

1. *Formulas (2.3) and (2.5).* In this case (5.1) reads as follows:

$$F(X_t) = F(X_0) + \int_0^t F'_-(X_s)\,dX_s - \frac{1}{2}\int_{\mathbb{R}} F'_-(x)\,d_x\ell_t^x \tag{5.2}$$

which (at least formally) coincides with the formula (2.5). To obtain (2.3) let us perform a formal partial integration in the final integral upon setting $u(x) = F'_-(x)$ and $dv(x) = d_x\ell_t^x$ so that $du(x) = dF'_-(x)$ and $v(x) = \ell_t^x$. The final integral in (5.2) then transforms as follows:

$$\int_{\mathbb{R}} F'_-(x)\,d_x\ell_t^x = F'_-(x)\,\ell_t^x\,\Big|_{x=-\infty}^{\infty} - \int_{\mathbb{R}} \ell_t^x\,dF'_-(x) = -\int_{\mathbb{R}} \ell_t^x\,dF'(x) \tag{5.3}$$

where we use that $\ell_t^{\pm\infty} \equiv 0$. Inserting (5.2) into (5.1) we formally obtain (2.3).

2. *Formula (2.9).* In this case (5.1) reads as follows:

$$F(X_t) = F(X_0) + \int_0^t F'(X_s)\,dX_s - \frac{1}{2}\int_{\mathbb{R}} F'(x)\,d_x\ell_t^x. \tag{5.4}$$

The final integral in (5.4) can be written as:

$$\int_{\mathbb{R}} F'(x)\,d_x\ell_t^x = \int_{\{|x|\leq\varepsilon\}} F'(x)\,d_x\ell_t^x + \int_{\{|x|>\varepsilon\}} F'(x)\,d_x\ell_t^x. \tag{5.5}$$

A formal partial integration in the first integral on the right-hand-side upon setting $u(x) = F'(x)$ and $dv(x) = d_x\ell_t^x$ so that $du(x) = dF'(x)$ and $v(x) = \ell_t^x$ yields:

$$\int_{\{|x|\leq\varepsilon\}} F'(x)\,d_x\ell_t^x = \Big(\int_{(-\varepsilon,0)} + \int_{\{0\}} + \int_{(0,\varepsilon)}\Big) F'(x)\,d_x\ell_t^x \tag{5.6}$$
$$= F'(x)\,\ell_t^x\Big(\Big|_{x=-\varepsilon}^{0-} + \Big|_{x=0}^{0} + \Big|_{x=0+}^{\varepsilon}\Big) - \int_{(-\varepsilon,0)\cup(0,\varepsilon)} \ell_t^x\,dF'(x)$$
$$- \int_{\{0\}} \ell_t^x\,dF'(x) \to (-\ell_t^0)\,(F'(0+) - F'(0-))$$

as $\varepsilon \downarrow 0$. Moreover, using (3.29) and (2.14) we see that the second integral transforms as follows:

$$\int_{\{|x|>\varepsilon\}} F'(x)\, d_x\ell_t^x = -\int_{\{|x|>\varepsilon\}} F''(x)\, \ell_t^x\, dx \tag{5.7}$$

$$= -\int_0^t F''(X_s)\, I(|X_s|>\varepsilon)\, d\langle X,X\rangle_s .$$

Letting $\varepsilon \downarrow 0$ in (5.5) using (5.6) and (5.7), where $d\langle X,X\rangle_s = ds$ when X is standard Brownian motion, and inserting the resulting identity in (5.4), we formally obtain (2.9).

3. *Formula (2.11).* In this case (5.1) reads as follows:

$$F(t,X_t) = F(0,X_0) + \int_0^t F_t(s,X_s-)\, ds + \int_0^t F_x(s,X_s-)\, dX_s \tag{5.8}$$

$$-\frac{1}{2}\int_0^t \int_{I\!R} F_x(s,x-)\, d\ell_s^x .$$

The final integral in (5.9) can be written as:

$$\int_0^t \int_{I\!R} F_x(s,x-)\, d\ell_s^x = \int_0^t \int_{I\!R\setminus\{b(s)\}} F_x(s,x-)\, d\ell_s^x \tag{5.9}$$

$$+\int_{\{b(s)\}} F_x(s,x-)\, d_x\Big(\int_0^t d_s\ell_s^x\Big).$$

Using (3.29) and (2.15) in the first integral on the right-hand-side and a formal partial integration in the second integral upon setting $u(x) = F_x(s,x-)$ and $dv(x) = d_x(\int_0^t d_s\ell_s^x)$ so that $du(x) = d_xF_x(s,x-)$ and $v(x) = \int_0^t d_s\ell_s^x$, we get:

$$\int_0^t \int_{I\!R} F_x(s,x-)\, d\ell_s^x = -\int_0^t \Big(\int_{I\!R\setminus\{b(s)\}} F_{xx}(s,x) d_s\ell_s^x\Big) dx \tag{5.10}$$

$$+\Big(F_x(s,x-)\int_0^t d_s\ell_s^x\Big)\Big|_{x=b(s)}^{b(s)} - \int_{\{b(s)\}} \Big(\int_0^t d_s\ell_s^x\Big)\, d_xF_x(s,x-)$$

$$= -\int_0^t F_{xx}(s,X_s)\, I(X_s \neq b(s))\, d\langle X,X\rangle_s - \int_0^t \Big(F_x(s,b(s)+) - F_x(s,b(s)-)\Big)\, d_s\ell_s^b$$

upon a formal identification $\int_0^t d_s\ell_s^{b(s)} = \int_0^t d_s\ell_s^b$. Inserting the resulting identity (5.10) into (5.8) we formally obtain (2.11).

4. Finally, let us briefly examine a new case not covered by known formulas from Section 2. For this, let us consider the setting of (2.11) where instead of a function $b : I\!R_+ \to I\!R$ of time we are given a function $c : I\!R \to I\!R$ of space. Setting $C = \{\, (s,x) \in I\!R_+ \times I\!R \mid s < c(x)\,\}$ and $D = \{\, (s,x) \in I\!R_+ \times I\!R \mid s > c(x)\,\}$ suppose that a continuous function $F : I\!R_+ \times I\!R \to I\!R$ is given such that F is $C^{1,2}$ on $\overline{C}$ and F is $C^{1,2}$ on $\overline{D}$. The question then arises to write down a change-of-variable formula for $F(t,X_t)$.

In this case (5.1) reads as follows:

$$F(t,X_t) = F(0,X_0) + \int_0^t F_t(s-,X_s)\,ds + \int_0^t F_x(s-,X_s)\,dX_s \qquad (5.11)$$
$$- \frac{1}{2}\int_{I\!R}\int_0^t F_x(s-,x)\,d\ell_s^x.$$

The final integral in (5.11) can be written as:

$$\int_{I\!R}\int_0^t F_x(s-,x)\,d\ell_s^x = \int_{I\!R}\int_{[0,t]\setminus\{c(x)\}} F_x(s-,x)\,d\ell_s^x \qquad (5.12)$$
$$+ \int_{\{c(x)\}} F_x(s-,x)\,d_s\Big(\int_{I\!R} d_x\ell_s^x\Big).$$

Using (3.29) and (2.15) in the first integral on the right-hand-side and a formal partial integration in the second integral upon setting $u(s) = F_x(s-,x)$ and $dv(s) = d_s(\int_{I\!R} d_x\ell_s^x)$ so that $du(s) = d_sF_x(s-,x)$ and $v(s) = \int_{I\!R} d_x\ell_s^x$, we get:

$$\int_{I\!R}\int_0^t F_x(s-,x)\,d\ell_s^x = -\int_{I\!R}\Big(\int_{[0,t]\setminus\{c(x)\}} F_{xx}(s,x)d_sd\ell_s^x\Big)dx \qquad (5.13)$$
$$+ \Big(F_x(s-,x)\int_{I\!R} d_x\ell_s^x\Big)\Big|_{x=c(x)}^{c(x)} - \int_{\{c(x)\}}\Big(\int_{I\!R} d_x\ell_s^x\Big)\,d_sF_x(s-,x)$$
$$= -\int_0^t F_{xx}(s,X_s)\,I(s\neq c(X_s))\,d\langle X,X\rangle_s$$
$$- \int_{I\!R}\Big(F_x(c(x)+,x) - F_x(c(x)-,x)\Big)\,d_x\ell_c^x$$

upon a formal identification $\int_0^t d_x\ell_{c(x)}^x = \int_0^t d_x\ell_c^x$. To give sense to the latter integral introduce:

$$\ell_c^x = I\!P\!-\!\lim_{\varepsilon\downarrow 0}\frac{1}{\varepsilon}\int_{c^{-1}([0,t])} I(x\leq X_{c(x)}<x+\varepsilon)\,d\langle X,X\rangle_{c(x)}. \qquad (5.14)$$

Then the final integral in (5.13) may be interpreted as the integral with respect to $x \mapsto \ell_c^x$, and inserting (5.13) into (5.11), we obtain the following candidate for the change-of-variable formula:

$$F(t,X_t) = F(0,X_0) + \int_0^t F_t(s-,X_s)\,ds + \int_0^t F_x(s-,X_s)\,dX_s \qquad (5.15)$$
$$+ \frac{1}{2}\int_0^t F_{xx}(s,X_s)I(s\neq c(X_s))d\langle X,X\rangle_s$$
$$+ \frac{1}{2}\int_{I\!R}\Big(F_x(c(x)+,x) - F_x(c(x)-,x)\Big)\,d_x\ell_c^x$$

In order to view the final integral differently note that:

$$\ell_c^x = I\!P\!-\!\lim_{\varepsilon\downarrow 0}\frac{1}{\varepsilon}\int_0^t I(c^{-1}(s)\leq X_s<c^{-1}(s)+\varepsilon)\,d\langle X,X\rangle_s \qquad (5.16)$$

where $c^{-1}(s) = \{ x \in I\!R \mid c(x) = s \}$ and $I(c^{-1}(s) \le X_s < c^{-1}(s)+\varepsilon)$ denotes the corresponding sum of $I(x \le X_s < x+\varepsilon)$ over x running through $c^{-1}(s)$.

Similar candidate formulas can be obtained when b or c above is a curve in $I\!R_+ \times I\!R$ defined through a mapping $\gamma : [0,1] \to I\!R_+ \times I\!R$, but we will omit the details. It is an interesting problem to establish these formulas rigorously under natural conditions.

References

[1] N. Bouleau *and* M. Yor, *Sur la variation quadratique des temps locaux de certaines semimartingales.* C. R. Acad. Sci. Paris Sér. I Math **292** (1981), 491–494.

[2] A. S. Cherny, *Principal values of the integral functionals of Brownian motion: Existence, continuity and extension of Itô's formula.* Sém. Probab. 35, Lecture Notes in Math. **1755** (2001), 348–370.

[3] N. Eisenbaum, *Integration with respect to local time.* Potential Anal. **13** (2000), 303–328.

[4] H. Föllmer, P. Protter *and* A. N. Shiryayev, *Quadratic covariation and an extension of Itô's formula.* Bernoulli **1** (1995), 149–169.

[5] K. Itô, *Stochastic integral.* Proc. Imp. Acad. Tokyo **20** (1944), 519–524.

[6] I. Karatzas *and* S. E. Shreve, *Brownian Motion and Stochastic Calculus.* Springer-Verlag, New York, (1991).

[7] H. Kunita *and* S. Watanabe, *On square integrable martingales.* Nagoya Math. J. **30** (1967), 209–245.

[8] P. Lévy, *Processus stochastiques et mouvement brownien.* Gauthier-Villars, Paris, 1948.

[9] P. A. Meyer, P. A. *Un cours sur les intégrales stochastiques.* Sém. Probab. 10, Lecture Notes in Math. **511** (1976), 245–400.

[10] J. L. Pedersen *and* G. Peskir, *On nonlinear integral equations arising in problems of optimal stopping.* Research Report No. **426** (2002), Dept. Theoret. Statist. Aarhus, (17 pp).

[11] G. Peskir, *A change-of-variable formula with local time on curves.* Research Report No. **428** (2002), Dept. Theoret. Statist. Aarhus, (17 pp).

[12] D. Revuz *and* M. Yor, *Continuous Martingales and Brownian Motion.* Springer-Verlag, Berlin, 1999.

[13] H. Tanaka, *Note on continuous additive functionals of the 1-dimensional Brownian path.* Z. Wahrscheinlichkeitstheorie und Verw. Gebiete **1** (1963), 251–257.

[14] A. T. Wang, *Generalized Itô's formula and additive functionals of Brownian motion.* Z. Wahrscheinlichkeitstheorie und Verw. Gebiete **41** (1977), 153–159.

Raouf Ghomrasni, Department of Mathematical Sciences, University of Aarhus, Ny Munkegade, DK-8000 Aarhus, Denmark
E-mail address: raouf@imf.au.dk

Goran Peskir, Department of Mathematical Sciences, University of Aarhus, Ny Munkegade, DK-8000 Aarhus, Denmark
E-mail address: goran@imf.au.dk

Progress in Probability, Vol. 55, 193–202

Local Times on Curves and Surfaces

Michael B. Marcus

Abstract. Given a positive measure μ on R^n, a continuous additive functional of an R^n valued Lévy process X can be defined (heuristically) by

$$L_t^\mu = \lim_{\epsilon \to 0} \int \int_0^t f_\epsilon(X(s) - x)\, ds\, d\mu(x) \tag{0.1}$$

where f_ϵ is an approximate δ function at zero. If the support of μ is a curve in R^2 or a surface in R^n we can consider L_t^μ as the local time on the curve or surface. Given a family $\mathcal{M}$ of such measures, endowed with a metric, some sufficient conditions for the continuity of $\{L_t^\mu, (t, \mu) \in R_+ \times \mathcal{M}\}$ are obtained. This problem is related, in part, to the continuity of certain second order Gaussian chaos processes indexed by the functions in $\mathcal{M}$.

1. Introduction

Let $\{X(t), t \in R_+\}$ be a symmetric Lévy process. Heuristically, one may think of the local time of X at x as the function defined by

$$L_t^x = \lim_{\epsilon \to 0} \int_0^t f_\epsilon(X(s) - x)\, ds \tag{1.1}$$

where f_ϵ is an approximate δ function at zero and consider this stochastic process as the occupation density of X at x up to time t. Local times are very important, but they only exist for certain Lévy processes in R^1. (For symmetric stable processes in R^1 of index p, they exist only for $1 < p \leq 2$).

When local times do not exist, and even when they do, one can consider continuous additive functionals of X determined by measures on R^n. We may think of these as

$$L_t^\mu = \lim_{\epsilon \to 0} \int \int_0^t f_\epsilon(X(s) - x)\, ds\, d\mu(x) \tag{1.2}$$

where μ is a positive measure on R^n. In this case, depending on the measure, such limits exist for all Lévy processes in R^n, for all $n \geq 1$. For some family of measures $\mathcal{M}$ for which (1.2) exists, endowed with some topology, one may consider the question of the continuity of $L = \{L_t^\mu, (t, \mu) \in R^+ \times \mathcal{M}\}$. This is done by R. Bass and D. Khoshnevisan in [1] and by the author and J. Rosen in [2] and several

---

2000 *Mathematics Subject Classification.* Primary 60J55, 60J65; Secondary 60G15 .
Key words and phrases. local times, continuous additive functionals, Gaussian chaos .
This work was supported in part by grants from the National Science Foundation and PSC CUNY.

subsequent works. See [3] for further references. We now describe the results of [1] and [2] on the continuity of L.

Let $p_t(x-y) := p_t(x,y)$ denote the transition probability density function of X. Consider the α–potential density of X

$$u^\alpha(x-y) := \int_0^\infty e^{-\alpha t} p_t(x-y)\,dt. \tag{1.3}$$

In general one considers the 0–potential density when it exists, which is generally the case in what follows, with the notable exception of Brownian motion in R^2. In the latter case we consider the 1–potential density. Let

$$d(\mu,\nu) = \sup_{x\in R^n} \left| \int u^\alpha(x-y)\,d(\mu(y)-\nu(y)) \right|^{1/2}. \tag{1.4}$$

It follows from Proposition 2.1 and Theorem 2.2, [1] that $(\mu,t) \to L_t^\mu$ is continuous on $(\mathcal{M},d) \times (R^+,|\cdot|)$ if

$$\int_0^D \log N_d(\mathcal{M},\epsilon)\,d\epsilon < \infty \tag{1.5}$$

where $N_d(\mathcal{M},\epsilon)$ is the minimum number of balls in the metric or pseudo–metric d that covers $\mathcal{M}$ and D is the diameter of $\mathcal{M}$ with respect to d. (Actually this result is only proved in [1] for Brownian motion but the authors say that it holds for symmetric stable processes as well).

The results in [2] are more complicated to explain, and although they apply in R^n for $n \le 3$, they are only applicable to the concerns of this paper for $n \le 2$. Nevertheless they introduce an interesting class of second order Gaussian chaos processes. It follows easily from the Chapman–Kolmogorov equation that for symmetric Lévy process, $p_t(x-y)$ is positive definite. Therefore $u^\alpha(x-y)$ is positive definite. However, in R^2, $u^\alpha(0)=\infty$. We restrict ourselves to R^2 and consider

$$u_\delta^\alpha(x-y) = \int_\delta^\infty e^{-\alpha t} p_t(x-y)\,dt. \tag{1.6}$$

u_δ is a bounded positive definite function. Let $G_\delta = \{G_\delta(x), x\in R^2\}$ be a mean zero stationary Gaussian process with covariance $u_\delta^\alpha(x-y)$. For a given Lévy process with α–potential density u^α let $\mathcal{M}$ denote the set of positive measures μ on R^2 for which

$$\int\int \left(u^\alpha(x-y)\right)^2 d\mu(x)\,d\mu(y) < \infty. \tag{1.7}$$

For $\mu \in \mathcal{M}$

$$H(\mu) := \lim_{\delta\to 0} \int \left(G_\delta^2(x) - E\left(G_\delta(x)\right)^2\right) d\mu(x) \tag{1.8}$$

exists in L^2 and is a second order Gaussian chaos. We consider the process $H = \{H(\mu), \mu \in \mathcal{M}\}$. Let

$$\tilde{d}(\mu,\nu) = \left(\int\int \left(u^\alpha(x-y)\right)^2 d(\mu(x)-\nu(x))\,d(\mu(y)-\nu(y))\right)^{1/2}. \tag{1.9}$$

The main result of [2] is that if H is bounded or continuous on $(\mathcal{M}, \widetilde{d})$ then $(\mu, t) \to L_t^\mu$ is bounded or continuous on $(\mathcal{M}, \widetilde{d}) \times (R^+, |\cdot|)$. (Actually this holds for all Markov processes with continuous 1–potential density). It follows that

$$\int_0^D \log N_{\widetilde{d}}(\mathcal{M}, \epsilon)\, d\epsilon < \infty \tag{1.10}$$

is a sufficient condition for the continuity of L on $(\mathcal{M}, \widetilde{d})$ for the simple reason that it is a sufficient condition for the continuity of H on $(\mathcal{M}, \widetilde{d})$. But note that (1.10) is not necessarily a necessary condition for the continuity of H. The most interesting results in [2] deal with second order Gaussian chaos processes for which (1.10) with $\log N_{\widetilde{d}}(\mathcal{M}, \epsilon)$ replaced by $(\log N_{\widetilde{d}}(\mathcal{M}, \epsilon))^{1/2}$ implies the continuity of H. (Although these processes do not include the processes considered in this note).

We now explain what we mean by 'local times on curves and surfaces'. Let f_n be a real valued measurable function on $[-1,1]^{n-1}$. Let $\bar{x}_{n-1} = (x_1, \ldots, x_{n-1})$. We define a measure on $[-1,1]^n$ by

$$\mu_{f_n}(A) = \int_{[-1,1]^{n-1}} I_A(\bar{x}_{n-1}, f_n(\bar{x}_{n-1}))\, d\bar{x}_{n-1}. \tag{1.11}$$

We consider $f_2(x_1)$ a curve in R^2 and $f_n(\bar{x}_{n-1})$, $n \geq 3$, a surface in R^n. Let $\mathcal{F}$ denote a set of functions $\{f_n\}$ on R^{n-1}. Let $\widetilde{\mathcal{M}} = \{\mu_{f_n}, f_n \in \mathcal{F}\}$. Refer to (1.2). $L_t^{\mu_{f_n}}$ only increases when X is in the vicinity of f_n. This is what we call a local time on the curve or surface determined by f_n. We define $\widetilde{L} = \left\{L_t^{f_n}, (t, f_n) \in R^+ \times \mathcal{F}\right\}$ as $\left\{L_t^{\mu_{f_n}}, (t, \mu_{f_n}) \in R^+ \times \widetilde{\mathcal{M}}\right\}$.

Of course, the continuity results mentioned above apply to $\widetilde{L}$. The purpose of this note is to find bounds for the pseudo–metrics defined in (1.4) and (1.9) in terms of more familiar metrics on the functions that determine the curves and surfaces, namely

$$\|f - g\|_\infty := \sup_{x \in [-1,1]^{n-1}} |f(x) - g(x)|$$

and

$$\|f - g\|_2 := \left(\int_{x \in [-1,1]^{n-1}} |f(x) - g(x)|^2\, dx \right)^{1/2}.$$

We give these estimates in the next section and consider some examples in Section 3.

Remark 1.1. Here is a more precise description of L_t^μ in (1.2). Any continuous additive functional A_t of X can be associated with a positive σ-finite measure ν_A called the Revuz measure of A_t, defined by

$$\nu_A(g) = \lim_{t \to 0} \frac{1}{t} E^m\left(\int_0^t g(X_s)\, dA_s\right) \tag{1.12}$$

where m is Lebesgue measure, for all bounded continuous functions g. A_t is uniquely determined by ν_A. L_t^μ is the continuous additive functional of X with Revuz measure μ. For our purposes it is enough to note that a sufficient condition for the existence of the limit in (1.12) is that

$$\sup_x \int u^\alpha(x-y)\, d\mu(y) < \infty$$

for some $\alpha \geq 0$.

Finally note that we can also define a second order Gaussian chaos process $\{H(f_2), f_2 \in \mathcal{F}\}$ as $\left\{H(\mu_{f_2}), \mu_{f_2} \in \widetilde{\mathcal{M}}\right\}$, without even mentioning Lévy processes. It would be very interesting to have a better understanding of the continuity properties of these processes.

2. Bounds for $\mathbf{d}(\mu,\nu)$ and $\widetilde{\mathbf{d}}(\mu,\nu)$

We say that a function $h : R_+ \to R_+$ is in class B if it is assymptotic to a decreasing function at the origin and satisfies

$$u|h'(u)| \leq Kh(u) \tag{2.1}$$

for $u > 0$, where K is a constant.

Lemma 2.1. *Let Φ be a function in class B, then for $\gamma \geq 0$*

$$\Phi(\alpha^2+\beta^2) - \Phi(\alpha^2+(\beta+\gamma)^2) \leq K\Phi(\alpha^2)\left(\frac{\gamma}{\alpha}+\frac{\gamma^2}{\alpha^2}\right). \tag{2.2}$$

Proof. Note that for $u, v \geq 0$

$$\begin{aligned} \Phi(u) - \Phi(u+v) &= \int_u^{u+v} |\Phi'(s)|\, ds \\ &\leq K\int_u^{u+v} \frac{\Phi(s)}{s}\, ds \\ &\leq K\frac{\Phi(u)v}{u}. \end{aligned} \tag{2.3}$$

Let $u = \alpha^2+\beta^2$ and $v = 2\beta\gamma+\gamma^2$ and note that

$$\frac{2\beta\gamma+\gamma^2}{\alpha^2+\beta^2} \leq \frac{\gamma}{\alpha}+\frac{\gamma^2}{\alpha^2} \tag{2.4}$$

since $2\beta\alpha \leq \alpha^2+\beta^2$. Equation (2.2) follows from (2.3) and (2.4). □

Let Ψ be a function in class B and let μ_{f_n} be as defined in (1.11). We define two pseudo–norms on the measures μ_{f_n}.

$$\|\mu_{f_n}\| = \sup_{\bar{x}_n \subset [\ 1,1]^n} \left|\int_T \Psi\left(|\bar{x}_n - \bar{y}_n|\right)\, d\mu_{f_n}(\bar{y}_n)\right|^{1/2} \tag{2.5}$$

and

$$\|\!|\mu_{f_n}\|\!| = \left| \iint_{T\times T} \Psi^2\left(|\bar{x}_n - \bar{y}_n|\right) d\mu_{f_n}(\bar{x}_n)\, d\mu_{f_n}(\bar{y}_n) \right|^{1/2} \tag{2.6}$$

where $T = [-1,1]^n$. When Ψ is also an α–potential density of a symmetric Lévy process, which is the case in what follows, $d(\mu,\nu) = \|\mu_{f_n} - \mu_{g_n}\|$ and $\widetilde{d}(\mu,\nu) = \|\!|\mu_{f_n} - \mu_{g_n}\|\!|$.

Lemma 2.2. *Assume that Ψ is in class B and is regularly varying at zero with index $p-n$, where $1 < p < 2$. I. e. $\Psi(u) = u^{p-n}L(u)$, where L is a slowly varying function. Then*

$$\|\mu_{f_n} - \mu_{g_n}\| \le C\left(\|f_n - g_n\|_\infty^{p-1} L(\|f_n - g_n\|_\infty)\right)^{1/2} \tag{2.7}$$

for some constant C, depending on n, but independent of f_n and g_n.

Let $\Psi_1(u) = u^{2-n}$ for $n \ge 3$, then

$$\|\mu_{f_n} - \mu_{g_n}\| \le C\left(\|f_n - g_n\|_\infty \log\left(2 + \frac{2}{\|f_n - g_n\|_\infty}\right)\right)^{1/2}. \tag{2.8}$$

Let $\Psi_2(u) = \log\left(2 + \dfrac{2}{u}\right)$, then

$$\|\mu_{f_2} - \mu_{g_2}\| \le C\|f_2 - g_2\|_\infty^{1/2} \log\left(2 + \frac{2}{\|f_2 - g_2\|_\infty}\right). \tag{2.9}$$

Proof. Let $\Phi(u) = \Psi(\sqrt{u})$. Thus $|\Phi'(u)| \le \dfrac{K\Phi(u)}{2u}$, so Φ is also in class B. We have

$$\begin{aligned}\|\mu_{f_n} - \mu_{g_n}\|^2 = \int_{[-1,1]^{n-1}} &\left[\Phi\left(|\bar{x}_{n-1} - \bar{y}_{n-1}|^2 + |x_n - f_n(\bar{y}_{n-1})|^2\right)\right. \\ &\left. - \Phi\left(|\bar{x}_{n-1} - \bar{y}_{n-1}|^2 + |x_n - g_n(\bar{y}_{n-1})|^2\right)\right] d\bar{y}_{n-1}\end{aligned} \tag{2.10}$$

We consider this integral over two regions, $I = \{\bar{y}_{n-1} : \sum_{i=1}^{n-1} |x_i - y_i|^2 \le \delta^2\}$ and II, its complement in $[-1,1]^{n-1}$. The integral of the right–hand side of (2.10) over I is obviously bounded by

$$2\int_I \Psi\left(|\bar{x}_{n-1} - \bar{y}_{n-1}|\right) d\bar{y}_{n-1} \le C\Psi(\delta)\delta^{n-1}. \tag{2.11}$$

To get this inequality we set $y_i = x_i + s_i$, $i = 1,\ldots,n-1$ and use radial coordinates and properties of regular variation.

We now consider the integral in (2.10) over II. Using the triangle inequality we see that it is

$$\begin{aligned}\le \int_{II} &\left[\Phi\left(|\bar{x}_{n-1} - \bar{y}_{n-1}|^2 + |x_n - f_n(\bar{y}_{n-1})|^2\right)\right. \\ &\left. - \Phi\left(|\bar{x}_{n-1} - \bar{y}_{n-1}|^2 + \big||x_n - f_n(\bar{y}_{n-1})| + |f_n(\bar{y}_{n-1}) + g_n(\bar{y}_{n-1})|\big|^2\right)\right] d\bar{y}_{n-1}.\end{aligned} \tag{2.12}$$

Using Lemma 2.1 with $\alpha = |\bar{x}_{n-1} - \bar{y}_{n-1}|$, $\beta = |x_n - f_n(\bar{y}_{n-1})|$ and $\gamma = |f_n(\bar{y}_{n-1}) - g_n(\bar{y}_{n-1})|$ we see that the integral in (2.12)

$$\begin{aligned} &\leq C \int_{II} \Psi\left(|\bar{x}_{n-1} - \bar{y}_{n-1}|\right) \\ &\left(\frac{|f_n(\bar{y}_{n-1}) - g_n(\bar{y}_{n-1})|}{|\bar{x}_{n-1} - \bar{y}_{n-1}|} + \frac{|f_n(\bar{y}_{n-1}) - g_n(\bar{y}_{n-1})|^2}{|\bar{x}_{n-1} - \bar{y}_{n-1}|^2} \right) d\bar{y}_{n-1} \quad (2.13) \\ &\leq C\left(\Psi(\delta)\delta^{n-2}\|f_n - g_n\|_\infty + \Psi(\delta)\delta^{n-3}\|f_n - g_n\|_\infty^2\right). \end{aligned}$$

We get (2.7) from (2.11) and (2.13) by taking $\delta = \|f_n - g_n\|_\infty$.

We now consider (2.8). It is easy to see that Ψ_1 is in class B and that

$$2\int_I \Psi_1\left(|\bar{x}_{n-1} - \bar{y}_{n-1}|\right) d\bar{y}_{n-1} \leq C\delta. \tag{2.14}$$

Also, using (2.12) and (2.13) we see that the integral in (2.10) over II

$$\begin{aligned} &\leq C\left(\|f_n - g_n\|_\infty \int_{s>\delta} \Psi(s)s^{n-3}\, ds + \|f - g\|_\infty^2 \int_{s>\delta} \Psi(s)s^{n-4}\, ds \right) \quad (2.15) \\ &\leq C\left(\|f_n - g_n\|_\infty \log\left(2 + \frac{2}{\delta}\right) + \frac{\|f_n - g_n\|_\infty^2}{\delta} \right). \end{aligned}$$

We get (2.8) from (2.14) and (2.15) by taking $\delta = \|f_n - g_n\|_\infty$.

We now consider (2.9). It is easy to see that Ψ_2 is in class B and that

$$2\int_I \Psi_2\left(|x_1 - y_1|\right) dy_1 \leq C\delta\Psi_2(\delta). \tag{2.16}$$

Also, using (2.12) and (2.13) we see that the integral in (2.10) over II

$$\begin{aligned} &\leq C\left(\|f_2 - g_2\|_\infty \int_{s>\delta} \frac{\Psi_2(s)}{s}\, ds + \|f_2 - g_2\|_\infty^2 \int_{s>\delta} \frac{\Psi_2(s)}{s^2}\, ds \right) \quad (2.17) \\ &\leq C\left(\|f_2 - g_2\|_\infty \Psi_2^2(\delta) + \frac{\|f_2 - g_2\|_\infty^2}{\delta}\Psi_2(\delta) \right). \end{aligned}$$

We get (2.8) from (2.16) and (2.17) by taking $\delta = \|f_2 - g_2\|_\infty$. □

The bounds in (2.8) and (2.9) do not depend on n. In the next lemma we see that this is not the case for the L^2 norm.

Lemma 2.3. *The following three expressions can be used on the right-hand sides of (2.7)–(2.9) respectively to give bounds in terms of the L_2 norm.*

$$C\left(\|f_n - g_n\|_2^{2(p-1)/(n+1)} L\left(\|f_n - g_n\|_2^{2(p-1)/(n+1)}\right)\right)^{1/2} \tag{2.18}$$

$$C\|f_n - g_n\|_2^{1/(n+1)} \tag{2.19}$$

and

$$C\left(\|f_n - g_n\|_2^{2/3} \log\left(2 + \frac{2}{\|f_2 - g_2\|_2}\right)\right)^{1/2}. \tag{2.20}$$

Proof. We follow the proof of Lemma 2.2. The inequality in (2.11) remains unchanged but we use a different estimate in (2.13) to express it in terms of $\|f-g\|_2$. We use the Schwarz inequality on the first integral in (2.13) and in the second we bound $\Psi(|\bar{x}_{n-1}-\bar{y}_{n-1}|)\,|\bar{x}_{n-1}-\bar{y}_{n-1}|^2$ by $C\psi(\delta)/\delta^2$. Consequently, in place of the last line in (2.13) we get

$$C\Psi(\delta)\delta^{n-1}\left(\frac{\|f-g\|_2}{\delta^{(n+1)/2}}+\frac{\|f-g\|_2^2}{\delta^{n+1}}\right). \tag{2.21}$$

We get (2.18) by taking $\delta=\|f-g\|_2^{2/(n+1)}$.

This same argument with $p=2$ gives (2.19) and (2.20) because in these cases, in contrast to the proof of Lemma 2.2, we don't integrate functions which are regularly varying with index minus one at the origin. I.e. (2.18) is also valid with $p=2$ with the slowly varying function equal to one for $n\geq 3$ and equal to the log when $n=2$. □

We now obtain bounds for $\|\!|\mu_f-\mu_g|\!\|$ for the functions Ψ considered in Lemma 2.2. We see immediately that (2.6) can be finite when $f_n\equiv 0$ only if $n=2$. Thus we restrict ourselves to this case.

Lemma 2.4. *Assume that Ψ is in class B and is regularly varying at zero with index $p-2$, where $1<p<2$. I. e. $\Psi(u)=u^{p-2}L(u)$, where L is a slowly varying function. Then*

$$\|\!|\mu_{f_2}-\mu_{g_2}|\!\|\leq C\left(\|f_2-g_2\|_2^{2p-3}L(\|f_2-g_2\|_2)\right)^{1/2} \tag{2.22}$$

for some constant C independent of f_2 and g_2.

Let $\Psi_2(u)=\log\left(2+\dfrac{2}{u}\right)$, then

$$\|\!|\mu_{f_2}-\mu_{g_2}|\!\|\leq C\|f_2-g_2\|_2^{1/2}\log^{3/2}\left(2+\frac{2}{\|f_2-g_2\|_2}\right). \tag{2.23}$$

Proof. We use the notation of lemma 2.2. Since Φ is in class B so is Φ^2. For real valued functions h and k let

$$\widetilde{\Phi}_{h,k}:=\Phi^2\left(|x_1-y_1|^2+|h(x_1)-k(y_1)|^2\right). \tag{2.24}$$

In this notation

$$\|\!|\mu_f-\mu_g|\!\|^2=\int_{-1}^{1}\int_{-1}^{1}\left(\widetilde{\Phi}_{f,f}+\widetilde{\Phi}_{g,g}-\widetilde{\Phi}_{f,g}-\widetilde{\Phi}_{g,f}\right)dx_1\,dy_1 \tag{2.25}$$

where we set $f_2=f$ and $g_2=g$. As in (2.12) and (2.13)

$$\begin{aligned}|\widetilde{\Phi}_{f,f}-\widetilde{\Phi}_{f,g}|&\leq\big[\Phi^2\left(|x_1-y_1|^2+|f(x_1)-f(y_1)|^2\right)\\&\quad-\Phi^2\left(|x_1-y_1|^2+\big||f(x_1)-g(y_1)|+|f(y_1)+g(y_1)|\big|^2\right)\big] \qquad (2.26)\\&\leq C\Psi^2(|x_1-y_1|)\left(\frac{|f(y_1)-g(y_1)|}{|x_1-y_1|}+\frac{|f(y_1)-g(y_1)|^2}{|x_1-y_1|^2}\right)\\&:=H_{f,g}(x_1,y_1).\end{aligned} \tag{2.27}$$

Clearly, the same bound holds for $|\widetilde{\Phi}_{g,g} - \widetilde{\Phi}_{g,f}|$. Therefore

$$\begin{aligned}\|\mu_f - \mu_g\|^2 &\le 4\int_{-1}^{1}\int_{|x_1-y_1|\le\delta} \Psi^2(|x_1-y_1|)\,dx_1\,dy_1 \\ &\quad +2\int_{-1}^{1}\int_{|x_1-y_1|\ge\delta} H_{f,g}(x_1,y_1)\,dx_1\,dy_1. \qquad (2.28)\\ &\le C\left(\delta\Psi^2(\delta) + \int_{s\ge\delta}\frac{\Psi^2(s)}{s}\,ds\|f-g\|_1 + \int_{s\ge\delta}\frac{\Psi^2(s)}{s^2}\,ds\|f-g\|_2^2\right).\end{aligned}$$

For $n \ge 3$ this last line

$$\le C\Psi_2^2(\delta)\left(\delta + \|f-g\|_2 + \frac{\|f-g\|_2^2}{\delta}\right) \qquad (2.29)$$

and we get (2.22) by taking $\delta = \|f-g\|_2$.

For $n = 2$ the last line in (2.28)

$$\le C\Psi_3^2(\delta)\left(\delta + \Psi_3(\delta)\|f-g\|_2 + \frac{\|f-g\|_2^2}{\delta}\right) \qquad (2.30)$$

since Ψ_3 is a log. We get (2.23) by taking $\delta = \|f-g\|_2$. □

Remark 2.5. Since the domain of f_2 and g_2 is $[-1,1]$, $\|f-g\|_2 \le \sqrt{2}\|f-g\|_\infty$. Therefore the bounds in Lemma 2.4 are also valid with $\|f-g\|_2$ replaced by $\|f-g\|_\infty$. However, we are not able to replace the right–hand sides of (2.22) and (2.23) with smaller functions of the larger norm. Examining these results we see that the bounds with respect to the sup norm are strictly smaller for $\|\mu_f - \mu_g\|$ than for $\|\mu_f - \mu_g\|$. However, the opposite is true for bounds with respect to the L^2 norm, when $n = 2$ and $7/4 < p \le 2$. This is not insignificant because $p = 2$ gives the result for Brownian motion on R^2.

3. Examples

Let T be a set of functions on $[-1,1]^n$. Let $\tau(f,g)$ and $\rho(f,g)$ be pseudo–metrics on $T \times T$. Suppose that

$$\tau(f,g) \le (\rho(f,g))^\gamma \qquad \forall f,g \in T \qquad (3.1)$$

then

$$\begin{aligned}\int_0^\infty \log N_\tau(T,\epsilon)\,d\epsilon &\le \int_0^\infty \log N_\rho(T,\epsilon^{1/\gamma})\,d\epsilon \qquad (3.2)\\ &= \gamma\int_0^\infty \frac{1}{\epsilon^{1-\gamma}}\log N_\rho(T,\epsilon)\,d\epsilon.\end{aligned}$$

Therefore the first integral in (3.2) is finite if

$$\log N_\rho(T,\epsilon) \le Const.\left(\frac{1}{\epsilon}\right)^\eta \qquad (3.3)$$

and $\eta < \gamma$. We use this observation together with the estimates in Section 2 to give examples of local times which are continuous over families of curves in R^2. Note that for symmetric stable processes on R^2, $u^0(x,y) = |x-y|^{p-2}$, $1 < p < 2$. For Brownian motion on R^2, $u^1(x,y)$ is in class B and $u^1(x,y) \le C \log\left(2 + \frac{2}{|x-y|}\right)$ for $x, y \in [-1,1]$.

1. Let $\mathcal{F}_1 = \{g(x) + y\}$, $y \in [-1,1]$, where $g(x)$, $x \in [-1,1]$ is a measurable function. It is easy to see that $N_{\|\cdot\|_\infty}(\mathcal{F}_1, \epsilon) \le 3/\epsilon$. Therefore, by Lemma 2.2, $\left\{L_t^f, (t,f) \in R^+ \times \mathcal{F}_1\right\}$ is continuous on $(\mathcal{F}_1, \|\cdot\|_\infty)$ for all symmetric stable processes on R^2 with index $1 < p \le 2$. Note that g is simply a measurable function; no smoothness conditions are required.

2. Let $\mathcal{F}_2$ be all polynomials of degree d, $\sum_{k=0}^d a_k x^k$, $x \in [-1,1]$, $|a_k| \le 1$. In this case it is also easy to see that $\log N_{\|\cdot\|_\infty}(\mathcal{F}_2, \epsilon) \le d(\log d + \log 3/\epsilon)$. Therefore, we have the same conclusion as in the previous example.

3. Let $\{\phi_k(x)\}$, $x \in [-1,1]$, $|\phi_k(x)| \le 1$, be a sequence of measurable functions on $[-1,1]$. Let $\mathcal{F}_3$ be all infinite series $\sum_{k=0}^\infty a_k \phi_k(x)$, $x \in [0,1]$, $|a_k| \le k^{-\beta}$, where $\beta > 1$.

We now find an upper bound for $N_{\|\cdot\|_\infty}(\mathcal{F}_3, \epsilon)$. Let $\delta = M^{-\beta}$. Then

$$\begin{aligned} \|\sum_{k=0}^{\infty} (\delta \wedge |a_k|)(\operatorname{sgn} a_k)\phi_k(x)\|_\infty &\le \sum_{k=0}^{M-1} \frac{1}{M^\beta} + \sum_{k=M}^{\infty} \frac{1}{k^\beta} \\ &\le \frac{C_\beta}{M^{\beta-1}}. \end{aligned} \tag{3.4}$$

Note that all the terms in the series starting from the M–th term are in a $\|\cdot\|_\infty$ ball of radius $C'_\beta M^{1-\beta}$. This shows us that

$$\prod_{k=1}^{M} \frac{2M^\beta}{k^\beta} \tag{3.5}$$

$\|\cdot\|_\infty$ balls of radius $C_\beta M^{1-\beta}$ cover $\mathcal{F}_3$. Therefore

$$\begin{aligned} \log N_{\|\cdot\|_\infty}(\mathcal{F}_3, C_\beta M^{1-\beta}) &\le M \log 2 + \beta M \log M - \beta \sum_{k=1}^{M} \log k \\ &\le C''_\beta M \end{aligned} \tag{3.6}$$

and by a change of variables

$$\log N_{\|\cdot\|_\infty}(\mathcal{F}_3, \epsilon) \le C_\beta \left(\frac{1}{\epsilon}\right)^{1/(\beta-1)}. \tag{3.7}$$

Therefore, by Lemma 2.2, $L = \left\{L_t^f, (t,f) \in R^+ \times \mathcal{F}_3\right\}$ is continuous on $(\mathcal{F}_3, \|\cdot\|_\infty)$ for all symmetric stable processes on R^2 with index $1 < p \le 2$ as long as $1/(\beta-1) < (p-1)/2$, or $\beta > (p+1)/(p-1)$.

Now let $\{\phi_k(x)\}$, $x \in [-1,1]$, $|\phi_k(x)| \le 1$, be a real orthogonal sequence of functions with respect to Lebesgue measure on $[-1,1]$. We find an upper bound for $N_{\|\cdot\|_2}(\mathcal{F}_3,\epsilon)$ in this case. Again we take $\delta = M^{-\beta}$.

$$\|\sum_{k=0}^{\infty}(\delta \wedge |a_k|)(\operatorname{sgn} a_k)\phi_k(x)\|_2 \le \left(\sum_{k=0}^{M-1}\frac{1}{M^{2\beta}} + \sum_{k=M}^{\infty}\frac{1}{k^{2\beta}}\right)^{1/2}$$
$$\le \frac{C_\beta}{M^{\beta-1/2}}. \tag{3.8}$$

The number in (3.5) is clearly an upper bound for the number of $\|\cdot\|_2$ balls of radius $C_\beta M^{(1/2)-\beta}$ that cover $\mathcal{F}_3$. Therefore

$$\log N_{\|\cdot\|_2}(\mathcal{F}_3, C_\beta M^{(1/2)-\beta}) \le C''_\beta M$$

and by a change of variables

$$\log N_{\|\cdot\|_2}(\mathcal{F}_3,\epsilon) \le C_\beta\left(\frac{1}{\epsilon}\right)^{2/(2\beta-1)}. \tag{3.9}$$

Therefore, by Lemma 2.4, $\left\{L_t^f,(t,f)\in R^+\times\mathcal{F}_3\right\}$ is continuous on $(\mathcal{F}_3,\|\cdot\|_2)$ for all symmetric stable processes on R^2 with index $3/2 < p \le 2$ as long as $2/(2\beta-1) < (2p-3)/2$, or $\beta > (2p+1)/(2p-3)$.

If $\phi_k(x) = \sin 2\pi kx$ the functions in $f \in \mathcal{F}_3$ are quite smooth. For example, when $\beta > 3$ they are twice differentiable. However, if $\phi_k(x) = \sin \pi 2^k x$ the functions in $f \in \mathcal{F}_3$ are not smooth at all. Thus, what gives us continuity of L_t^f over a family of functions f is not the smoothness of the functions but that, in some sense, the family of functions is not too large.

These example apply to a much lager class of processes than symmetric stable processes. In [3] we discuss a class of Lévy processes which we call stable mixtures. These processes have 1–potential densities which are in class B.

References

[1] R. Bass and D. Khoshnevisan, *Local times on curves and uniform invariance principles*, Stochastic Process. Appl. **92** (1992) 465–492.

[2] M. B. Marcus and J. Rosen, *Gaussian chaos and sample path properties of additive functionals of symmetric Markov processes*, Ann. Probab. **24** (1996) 1130–1177.

[3] M. B. Marcus and J. Rosen, *Renormalized self-intersection local times and Wick power chaos processes*, Mem. Amer. Math. Soc. **142**, (1999) No. 675.

Department of Mathematics
City College
New York, NY 10031, USA
E-mail address: mbmarcus@earthlink.net

Part 5

Large and Small Deviations

Progress in Probability, Vol. 55, 205–223

Large Deviations of Empirical Processes

Miguel A. Arcones

Abstract. We give necessary and sufficient conditions for the large deviations of empirical processes and of Banach space valued random vectors. We also consider the large deviations of partial sums processes. The main tool used is an isoperimetric inequality for empirical processes due to Talagrand.

1. Introduction

We study the (LDP) large deviation principle for different types of sequences of empirical processes $\{U_n(t) : t \in T\}$, where T is an index set. General references on large deviations are Bahadur [3], Varadhan [24], Deuschel and Stroock [10] and Dembo and Zeitouni [9]. We consider stochastic processes as elements of $l_\infty(T)$, where T is an index set. $l_\infty(T)$ is the Banach space consisting of the bounded functions defined in T with the norm $\|x\|_\infty = \sup_{t\in T} |x(t)|$. We will use the following definition.

Definition 1.1. Given a sequence of stochastic processes $\{U_n(t) : t \in T\}$, a sequence of positive numbers $\{\epsilon_n\}_{n=1}^{\infty}$ such that $\epsilon_n \to 0$, and a function $I : l_\infty(T) \to [0, \infty]$, we say that $\{U_n(t) : t \in T\}$ satisfies the LDP with speed ϵ_n^{-1} and with good rate function I if:

(i) For each $0 \le c < \infty$, $\{z \in l_\infty(T) : I(z) \le c\}$ is a compact set of $l_\infty(T)$.

(ii) For each set $A \in l_\infty(T)$,

$$-I(A^o) \le \liminf_{n\to\infty} \epsilon_n \log(\Pr{}_*\{\{U_n(t) : t \in T\} \in A\})$$

and

$$\limsup_{n\to\infty} \epsilon_n \log(\Pr{}^*\{\{U_n(t) : t \in T\} \in A\}) \le -I(\bar{A}),$$

where for $B \subset l_\infty(T)$, $I(B) = \inf\{I(x) : x \in B\}$.

It was shown in Arcones [1], that this definition is equivalent to the large deviations of the finite dimension distributions plus an asymptotic equicontinuity condition. Thus, large deviations can be studied similarly to the weak convergence of empirical processes.

2000 *Mathematics Subject Classification.* Primary 62E20; Secondary 62F12.
Key words and phrases. Large deviations, empirical processes.

In Section 2, we present necessary and sufficient conditions for the large deviations of empirical processes and of sums of Banach space valued i.i.d.r.v.'s. If $\{X_j\}_{j=1}^{\infty}$ is a sequence of real i.i.d.r.v.'s, it follows from the results in Cramér [8] and Chernoff [7] that $n^{-1}\sum_{j=1}^{n} X_j$ satisfies the LDP with a good rate with speed n if and only if for some $\lambda > 0$ $E[e^{\lambda|X_1|}] < \infty$. This is also true for r.v.'s with values in a finite dimensional vector space (see Corollary 6.1.6 in Dembo and Zeitouni, 1999). Given a sequence $\{X_i\}$ of B–valued i.i.d.r.v.'s, where B is a separable Banach space, Sethuraman [20] (Theorem 7) and Donsker and Varadhan [12] (Theorem 5.3) showed that if for each $\lambda > 0$ $E[e^{\lambda|X_1|}] < \infty$, then the LDP holds for $n^{-1}\sum_{j=1}^{n} X_j$ with speed n and with rate function

$$I(x) = \sup\{f(x) - \log(E[e^{f(X)}]) : f \in B^*\},$$

where X is a copy of X_1 and B^* is the dual of B. We obtain that in the previous situation the LDP holds for $n^{-1}\sum_{j=1}^{n} X_j$ with speed n and a good rate function if and only if there exists a $\lambda > 0$ such that $E[e^{\lambda|X_1|}] < \infty$; and for each $\lambda > 0$ there exists a $\eta > 0$ such that $E[e^{\lambda W^{(\eta)}}] < \infty$, where

$$W^{(\eta)} = \sup\{|f_1(X) - f_2(X)| : f_1, f_2 \in B_1^*,\ E[|f_1(X) - f_2(X)|] \le \eta\},$$

where B_1^* is the unit ball of B^*. As a corollary, we obtain that when B is a Hilbert space $(H, <\cdot>)$, the LDP holds for $n^{-1}\sum_{j=1}^{n} X_j$ with speed n and a good rate function if and only if there exists a $\lambda > 0$ such that $E[\exp(\lambda|X|)] < \infty$; and for each $\lambda > 0$, there exists an integer m such that $E[\exp(\lambda|X^{(m)}|)] < \infty$, where $X^{(m)} = \sum_{k=m+1}^{\infty} < X, h_k > h_k$ and $\{h_k\}$ is an orthogonal basis of H. We also prove that the stochastic process $\{n^{-1}\sum_{j=1}^{[nt]} X_j : 0 \le t \le 1\}$ satisfies the LDP in $l_\infty([0,1], B)$ with speed n and a good rate function if and only if for each $\lambda > 0$ $E[e^{\lambda|X|}] < \infty$. Here, $l_\infty([0,1], B)$ denotes the Banach space consisting of the bounded functions from $[0,1]$ into B with the norm $\|x\|_{\infty,[0,1],B} = \sup_{0\le t\le 1} |x(t)|$.

We will obtain the previous results from characterizations of the LDP of empirical processes. The study of the large deviations of empirical processes started with Sethuraman [20, 21, 22], where sufficient conditions were given such that for each $\epsilon > 0$, the following limit exists,

$$\lim_{n\to\infty} n^{-1}\log(\Pr\{\sup_{f\in\mathcal{F}} n^{-1}|\sum_{j=1}^{n}(f(X_j) - E[f(X_j)]) \ge \epsilon\}),$$

where $\{X_j\}_{j=1}^{\infty}$ is sequence of i.i.d.r.v.'s with values in separable compact metric space S and $\mathcal{F}$ is a collection of functions on S.

Another paper worth mentioning, but unrelated with the results in the present paper, is Bahadur and Zabell [4].

We will dealing with large deviations for empirical processes. Many other authors have considered large deviations for empirical measures (see for example, Groeneboom, Oosterhoff and Ruymgaart [16] and the references there).

c will denote an universal constant that may vary from line to line. We will use the usual multivariate notation. For example, given $u = (u_1, \dots, u_d)' \in \mathbb{R}^d$ and $v = (v_1, \dots, v_d)' \in \mathbb{R}^d$, $u'v = \sum_{j=1}^d u_j v_j$ and $|u| = (\sum_{j=1}^n u_j^2)^{1/2}$. Given a sequence of real numbers $a = \{a_k\}$, we denote $|a|_\infty = \sup_{k\geq 1} |a_k|$ and $|a|_p = (\sum_{k=1}^\infty |a_k|^p)^{1/p}$.

2. Large deviations of empirical processes

We consider the LDP for general triangular arrays of empirical processes. Let $(\Omega_n, \mathcal{A}_n, Q_n)$ be a sequence of probability spaces. Let $(S_{n,j}, \mathcal{S}_{n,j})$ be measurable spaces for $1 \leq j \leq k_n$, where $\{k_n\}_{n=1}^\infty$ is a sequence of positive integers converging to infinity. Let $\{X_{n,j} : 1 \leq j \leq k_n\}$ be $S_{n,j}$-valued independent r.v.'s defined on Ω_n. To avoid measurability problems, we assume that $\Omega_n = \prod_{j=1}^{k_n} S_{n,j}$, $\mathcal{A}_n = \prod_{j=1}^{k_n} \mathcal{S}_{n,j}$ and $Q_n = \prod_{j=1}^{k_n} \mathcal{L}(X_{n,j})$. Let $f_{n,j}(\cdot, t) : S_{n,j} \to \mathbb{R}$ be a measurable function for each $1 \leq j \leq k_n$, each $n \geq 1$ and each $t \in T$. Let $U_n(t) := \sum_{j=1}^{k_n} f_{n,j}(X_{n,j}, t)$. We also will assume that the probabilities of $\{U_n(t) : t \in T\}$ are determined by a countable set T_0. Alternatively, we could assume that for each $1 \leq j \leq k_n$, $\{f_{n,j}(x,t) : t \in T\}$ is an image admissible Suslin class of functions (see page 80 in Dudley [13]).

First, we present a couple of lemmas that we will need later on.

Lemma 2.1. *Under the previous notation, let $\{\epsilon_n\}$ be sequence of positive numbers converging to zero. Let $0 < c_1, c_2, M_1, M_2 < \infty$. Suppose that*

$$\limsup_{n\to\infty} \epsilon_n \log\Big(\Pr\{\sup_{t\in T} |\sum_{j=1}^{k_n} f_{n,j}(X_{n,j}, t)| \geq M_1\}\Big) \leq -c_1$$

and

$$\liminf_{n\to\infty} \min_{1\leq j\leq k_n} \epsilon_n \log\Big(\Pr\{\sup_{t\in T} |f_{n,j}(X_{n,j}, t)| \leq M_2\}\Big) \geq -c_2.$$

Then,

$$\limsup_{n\to\infty} \epsilon_n \log\Big(\sum_{j=1}^{k_n} \Pr\{\sup_{t\in T} |f_{n,j}(X_{n,j}, t)| \geq 2M_1 + M_2\}\Big) \leq -(c_1 - c_2).$$

Proof. Let $0 < c_1' < c_1$ and let $c_2' > c_2$. For n large enough,

$$\Pr\{\sup_{t\in T} |\sum_{j=1}^{k_n} f_{n,j}(X_{n,j}, t)| \geq M_1\} \leq e^{-c_1'\epsilon_n^{-1}}.$$

Let $\{X_{n,j}' : 1 \leq j \leq k_n, 1 \leq n\}$ be an independent copy of $\{X_{n,j} : 1 \leq j \leq k_n, 1 \leq n\}$. Then, for n large enough,

$$\Pr\{\sup_{t\in T} |\sum_{j=1}^{k_n} (f_{n,j}(X_{n,j}, t) - f_{n,j}(X_{n,j}', t))| \geq 2M_1\} \leq 2e^{-c_1'\epsilon_n^{-1}}.$$

By the Lévy inequality (see for example Proposition 2.3 in Ledoux and Talagrand [18]),

$$\Pr\{\max_{1\le j\le k_n}\sup_{t\in T}|f_{n,j}(X_{n,j},t)-f_{n,j}(X'_{n,j},t)|\ge 2M_1\}\le 4e^{-c'_1\epsilon_n^{-1}}.$$

This inequality and Lemma 2.6 in Ledoux and Talagrand [18] imply that for n large enough,

$$\sum_{j=1}^{k_n}\Pr\{\sup_{t\in T}|f_{n,j}(X_{n,j},t)-f_{n,j}(X'_{n,j},t)|\ge 2M_1\}\le 8e^{-c'_1\epsilon_n^{-1}}.$$

We also have that for n large enough and each $1\le j\le k_n$,

$$\begin{aligned}
&e^{-\epsilon_n^{-1}c'_2}\Pr\{\sup_{t\in T}|f_{n,j}(X_{n,j},t)|\ge 2M_1+M_2\}\\
&\quad\le \Pr\{\sup_{t\in T}|f_{n,j}(X_{n,j},t)|\ge 2M_1+M_2\}\Pr\{\sup_{t\in T}|f_{n,j}(X'_{n,j},t)|\le M_2\}\\
&\quad\le \Pr\{\sup_{t\in T}|f_{n,j}(X_{n,j},t)-f_{n,j}(X'_{n,j},t)|\ge 2M_1\}.
\end{aligned}$$

Thus, for n large enough,

$$\sum_{j=1}^{k_n}\Pr\{\sup_{t\in T}|f(X_{n,j},t)|\ge 2M_1+M_2\}\le 8e^{-(c'_1-c'_2)\epsilon_n^{-1}}.$$

This implies the claim. □

Lemma 2.2. *Under the notation in Lemma 2.1, let d be a pseudometric in T and let v be a function on T. Suppose that:*

(i) (T,d) is totally bounded.

(ii) For each $\tau>0$,

$$\lim_{\delta\to 0}\limsup_{n\to\infty}\Pr\{\sup_{d(s,t)\le\delta}|U_n(s)-U_n(t)|\ge\tau\}=0.$$

(iii) For each $t\in T$, $U_n(t)\xrightarrow{\Pr}v(t)$.

(iv) $v(t)$ is uniformly continuous in (T,d).

Then,

$$\sup_{t\in T}|U_n(t)-v(t)|\xrightarrow{\Pr}0.$$

Proof. We have to prove that for each $\tau>0$,

$$\limsup_{n\to\infty}\Pr\{\sup_{t\in T}|U_n(t)-v(t)|\ge\tau\}\le\tau.$$

Take $\delta>0$ such that $\sup_{d(s,t)\le\delta}|v(s)-v(t)|\le\tau/3$ and

$$\limsup_{n\to\infty}\Pr\{\sup_{d(s,t)\le\delta}|U_n(s)-U_n(t)|\ge\tau/3\}\le\tau.$$

Hence,

$$\limsup_{n\to\infty}\Pr\{\sup_{d(s,t)\le\delta}|U_n(s)-v(s)-(U_n(t)-v(t))|\ge 2\tau/3\}\le\tau. \tag{2.1}$$

Take a function $\pi : T \to T$ with finite range such that $\sup_{t\in T} d(t,\pi(t)) \leq \delta$. By condition (iii),

$$\lim_{n\to\infty} \Pr\{\sup_{t\in T} |U_n(\pi(t)) - v(\pi(t))| \geq \tau/3\} = 0.$$

By (2.1),

$$\limsup_{n\to\infty} \Pr\{\sup_{t\in T} |U_n(t) - v(t) - (U_n(\pi(t)) - v(\pi(t)))| \geq 2\tau/3\} \leq \tau.$$

The two last estimations imply the claim. □

We also will need the following exponential inequality for empirical processes

Theorem 2.3. *(Talagrand [23], Theorem 1.4). With the above notation, suppose that $E[f(X_i,t)] = 0$, for each $T \in T$ and each $1 \leq i \leq n$. Then, there exists a universal constant K such that for any $\tau > 0$,*

$$\Pr\{|Z - E[Z]| \geq \tau\} \leq K \exp\left(-\frac{\tau}{KC} \log\left(1 + \frac{\tau C}{\sigma^2 + CE[Z]}\right)\right),$$

where $Z = \sup_{t\in T} |\sum_{i=1}^n f(X_i,t)|$, $C = \sup_{t\in T} \sup_{1\leq i\leq n} \|f(X_i,t)\|_\infty$, and $\sigma^2 = \sup_{t\in T} \sum_{i=1}^n \mathrm{Var}(f(X_i,t))$.

To get the large deviations for the considered empirical processes, we use the following theorem:

Theorem 2.4. *(Arcones [1], Theorem 3.1) Under the notation above, let $\{\epsilon_n\}$ be a sequence of positive numbers that converges to zero. Let $I : l_\infty(T) \to [0,\infty]$ and let $I_{t_1,\dots,t_m} : \mathbb{R}^m \to [0,\infty]$ be a function, where $t_1,\dots,t_m \in T$. Let d be a pseudometric in T.*

Consider the conditions:

(a.1) (T,d) is totally bounded.

(a.2) For each $t_1,\dots,t_m \in T$, $(U_n(t_1),\dots,U_n(t_m))$ satisfies the LDP with speed ϵ_n and good rate function $I_{t_1,\dots,t_m}$.

(a.3) For each $\tau > 0$,

$$\lim_{\eta\to 0} \limsup_{n\to\infty} \epsilon_n \log \Pr{}^*\{\sup_{d(s,t)\leq\eta} |U_n(t) - U_n(s)| \geq \tau\} = -\infty.$$

(b.1) For each $0 \leq c < \infty$, $\{z \in l_\infty(T) : I(z) \leq c\}$ is a compact set in $l_\infty(T)$.

(b.2) For each $A \subset l_\infty(T)$,

$$-I(A^o) \leq \liminf_{n\to\infty} \epsilon_n \log \Pr{}_*\{U_n \in A\} \leq \limsup_{n\to\infty} \epsilon_n \log \Pr{}^*\{U_n \in A\} \leq -I(\bar{A}).$$

If the set of conditions (a) is satisfied, then the set of conditions (b) holds with

$$I(z) = \sup\{I_{t_1,\dots,t_m}(z(t_1),\dots,z(t_m)) : t_1,\dots,t_m \in T, m \geq 1\}.$$

If the set of conditions (b) is satisfied for some pseudometric d, then the set of conditions (a) holds with

$$I_{t_1,\dots,t_m}(u_1,\dots,u_m) = \inf\{I(z) : z \in l_\infty(T), (z(t_1),\dots,z(t_m)) = (u_1,\dots,u_m)\}$$

and the pseudometric $\rho(s,t)=\sum_{k=1}^{\infty} k^{-2}\min(\rho_k(s,t),1)$, *where* $\rho_k(s,t)=\sup\{|u_2-u_1| : I_{s,t}(u_1,u_2)\le k\}$.

To check condition (a.3) in the previous theorem, we will use the following lemma:

Lemma 2.5. *Under the previous notation, let* d *be a pseudometric in* T *such that* (T,d) *is totally bounded. Suppose that:*

(i) $\lim_{M\to\infty,\ \eta\to 0} \limsup_{n\to\infty} \epsilon_n \log(\sum_{j=1}^{k_n} \Pr\{F_{n,j}^{(\eta)}(X_{n,j})\ge M\}) = -\infty$,

where $F_{n,j}^{(\eta)}(x)=\sup_{d(s,t)\le\eta}|f_{n,j}(x,s)-f_{n,j}(x,t)|$.

(ii) For each $0<M,\lambda<\infty$,

$$\lim_{\substack{\eta\to 0\\ a\to\infty}} \limsup_{n\to\infty} \epsilon_n \log E[\exp(\epsilon_n^{-1}\lambda \sum_{j=1}^{k_n} F_{n,j}^{(\eta)}(X_{n,j}) I_{M\ge F_{n,j}^{(\eta)}(X_{n,j})\ge a\epsilon_n})]=0.$$

(iii) For each $a,\eta_0>0$,

$$\lim_{\eta\to 0}\limsup_{n\to\infty}\sup_{d(s,t)\le\eta}|\sum_{j=1}^{k_n} E[(f_{n,j}(X_{n,j},s)-f_{n,j}(X_{n,j},t))I_{F_{n,j}^{(\eta_0)}(X_{n,j})\le a\epsilon_n}]|=0.$$

(iv) For each $a,\eta_0>0$,

$$E[\sup_{t\in T}|\sum_{j=1}^{k_n}(f_{n,j}(X_{n,j},t)I_{F_{n,j}^{(\eta_0)}(X_{n,j})\le a\epsilon_n}-E[f_{n,j}(X_{n,j},t)I_{F_{n,j}^{(\eta_0)}(X_{n,j})\le a\epsilon_n}])|]\to 0.$$

(v) For each $a,\eta_0>0$,

$$\lim_{\eta\to 0}\limsup_{n\to\infty}\sup_{d(s,t)\le\eta}\epsilon_n^{-1}\sum_{j=1}^{k_n}\mathrm{Var}((f_{n,j}(X_{n,j},s)-f_{n,j}(X_{n,j},t))I_{F_{n,j}^{(\eta_0)}(X_{n,j})\le a\epsilon_n})=0.$$

Then, for each $\tau,M>0$,

$$\lim_{\eta\to 0}\limsup_{n\to\infty}\epsilon_n\log\Pr\{\sup_{d(s,t)\le\eta}|\sum_{j=1}^{k_n}(f_{n,j}(X_{n,j},s)-f_{n,j}(X_{n,j},t))|\ge\tau\}=-\infty,$$

and

$$\lim_{\eta\to 0}\limsup_{n\to\infty}\epsilon_n\log\Pr\{\sup_{d(s,t)\le\eta}|\sum_{j=1}^{k_n}(f_{n,j}(X_{n,j},s)-f_{n,j}(X_{n,j},t)$$
$$-E[(f_{n,j}(X_{n,j},s)-f_{n,j}(X_{n,j},t))I_{F_{n,j}^{(\eta)}(X_{n,j})\le M}])|\ge\tau\}=-\infty.$$

Proof. Note that by conditions (ii) and (iii), the two limits above are equivalent. By (i) and (ii), given $c>0$, we may take $\eta_0,\lambda,M,a>0$ such that $\lambda\tau\ge 6c$,

$$\limsup_{n\to\infty}\epsilon_n\log(\sum_{j=1}^{k_n}\Pr\{F_{n,j}^{(\eta_0)}(X_{n,j})\ge M\})\le -c$$

and

$$\limsup_{n\to\infty} \epsilon_n \log E[\exp(\epsilon_n^{-1}\lambda \sum_{j=1}^{k_n} F_{n,j}^{(\eta_0)}(X_{n,j}) I_{M > F_{n,j}^{(\eta_0)}(X_{n,j}) \geq a\epsilon_n})] \leq 6^{-1}\lambda\tau.$$

Then,

$$\epsilon_n \log(\Pr\{\sup_{d(s,t)\leq \eta_0} |\sum_{j=1}^{k_n} (f_{n,j}(X_{n,j},s) - f_{n,j}(X_{n,j},t)) I_{F_{n,j}^{(\eta_0)}(X_{n,j})\geq M}| \geq 3^{-1}\tau\})$$

$$\leq \epsilon_n \log(\sum_{j=1}^{k_n} \Pr\{F_{n,j}^{(\eta_0)}(X_{n,j}) \geq M\}) \leq -c.$$

We also have that

$$\Pr\{\sup_{d(s,t)\leq \eta_0} |\sum_{j=1}^{k_n} (f_{n,j}(X_{n,j},s) - f_{n,j}(X_{n,j},t)) I_{M > F_{n,j}^{(\eta_0)}(X_{n,j}) \geq a\epsilon_n}| \geq 3^{-1}\tau\}$$

$$\leq \Pr\{\sum_{j=1}^{k_n} F_{n,j}^{(\eta_0)} I_{M > F_{n,j}^{(\eta_0)}(X_{n,j}) \geq a\epsilon_n} \geq 3^{-1}\tau\}$$

$$\leq e^{-3^{-1}\epsilon_n^{-1}\lambda\tau} E[\exp(\epsilon_n^{-1}\lambda \sum_{j=1}^{k_n} F_{n,j}^{(\eta_0)}(X_{n,j}) I_{M > F_{n,j}^{(\eta_0)}(X_{n,j}) \geq a\epsilon_n})] \leq e^{-c\epsilon_n^{-1}}.$$

Hence,

$$\limsup_{n\to\infty} \epsilon_n \log(\Pr\{\sup_{d(s,t)\leq \eta_0} |\sum_{j=1}^{k_n} (f_{n,j}(X_{n,j},s) - f_{n,j}(X_{n,j},t)) \times I_{F_{n,j}^{(\eta_0)}(X_{n,j}) \geq a\epsilon_n}| \geq (2/3)\tau\}) \leq -c.$$

By previous estimations and condition (iii), it suffices to consider $\sup_{d(s,t)\leq\eta} |Z_n(s) - Z_n(t)|$, where

$$Z_n(t) = \sum_{j=1}^{k_n} (f_{n,j}(X_{n,j},t) I_{F_{n,j}^{(\eta_0)}(X_{n,j}) < a\epsilon_n} - E[f_{n,j}(X_{n,j},t) I_{F_{n,j}^{(\eta_0)}(X_{n,j}) < a\epsilon_n}]).$$

It is easy to see that by Theorem 2.3, conditions (iv) and (v) imply that

$$\lim_{\eta\to 0} \limsup_{n\to\infty} \epsilon_n \log(\Pr\{|\sup_{d(s,t)\leq\eta} |Z_n(s) - Z_n(t)| - E[\sup_{d(s,t)\leq\eta} |Z_n(s) - Z_n(t)|]| \geq 3^{-1}\tau\}) = -\infty.$$

Hence, the claim follows. □

We will use some notation from convex analysis. We refer to Borwein and Lewis (2000) for the definitions of Fenchel conjugate and of closure of a convex function. Let

$$\Psi(x) = \begin{cases} x\log\left(\frac{x}{e}\right) + 1, & \text{if } x > 0; \\ 1 & \text{if } x = 0; \\ \infty & \text{if } x < 0. \end{cases} \tag{2.2}$$

Ψ is the Fenchel conjugate of the function $\Phi(x) = e^x - 1$: for each $x, y \in \mathbb{R}$,

$$xy \le \Psi(x) + \Phi(y),\ \Psi(x) = \sup_{y\in R}(xy - \Phi(y)),\ \Phi(y) = \sup_{x\in R}(xy - \Psi(x)).$$

We also will need the following lemmas:

Lemma 2.6. *Let X be a r.v. with values in a measurable space $(S, \mathcal{S})$. Let f be a measurable function in S. Then,*

$$\sup\{E[f(X)\gamma(X) - \Psi(\gamma(X))] : E[\gamma(X)] = 1\} = \log(E[\exp(f(X))]).$$

Proof. By the Jensen's inequality, for each γ with $E[\gamma(X)] = 1$,

$$\begin{aligned} E[f(X)\gamma(X) - \Psi(\gamma(X))] &= E[(f(X) - \log(\gamma(X)))\gamma(X)] \\ &\le \log(E[\exp(f(X) - \log(\gamma(X)))\gamma(X)]) \\ &= \log(E[\exp(f(X))]). \end{aligned}$$

Taking $\gamma(X) = P_M^{-1}e^{f(X)}I(|f(X)| \le M)$, where $0 < M < \infty$ and

$$P_M = E[e^{f(X)}I(|f(X)| \le M)],$$

we get that

$$\begin{aligned} &\sup\{E[f(X)\gamma(X) - \Psi(\gamma(X))] : E[\gamma(X)] = 1\} \\ &\quad \ge E[f(X)P_M^{-1}e^{f(X)}I(|f(X)| \le M) - \Psi(P_M^{-1}e^{f(X)}I(|f(X)| \le M))] \\ &\quad = \log(P_M), \end{aligned}$$

which tends to $\log(E[\exp(f(X))])$ as $M \to \infty$. □

Lemma 2.7. *Let X be a r.v. with values in a measurable space $(S, \mathcal{S})$. Let $f_1, \dots, f_m$ be measurable functions in S. Suppose that there exists a $\lambda > 0$ such that for each $1 \le j \le m$, $E[e^{\lambda|f_j(X)|}] < \infty$. Let*

$$I^{(1)}(u_1, \dots, u_m) = \sup\left\{\sum_{j=1}^m \lambda_j u_j - \log\left(E\left[\exp\left(\sum_{j=1}^m \lambda_j f_j(X)\right)\right]\right) : \lambda_1, \dots, \lambda_m \in \mathbb{R}\right\},$$

and let

$$\begin{aligned} I^{(2)}(u_1, \dots, u_m) = \inf\{E[\Psi(\gamma(X))] : E[\gamma(X)] = 1 \text{ and} \\ E[f_j(X)\gamma(X)] = u_j \text{ for each } 1 \le j \le m\}. \end{aligned}$$

Then, $I^{(1)}$ is a convex lower semicontinuous function; $I^{(2)}$ is a convex function; $I^{(1)}$ is the closure of the function $I^{(2)}$; and for each $u_1, \dots, u_m \in \mathbb{R}$,

$$I^{(1)}(u_1, \dots, u_m) \le I^{(2)}(u_1, \dots, u_m).$$

Proof. It is obvious that $I^{(1)}$ and $I^{(2)}$ are convex functions and $I^{(1)}$ is lower semi-continuous. By the Lemma 2.6, for each $u_1, \dots, u_m \in \mathbb{R}$, $\lambda_1, \dots, \lambda_m \in \mathbb{R}$, and each function γ with $E[\gamma(X)] = 1$ and $E[f_j(X)\gamma(X)] = u_j$ for each $1 \le j \le m$,

$$\begin{aligned}
&\sum_{j=1}^{m} \lambda_j u_j - \log\left(E\left[\exp\left(\sum_{j=1}^{m} \lambda_j f_j(X)\right)\right]\right) \\
&\qquad = E\left[\sum_{j=1}^{m} \lambda_j f_j(X)\gamma(X)\right] - \log\left(E\left[\exp\left(\sum_{j=1}^{m} \lambda_j f_j(X)\right)\right]\right) \\
&\qquad \le E[\Psi(\gamma(X))]
\end{aligned}$$

Hence, $I^{(1)}(u_1, \dots, u_m) \le I^{(2)}(u_1, \dots, u_m)$.

By Theorem 4.2.8 in Borwein and Lewis [6], to prove that $I^{(2)}$ is the closure of the function $I^{(1)}$, it suffices to prove that the Fenchel conjugates $I^{(1)}$ and $I^{(2)}$ agree. We have that for each $\lambda_1, \dots, \lambda_m \in \mathbb{R}$,

$$\begin{aligned}
&\sup\left\{\sum_{j=1}^{m} \lambda_j u_j - I^{(1)}(u_1, \dots, u_m) : u_1, \dots, u_m \in \mathbb{R}\right\} \\
&\qquad = \log\left(E\left[\exp\left(\sum_{j=1}^{m} \lambda_j f_j(X)\right)\right]\right),
\end{aligned}$$

and

$$\begin{aligned}
&\sup\{\sum_{j=1}^{m} \lambda_j u_j - I^{(2)}(u_1, \dots, u_m) : u_1, \dots, u_m \in \mathbb{R}\} \\
&= \sup\{\sum_{j=1}^{m} \lambda_j u_j - E[\Psi(\gamma(X))] : E[\gamma(X)] = 1, \\
&\qquad E[f_j(X)\gamma(X)] = u_j \text{ for each } 1 \le j \le m, u_1, \dots, u_m \in \mathbb{R}\} \\
&= \sup\{E[\sum_{j=1}^{m} \lambda_j f_j(X)\gamma(X) - \Psi(\gamma(X))] : E[\gamma(X)] = 1\} \\
&= \log\left(E\left[\exp\left(\sum_{j=1}^{m} \lambda_j f_j(X)\right)\right]\right),
\end{aligned}$$

by Lemma 2.6. □

By Theorem 5.2 in Donsker and Varadhan (1976), if for each $\lambda > 0$ and each $1 \le j \le m$, $E[e^{\lambda |f_j(X)|}] < \infty$, then the functions $I^{(1)}$ and $I^{(2)}$ in the previous lemma agree.

Next, we present necessary and sufficient conditions for the large deviations of empirical processes. The set–up for sums of i.i.d. r.v.'s is as follows. Let $(S, \mathcal{S}, \nu)$ be a probability space. Let $\Omega = S^{\mathbb{N}}$, $\mathcal{A} = \mathcal{S}^{\mathbb{N}}$, and $Q = \nu^{\mathbb{N}}$. Let X_n be the n–th projection from Ω into S. Then, $\{X_n\}_{n=1}^{\infty}$ is a sequence of i.i.d.r.v.'s with values in S.

Let $\{f(\cdot,t) : t \in T\}$ be an image admissible Suslin class of measurable functions from S into $\mathbb{R}$. We consider the LDP for $\{n^{-1}\sum_{j=1}^n f(X_j,t) : t \in T\}$. Sethuraman [20] got the large deviations for classical empirical processes, i.e the case $T = \mathbb{R}$ and $f(x,t) = I(x \le t)$, $t \in \mathbb{R}$. The large deviations for general empirical processes was considered by Wu [25]. He obtained necessary and sufficient conditions for a bounded set of functions Wu [25] (Theorem 1). But, for unbounded classes, the sufficient conditions in Theorem 4 in Wu [25] are not necessary. Next theorem gives necessary and sufficient conditions for the large deviations of empirical processes.

Theorem 2.8. *Suppose that* $\sup_{t\in T}|f(X,t)| < \infty$ *a.s. Then, the following sets of conditions ((a) and (b)) are equivalent:*

(a.1) (T,d) *is totally bounded, where* $d(s,t) = E[|f(X,s) - f(X,t)|]$.

(a.2) There exists a $\lambda > 0$ *such that*

$$E[\exp(\lambda F(X))] < \infty,$$

where $F(x) = \sup_{t\in T}|f(x,t)|$.

(a.3) For each $\lambda > 0$, *there exists a* $\eta > 0$ *such that* $E[\exp(\lambda F^{(\eta)}(X))] < \infty$, *where* $F^{(\eta)}(x) = \sup_{d(s,t)\le\eta}|f(x,s) - f(x,t)|$.

(a.4) $\sup_{t\in T}|n^{-1}\sum_{j=1}^n (f(X_j,t) - E[f(X_j,t)])| \overset{\Pr}{\to} 0$.

(b) $\{n^{-1}\sum_{j=1}^n f(X_j,t) : t \in T\}$ *satisfies the large deviation principle in* $l_\infty(T)$ *with speed* n *and a good rate.*

Moreover, the rate function is given by

$$I(z) = \sup\{I_{t_1,\dots,t_m}(z(t_1),\dots,z(t_m)) : t_1,\dots,t_m \in T, m \ge 1\},$$

where

$$I_{t_1,\dots,t_m}(u_1,\dots,u_m) = \sup\Big\{\sum_{j=1}^m \lambda_j u_j - \log\Big(E\Big[\exp\Big(\sum_{j=1}^m \lambda_j f(X,t_j)\Big)\Big]\Big) : \lambda_1,\dots,\lambda_m \in \mathbb{R}\Big\}. \tag{2.3}$$

Proof. Assume the set of conditions (a). We apply Theorem 2.4. Condition (a.1) in Theorem 2.4 is obviously satisfied. Condition (a.2) in Theorem 2.4 follows from the Cramér–Chernoff theorem (see Corollary 6.1.6 in Dembo and Zeitouni, 1999). Condition (a.3) in Theorem 2.4 follows from Lemma 2.5.

Assume (b). Since we have a good rate, for each $0 < c < \infty$, there exists $M < \infty$ such that

$$\limsup_{n\to\infty} n^{-1}\log(\Pr\{\sup_{t\in T}|n^{-1}\sum_{j=1}^n f(X_j,t)| \ge M\}) \le -c.$$

By Lemma 2.1, for each $\delta > 0$,

$$\limsup_{n\to\infty} n^{-1}\log(n\Pr\{n^{-1}F(X) \ge 2M + \delta\}) \le -c,$$

is easy to see that this implies $E[e^{\lambda F(X)}] < \infty$, for each $0 < \lambda < c(2M+\delta)^{-1}$. So, condition (a.2) holds. Besides the rate function for the finite dimensional distributions is given by (2.3). Hence, by Theorem 2.4, for each $k \geq 1$, $(T, \rho_k^{(1)})$ is totally bounded, where

$$\rho_k^{(1)}(s,t) = \sup\{|u_2 - u_1| : I_{s,t}^{(1)}(u_1,u_2) \leq k\},$$

and

$$I_{s,t}^{(1)}(u_1,u_2) = \sup\{\lambda_1 u_1 + \lambda_2 u_2 - \log(E[\exp(\lambda_1 f(X,s) + \lambda_2 f(X,t))]) : \lambda_1, \lambda_2 \in \mathbb{R}\}.$$

By Lemma 2.7, for each $s, t \in T$, and each $k > 0$, $\rho_k^{(2)}(s,t) \leq \rho_k^{(1)}(s,t)$, where

$$\rho_k^{(2)}(s,t) = \sup\{|E[\gamma(X)(f(X,t) - f(X,s))]| : E[\gamma(X)] = 1, E[\Psi(\gamma(X))] \leq k\}. \quad (2.4)$$

Hence, for each $k > 0$, $(T, \rho_k^{(2)})$ is totally bounded. Given $1 > \delta > 0$, there exists a $b > \delta$ such that $E[F(X) I_{F(X) \geq 2^{-1} b}] < 2^{-2}\delta$. Hence, for each $s, t \in T$,

$$E[|f(X,t) - f(X,s)| I_{|f(X,t)-f(X,s)| \geq b}] \leq 2^{-1}\delta. \quad (2.5)$$

Take $k_0 > \Psi(4b\delta^{-1})$. Given $s, t, \in T$ with $\rho_{k_0}^{(2)}(s,t) < 2^{-2}\delta$. We define

$$\gamma(x) = a^{-1}(f(x,t) - f(x,s)) I_{b > f(x,t) - f(x,s) > 0},$$

where

$$a = E[(f(X,t) - f(X,s)) I_{b > f(X,t) - f(X,s) > 0}].$$

If $a^{-1} \leq 4\delta^{-1}$, then $E[\Psi(\gamma(X))] \leq \Psi(\delta^{-1} 4b)$. So,

$$\begin{aligned} &E[(f(X,t) - f(X,s)) I_{b > f(X,t) - f(X,s) > 0}] \\ &\quad \leq a^{-1} E[(f(X,t) - f(X,s))^2 I_{b > f(X,t) - f(X,s) > 0}] \\ &\quad = E[(f(X,t) - f(X,s))\gamma(X)] \leq 2^{-2}\delta. \end{aligned} \quad (2.6)$$

If $a^{-1} > 4\delta^{-1}$, then (2.5) holds obviously. Combining (2.4) and (2.5), we get that if $\rho_{k_0}^{(2)}(s,t) < 2^{-2}\delta$ then $d(s,t) \leq \delta$. Therefore, (T, d) is totally bounded, that is (a.1) holds.

Since (T, d) is totally bounded, by Theorem 2.4, for each $\tau, c < \infty$, there exists $0 < \eta < \infty$ such that

$$\limsup_{n \to \infty} n^{-1} \log \Pr\{\sup_{d(s,t) \leq \eta} |n^{-1} \sum_{j=1}^{n} f(X_j, t)| \geq \tau\}) \leq -c.$$

Using a previous argument, this limit and Lemma 2.1 implies that $E[e^{\lambda F^{(\eta)}(X)}] < \infty$, for each $0 < \lambda < c 2^{-1} \tau^{-1}$. So, condition (a.3) holds.

Condition (a.4) follows from Lemma 2.2. □

In condition (a.1) in the theorem above, we may use $d(s,t) = (E[|f(X,s) - f(X,t)|^p])^{1/p}$, for any $p \geq 1$, or $d(s,t) = E[|f(X,s) - f(X,t)| \wedge 1]$.

It is not sufficient to have that for some $\lambda > 0$, $E[\exp(\lambda F(X))] < \infty$ to have the large deviations for empirical processes. Let $T = \{0, 1, 2, \dots\}$ and let $\{\xi_n\}_{n=1}^{\infty}$

be a sequence of symmetric independent r.v.'s, with $\Pr\{|\xi_0| = 0\} = 1$, and for $n \geq 1$, $\Pr\{|\xi_n| \geq t\} = e^{-t}$ for each $t \geq n$ and $\Pr\{\xi_n = 0\} = 1 - e^{-n}$. Then, for each $0 < \lambda < 1$, $E[\exp(\lambda \sup_{n \geq 0} |\xi_n|)] < \infty$. However, condition (a.3) in Theorem 2.8 does not hold. We have that then $E[(\xi_n - \xi_0)^2] = 4(n_0+1)e^{-n_0}$. For each $\eta > 0$, if n_0 is an integer with $4(n_0+1)e^{-n_0} \leq \eta^2$, then $\sup_{m,n,d(m,n) \leq \eta} |\xi_m - \xi_n| \geq \sup_{n \geq n_0} |\xi_n|$. However, for $\lambda > 1$ and any n_0, $E[\exp(\lambda \sup_{n \geq n_0} |\xi_n|)] = \infty$.

Conditions (a.2) and (a.3) hold if for each $\lambda > 0$ $E[\exp(\lambda F(X))] < \infty$. However, however there are empirical processes for which the large deviations hold, but it is not true that for each $\lambda > 0$ $E[\exp(\lambda F(X))] < \infty$. Let $\{\xi_k\}_{k=1}^{\infty}$ be a sequence of symmetric i.i.d.r.v.'s with $\Pr\{|\xi_k| \geq t\} = e^{-t/a_k}$ for each $t > 0$ and each $k \geq 1$, where $\{a_k\}$ is a sequence of positive numbers such that $\sum_{k=1}^{\infty} a_k^2 < \infty$. Then, there exists a r.v. X and functions $f(x,k)$ such that for each $k \geq 1$ $f(X,k) = \xi_k$. Then, (a.1)–(a.4) in Theorem 2.8 hold, but for $\lambda > \sup_{k \geq 1} a_k$, $E[\exp(\lambda \sup_{k \geq 1} |\xi_k|)] = \infty$.

By Theorem 4.2 in Arcones [2], if for each $\lambda \in \mathbb{R}$ and each $t \in T$, $E[\exp(\lambda f(X,t))] < \infty$, then the rate function in the previous theorem is given by

$$I(z) = \inf\{E[\Psi(\gamma(X))] : E[\gamma(X)] = 1 \text{ and } z(t) = E[\gamma(X) f(X,t)] \text{ for each } t \in T\},$$

where Ψ is as in (2.2).

By Corollary 3.4 in Arcones [1], the previous theorem gives necessary and sufficient conditions for the LDP for Banach space values r.v.'s:

Corollary 2.9. *Let $\{X_j\}_{j=1}^{\infty}$ be a sequence of i.i.d.r.v.'s with values in a separable Banach space B. Then, the following conditions are equivalent:*

(a.1) There exists a $\lambda > 0$ such that

$$E[\exp(\lambda |X|)] < \infty.$$

(a.2) For each $\lambda > 0$, there exists a $\eta > 0$ such that

$$E[\exp(\lambda W^{(\eta)})] < \infty,$$

where

$$W^{(\eta)} = \sup\{|f_1(X) - f_2(X)| : f_1, f_2 \in B_1^*, E[|f_1(X) - f_2(X)|] \leq \eta\}.$$

(b) $\{n^{-1} \sum_{j=1}^{n} X_j\}$ satisfies the LDP in B with speed n.

Proof. By Theorem 2.5, it suffices to show that (a.1) and (a.2) imply that $\{f(X) : f \in B_1^*\}$ is totally bounded in L_1. Conditions (a.1) and (a.2) imply that $E[|X|] < \infty$. Hence, given $\epsilon > 0$, there exists a r.v. $Y = \sum_{j=1}^{m} x_j I(E_j)$ such that $E[|X - Y|] < \epsilon$, where $x_j \in B$ and $E_1, \ldots, E_m$ are disjoint Borel sets. It is easy to see that $\{f(Y) : f \in B_1^*\}$ is totally bounded in L_1. $\square$

If B is a finite dimensional space, then (a.1) in the previous corollary implies (a.2). So, for a finite dimensional Banach space, (a.1) and (b) are equivalent.

The previous theorem relaxes the conditions in Donsker and Varadhan [12]. Let $\{X_j\}_{j=1}^{\infty}$ be a sequence of i.i.d r.v.'s with $\Pr\{|X_j| > t\} = e^{-t}$, for each $t > 0$ and each $1 \leq j$. Then, $f(X_j, t) = |X_j - t|, 0 \leq t \leq 1$, defines a r.v. with values in $C[0,1]$ with the uniform norm. It is easy to see that Corollary 2.9 applies to this example. However, it is not true that for each $\lambda > 0$, $E[\exp(\lambda \sup_{0 \leq t \leq 1} |X - t|)] < \infty$. So, the theorem by Donsker and Varadhan [12] does not apply to this case. Jiang, Bhaskara Rao and Wang [17] obtained another type of conditions for the large deviations for Banach space random variables. Their conditions are more difficult to check.

The previous corollary gives the following for Hilbert space valued r.v.'s

Corollary 2.10. *Let $\{X_j\}_{j=1}^{\infty}$ be a sequence of i.i.d.r.v.'s with values in a separable Hilbert space $(H, < \cdot >)$. Then, the following conditions are equivalent:*

(a.1) There exists a $\lambda > 0$ such that

$$E[\exp(\lambda |X|)] < \infty.$$

(a.2) For each $\lambda > 0$, there exists an integer m such that

$$E[\exp(\lambda |X^{(m)}|))] < \infty,$$

where $X^{(m)} = \sum_{k=m+1}^{\infty} < X, h_k > h_k$ and $\{h_k\}$ is an orthogonal basis of H.

(b) $\{n^{-1} \sum_{j=1}^{n} X_j\}$ satisfies the LDP in B with speed n.

Proof. We show that under (a.1), (a.2) above is equivalent to (a.2) in Corollary 2.9.

Suppose that (a.2) in Corollary 2.10 holds. Then, given $\lambda > 0$, there exists an integer m such that $E[e^{4\lambda |X^{(m)}|}] < \infty$. Let $Y^{(m)} = X - X^{(m)} = \sum_{k=1}^{m} < X, h_k > h_k$. $Y^{(m)}$ is a finite dimensional r.v. So, there exists a $\eta > 0$ such that $E[e^{2\lambda V^{(m,\eta)}}] < \infty$, where

$$V^{(m,\eta)} = \sup\{|f_1(x) - f_2(x)| : f_1, f_2 \in B_1^*, E[|f_1(X) - f_2(X)|] \leq \eta\}.$$

Since $W^{\eta} \leq V^{(m,\eta)} + 2|X^{(m)}|$, (a.2) in Corollary 2.9 follows.

Suppose that (a.2) in Corollary 2.9 holds. For each $\lambda > 0$, there exists a $\eta > 0$ such that $E[\exp(\lambda W^{(\eta)})] < \infty$. Take an integer m such that $E[|X^{(m)}|] \leq \eta$, Then, for any $f_1 \in B_1^*$ with $f_1(h_j) = 0$, for each $1 \leq j \leq m$, and $f_2 = 0$, we have $E[|f_1(X) - f_2(X)| \leq E[|X^{(m)}|] \leq \eta$. So, $|X^{(m)}| \geq W^{(\eta)}$ and (a.2) in Corollary 2.10 follows. □

A similar result holds for r.v.'s with values in l_p, $p \geq 1$. In this case, (a.2) in Corollary 2.9 can be substituted by

(a.2)' For each $\lambda > 0$, there exists an integer m such that

$$E[\exp(\lambda |X^{(m)}|)] < \infty,$$

where $X = (Y^{(1)}, Y^{(2)}, \ldots,)$ and $X^{(m)} = (0, \ldots, 0, Y^{(m+1)}, Y^{(m+2)}, \ldots,)$.

We must notice the conditions above are sort of compactness conditions. A set K of a separable Hilbert space is compact if and only if it is closed, bounded

and

$$\lim_{m\to\infty} \sup_{x\in K} \sum_{k=m+1}^{\infty} |<x,h_k>|^2 = 0,$$

where $\{h_k\}$ is an orthogonal basis of H. For $p \geq 1$, a set K of l_p is compact if and only if it is closed, bounded and $\lim_{m\to\infty} \sup_{x\in K} \sum_{k=m+1}^{\infty} |x^{(k)}|^p = 0$, where $x = (x^{(1)}, x^{(2)}, \dots)$ (see for example page 6 in Diestel [11]).

Let $\{\xi_n\}_{n=1}^{\infty}$ be a sequence of symmetric independent r.v.'s, with $\Pr\{|\xi_n| \geq t\} = e^{-t}$ for each $t \geq n$ and $\Pr\{\xi_n = 0\} = 1 - e^{-n}$. Then, for each $p \geq 1$, $X = (\xi_1, \xi_2, \dots,)$ is a r.v. with values in l_p such that for each $0 < \lambda < 1$, $E[\exp(\lambda |X|_p)] < \infty$, where $|\cdot|_p$ is the l_p norm. However, X does not satisfy (a.2)'.

Our methods also apply to partial sums processes. First, we consider the case of a unique function.

Theorem 2.11. *Let $\{X_j\}_{j=1}^{\infty}$ be a sequence of i.i.d.r.v.'s. Then, the following conditions are equivalent:*

(a) For each $\lambda \in \mathbb{R}$, $E[\exp(\lambda X)] < \infty$.

(b) $\{n^{-1} \sum_{j=1}^{[nt]} X_j : 0 \leq t \leq 1\}$ satisfies the large deviation principle in $l_\infty([0,1])$ with speed n and a good rate.

Moreover, the rate function is given by

$$I(z) = \begin{cases} \int_0^1 \Psi_1(z'(t))\,dt, & \text{if } z(0) = 0 \\ & \text{and } z(t) \text{ is absolutely continuous,} \\ \infty & \text{else,} \end{cases} \tag{2.7}$$

where $\Psi_1(x) = \sup_y (xy - \Phi_1(y))$ and $\Phi_1(y) = \log(E[e^{yX}])$.

Proof. Let $U_n(t) = n^{-1} \sum_{j=1}^{[nt]} X_j$. Assume condition (a). We apply Theorem 2.4 with $d(s,t) = |s-t|$. Obviously $([0,1], d)$ is totally bounded. Given $0 \leq t_1 \leq \cdots \leq t_m \leq 1$, we have that

$$\sum_{j=1}^{m} \sum_{i=1}^{[nt_j]} \lambda_j X_i = \sum_{j=1}^{m} \sum_{p=1}^{j} \sum_{i=[nt_{p-1}]+1}^{[nt_p]} \lambda_j X_i \tag{2.8}$$

$$= \sum_{p=1}^{m} \sum_{j=p}^{m} \sum_{i=[nt_{p-1}]+1}^{[nt_p]} \lambda_j X_i = \sum_{p=1}^{m} \sum_{i=[nt_{p-1}]+1}^{[nt_p]} \sum_{j=p}^{m} \lambda_j X_i,$$

where $t_0 = 0$.

Hence,

$$n^{-1}\log E[\exp(\sum_{j=1}^{m}\sum_{i=1}^{[nt_j]}\lambda_j X_i)] \tag{2.9}$$

$$= n^{-1}\sum_{p=1}^{m}\sum_{i=[nt_{p-1}]+1}^{[nt_p]}\log E[\exp(\sum_{j=p}^{m}\lambda_j X)]$$

$$\to \sum_{p=1}^{m}(t_p - t_{p-1})\log E[\exp(\sum_{j=p}^{m}\lambda_j X)] = \int_0^1 \Phi_1(\sum_{j=1}^{m}\lambda_j I_{0\le s\le t_j})\, ds.$$

This limit and Theorem II.2 in Ellis [14] imply condition (a.2) in Theorem 2.4. To check condition (a.3) in Theorem 2.4, given $c, \tau > 0$, take $\lambda > 0$ such that $\lambda > 2^3 c\tau^{-1}$ and take an integer

$$m > \max\left(c^{-1}\log(E[\exp(\lambda|X|)]), 2^3\tau^{-1}E[|X|]\right).$$

Let $\{X_i'\}$ be a independent copy of $\{X_i\}$. Let $s_j = m^{-1}j$, for $0 \le j \le m$. Let $\pi(s) = s_j$ if $s_{j-1} \le s < s_j$ for some $j = 1, \dots, m-1$. Let $\pi(s) = s_m$ if $s_{m-1} \le s \le s_m$. By symmetrization (see Lemma 1.2.1 in Giné and Zinn [15]) and the Lévy inequality

$$\Pr\{\sup_{0\le s\le 1}|U_n(\pi(s)) - U_n(s)| \ge \tau\}$$

$$\le m\max_{1\le j\le m}\Pr\{\sup_{s_{j-1}\le s\le s_{j-1}}|\sum_{i=[ns_{j-1}]+1}^{[ns]}X_i| \ge n\tau\}$$

$$\le 2m\max_{1\le j\le m}\Pr\{\sup_{s_{j-1}\le s\le s_{j-1}}|\sum_{i=[ns_{j-1}]+1}^{[ns]}(X_i - X_i')| \ge 2^{-1}n\tau\}$$

$$\le 4m\Pr\{|\sum_{i=1}^{[m^{-1}n]+2}(X_i - X_i')| \ge 2^{-1}\tau n\}$$

$$\le 8m\Pr\{|\sum_{i=1}^{[m^{-1}n]+2}X_i| \ge 2^{-2}\tau n\}$$

$$\le 8me^{-\lambda 2^{-2}\tau n}(E[\exp(\lambda|X|)])^{[m^{-1}n]+2},$$

for n large enough. Observe that we may symmetrize because

$$E[n^{-1}|\sum_{i=[ns_{j-1}]+1}^{[ns]}X_i|] \le n^{-1}([m^{-1}n]+2)E[|X|] \le 2^{-3}\tau,$$

for n large enough.

Hence,

$$\limsup_{n\to\infty} n^{-1}\log(\Pr\{\sup_{0\le s\le 1}|U_n(\pi(s))-U_n(s)|\ge\tau\})$$
$$\le -\lambda 2^{-2}\tau+m^{-1}\log(E[\exp(\lambda|X|)])\le -c.$$

Therefore, (b) follows.

Assume (b). By Theorem 2.4, there exists a pseudometric ρ in $[0,1]$ such that $([0,1]),\rho)$ is totally bounded and for each $\tau>0$, there exists a $\delta>0$ such that

$$\limsup_{n\to\infty} n^{-1}\log\Pr\{\sup_{\substack{0\le s,t\le 1\\ \rho(s,t)\le\delta}}|U_n(s)-U_n(t)|\ge\tau\}\le -1.$$

Take $0\le s<t\le 1$ such that $\rho(s,t)\le\delta$. Then,

$$\limsup_{n\to\infty} n^{-1}\log\Pr\{|\sum_{i=[ns]+1}^{[nt]} X_i|\ge\tau n\}\le -1.$$

By Lemma 2.1, this implies that $E[e^{\lambda X}]<\infty$, for each $|\lambda|<\tau^{-1}$. So, condition (a) holds.

The results in Section 4 in Arcones [2] give that the rate function is the one claimed. □

Large deviations for partial sums processes in a similar set–up to the previous one have been considered by Borovkov [5] and Mogulskii [19].

For partial sums of empirical processes, we present the following theorem:

Theorem 2.12. *Let $\{X_j\}$ be a sequence of i.i.d.r.v.'s with values in a measurable space $(S,\mathcal{S})$. Let $\{f(x,t): t\in T\}$ be a class of measurable functions defined on S such that $\sup_{t\in T}|f(X,t)|<\infty$ a.s. Let $U_n(s,t)=n^{-1}\sum_{j=1}^{[ns]} f(X_j,t)$. Then, the following sets of conditions (a) and (b) are equivalent:*

(a.1) (T,d) is totally bounded, where $d(s,t)=E[|f(X,s)-f(X,t)|]$.

(a.2) For each $\lambda>0$,

$$E[\exp(\lambda F(X))]<\infty,$$

where $F(x)=\sup_{t\in T}|f(x,t)|$.

(a.3) $\sup_{t\in T}|n^{-1}\sum_{j=1}^{n}(f(X_j,t)-E[f(X_j,t)])|\overset{\Pr}{\to}0$.

(b) $\{U_n(s,t): s\in[0,1], t\in T\}$ satisfies the large deviation principle in $l_\infty([0,1]\times T)$ with speed n and a good rate.

Moreover, the rate function is given by

$$I(z)=\inf\{\textstyle\int_0^1 E[\Psi(\gamma(X,u))]\,du: E[\gamma(X,u)]=1 \text{ for each } 0\le u\le 1, \quad (2.10)$$
$$\text{and } z(s,t)=\textstyle\int_0^s E[f(X,t)\gamma(X,u)]\,du \text{ for each } (s,t)\in[0,1]\times T\}.$$

Proof. The proof of (a) is equivalent to (b) is similar to that of Theorem 2.8 and it is omitted. The new part is the form of the rate function. By the argument in (2.7) for $0 \leq s_1 \leq \cdots \leq s_m \leq 1$ and $t_1, \ldots, t_m \in T$,

$$\sum_{j=1}^{m}\sum_{k=1}^{m}\lambda_{j,k}\sum_{i=1}^{[ns_j]} f(X_i,t_k) = \sum_{p=1}^{m}\sum_{i=[s_{p-1}]+1}^{[ns_p]}\sum_{j=p}^{m}\sum_{k=1}^{m}\lambda_{j,k} f(X_i,t_k),$$

where $s_0 = 0$. Hence,

$$\begin{aligned}
&n^{-1}\log(E[\exp(\sum_{j=1}^{m}\sum_{k=1}^{m}\lambda_{j,k}\sum_{i=1}^{[ns_j]} f(X_i,t_k))]) \\
&= n^{-1}\sum_{p=1}^{m}\sum_{i=[s_{p-1}]+1}^{[ns_p]}\log(E[\exp(\sum_{j=p}^{m}\sum_{k=1}^{m}\lambda_{j,k} f(X_i,t_k))]) \\
&\to \sum_{p=1}^{m}(s_p - s_{p-1})\log\Big(E\Big[\exp\Big(\sum_{j=p}^{m}\sum_{k=1}^{m}\lambda_{j,k} f(X,t_k)\Big)\Big]\Big)\,du \\
&= \int_0^1 \log\Big(E\Big[\exp\Big(\sum_{j=1}^{m}\sum_{k=1}^{m}\lambda_{j,k} I(0 \leq u \leq s_j) f(X,t_k)\Big)\Big]\Big)\,du.
\end{aligned}$$

Hence, by Theorem II.2 in Ellis [14], the rate function for the LDP of

$$\{(U_n(s_1,t_1),\ldots,U_n(s_1,t_m),\ldots,U_n(s_m,t_1),\ldots,U_n(s_m,t_m))\}$$

is given by

$$\begin{aligned}
\sup\Big\{&\sum_{j=1}^{m}\sum_{k=1}^{m}\lambda_{j,k}u_{j,k} \\
&- \int_0^1 \log\Big(E\Big[\exp\Big(\sum_{j=1}^{m}\sum_{k=1}^{m}\lambda_{j,k} I(0 \leq u \leq s_j) f(X,t_k)\Big)\Big]\Big)\,du : \lambda_{j,k} \in \mathbb{R}\Big\}.
\end{aligned}$$

The arguments in Lemma 2.7 give that

$$\begin{aligned}
&I_{(s_1,t_1),\ldots,(s_m,t_m)}(u_{1,1},\ldots,u_{m,m}) \\
&:= \inf\Big\{\sum_{p=1}^{m}(s_p - s_{p-1})E[\Psi(\gamma_p(X))] : E[\gamma_j(X)] = 1 \text{ for each } 1 \leq j \leq m\} \\
&\qquad \text{and } \sum_{p=1}^{j}(s_p - s_{p-1})E[f(X,t_k)\gamma_p(X)]\,du = u_{j,k} \text{ for each } 1 \leq j,k \leq m\Big\} \\
&= \inf\Big\{\int_0^1 E[\Psi(\gamma(u,X))]\,du : E[\gamma(X,u)] = 1 \text{ for each } 0 \leq u \leq 1, \\
&\qquad \text{and } \int_0^{s_j} E[f(X,t_k)\gamma(X,u)]\,du = u_{j,k} \text{ for each } 1 \leq j,k \leq m\Big\}.
\end{aligned}$$

It is easy to that the methods in Section 4 in Arcones [2], give that

$$\sup\{I_{(s_1,t_1),\dots,(s_m,t_m)}(z(s_1,t_1),\dots,z(s_m,t_m)) : s_i \in [0,1], t_i \in T, \text{ for } 1 \le i \le m\},$$

is the rate in (2.9). □

For Banach space r.v.'s, we have that:

Corollary 2.13. *Let* $\{X_j\}_{j=1}^{\infty}$ *be a sequence of i.i.d.r.v.'s with values in a separable Banach space* B. $l_\infty([0,1], B)$ *is the Banach space consisting by the bounded functions from* $[0,1]$ *into* B *with the norm* $\|x\|_{\infty,T,B} = \sup_{t\in T} |x(t)|$. *Then, the following conditions are equivalent:*

(a) For each $\lambda > 0$ $E[\exp(\lambda|X|)] < \infty$.

(b) $\{n^{-1}\sum_{j=1}^{[nt]} X_j : 0 \le t \le 1\}$ *satisfies the LDP in* $l_\infty([0,1], B)$ *with speed* n *and a good rate function.*

Moreover, the rate function is

$$I(z) = \inf\Big\{ \int_0^1 E[\Psi(\gamma(X,u))]\,du : E[\gamma(X,u)] = 1 \text{ for each } 0 \le u \le 1,$$
$$\text{and } z(t) = \int_0^t E[\gamma(X,u)X]\,du \text{ for each } t \in [0,1]\Big\}.$$

References

[1] M. A. Arcones, *The large deviation principle for stochastic processes I.* To appear in Theor. Probab. Applic. (2002).

[2] M. A. Arcones, *The large deviation principle for stochastic processes II.* To appear in Theor. Probab. Applic. (2002).

[3] R. R. Bahadur, *Some Limit Theorems in Statistics.* SIAM, Philadelphia, PA. 1971.

[4] R. R. Bahadur, and S. L. Zabell, *Large deviations of the sample mean in general vector spaces.* Ann. Probab. **57** (1979), 587-621.

[5] A. A. Borovkov, *Boundary–value problems for random walks and large deviations in function spaces.* Theor. Probab. Appl. **12** (1967), 575–595.

[6] J. M. Borwein and A. S. Lewis, *Convex Analysis and Nonlinear Optimization: Theory and Examples.* Springer–Verlag, New York. 2000.

[7] H. Chernoff, *A measure of asymptotic efficiency for tests of hypothesis based on the sum of observations.* Ann. Math. Statist. **23** (1952), 493–507.

[8] H. Cramér, *Sur un nouveau théorèm limite de la théorie des probabilities.* Actualités Sci. Ind. **736** (1938), 5–23.

[9] A. Dembo, and O. Zeitouni, (1999). *Large Deviations Techniques and Applications.* Second Edition, Springer, New York.

[10] J. D. Deuschel, and D. W. Stroock, *Large Deviations.* Academic Press, Inc., Boston, MA. 1989.

[11] J. Diestel, *Sequences and Series in Banach Spaces.* Springer–Verlag, New York. 1984.

[12] M. D. Donsker, and S. R. S. Varadhan, *Asymptotic evaluation of certain Markov process expectations for large time III.* Comm. Pure Appl. Math. **29** (1976), 389–461.

[13] R. M. Dudley, *Uniform Central Limit Theorems.* Cambridge University Press, Cambridge. 1999.

[14] R. S. Ellis, *Large deviations for a general class of random vectors.* Ann. Probab. **12** (1984), 1–12.

[15] E. Giné, and J. Zinn, Lectures on the central limit theorem for empirical processes. Lect. Notes in Math. **1221** (1986), 50–112. Springer–Verlag, New York.

[16] P. Groeneboom; J. Oosterhoff and F. H. Ruymgaart, Large deviation theorems for empirical probability measures. Ann. Probab. **7** (1979), 553–586.

[17] T. Jiang; M. Bhaskara Rao, and X. P. Wang, *Large deviations for moving average processes.* Stoch. Proc. Appl. **59** (1995), 309–320.

[18] M. Ledoux, and M. Talagrand, *Probability in Banach Spaces.* Springer–Verlag, New York. 1991.

[19] A. A. Mogulskii, *Large deviations for trajectories of multidimensional random walks.* Theor. Probab. Appl. **21** (1976), 300–315.

[20] J. Sethuraman, *On the probability of large deviations of families of sample means.* Ann. Math. Statist. **35** (1964), 1964 1304–1316.

[21] J. Sethuraman, *On the probability of large deviations of of mean for random variables in* $D[0,1]$. Ann. Math. Statist. **36** (1965), 280–285.

[22] J. Sethuraman, *Corrections to: "On the probability of large deviations of families of sample means".* Ann. Math. Statist. **41** (1970), 1376–1380.

[23] M. Talagrand, *New concentration inequalities in product spaces.* Inv. Math. **126** (1996), 505–563.

[24] S. R. S. Varadhan, *Large Deviations and Applications.* SIAM, Philadelphia, Pennsylvania. (1984).

[25] L. Wu, *Large deviations, moderate deviations and LIL for empirical processes.* Ann. Probab. **22** (1994), 17–27.

Department of Mathematical Sciences
Binghamton University
Binghamton, NY 13902, USA
E-mail address: arcones@math.binghamton.edu

Progress in Probability, Vol. 55, 225–238

Small Deviation Estimates for Some Additive Processes

Xia Chen and Wenbo V. Li

Abstract. We study the small deviation probabilities for real valued additive processes. This naturally leads to the small deviation for the corresponding range process. Our general results can be applied to a wide range of additive processes generated from fractional Brownian motions, stable processes, Brownian sheets, etc. As an application, limit inf type LIL are proved for additive stable processes.

1. Introduction

Let $X_j(t)$, $1 \leq j \leq d$, be independent copies of a given real valued stochastic process $\{X(t), t \in E\}$ with index set E and $X(t_0) = 0$ for some $t_0 \in E$. Define the corresponding additive process

$$\mathbb{X}(\mathbf{t}) = \mathbb{X}(t_1, \cdots, t_d) = \sum_{j=1}^{d} X_j(t_j), \quad \mathbf{t} = (t_1, \cdots, t_d) \in E^d.$$

There are various motivations for the study of the additive process $X(\mathbf{t})$, $\mathbf{t} \in E^d$, and it has been active investigated recently from different points of view, see Khoshnevisan, Xiao, and Zhong (2002a,b) for detailed discussion and the bibliography for further works in this area. First of all, additive processes play a role in the study of other more interesting multiparameter processes. For example, locally and with time suitable rescaled, the Brownian sheet closely resembles additive Brownian motion, see Dalang and Walsh (1993a,b), Dalang and Mountford (2002). They also arise in the theory of intersections and selfintersections of Brownian processes; see Khoshnevisan and Xiao (2001) and Chen and Li (2002). Moreover, recent progress has shown that additive processes are more amenable to analysis, as we will also see in this paper.

The main objective of this paper is a study of the small deviation probabilities for real valued additive processes. This naturally leads to the small deviation for

Received by the editors December 27, 2002.
2000 *Mathematics Subject Classification.* Primary 60G60; Secondary 60G17, 60F15.
Key words and phrases. Small deviation, Additive process, Stable process.
Supported in part by NSF Grant DMS-0102238 and DMS-0204513.

the corresponding range process. Our general results given in Theorem 2.1 and 3.1 can be applied to a wide range of additive processes generated from fractional Brownian motions, stable processes, Brownian sheets, etc. As an application, limit inf type LIL are proved for additive stable processes.

It is useful and important to view our main results of this paper, under the sup-norm for various additive processes, as a part of a bigger picture. The small deviation/ball probability studies the behavior of

$$\log \mu(x : \|x\| \leq \varepsilon) = -\phi(\varepsilon) \quad \text{as} \quad \varepsilon \to 0 . \tag{1.1}$$

for a given measure μ and a norm $\|\cdot\|$. In the literature, small deviation probabilities of various types are studied and applied to many problems of interest under different names such as small ball probability, lower tail behaviors, two sided boundary crossing probability and exit time.

For a Gaussian measure and any norm on a separable Banach space, there is a precise link, discovered in Kuelbs and Li (1993) and completed in Li and Linde (1999), between the function $\phi(\varepsilon)$ and the metric entropy of the unit ball of the reproducing kernel Hilbert space generated by μ. This powerful connection allows the use of tools and results from functional analysis and approximation theory to estimate small ball probabilities. The survey paper of Li and Shao (2001) on small ball probabilities for Gaussian processes, together with its extended references, covers much of the recent progress in this area. In particular, various applications and connections with other areas of probability and analysis are discussed.

For many other important processes such as Markov processes and additive processes, there is no general result available unless the process and the norm have the correct scaling (or self-similar) property. In that case (1.1) can be rewritten in terms of the first exit time of certain region and certain general results are known. For example, in the case of stable processes, the problems are related to the large deviation for occupation measures developed by Donsker and Varadhan (1977).

It is somewhat surprising that we are able to find the exact small deviation constants for various additive processes since the main results in many works in this area determine only the asymptotic behavior in (1.1) up to some constant factor in front of the rate. As far as we know, this is the first time that explicit constants are found for non-trivial multiparameter processes under the sup-norm.

The remaining of the paper is organized as follows. Section 2 contains the small deviations for the range process. Various remarks and examples are also given.

In Section 3, we obtain the small deviation for additive processes. The proof we present is much simpler than our original one but it is strictly based on one dimensional structure.

Section 4 establishes, as an application of our probability estimates, limit inf type LIL for additive stable processes. The key idea for the proof of the upper bound essentially comes from Kuelbs (1981). For the additive fractional Brownian motion and the additive fractional integrated Brownian motion, limit inf type LIL's are formulated

In Section 5, we first exam some related additive type multiparameter processes generated by a single copy. Amongst other implications, these results show that their small deviation constants are different from the additive case by a factor of d, which is the number of independent copies needed in the additive process. The corresponding limit inf type LIL is also given. Finally, we generalize our small deviation estimates to additive type processes with sums of independent processes which are not necessarily copies of each other.

Acknowledgment

The authors are grateful to Davar Khoshnevisan and Yimin Xiao for helpful discussions.

2. Small Deviations for Range

In this section, we first present a general relation between the small deviation behaviors of the sup-norm and the range. The basic observation is that the range is about twice the sup-norm when they take small values.

Theorem 2.1. *Let $X(t)$, $t \in E$, be a real valued stochastic process with index set E and $X(t_0) = 0$ for some $t_0 \in E$. Assume the process satisfies the shift inequality*

$$\mathbb{P}\left(\sup_{t\in E}|X(t) - x| \le \varepsilon\right) \le \mathbb{P}\left(\sup_{t\in E}|X(t)| \le \varepsilon\right) \tag{2.1}$$

for any $x \in \mathbb{R}$, $\varepsilon > 0$, and has the small ball property

$$\lim_{\varepsilon\to 0^+} \varepsilon^\beta \log \mathbb{P}\left(\sup_{t\in E}|X(t)| \le \varepsilon\right) = -A_\beta \tag{2.2}$$

where $\beta > 0$ and $0 < A_\beta < \infty$ is a constant. Then for the range

$$R = \sup_{s,t\in E}|X(t) - X(s)| = \sup_{t\in E} X(t) - \inf_{t\in E} X(t),$$

we have

$$\lim_{\varepsilon\to 0^+} \varepsilon^\beta \log \mathbb{P}(R \le \varepsilon) = -2^\beta A_\beta \tag{2.3}$$

More general, for any fixed constants $a, b \ge 0$,

$$\lim_{\varepsilon\to 0^+} \varepsilon^\beta \log \mathbb{P}(aR + bM \le \varepsilon) = -(2a + b)^\beta A_\beta \tag{2.4}$$

where $M = \sup_{t\in E}|X(t)|$.

Before we prove the result, a few remarks and examples are needed. First, the shift inequality (2.1) holds for all centered Gaussian processes since it is a special case of Anderson's inequality, and the small ball property (2.2) is satisfied by various Gaussian processes. See the survey of Li and Shao (2001) for more details. Of particular interests to us in this paper are fractional Brownian motion

$B_H(t)$ with $B_H(0) = 0$ and index parameter $H \in (0,1)$, and fractional integrated Brownian motion

$$W_\gamma(t) = \frac{1}{\Gamma(\gamma+1)} \int_0^t (t-s)^\gamma dW(s) \quad \gamma > -1/2.$$

where $W_0(t) = W(t) = B_{1/2}(t)$ is the standard Brownian motion. To be more precise, $\{B_H(t), t \geq 0\}$ is a Gaussian process with mean zero and covariance function

$$\mathbb{E}\, B_H(t) B_H(s) = \frac{1}{2}(|s|^{2H} + |t|^{2H} - |s-t|^{2H}), \quad 0 < H < 1.$$

The small ball property (2.2) was proved in Li and Linde (1998), namely,

$$\lim_{\varepsilon \to 0} \varepsilon^{2/(2\gamma+1)} \log \mathbb{P}\left(\sup_{0 \leq t \leq 1} |W_\gamma(t)| \leq \varepsilon \right) = -k_\gamma \tag{2.5}$$

for any $\gamma > -1/2$ and

$$\lim_{\varepsilon \to 0} \varepsilon^{1/H} \log \mathbb{P}\left(\sup_{0 \leq t \leq 1} |B_H(t)| \leq \varepsilon \right) = -C_H \tag{2.6}$$

where $k_\gamma, C_H \in (0, \infty)$ are given by

$$\begin{aligned} k_\gamma &= -\inf_{\varepsilon > 0} \varepsilon^{2/(2\gamma+1)} \log \mathbb{P}\left(\sup_{0 \leq t \leq 1} |W_\gamma(t)| \leq \varepsilon \right) \\ C_\gamma &= k_{H-1/2} \cdot (\Gamma(H+1/2))^{1/H} \\ &\quad \cdot \left((2H)^{-1} + \int_{-\infty}^0 ((1-s)^{H-1/2} - (-s)^{H-1/2})^2 ds \right)^{-1/(2H)} \end{aligned}$$

The existence of the constant in (2.6) is also proved in Shao (1999) by developing a weaker form of correlation inequality. In the Brownian motion case, i.e. $\gamma = 0$, it is well known that $k_0 = C_{1/2} = \pi^2/8$.

Second, the symmetric α-stable processes $S_\alpha(t)$ with $S_\alpha(0) = 0$, $0 < \alpha \leq 2$, is covered by Theorem 2.1. The shift inequality (2.1) for $S_\alpha(t)$ is easy to prove when the process is viewed as a mixture of Gaussian, see, e.g. Chen, Kuelbs and Li (2000). The small ball property (2.2) is well known and more precisely,

$$\lim_{\varepsilon \to 0+} \varepsilon^\alpha \log \mathbb{P}\left(\sup_{0 \leq t \leq 1} |S_\alpha(t)| \leq \varepsilon \right) = -\lambda_\alpha \tag{2.7}$$

where $\lambda_\alpha > 0$ is the principle Dirichlet eigenvalue for the fractional Laplacian operator associated with $S_\alpha(t)$ in the interval $[-1,1]$. It should be pointed out that (2.7) can be equivalently stated as

$$\lim_{t \to \infty} \frac{1}{t} \mathbb{P}(\tau > t) = -\lambda_\alpha$$

due to scaling property of $S_\alpha(t)$, where

$$\tau = \inf\{s : |S_\alpha(s)| > 1\}$$

is the first exit time of the interval domain $[-1,1]$. Little seems to be known concerning the explicit value of λ_α, $0<\alpha<2$ despite the often appearances of this constant in other problems. The best known bounds to date are

$$\Gamma(\alpha+1)\le\lambda_\alpha\le\Gamma(\frac{\alpha}{2}+1)\Gamma(\alpha+\frac{3}{2})\Big/\Gamma(\frac{\alpha+3}{2}),\quad 0<\alpha<2 \tag{2.8}$$

and it is a challenge to find more explicit expression for λ_α than the well known variation ones. For more information, see a recent survey of Li and Linde (2002). The existence of a constant λ_α in (2.7) and the result (2.3) are proved in Mogul'skii (1974) based on scaling and independent increment properties of symmetric α-stable processes.

Third, the exact distributions in infinite series forms for the range of Brownian motion and Brownian bridge were first found in Feller (1951) by observing the connection between the joint distribution function of the maximum and minimum of $W(t)$ for $0\le t\le 1$ and a certain distribution function arising in the Kolmogorov–Smirnov theorem on empirical distributions. In particular, the following more precise results are known:

$$\mathbb{P}\left(\sup_{0\le s,t\le 1}|W(t)-W(s)|\le\varepsilon\right)\sim\frac{8}{\varepsilon^2}\exp\left(-\frac{\pi^2}{2\varepsilon^2}\right)$$

as $\varepsilon\to\infty$, and this should be compared with the well known fact

$$\mathbb{P}\left(\sup_{0\le t\le 1}|W(t)|\le\varepsilon\right)\sim\frac{4}{\pi}\exp\left(-\frac{\pi^2}{8\varepsilon^2}\right).$$

Fourth, in the case of standard two dimensional Brownian sheet $W(s,t)$ and its tied (pinned) down variants, the correct decay rate for the sup-norm is known but not the existence of the constant, see Li and Shao (2001) for a brief history of the problem and open conjectures in three and higher dimension. Nevertheless, the proof for Theorem 2.1 can be used to obtain, as $\varepsilon\to 0$,

$$\begin{aligned}&\log\mathbb{P}\left(\sup_{0\le s,s',t,t'\le 1}|W(s,t)-W(s',t')|\le\varepsilon\right)\\ \approx\ &\log\mathbb{P}\left(\sup_{0\le s,t\le 1}|W(s,t)|\le\varepsilon\right)\\ \approx\ &\frac{1}{\varepsilon^2}\left(\log\frac{1}{\varepsilon}\right)^3\end{aligned}$$

where $\approx$ means both side the ratio differ by at most some constants above and below.

Fifth, our results also apply to the range of additive processes discussed in the next section, see (3.2) in Theorem 3.1. Finally we can present the proof of Theorem 2.1.

Proof. The lower bound of (2.3) follows easily from (2.2) and the fact

$$R=\sup_{s,t\in E}|X(t)-X(s)|\le 2\sup_{t\in E}|X(t)|=2M.$$

For the upper bound, fix a large integer N, we have by using $X(t_0) = 0$,

$$
\begin{aligned}
\mathbb{P}(R \le \varepsilon) &= \sum_{k=0}^{N-1} \mathbb{P}\left(\sup_{t\in E} X(t) - \inf_{t\in E} X(t) \le \varepsilon,\ \frac{k}{N}\varepsilon \le \sup_{t\in E} X(t) \le \frac{k+1}{N}\varepsilon\right) \\
&\le \sum_{k=0}^{N-1} \mathbb{P}\left(\inf_{t\in E} X(t) \ge -\frac{N-k}{N}\varepsilon,\ \frac{k}{N}\varepsilon \le \sup_{t\in E} X(t) \le \frac{k+1}{N}\varepsilon\right) \\
&\le \sum_{k=0}^{N-1} \mathbb{P}\left(\sup_{t\in E} \left|X(t) - \frac{2k-N+1}{2N}\varepsilon\right| \le \frac{N+1}{2N}\varepsilon\right) \\
&\le N \cdot \mathbb{P}\left(\sup_{t\in E} |X(t)| \le \frac{N+1}{2N}\varepsilon\right) = N \cdot \mathbb{P}\left(M \le \frac{N+1}{2N}\varepsilon\right)
\end{aligned}
$$

where the last step follows from the shift inequality (2.1). Thus from the small ball property (2.2),

$$
\begin{aligned}
\limsup_{\varepsilon\to 0} \varepsilon^\beta \log \mathbb{P}(R \le \varepsilon) &\le \limsup_{\varepsilon\to 0} \varepsilon^\beta \log \mathbb{P}\left(\sup_{t\in E} |X(t)| \le \frac{N+1}{2N}\varepsilon\right) \\
&= -\left(\frac{2N}{N+1}\right)^\beta A_\beta
\end{aligned}
$$

Taking $N \to \infty$, we obtain (2.3).

The result (2.4) follows from the relation

$$(\gamma + \beta/2)R \le \gamma R + \beta M \le (2\gamma + \beta)M$$

and the tight estimates (2.2) and (2.3). □

3. Small Deviations for Additive Processes

We present our main result here. Some generalizations are given in the last section. An inspection of our arguments reveals that the special structure of additive processes plays a very important role in our derivations.

Theorem 3.1. *Let $X_j(t)$ be independent copies of a stochastic process $X(t)$ index by E. Assume the conditions in Theorem 2.1 are satisfied. Then for the additive process $\mathbb{X}(\mathbf{t}) = \sum_{j=1}^d X_j(t_j)$, $\mathbf{t} \in E^d$,*

$$\lim_{\varepsilon\to 0+} \varepsilon^\beta \log \mathbb{P}\left(\sup_{\mathbf{t}\in E^d} |\mathbb{X}(\mathbf{t})| \le \varepsilon\right) = -d^{\beta+1} A_\beta \tag{3.1}$$

and

$$\lim_{\varepsilon\to 0+} \varepsilon^\beta \log \mathbb{P}\left(\sup_{\mathbf{t},\mathbf{s}\in E^d} |\mathbb{X}(\mathbf{t}) - \mathbb{X}(\mathbf{s})| \le \varepsilon\right) = -2^\beta d^{\beta+1} A_\beta \tag{3.2}$$

where A_β is the small ball constant for $X(t)$ given in (2.2).

Proof. We only need to show (3.1) since (3.2) follows from Theorem 2.1 and (3.1). First note that

$$\sup_{\mathbf{t}\in E^d} |\mathbb{X}(t)| \le \sup_{\mathbf{t}\in E^d} \sum_{j=1}^d |X_j(t_j)| = \sum_{j=1}^d \sup_{t\in E} |X_j(t)|.$$

Thus

$$\begin{aligned}
\mathbb{P}\left(\sup_{\mathbf{t}\in E^d} |\mathbb{X}(t)| \le \varepsilon\right) &\ge \mathbb{P}\left(\sum_{j=1}^d \sup_{t\in E} |X_j(t)| \le \varepsilon\right) \\
&\ge \mathbb{P}\left(\max_{1\le j\le d} \sup_{t\in E} |X_j(t)| \le d^{-1}\varepsilon\right) \\
&= \prod_{j=1}^d \mathbb{P}\left(\sup_{t\in E} |X_j(t)| \le d^{-1}\varepsilon\right) \\
&= \mathbb{P}^d\left(\sup_{t\in E} |X(t)| \le d^{-1}\varepsilon\right)
\end{aligned}$$

and we obtain the lower bound

$$\begin{aligned}
\liminf_{\varepsilon\to 0+} \varepsilon^\beta \log \mathbb{P}\left(\sup_{\mathbf{t}\in E^d} |\mathbb{X}(t)| \le \varepsilon\right) &\ge d\cdot \liminf_{\varepsilon\to 0+} \varepsilon^\beta \log \mathbb{P}\left(\sup_{t\in E} |X(t)| \le d^{-1}\varepsilon\right) \\
&= -d^{\beta+1} A_\beta
\end{aligned}$$

where the last line follows from (2.2).

For the upper bound, consider the range process $R_j(t)$ of X_j over the index set E, that is,

$$R_j = \sup_{s,t\in E} |X_j(t) - X_j(s)| = \sup_{t\in E} X_j(t) - \inf_{t\in E} X_j(t).$$

Then

$$\begin{aligned}
\sum_{j=1}^d R_j &= \sum_{j=1}^d \left(\sup_{t\in E} X_j(t) - \inf_{t\in E} X_j(t)\right) \\
&= \sum_{j=1}^d \sup_{t\in E} X_j(t) - \sum_{j=1}^d \inf_{t\in E} X_j(t) \\
&= \sup_{\mathbf{t}\in E^d} \sum_{j=1}^d X_j(t_j) + \sup_{\mathbf{t}\in E^d} \sum_{j=1}^d (-X_j(t_j)) \\
&\le \sup_{\mathbf{t}\in E^d} \left|\sum_{j=1}^d X_j(t_j)\right| + \sup_{\mathbf{t}\in E^d} \left|\sum_{j=1}^d X_j(t_j)\right| \\
&= 2 \sup_{\mathbf{t}\in E^d} |\mathbb{X}(t)|
\end{aligned}$$

Thus we have

$$\mathbb{P}\left(\sup_{\mathbf{t}\in E^d}|\mathbb{X}(t)|\le\varepsilon\right)\le\mathbb{P}\left(\sum_{j=1}^d R_j\le 2\varepsilon\right)$$

and by exponential Chebyshev inequality, for any $\lambda>0$,

$$\begin{aligned}\mathbb{P}\left(\sum_{j=1}^d R_j\le 2\varepsilon\right) &= \mathbb{P}\left(e^{-\lambda\sum_{j=1}^d R_j}\ge e^{-2\lambda\varepsilon}\right)\\ &\le e^{2\lambda\varepsilon}\cdot\mathbb{E}\,e^{-\lambda\sum_{j=1}^d R_j}\\ &= e^{2\lambda\varepsilon}\cdot\left(\mathbb{E}\,e^{-\lambda R}\right)^d. \qquad (3.3)\end{aligned}$$

Now it follows from (2.3) and Tauberian's theorem, see Li and Shao (2001, p547),

$$\lim_{\lambda\to\infty}\lambda^{-\beta/(\beta+1)}\log\mathbb{E}\,e^{-\lambda R}=-(\beta+1)(2/\beta)^{\beta/(\beta+1)}A_\beta^{1/(\beta+1)}$$

Taking $\lambda=2^{-1}\beta d^{\beta+1}A_\beta\varepsilon^{-(\beta+1)}$ in (3.3), we see that

$$\limsup_{\varepsilon\to 0^+}\varepsilon^\beta\log\mathbb{P}\left(\sup_{\mathbf{t}\in E^d}|\mathbb{X}(t)|\le\varepsilon\right)\le -d^{\beta+1}A_\beta.$$

Note that we know in fact that

$$\lim_{\varepsilon\to 0^+}\varepsilon^\beta\log\mathbb{P}\left(\sum_{j=1}^d R_j\le\varepsilon\right)=-d^{\beta+1}2^\beta A_\beta$$

from Lemma 2 in Li (2001). Here we try to stress the direct and relative easy upper bound estimates. Put out upper and lower bound together, we finish our proof of Theorem 3.1. □

4. Limit Theorems for additive stable processes

Consider the additive α-stable processes

$$\mathbb{S}_\alpha(\mathbf{t})=\sum_{j=1}^d S_j(t_j)$$

on $[0,\infty)^d$ constructed from α-stable processes $S_j(t)$.

Theorem 4.1.

$$\liminf_{T\to\infty}\left(T^{-1}\log\log T\right)^{1/\alpha}\sup_{\mathbf{t}\in[0,T]^d}|\mathbb{S}_\alpha(\mathbf{t})|=(d^{\alpha+1}\lambda_\alpha)^{1/\alpha}\qquad a.s.$$

Proof. According to Theorem 3.1 and small ball estimate (2.7),

$$\lim_{\varepsilon\to 0}\varepsilon^\alpha\log\mathbb{P}\left(\sup_{\mathbf{t}\in[0,1]^d}|\mathbb{S}_\alpha(\mathbf{t})|\le\varepsilon\right)=-d^{\alpha+1}\lambda_\alpha.$$

Given $\theta > 1$, let $T_k = \theta^k$, $k \geq 1$. For any $\lambda < (d^{\alpha+1}\lambda_\alpha)^{1/\alpha}$, using the scaling property and the above estimate

$$\sum_{k\geq 1} \mathbb{P}\left(\sup_{\mathbf{t}\in[0,T_k]^d} |\mathbb{S}_\alpha(\mathbf{t})| \leq \lambda \left(T_k(\log\log T_k)^{-1}\right)^{1/\alpha}\right)$$

$$= \sum_{k\geq 1} \mathbb{P}\left(\sup_{\mathbf{t}\in[0,1]^d} |\mathbb{S}_\alpha(\mathbf{t})| \leq \lambda(\log\log T_k)^{-1/\alpha}\right) < \infty.$$

Hence by the Borel-Cantelli lemma,

$$\liminf_{k\to\infty} \left(T_k^{-1}\log\log T_k\right)^{1/\alpha} \sup_{\mathbf{t}\in[0,T_k]^d} |\mathbb{S}_\alpha(\mathbf{t})| \geq \lambda \qquad a.s.$$

Now for any $T_k \leq T \leq T_{k+1}$,

$$\left(T^{-1}\log\log T\right)^{1/\alpha} \sup_{\mathbf{t}\in[0,T]^d} |\mathbb{S}_\alpha(\mathbf{t})|$$

$$\geq \left(\theta^{-1/\alpha} + o(1)\right)\left(T_k^{-1}\log\log T_k\right)^{1/\alpha} \sup_{\mathbf{t}\in[0,T_k]^d} |\mathbb{S}_\alpha(\mathbf{t})|.$$

Thus

$$\liminf_{T\to\infty} \left(T^{-1}\log\log T\right)^{1/\alpha} \sup_{\mathbf{t}\in[0,T]^d} |\mathbb{S}_\alpha(\mathbf{t})| \geq \theta^{1/\alpha}\lambda \qquad a.s.$$

Letting $\lambda \to (d^{\alpha+1}\lambda_\alpha)^{1/\alpha}$ and $\theta \to 1$ proves the lower bound.

The idea for the proof of the upper bound essentially comes from Kuelbs (1981). We take $T_k = 2^k$. Let $\lambda > (d^{\alpha+1}\lambda_\alpha)^{1/\alpha}$ and $\delta > 0$ be fixed. Choose $j \geq 1$, independent of k, so that $T_{k+j} \geq \delta^{-1}T_k$ and

$$\left(T_{k+j}^{-1}\log\log T_{k+j}\right)^{1/\alpha} < \delta\left(T_k^{-1}\log\log T_{k+1}\right)^{1/\alpha}, \qquad \forall k \geq 1. \tag{4.1}$$

Next define the events

$$D_k \equiv \Big\{ \left(T^{-1}\log\log T\right)^{1/\alpha} \sup_{\mathbf{t}\in[0,T]^d} |\mathbb{S}_\alpha(\mathbf{t})| > \lambda \text{ for all } T \geq T_{k+j},$$
$$\left(T_k^{-1}\log\log T_k\right)^{1/\alpha} \sup_{\mathbf{t}\in[0,T_k]^d} |\mathbb{S}_\alpha(\mathbf{t})| \leq \lambda\Big\}.$$

Then from (4.1),

$$\Big\{ \left(T^{-1}\log\log T\right)^{1/\alpha} \sup_{\mathbf{t}\in[T_k,T]^d} |\mathbb{S}_\alpha(\mathbf{t}) - \mathbb{S}_\alpha(\mathbf{T_k})| > (1+\delta)\lambda \text{ for all } T \geq T_{k+j},$$
$$\left(T_k^{-1}\log\log T_k\right)^{1/\alpha} \sup_{\mathbf{t}\in[0,T_k]^d} |\mathbb{S}_\alpha(\mathbf{t})| \leq \lambda\Big\} \subset D_k$$

where $\mathbf{T_k} = (T_k, \cdots, T_k)$.

Hence by independence of increment, stationarity, and scaling

$$\mathbb{P}(D_k) \geq p_k(\lambda) \cdot \mathbb{P}\left(\phi(T) \sup_{\mathbf{t}\in[0,T-T_k]^d} |\mathbb{S}_\alpha(\mathbf{t})| > (1+\delta)\lambda \text{ for all } T \geq T_{k+j}\right)$$

$$\geq p_k(\lambda) \cdot \mathbb{P}\left(\phi(T) \sup_{\mathbf{t}\in[0,(1-\delta)T]^d} |\mathbb{S}_\alpha(\mathbf{t})| > (1+\delta)\lambda \text{ for all } T \geq T_{k+j}\right)$$

$$= p_k(\lambda) \cdot \mathbb{P}\left(\phi(T) \sup_{\mathbf{t}\in[0,T]^d} |\mathbb{S}_\alpha(\mathbf{t})| > (1-\delta)^{-1/\alpha}(1+\delta)\lambda \text{ for all } T \geq T_{k+j}\right)$$

where

$$p_k(\lambda) = \mathbb{P}\left(\left(T_k^{-1}\log\log T_k\right)^{1/\alpha} \sup_{\mathbf{t}\in[0,T_k]^d} |\mathbb{S}_\alpha(\mathbf{t})| \leq \lambda\right)$$

and

$$\phi(T) = \left(T^{-1}\log\log T\right)^{1/\alpha}.$$

Hence, for any integer $N \geq 1$, as long as $T_k \geq N$, i.e. $k \geq \log N$,

$$\mathbb{P}(D_k) \geq p_k(\lambda)$$
$$\cdot \mathbb{P}\left(\left(T^{-1}\log\log T\right)^{1/\alpha} \sup_{\mathbf{t}\in[0,T]^d} |\mathbb{S}_\alpha(\mathbf{t})| > (1-\delta)^{-1/\alpha}(1+\delta)\lambda \text{ for all } T \geq N\right).$$

On the other hand,

$$\sum_{k=1}^{\infty} \mathbb{P}(D_k) = \mathbb{E}\sum_{k=1}^{\infty} I_{D_k} \leq j$$

since among $\{D_k;\ \ k \geq 1\}$, at most j of them occur. Hence,

$$j \geq \sum_{k\geq \log N} p_k(\lambda)$$
$$\cdot \mathbb{P}\left(\left(T^{-1}\log\log T\right)^{1/\alpha} \sup_{\mathbf{t}\in[0,T]^d} |\mathbb{S}_\alpha(\mathbf{t})| > (1-\delta)^{1/\alpha}(1+\delta)\lambda \text{ for all } T \geq N\right)$$

Notice that by scaling, for $\lambda > (d^{\alpha+1}\lambda_\alpha)^{1/\alpha}$,

$$\sum_{k\geq \log N} p_k(\lambda) = \mathbb{P}\left(\sup_{\mathbf{t}\in[0,T_k]^d} |\mathbb{S}_\alpha(\mathbf{t})| \leq \lambda\left(T_k(\log\log T_k)^{-1}\right)^{1/\alpha}\right) = \infty.$$

We must have for all N

$$\mathbb{P}\left(\left(T^{-1}\log\log T\right)^{1/\alpha} \sup_{\mathbf{t}\in[0,T]^d} |\mathbb{S}_\alpha(\mathbf{t})| > (1-\delta)^{-1/\alpha}(1+\delta)\lambda \text{ for all } T \geq N\right) = 0.$$

Hence

$$\liminf_{T\to\infty} \left(T^{-1}\log\log T\right)^{1/\alpha} \sup_{\mathbf{t}\in[0,T]^d} |\mathbb{S}_\alpha(\mathbf{t})| \leq (1-\delta)^{-1/\alpha}(1+\delta)\lambda \quad a.s.$$

Let $\delta \to 0$ and $\lambda \to (d^{\alpha+1}\lambda_\alpha)^{1/\alpha}$ we obtain the desired upper bound and finished the proof. □

Next we formulate limiting behaviors for some additive Gaussian processes. First, consider the additive fractional Brownian motions

$$\mathbb{B}_H(\mathbf{t}) = \sum_{j=1}^{d} B_j(t_j)$$

on $[0,\infty)^d$ constructed from fractional Brownian motions $B_j(t)$ with index parameter $H \in (0,1)$. It is nature to expect

$$\liminf_{T\to\infty} \left(T^{-1}\log\log T\right)^H \sup_{\mathbf{t}\in[0,T]^d} |\mathbb{B}_H(\mathbf{t})| = -\left(d^{1+1/H}C_H\right)^H \tag{4.2}$$

based on the small deviation estimate

$$\lim_{\varepsilon\to 0} \varepsilon^{1/H} \log \mathbb{P}\left(\sup_{\mathbf{t}\in[0,1]^d} |\mathbb{B}_H(\mathbf{t})| \le \varepsilon\right) = -d^{1+1/H}C_H.$$

from (2.6) and Theorem 3.1. Similarly, for the additive fractional integrated Brownian motions

$$\mathbb{W}_\gamma(\mathbf{t}) = \sum_{j=1}^{d} W_j(t_j)$$

on $[0,\infty)^d$ constructed from fractional Brownian motions $W_j(t)$ with index parameter $\gamma > -1/2$. It is nature to expect

$$\liminf_{T\to\infty} \left(T^{-1}\log\log T\right)^{(2\gamma+1)/2} \sup_{\mathbf{t}\in[0,T]^d} |\mathbb{W}_\gamma(\mathbf{t})| = -\left(d^{1+2/(2\gamma+1)}k_\gamma\right)^{(2\gamma+1)/2} \tag{4.3}$$

based on the small deviation estimate

$$\lim_{\varepsilon\to 0} \varepsilon^{2/(2\gamma+1)} \log \mathbb{P}\left(\sup_{\mathbf{t}\in[0,1]^d} |\mathbb{W}_\gamma(\mathbf{t})| \le \varepsilon\right) = -d^{1+2/(2\gamma+1)}k_\gamma$$

from (2.5) and Theorem 3.1. The lower bounds for (4.2) and (4.3) follows easily from standard arguments given in the proof of Theorem 4.1. For the upper bounds in (4.2) and (4.3), we believe that detailed proofs can be obtained, but we will not go further in this direction since we do not have a nice and instructive arguments.

5. Some related multi-parameter processes

The additive process in the early sections requires copies of independent process $X(t)$ on E, and each addition sign can be changed to minus if $X(t)$ is symmetric. If the same process $X(t)$ is used with plus/minus sign, then we obtain some related multi-parameter processes. Using symmetry, these processes can be represented as

$$\mathbb{X}_{m,d}(\mathbf{t}) = \sum_{i=1}^{d-m} X(t_i) - \sum_{i=d-m+1}^{d} X(t_i), \quad \mathbf{t}\in E^d, \quad m \le d/2.$$

Next observe that the supremum norm of the process $\mathbb{X}_{m,d}(\mathbf{t})$ is simple. Indeed, it is easy to see

$$\sup_{\mathbf{t}\in E^d} |\mathbb{X}_{m,d}(\mathbf{t})| = \sup_{0\le s,t,u\le 1} |m(X(t)-X(s)) + (d-2m)X(u)| = mR + (d-2m)M.$$

and thus the small ball estimates follow from (2.4). More precisely,

$$\lim_{\varepsilon\to 0} \varepsilon^\beta \log \mathbb{P}\left(\sup_{\mathbf{t}\in E^d} |\mathbb{X}_{m,d}(\mathbf{t})| \le \varepsilon\right) = -d^\beta A_\beta. \tag{5.1}$$

Theorem 5.1. *Let $\mathbb{S}_{m,d}(\mathbf{t}), t \in [0,\infty)^d$ be the multi-parameter process generated by the same α-stable process $S_\alpha(t), t\ge 0$. Then*

$$\liminf_{T\to\infty} \left(T^{-1}\log\log T\right)^{1/\alpha} \sup_{\mathbf{t}\in[0,T]^d} |\mathbb{S}_{m,d}(\mathbf{t})| = -d\lambda_\alpha^{1/\alpha}.$$

In particular, for the range process $R_\alpha(T) = \sup_{s,t\in[0,T]} |S_\alpha(t) - S_\alpha(s)|$ of α-stable processes $S_\alpha(t)$,

$$\liminf_{T\to\infty} \left(T^{-1}\log\log T\right)^{1/\alpha} R_\alpha(T) = -2\lambda_\alpha^{1/\alpha}.$$

Proof. The proof of Theorem 5.1 follows from the same argument given in the proof of Theorem 4.1 and we omit the details. □

Next we consider a generalization of the additive processes in the early sections by requiring additions of independent processes $Y_j(t)$ on E, which are not necessarily copies of each other. These cover processes such as $\sum_{j=1}^d \pm X_j(t_j)$ where $X_j(t)$ are independent copies given in Theorem 3.1. Note that $X(t)$ need not be symmetric.

Theorem 5.2. *Let $Y_j(t)$, $1\le j\le d$, be independent stochastic processes index by E. Assume the conditions in Theorem 2.1 are satisfied for each $Y_j(t)$ with the small ball properties*

$$\lim_{\varepsilon\to 0+} \varepsilon^{\beta_j} \log \mathbb{P}\left(\sup_{t\in E} |Y_j(t)| \le \varepsilon\right) = -A_j, \quad 0 < A_j < \infty, \quad 1\le j\le d$$

and for a fixed m, $1\le m\le d$,

$$\beta = \beta_1 = \beta_2 = \cdots = \beta_m > \beta_{m+1} \ge \beta_{m+2} \ge \cdots \ge \beta_d \ge 0.$$

Then for the additive type process $\mathbb{Y}(\mathbf{t}) = \sum_{j=1}^d Y_j(t_j)$, $\mathbf{t}\in E^d$,

$$\lim_{\varepsilon\to 0+} \varepsilon^\beta \log \mathbb{P}\left(\sup_{\mathbf{t}\in E^d} |\mathbb{Y}(\mathbf{t})| \le \varepsilon\right) = -\left(\sum_{j=1}^m A_j^{1/(1+\beta)}\right)^{1+\beta} \tag{5.2}$$

and

$$\lim_{\varepsilon\to 0+} \varepsilon^\beta \log \mathbb{P}\left(\sup_{\mathbf{t},\mathbf{s}\in E^d} |\mathbb{Y}(\mathbf{t}) - \mathbb{Y}(\mathbf{s})| \le \varepsilon\right) = -2^\beta \left(\sum_{j=1}^m A_j^{1/(1+\beta)}\right)^{1+\beta}. \tag{5.3}$$

Proof. We only need to show (5.2) since (5.3) follows from Theorem 2.1 and (5.2). The proof of (5.2) follows more or less the arguments given for (3.1). Here we only point out the additional differences. For the lower bound, fix $\delta > 0$ small and set $A = \sum_{j=1}^m A_j^{1/(1+\beta)}$. Then

$$\begin{aligned}
\mathbb{P}\left(\sup_{\mathbf{t}\in E^d} |\mathbb{Y}(t)| \le \varepsilon\right) &\ge \left(\sum_{j=1}^d \sup_{t\in E} |Y_j(t)| \le \varepsilon\right) \\
&\ge \mathbb{P}\left(\max_{1\le j\le m} \sup_{t\in E} |Y_j(t)| \le A_j^{1/(1+\beta)}(A+d\delta)^{-1}\varepsilon,\right. \\
&\qquad \left.\max_{m<j\le d} \sup_{t\in E} |Y_j(t)| \le \delta(A+d\delta)^{-1}\varepsilon\right) \\
&= \prod_{j=1}^m \mathbb{P}\left(\sup_{t\in E} |Y_j(t)| \le A_j^{1/(1+\beta)}(A+d\delta)^{-1}\varepsilon\right) \\
&\qquad \cdot \prod_{j=m+1}^d \mathbb{P}\left(\sup_{t\in E} |Y_j(t)| \le \delta(A+d\delta)^{-1}\varepsilon\right)
\end{aligned}$$

and hence

$$\begin{aligned}
&\liminf_{\varepsilon\to 0^+} \varepsilon^\beta \log \mathbb{P}\left(\sup_{\mathbf{t}\in E^d} |\mathbb{Y}(t)| \le \varepsilon\right) \\
&\ge \sum_{j=1}^m \liminf_{\varepsilon\to 0^+} \varepsilon^\beta \log \mathbb{P}\left(\sup_{t\in E} |Y_j(t)| \le A_j^{1/(1+\beta)}(A+d\delta)^{-1}\varepsilon\right) \\
&\quad + \sum_{j=m+1}^d \liminf_{\varepsilon\to 0^+} \varepsilon^\beta \log \mathbb{P}\left(\sup_{t\in E} |Y_j(t)| \le \delta(A+d\delta)^{-1}\varepsilon\right) \\
&= -\sum_{j=1}^m A_j \left(A_j^{1/(1+\beta)}(A+d\delta)^{-1}\right)^{-\beta} \\
&= -A(A+d\delta)^\beta.
\end{aligned}$$

The lower bound follows by taking $\delta \to 0$.

For the upper bound, we use

$$\left(\sup_{\mathbf{t}\in E^d} |\mathbb{Y}(t)| \le \varepsilon\right) \le \left(\sum_{j=1}^d R_j \le 2\varepsilon\right) \le \left(\sum_{j=1}^m R_j \le 2\varepsilon\right)$$

and then Lemma 2 in Li (2001). □

References

[1] Chen, X., Kuelbs, J. and Li, W.V. (2000), A functional LIL for symmetric stable processes. *Ann. Probab.* **28**, 258–276.

[2] Chen, X. and Li, W.V. (2002), Large and moderate deviations for intersection local times, preprint.

[3] Dalang, R. and Mountford, T. (2002). Non-independence of excursions of the Brownian sheet and of additive Brownian motion, to appear in *Trans. Amer. Math. Soc.*

[4] Dalang, R. and Walsh, J. (1993a), Geography of the level sets of the Brownian sheet, *Prob. Th. Rel. Fields*, **96**, 153–176.

[5] Dalang, R. and Walsh, J.(1993b), The structure of a Brownian bubble, *Prob. Th. Rel. Fields*, **96**, 475–501.

[6] Donsker, M.D. and Varadhan, S.R.S. (1977). On laws of the iterated logarithm for local times. *Comm. Pure. Appl. Math.* **30**, 707-753.

[7] Feller, W. (1951), The asymptotic distribution of the range of sums of independent random variables, *Ann. Math. Stat.* **22**, 427-432.

[8] Khoshnevisan, D. and Xiao, Y. (2002), Level sets of additive L'evy process, *Ann. Prob.* (to appear).

[9] Khoshnevisan, D., Xiao, Y. and Zhong, Y. (2001). Local times of additive Lévy processes, I: regularity. (preprint).

[10] Khoshnevisan, D., Xiao, Y. and Zhong, Y. (2001). Measuring the range of an additive Lévy process. *Ann. Probab.* (to appear).

[11] Kuelbs, J. (1981), When is the cluster set of S_n/a_n empty? *Ann. Probab.* **9** 377–394.

[12] Kuelbs, J. and Li, W.V. (1993). Metric entropy and the small ball problem for Gaussian measures. *J. Funct. Anal.* **116** 133–157.

[13] Li, W.V. (2001), Small ball probabilities for Gaussian Markov processes under the L_p-norm, *Stoch. Processes and Their Appl.*, **92**, 87-102.

[14] Li, W.V. and Linde, W. (1999). Approximation, metric entropy and small ball estimates for Gaussian measures, *Ann. Probab.* **27**, 1556-1578.

[15] Li, W.V. and Linde, W. (2002), Recent developments on small ball probabilities for stable processes. *preprint.*

[16] Li, W.V. and Shao, Q.M. (2001). Gaussian Processes: Inequalities, Small Ball Probabilities and Applications. In: Stochastic Processes: Theory and Methods. Handbook of Statistics, Vol. 19 (2001), Edited by C.R. Rao and D. Shanbhag, 533-597.

[17] Mogul'skii, A.A. (1974). Small deviations in a space of trajectories. *Theor. Probab. Appl.* **19** , 726–736.

[18] Samorodnitsky, G. (1998). Lower tails of self-similar stable processes. *Bernoulli* **4**, 127-142.

[19] Shao, Q.M. (1999). A Gaussian correlation inequality and its applications to the existence of small ball constant, preprint.

[20] Taylor, S.J. (1967). Sample path properties of a transient stable process. *J. Math. Mech.* **16**, 1229-1246.

Department of Mathematics, University of Tennessee
Knoxville, TN 37996, USA
E-mail address: xchen@math.utk.edu

Department of Mathematical Sciences, University of Delaware
Newark, DE 19716, USA
E-mail address: wli@math.udel.edu

Part 6

Density Estimation

Progress in Probability, Vol. 55, 241–253

Convergence in Distribution of Self-Normalized Sup-Norms of Kernel Density Estimators

Evarist Giné[1], Vladimir Koltchinskii[2], and Lyudmila Sakhanenko

Abstract. Let f_n denote a kernel density estimator of a density f on the real line, for a bounded, compactly supported probability kernel. Under relatively weak smoothness conditions on f and K it is proved, for every $0 < \beta < 1/2$, that the sequence

$$\hat{A}_n \left(\frac{\sqrt{nh_n}}{\|K\|_2 \|f_n\|_\infty^{1/2-\beta}} \sup_{t \in \hat{D}_{a_n}} \frac{|f_n(t) - f(t)|}{f_n^\beta(t)} - \hat{A}_n \right)$$

converges in distribution to the double exponential law. Here $\hat{A}_n$ is constructed from the sample, $a_n \to \infty$ as a power of n and $\hat{D}_{a_n} = \{t : f_n(t) \geq a_n^{-1}\}$. Thus, this result provides distribution free asymptotic confidence bands for densities on the real line.

1. Introduction

Let X, X_i, $i \in \mathbb{N}$, be i.i.d. real random variables with density f, let K be a kernel and $\{h_n\}$, the bandwidths, be a monotonically decreasing sequence such that $h_n \to 0$ and $nh_n \to \infty$, and let

$$f_n(t) := \frac{1}{nh_n} \sum_{i=1}^{n} K\left(\frac{t - X_i}{h_n}\right), \quad n \in \mathbb{N}, \tag{1.1}$$

be the corresponding kernel density estimator. The object of this note is to obtain a distributional limit theorem for weighted supremum norms of $|f_n - f|$, suitably centered and normalized, so that the weights, the centering and the normalization depend only on the data, the intervals over which the supremum is taken increases to the whole line, and the limit is distribution free. Obviously, such a result can be used for the construction of asymptotic confidence bands for f. However, speed of convergence, which would be nice to have, will not be considered.

2000 *Mathematics Subject Classification.* Primary 62G07; Secondary 62G20, 60F15.
Key words and phrases. kernel density estimates, convergence in distribution, self-normalized sup norms, (asymptotic) confidence bands for densities.
[1]Research partially supported by NSF Grant No. DMS-0070382. [2]Research partially supported by NSA Grant No. MDA904-02-1-0075.

This research builds upon recent previous work by the same authors in [2], where we obtained convergence in distribution to the double exponential distribution of statistics of the form

$$A_n\left(\|\Psi(t)(f_n(t)-Ef_n(t))\|_\infty - A_n\right), \tag{1.2}$$

thus extending to suprema over the whole line the classical result of Bickel and Rosenblatt ([1]). In that article we actually characterized such convergence by necessary and sufficient conditions in terms of the tail probabilities of $\Psi(X)$. Implicit in that work, and made explicit below, is the observation that if the n-th sup is taken over the set $\{t : f(t) \geq (nh_n)^{-\theta}\}$ for $\theta \in (0,1)$, then the tail probability condition becomes superfluous and the result holds for all densities satisfying some not too restrictive smoothness properties. However A_n still depends on Ψ which in turn must satisfy a condition related to f (namely, $\Psi(x)f(x)^\beta \leq c$ for some $0 < \beta < 1/2$). Our main result below shows that one can replace A_n by a function of the data, that the weights can also be made solely dependent on the data, and that the sup can be taken over the data-dependent set $\{t : f_n(t) \geq (nh_n)^{-\theta}\}$.

This article is far from being self-contained: it relies instead on results and proofs from [3] and [2], and the methods used originate as well with these papers, particularly the latter. What ultimately make these results possible are adaptations of classical work on the asymptotic distribution of weighted suprema of stationary Gaussian processes (Theorem 2 in [2]) in combination with the KMT approximation of the empirical process, and estimates of moments and tail probabilities of suprema of empirical processes over certain classes of functions (see [3] and [2]).

Next we describe (and so, we get out of the way) the general assumptions for the main theorem below. We begin with hypotheses on the density, that in our case will coincide with hypotheses on the weights as well because we will take our weights to be $f^{-\beta}$ and $f_n^{-\beta}$ for some $0 < \beta < 1/2$ in the main result. These hypotheses are:

(D) f is a non-vanishing bounded, continuous, piecewise monotone density on $\mathbb{R}$ satisfying the following two conditions: a) for every $\delta > 0$ there exist $c \in (0, \infty)$ and $h_\delta > 0$ such that for all $|\tau| \leq h_\delta$ and all $t \in \mathbb{R}$,

$$c^{-1} f^\delta(t) \leq \frac{f(t+\tau)}{f(t)} \leq c f^{-\delta}(t),$$

and b) for all $r > 0$,

$$\sup_{f(t)\geq h^r, |\tau|\leq h} \left|\frac{f(t+\tau)}{f(t)} - 1\right| = o\left(\frac{1}{|\log h|}\right) \quad \text{as} \quad h \to 0.$$

Note that symmetric exponential as well as normal densities satisfy this condition. For a discussion on this type of conditions see [3].

The hypotheses on the kernel are:

(K$_\alpha$) K is a probability kernel of bounded variation, with support contained in $[-1/2, 1/2]$, and such that the function

$$r(t) = \frac{\int_{-\infty}^{\infty} K(u)K(u+t)du}{\|K\|_2^2}$$

satisfies that

$$r(t) = 1 - C|t|^\alpha + o\left(|t|^\alpha\right) \quad \text{as } t \to 0,$$

for some $0 < \alpha \le 2$, and $\sup_{|t|>\varepsilon} |r(t)| < 1$ for all $\varepsilon > 0$. We set $\kappa := \|K\|_\infty$.

Finally, here is the hypothesis on the bandwidths h_n:

(H) h_t, $t \ge 1$, is regularly varying at infinity with exponent $-\tau$ for some $\tau \in (0,1)$, $h_t \searrow 0$ and $th_t \nearrow \infty$ as $t \nearrow \infty$.

Taking $\Psi = f^{-\beta}$ for some $0 < \beta < 1/2$, these conditions imply all the general assumptions (K, F, UH, D, W, WD and H) in [2].

The results in the present article complement Theorem 6 of [2] in substantial ways. It is thus worthwhile to conclude this introduction with a statement of this theorem. For simplicity we state it only in the case of non-vanishing densities (this is not a harmless simplification, but we make it anyway). That theorem deals with weighted suprema, and here are the assumptions on the weight function $\Psi : \mathbb{R} \mapsto \mathbb{R}_+$:

(W.a): Ψ is strictly positive, continuous and piecewise monotone.

(W.b): For all $\delta > 0$ there exist $c \in (0, \infty)$ and $h_0 > 0$ such that, for all $|y| \le h_0$ and all $x \in \mathbb{R}$,

$$\frac{1}{c}\Psi^{-\delta}(x) \le \frac{\Psi(x+y)}{\Psi(x)} \le c\Psi^\delta(x).$$

(W.c): For all $r > 0$,

$$\sup_{\Psi(x) \le h^{-r},\ |y| \le h} \left| \frac{\Psi(x+y)}{\Psi(x)} - 1 \right| = o\left(\frac{1}{|\log h|}\right) \quad \text{as } h \to 0.$$

(W.D)$_\beta$: For a fixed $\beta \in (0,2)$, $\|f^\beta\|_{\Psi,\infty} := \sup_{t\in\mathbb{R}} \Psi(t) f^\beta(t) < \infty$.

The following quantities are needed in order to define the norming and centering constants A_n: for $\alpha \in (0,2]$, the same α from hypothesis (K$_\alpha$), we define the functions

$$\Psi_\alpha(u) := \sqrt{\frac{2}{\pi}} C^{1/\alpha} H_\alpha u^{2/\alpha - 1} e^{-u^2/2}, \quad u \ge 0, \tag{1.3}$$

where the constants H_α are absolute and are related to the asymptotic theory of stationary Gaussian processes (see [2]), with $H_2 = \sqrt{\pi}$, and

$$\Lambda_\alpha(u) := \int_{-\infty}^{\infty} \Psi_\alpha\left(\frac{u}{\Psi(y) f^{1/2}(y)}\right) dy, \quad u > 0. \tag{1.4}$$

With these definitions we also assume that the weight function Ψ and the density f satisfy

(w.2): $\Lambda_\alpha(u_0) < \infty$ for some $0 < u_0 < \infty$ (hence for all $u \ge u_0$).

Finally, we define the norming and centering constants A_n: if $A_T > 0$ is a function of $T > 0$ satisfying the relation

$$\lim_{T\to\infty} T\Lambda_\alpha(A_T) = 1, \tag{1.5}$$

and h_n are the bandwidths, then

$$A_n := A_{h_n^{-1}}. \tag{1.6}$$

It can be proved (see [2]) that A_T is, up to multiplicative constants, of the order of $\sqrt{\log T}$ as $T \to \infty$. We then have:

Theorem 0. (Special case of Theorem 6 in [2].) *Assume the conditions (D), (K_α) for some $\alpha \in (0, 2]$, (H), (W.a)-(W.c), $(WD)_\beta$ for some $\beta \in (0, 1/2)$, and (w.2), and let Ψ be normalized so that $\|\Psi f^{1/2}\|_\infty = 1$. Then, the condition*

$$\lim_{n\to\infty} n \Pr\left\{\Psi(X) > \sqrt{nh_n|\log h_n|}\right\} = 0,$$

is necessary and sufficient for

$$\lim_{n\to\infty} \Pr\left\{A_n\left(\frac{\sqrt{nh_n}}{\|K\|_2}\sup_{t\in\mathbb{R}}|\Psi(t)(f_n(t) - Ef_n(t))| - A_n\right) \le x\right\} = \exp\{-e^{-x}\}$$

for all $x \in \mathbb{R}$.

2. The limit theorem

We still need some extra definitions and hypotheses. For any $a > 0$ we set

$$D_a := \left\{t : f(t) \ge a^{-1}\right\}, \; \hat{D}_a := \left\{t : f_n(t) \ge a^{-1}\right\}. \tag{2.1}$$

Taking

$$a_n = O\left((nh_n)^\theta\right), \tag{2.2}$$

for some $\theta \in (0, 1)$, we assume that for all $0 < \gamma < 1/2$

$$\sup_{t\in\mathbb{R}} \left(a_n^\gamma \wedge f^{-\gamma}(t)\right) \mathrm{Bias}_n(t) = o\left((nh_n|\log h_n|)^{-1/2}\right) \text{ as } n \to \infty \tag{2.3}$$

and also that for some $p < 1$

$$\int_{\mathbb{R}} f^p(t)dt < +\infty. \tag{2.4}$$

Here,

$$\mathrm{Bias}_n(t) := |\mathbb{E}f_n(t) - f(t)|.$$

It is well known and easy to check that if K is a probability kernel supported by $[-1/2, 1/2]$ and

$$\omega_f(t; \delta) := \sup\left\{|f(t_1) - f(t_2)| : t_1, t_2 \in (t - \delta, t + \delta)\right\}.$$

is the modulus of continuity of f, then $\mathrm{Bias}_n(t) \le \omega_f(t; h_n)$, and if moreover $\int_{-1/2}^{1/2} uK(u)du = 0$, $v = \int_{-1/2}^{1/2} u^2K(u)du$ and f is twice continuously differentiable then $\mathrm{Bias}_n(t) \le 2^{-1}v(f''(t) + \omega_{f''}(t; h_n))h_n^2$. This is the extent to which we will deal with the bias. Finally, we let $\beta \in (0, 1/2)$ and let $\hat{A}_n$ be defined as the solution of the equation

$$\int_{\hat{D}_{a_n}} \Psi_\alpha \left(\frac{\hat{A}_n \|f_n\|_\infty^{1/2-\beta}}{f_n^{1/2-\beta}(y)} \right) dy = h_n, \tag{2.5}$$

where Ψ_α is defined by (1.3) if the kernel K satisfies (K_α). The result we are going to prove is as follows:

Theorem 1. *Under assumptions (D), (K_α) for some $0 < \alpha \leq 2$, (H), and (2.3) and (2.4) above, with $\beta \in (0, 1/2)$ and a_n, $\hat{D}_a$ and $\hat{A}_n$ as defined respectively in (2.2), (2.1) and (2.5), we have*

$$\lim_{n\to\infty} \Pr\left\{ \hat{A}_n \left(\frac{\sqrt{nh_n}}{\|K\|_2 \|f_n\|_\infty^{1/2-\beta}} \sup_{t \in \hat{D}_{a_n}} \frac{|f_n(t) - f(t)|}{f_n^\beta(t)} - \hat{A}_n \right) \leq x \right\} = \exp\{-e^{-x}\}. \tag{2.6}$$

for all $x \in \mathbb{R}$.

The proof will be based on Theorem 2 below and two lemmas.

Lemma 1. *Under the assumptions of Theorem 1, for any $\gamma \in (0, 1/2)$, the sequence*

$$\sqrt{\frac{nh_n}{|\log h_n|}} \sup_{t \in \mathbb{R}} \frac{|f_n(t) - f(t)|}{f^\gamma(t) \vee a_n^{-\gamma}}, \quad n \in \mathbb{N}, \tag{2.7}$$

is stochastically bounded.

Proof (Sketch). By the bias condition (2.3), we can eliminate the bias, that is, we can replace f by $\mathbb{E}f_n$ above, and then, the lemma follows along the steps in the proof of Theorem 2.1 in [3], with simplifications due to the particular form of the weights and to the fact that we are only proving tightness. The main difference with that theorem is that having a_n^γ in the denominator above allows us to obtain tightness without imposing any conditions on the tail of $\Psi(x) = f^{-\gamma}(x)$. The sup over $\mathbb{R}$ is decomposed into sups over sets A_n, B_n, C_n defined as follows:

$$A_n := \{t \in \mathbb{R} : f(t) < h_n^r\},$$

where r is chosen large enough, and, with $\varepsilon_n := 1/\log n$,

$$B_n := \left\{ t \in \mathbb{R} : h_n^r \leq f(t) \leq \varepsilon_n \left(\frac{|\log h_n|}{nh_n} \right)^{1/(2(1-\gamma))} \right\},$$

and

$$C_n := \left\{ t \in \mathbb{R} : f(t) > \varepsilon_n \left(\frac{|\log h_n|}{nh_n} \right)^{1/(2(1-\gamma))} \right\}.$$

Exactly as in [3], we can eliminate the centering for the sups over A_n and B_n. For the uncentered sup over A_n we proceed as follows. By (H) we can choose $\delta \in (0, 1)$ and $r > 0$ such that, with $p < 1$ as in (2.4), $nh_n^{r(1-\delta)(1-p)} \to 0$, and by (D), there exist $c < \infty$ and $n_0 < \infty$ such that $t \in A_n$ and $|s - t| \leq h_n/2$

imply that $f(s) \le cf^{1-\delta}(t) < ch_n^{r(1-\delta)}$ for all $n > n_0$. Therefore, setting $\lambda_n := nh_n|\log h_n|^{1/2}$, we have that for these values of n, for p as in (2.4) and for some $C < \infty$,

$$\Pr\left\{\sup_{t\in A_n} \frac{f^{-\gamma}(t)\sum_{i=1}^n K((t-X_i)/h_n)}{\lambda_n} > \varepsilon\right\} \le n\Pr\{f(X) \le ch_n^r\}$$
$$= n\int_{\{t:f(t)\le ch_n^r\}} f(t)dt$$
$$\le Cnh_n^{r(1-\delta)(1-p)}\int_{\mathbb{R}} f^p(t)dt \to 0.$$

On the uncentered sups over B_n, following the proof of Theorem 2.1 in [3] (more concretely, the proof of (2.11) there), we can establish that for any $c > 1$ and large enough n

$$\sup_{t\in B_n} \frac{(f^{-\gamma}(t)\wedge a_n^\gamma)\sum_{i=1}^n K((t-X_i)/h_n)}{\lambda_n} \le \max_{1\le i\le n} \frac{c\kappa(f^{-\gamma}(X_i)\wedge a_n^\gamma)}{\lambda_n}$$
$$+ \max_{1\le j\le n} \frac{c\kappa(f^{-\gamma}(X_j)\wedge a_n^\gamma)I_j\sum_{1\le i\le n, i\ne j} I(|X_i - X_j| \le h_n)}{\lambda_n}, \tag{2.8}$$

where

$$I_j = I_{n,j} := I\left(c^{-1}h_n^r \le f(t) \le c\varepsilon_n\left(\frac{|\log h_n|}{nh_n}\right)^{1/(2(1-\gamma))}\right).$$

Since $a_n^\gamma = o(\lambda_n)$, the first maximum at the right hand side of (2.8) tends to 0. To bound the second maximum, let P_j be conditional probability given X_j. Then, by a standard bound for binomial probabilities,

$$P_j\left\{\frac{(f^{-\gamma}(X_i)\wedge a_n^\gamma)I_j\sum_{1\le i\le n, i\ne j} I(|X_i - X_j| \le h_n)}{\lambda_n} \ge \varepsilon\right\}$$
$$\le \left(\frac{(n-1)ep_j(f^{-\gamma}(X_j)\wedge a_n^\gamma)}{\lambda_n\varepsilon}\right)^{(\varepsilon\lambda_n/(f^{-\gamma}(X_j)\wedge a_n^\gamma))\vee 1}, \tag{2.9}$$

where, by condition (D),

$$2c^{-1}h_nf(X_j) \le p_j := P_j\{|X - X_j| \le h_n\} \le 2ch_nf(X_j) \tag{2.10}$$

(provided that $I_j = 1$). Using (2.10), we can bound (2.9) further by

$$\left(\frac{2ecnh_nf^{1-\gamma}(X_j)}{\lambda_n\varepsilon}\right)^{(\varepsilon\lambda_n/a_n^\gamma)\vee 1},$$

and if $I_j = 1$ (otherwise the conditional probability in question is 0) by

$$\left(\frac{2ec^{2-\gamma}nh_n\varepsilon_n^{1-\gamma}|\log h_n|^{1/2}}{(nh_n)^{1/2}\lambda_n\varepsilon}\right)^{(\varepsilon\lambda_n/a_n^\gamma)\vee 1} \le \left(\frac{C_1\varepsilon_n^{1-\gamma}}{\varepsilon}\right)^{(\varepsilon\lambda_n/a_n^\gamma)\vee 1}$$

with some $C_1 < \infty$ (and all n large enough). Since $\varepsilon_n \to 0$ and $\lambda_n/a_n^\gamma \to \infty$ faster than a power of n, we can conclude that

$$\max_{1\le j\le n} P_j\left\{\frac{(f^{-\gamma}(X_i)\wedge a_n^\gamma)I_j\sum_{1\le i\le n, i\ne j} I(|X_i - X_j| \le h_n)}{\lambda_n} \ge \varepsilon\right\}$$

$$\le \left(\frac{C_1\varepsilon_n^{1-\gamma}}{\varepsilon}\right)^{\varepsilon\lambda_n/a_n^\gamma} = O(n^{-A})$$

for any $A > 0$. This implies that

$$\Pr\left\{\max_{1\le j\le n} \frac{(f^{-\gamma}(X_j)\wedge a_n^\gamma)I_j\sum_{1\le i\le n, i\ne j} I(|X_i - X_j| \le h_n)}{\lambda_n} \ge \varepsilon\right\}$$

$$\le \sum_{j=1}^n \mathbb{E}I_j P_j\left\{\frac{f^{-\gamma}(X_j)\wedge a_n^\gamma I_j\sum_{1\le i\le n, i\ne j} I(|X_i - X_j| \le h_n)}{\lambda_n} \ge \varepsilon\right\}$$

$$\le nO(n^{-A}) \to 0$$

for $A > 1$.

The centered suprema over C_n can be handled as in the proof of Theorem 2.1 in [3], which, for this part, does not require the tail assumption of that theorem, and they are stochastically bounded even if we replace the weights $(f^\gamma \vee a_n^{-\gamma})^{-1}$ by $f^{-\gamma}$. □

The next result is implicit in [2] and constitutes a nice complement to Theorem 6 there (or Theorem 0 above) when no assumptions are made on the tail probabilities of X.

Theorem 2. *Assume the hypotheses of Theorem 1, and let $\Psi(t)$ be a piecewise monotone function that satisfies the hypotheses (W.a) – (W.c) from [3] for $B_f = \mathbb{R}$, and, moreover, $\|\Psi f^\beta\|_\infty < \infty$ for some $0 < \beta < 1/2$, normalized so that $\|\Psi f^{1/2}\|_\infty = 1$. Let A_n be constants satisfying*

$$\int_{-\infty}^{\infty} \frac{1}{h_n}\Psi_\alpha\left(\frac{A_n}{\Psi(y)f^{1/2}(y)}\right)dy \to 1.$$

Then,

$$\lim_{n\to\infty} \Pr\left\{A_n\left(\frac{\sqrt{nh_n}}{\|K\|_2}\sup_{t\in D_{a_n}} |\Psi(t)(f_n(t) - f(t))| - A_n\right) \le x\right\} = e^{-e^{-x}} \quad (2.11)$$

for all $x \in \mathbb{R}$. If the bias condition (2.3) is not assumed, then $f(t)$ must be replaced by $\mathbb{E}f_n(t)$ in (2.11).

Proof. Recall the definition of the sets A_n, B_n and C_n in Lemma 1 for $\gamma = \beta$. (Although A_n is used for constants and for sets, it is clear what is meant by the context and there is no possibility of confusion.) Let

$$\tilde{D}_a := D_a \cap \{t : |t| \le a\},\ C_{n,a} := C_n \cap \tilde{D}_a^c.$$

Since $\Psi \le cf^{-\beta}$ and, on D_{a_n}, $f^\beta(t) \vee a_n^{-\beta} = f^\beta(t)$, it follows from the proof of Lemma 1 that

$$\lim_{n\to\infty} \sqrt{\frac{nh_n}{|\log h_n|}} \sup_{t\in A_n\cap D_{a_n}} |\Psi(t)(f_n(t) - \mathbb{E}f_n(t))| = 0 \text{ in pr.},$$

$$\lim_{n\to\infty} \sqrt{\frac{nh_n}{|\log h_n|}} \sup_{t\in B_n\cap D_{a_n}} |\Psi(t)(f_n(t) - \mathbb{E}f_n(t))| = 0 \text{ in pr.}$$

and it also follows from the proof of Theorem 2.1 in [3] that

$$\lim_{a\to\infty} \limsup_{n\to\infty} \mathbb{E}\sqrt{\frac{nh_n}{|\log h_n|}} \sup_{t\in C_{n,a}\cap D_{a_n}} |\Psi(t)(f_n(t) - \mathbb{E}f_n(t))| = 0$$

(see (2.20) there). By Lemma 5 in [2], for all $x \in \mathbb{R}$,

$$\limsup \frac{A_n + x/A_n}{\sqrt{|\log h_n|}} < \infty,$$

and therefore, replacing $\sqrt{\log h_n^{-1}}$ by $A_n + x/A_n$ in the previous limits, we get

$$\lim_{n\to\infty} \Pr\left\{A_n\left(\frac{\sqrt{nh_n}}{\|K\|_2} \sup_{t\in A_n\cap D_{a_n}} |\Psi(t)(f_n(t) - \mathbb{E}f_n(t))| - A_n\right) \ge x\right\} = 0, \quad (2.12)$$

$$\lim_{n\to\infty} \Pr\left\{A_n\left(\frac{\sqrt{nh_n}}{\|K\|_2} \sup_{t\in B_n\cap D_{a_n}} |\Psi(t)(f_n(t) - \mathbb{E}f_n(t))| - A_n\right) \ge x\right\} = 0 \quad (2.13)$$

and

$$\lim_{a\to\infty} \limsup_{n\to\infty} \Pr\left\{A_n\left(\frac{\sqrt{nh_n}}{\|K\|_2} \sup_{t\in C_{n,a}\cap D_{a_n}} |\Psi(t)(f_n(t) - \mathbb{E}f_n(t))| - A_n\right) \ge x\right\} = 0 \quad (2.14)$$

for all $x \in \mathbb{R}$. As in (4.12) in the proof of Theorem 6 in [3], we get for all a large enough and all $x \in \mathbb{R}$,

$$\lim_{n\to\infty} \Pr\left\{A_n\left(\frac{\sqrt{nh_n}}{\|K\|_2} \sup_{t\in \tilde{D}_a} |\Psi(t)(f_n(t) - \mathbb{E}f_n(t))| - A_n\right) \le x\right\} = e^{-e^{-x}}.$$

This together with (2.12)–(2.14) give

$$\lim_{n\to\infty} \Pr\left\{A_n\left(\frac{\sqrt{nh_n}}{\|K\|_2} \sup_{t\in D_{a_n}} |\Psi(t)(f_n(t) - \mathbb{E}f_n(t))| - A_n\right) \le x\right\} = e^{-e^{-x}}.$$

The limit (2.11) is now a consequence of this limit and the bias condition (2.3). □

For simplicity, we have stated Theorem 2 for densities that do not vanish anywhere, but it is true as well (and with the same proof) for densities which may vanish outside an open set, as in [2].

We will use Theorem 2 for the weight function $\Psi(t) = \|f^{1/2-\beta}\|_\infty f^{-\beta}(t)$.

Lemma 2. *Under the assumptions of Theorem 1, with* $0 < \delta < (1-\theta)/2$, *and* $\varepsilon_n := (nh_n)^{-\delta}$, *we have*

$$\sup_{t\in D_{a_n}\cup \hat{D}_{a_n}} \left|\frac{f_n(t)}{f(t)} - 1\right| = o_P(\varepsilon_n), \tag{2.15}$$

which also implies

$$\sup_{t\in D_{a_n}\cup \hat{D}_{a_n}} \left|\frac{f(t)}{f_n(t)} - 1\right| = o_P(\varepsilon_n) \tag{2.16}$$

and

$$\sup_{t\in D_{a_n}\cup \hat{D}_{a_n}} \left|\left(\frac{f}{f_n}\right)^{\eta}(t) - 1\right| = o_P(\varepsilon_n) \text{ and } \sup_{t\in D_{a_n}\cup \hat{D}_{a_n}} \left|\left(\frac{f_n}{f}\right)^{\eta}(t) - 1\right| = o_P(\varepsilon_n) \tag{2.17}$$

for all $\eta \in (0,1)$.

Proof. We take $\gamma > 0$ such that

$$\frac{1}{2} < 1-\gamma < \frac{1}{2\theta} - \frac{\delta}{\theta}.$$

By Lemma 1,

$$\begin{aligned}
\sup_{t\in D_{a_n}} \left|\frac{f_n(t)}{f(t)} - 1\right| &\le \sup_{t\in D_{a_n}} \frac{|f_n(t)-f(t)|}{f^{\gamma}(t)} \sup_{t\in D_{a_n}} f^{-(1-\gamma)}(t) \\
&\le \sup_{t\in\mathbb{R}} \frac{|f_n(t)-f(t)|}{f^{\gamma}(t)\vee a_n^{-\gamma}} \times O\left((nh_n)^{\theta(1-\gamma)}\right) \\
&= O_P\left(\sqrt{\frac{|\log h_n|}{nh_n}}\right) \times O\left((nh_n)^{\theta(1-\gamma)}\right) \\
&= O_P\left((nh_n)^{\theta(1-\gamma)-1/2}|\log h_n|^{1/2}\right) \\
&= o_P\left((nh_n)^{-\delta}\right) = o_p(\varepsilon_n).
\end{aligned} \tag{2.18}$$

Lemma 1 and the same calculation that leads to (2.18) also give

$$\begin{aligned}
\sup_{t\in \hat{D}_{a_n}, f_n(t)\ge f(t)} \left|1 - \frac{f(t)}{f_n(t)}\right| &= \sup_{t: f_n(t)\ge f(t)\vee a_n^{-1}} \frac{|f_n(t)-f(t)|}{f_n(t)} \\
&\le \sup_{t\in\mathbb{R}} \frac{|f_n(t)-f(t)|}{f^{\gamma}(t)\vee a_n^{-\gamma}} \sup_{t\in \hat{D}_{a_n}} f_n^{-(1-\gamma)}(t) \\
&\le \sup_{t\in\mathbb{R}} \frac{|f_n(t)-f(t)|}{f^{\gamma}(t)\vee a_n^{-\gamma}} \times O((nh_n)^{\theta(1-\gamma)}) \\
&\le o_P((nh_n)^{-\delta}),
\end{aligned}$$

in particular, with probability tending to 1,

$$\sup_{t\in\hat{D}_{a_n}, f_n(t)\geq f(t)} \frac{f_n(t)}{f(t)} \leq (1-\varepsilon_n)^{-1}.$$

This and (2.18) show that

$$\sup_{t\in\hat{D}_{a_n}} \left|\frac{f_n(t)}{f(t)} - 1\right| \leq \max\left(\sup_{t\in\hat{D}_{a_n}, f_n(t)\geq f(t)} \frac{|f_n(t)-f(t)|}{f_n(t)}\frac{f_n(t)}{f(t)}, \sup_{t\in D_{a_n}} \frac{|f_n(t)-f(t)|}{f(t)}\right)$$

$$= o_P\left(\frac{\varepsilon_n}{1-\varepsilon_n}\right) = o_P(\varepsilon_n), \tag{2.19}$$

and (2.15) follows from (2.18) and (2.19). Obviously, (2.16) is equivalent to (2.15). Finally, the inequality

$$|(1+x)^\eta - 1| \leq \eta(1-|x|)^{\eta-1}|x|, \; |x|<1,$$

which holds for $\eta \in (0,1)$, and (2.15)-(2.16) imply

$$\sup_{t\in D_{a_n}\cup\hat{D}_{a_n}} \left|\left(\frac{f}{f_n}\right)^\eta(t) - 1\right| = \sup_{t\in D_{a_n}\cup\hat{D}_{a_n}} \left|\left(1+\frac{f-f_n}{f_n}\right)^\eta(t) - 1\right| = o_P(\varepsilon_n)$$

and similarly

$$\sup_{t\in D_{a_n}\cup\hat{D}_{a_n}} \left|\left(\frac{f_n}{f}\right)^\eta(t) - 1\right| = o_P(\varepsilon_n).$$

□

Proof of Theorem 1. Let ε_n be as in Lemma 2. Note that if we define

$$E_n := \left\{\forall\, t\in D_{a_n}\; 1-\varepsilon_n < \frac{f_n(t)}{f(t)} < 1+\varepsilon_n \text{ and } 1-\varepsilon_n < \frac{f(t)}{f_n(t)} < 1+\varepsilon_n\right\}, \tag{2.20}$$

then Lemma 2 implies that

$$\Pr(E_n) \to 1 \quad\text{as}\quad n\to\infty. \tag{2.21}$$

Moreover, on the event E_n, the following inclusions hold:

$$D_{a_n(1-\varepsilon_n)} \subset \hat{D}_{a_n} \subset D_{a_n(1+\varepsilon_n)}. \tag{2.22}$$

To prove convergence in distribution of

$$\hat{A}_n\left(\frac{\sqrt{nh_n}}{\|K\|_2\|f_n\|_\infty^{1/2-\beta}} \sup_{t\in\hat{D}_{a_n}} \frac{|f_n(t)-f(t)|}{f_n^\beta(t)} - \hat{A}_n\right)$$

to the double exponential law, where $\hat{A}_n$ is defined in (2.5), we begin with the decomposition

$$\frac{f_n(t)-f(t)}{\|f_n\|_\infty^{1/2-\beta} f_n^\beta(t)} = \frac{f_n(t)-f(t)}{\|f\|_\infty^{1/2-\beta} f^\beta(t)} + \frac{f_n(t)-f(t)}{\|f\|_\infty^{1/2-\beta} f^\beta(t)} \left[\frac{\|f\|_\infty^{1/2-\beta}}{\|f_n\|_\infty^{1/2-\beta}} \frac{f^\beta(t)}{f_n^\beta(t)} - 1 \right]. \tag{2.23}$$

Now,

$$\left| 1 - \frac{\|f_n\|_\infty}{\|f\|_\infty} \right| \le \frac{\|f_n - f\|_\infty}{\|f\|_\infty} = O_P\left(\sqrt{\frac{|\log h_n|}{nh_n}} \right) = o_P(\varepsilon_n),$$

for instance, by Lemma 1. Hence, proceeding as in the proof (2.17) in Lemma 2, we also have

$$\left| \frac{\|f\|_\infty^{1/2-\beta}}{\|f_n\|_\infty^{1/2-\beta}} - 1 \right| = o_P(\varepsilon_n),$$

which, together with (2.17) for $\eta = \beta$ (note $|ab-1| \le |a-1||b| + |b-1|$) gives

$$\sup_{t \in D_{a_n} \cup \hat{D}_{a_n}} \left| \frac{\|f\|_\infty^{1/2-\beta}}{\|f_n\|_\infty^{1/2-\beta}} \frac{f^\beta(t)}{f_n^\beta(t)} - 1 \right| = o_P(\varepsilon_n). \tag{2.24}$$

By the inclusions (2.22), we have that on E_n,

$$\sup_{t \in \hat{D}_{a_n}} \frac{|f_n(t)-f(t)|}{f^\beta(t)} \le \sup_{t \in D_{a_n(1+\varepsilon_n)}} \frac{|f_n(t)-f(t)|}{f^\beta(t)} \le (1+\varepsilon_n)^\beta \sup_{t \in \mathbb{R}} \frac{|f_n(t)-f(t)|}{f^\beta(t) \vee a_n^{-\beta}},$$

hence, by Lemma 1 and (2.21),

$$\sqrt{\frac{nh_n}{|\log h_n|}} \sup_{t \in \hat{D}_{a_n}} \frac{|f_n(t)-f(t)|}{f^\beta(t)} = O_P(1).$$

This and (2.24) then give

$$\sqrt{\frac{nh_n}{|\log h_n|}} \sup_{t \in \hat{D}_{a_n}} \frac{|f_n(t)-f(t)|}{f^\beta(t)} \left[\frac{\|f\|_\infty^{1/2-\beta}}{\|f_n\|_\infty^{1/2-\beta}} \frac{f^\beta(t)}{f_n^\beta(t)} - 1 \right] = o_P(\varepsilon_n). \tag{2.25}$$

We will see below that $\hat{A}_n = O_P(\sqrt{\log n})$, so that, since ε_n tends to zero as a negative power of n, (2.25) shows that we can ignore the second term in the decomposition (2.23) and that it suffices to show that the random variables

$$\hat{A}_n \left(\frac{\sqrt{nh_n}}{\|K\|_2 \|f\|_\infty^{1/2-\beta}} \sup_{t \in \hat{D}_{a_n}} \frac{|f_n(t)-f(t)|}{f^\beta(t)} - \hat{A}_n \right)$$

converge in distribution to the double exponential law. That is, we must show that we can replace A_n by $\hat{A}_n$ in Theorem 2. To this end we argue as in the proof of Theorem 10 in [2]. First, by Theorem 2 above and Lemma 3 in [2], for any sequence A_n satisfying

$$h_n^{-1} \int_{D_{a_n}} \Psi_\alpha\left(\frac{A_n \|f\|_\infty^{1/2-\beta}}{f^{1/2-\beta}(y)} \right) dy \to 1 \text{ as } n \to \infty, \tag{2.26}$$

we have that

$$\lim_{n\to\infty} \Pr\left\{ A_n \left(\frac{\sqrt{nh_n}}{\|K\|_2 \|f\|_\infty^{1/2-\beta}} \sup_{t\in D_{a_n}} |f^{-\beta}(t)(f_n(t) - f(t))| - A_n \right) \le x \right\} = e^{-e^{-x}}. \tag{2.27}$$

for all $x \in \mathbb{R}$, and D_{a_n} in (2.26) can be replaced by $D_{a_n(1-\varepsilon_n)}$ or $D_{a_n(1+\varepsilon_n)}$. If A_n and $\bar{A}_n$ are two sequences satisfying (2.26), then they also satisfy the limit (2.27), and it follows that

$$\left| \frac{\bar{A}_n}{A_n} - 1 \right| = o(A_n^{-2}) \quad \text{as } n \to \infty. \tag{2.28}$$

On the other hand, if (2.28) holds, then A_n in (2.27) can be replaced by $\bar{A}_n$ (even if $\bar{A}_n$ is a sequence of random variables and (2.28) holds in probability, which is the case below).

Next, by Lemma 2 and arguing as in the proof of (2.24), we have

$$\sup_{t\in D_{a_n}\cup \hat{D}_{a_n}} \left| \left(\frac{\|f\|_\infty}{\|f_n\|_\infty} \frac{f_n(t)}{f(t)} \right)^{1/2-\beta} - 1 \right| = o_P(\varepsilon_n). \tag{2.29}$$

Since also the inclusion (2.22) holds on the event E_n, we have on an event $\tilde{E}_n$ whose probability tends to 1

$$\int_{D_{a_n(1-\varepsilon_n)}} \Psi_\alpha\left(\frac{\hat{A}_n \|f\|_\infty^{1/2-\beta}}{(1-\varepsilon_n) f^{1/2-\beta}(y)} \right) dy \le \int_{\hat{D}_{a_n}} \Psi_\alpha\left(\frac{\hat{A}_n \|f_n\|_\infty^{1/2-\beta}}{f_n^{1/2-\beta}(y)} \right) dy = h_n \tag{2.30}$$

$$\le \int_{D_{a_n(1+\varepsilon_n)}} \Psi_\alpha\left(\frac{\hat{A}_n \|f\|_\infty^{1/2-\beta}}{(1+\varepsilon_n) f^{1/2-\beta}(y)} \right) dy.$$

Defining now A_n^- and A_n^+ as the solutions of the equations

$$\int_{D_{a_n(1-\varepsilon_n)}} \Psi_\alpha\left(\frac{A_n^- \|f\|_\infty^{1/2-\beta}}{f^{1/2-\beta}(y)} \right) dy = h_n$$

and

$$\int_{D_{a_n(1+\varepsilon_n)}} \Psi_\alpha\left(\frac{A_n^+ \|f\|_\infty^{1/2-\beta}}{f^{1/2-\beta}(y)} \right) dy = h_n$$

we derive from (2.30) (using the monotonicity of the functions involved) that on the event $\tilde{E}_n$

$$A_n^-(1-\varepsilon_n) \le \hat{A}_n \le A_n^+(1+\varepsilon_n). \tag{2.31}$$

Since we can replace A_n in (2.27) by A_n^+ or by A_n^-, we have

$$\left| \frac{A_n^+}{A_n} - 1 \right| = o(A_n^{-2}) \quad \text{and} \quad \left| \frac{A_n^-}{A_n} - 1 \right| = o(A_n^{-2}) \quad \text{as } n \to \infty.$$

By Lemma 5 in [2], A_n is of the order $|\log h_n|^{1/2}$ and also $\varepsilon_n = o(\frac{1}{|\log h_n|})$. It follows that

$$\left|\frac{\hat{A}_n}{A_n} - 1\right| = o_P(A_n^{-2}) \quad \text{as} \quad n \to \infty.$$

This implies that A_n can be replaced by $\hat{A}_n$ in (2.27), which, together with (2.25) and the fact that $\hat{A}_n = O_P\left(\sqrt{|\log h_n|}\right)$, completes the proof of the theorem. □

For a result analogous to Theorem 1 in the case $\beta = 0$, see Corollary 11 in [2]. In that corollary, the assumption on the modulus of continuity of f has the only effect of handling the bias, and can be replaced by a weaker hypothesis.

In the Bickel-Rosenblatt theorem ([1]), corresponding to the case $\beta = 1/2$, the constants A_n do not depend on f, and their theorem with $\mathbb{E}f_n$ replaced by f already provides asymptotic confidence bands for f, which moreover simplify by further replacing the weight $\sqrt{f}$ by $\sqrt{f_n}$, as these authors did; Lemma 2 above provides an explicit justification for this last replacement. Our Theorem 1 can be used to extend to the case $\beta \neq 1/2$ and to the whole line the corollary on confidence bands in the Introduction of [1], which corresponds to $\beta = 1/2$ and bounded intervals.

References

[1] P. J. Bickel and M. Rosenblatt, *On some global measures of the deviations of density function estimates.* Ann. Statist. **1** (1973), 86–118.

[2] E. Giné, V. I. Koltchinskii and L. Sakhanenko, *Kernel density estimators: convergence in distribution for weighted sup norms.* 2002, to appear. Can be downloaded from http://www.math.uconn.edu/~gine/Publications/index.html.

[3] E. Giné, V. I. Koltchinskii and J. Zinn, *Weighted uniform consistency of kernel density estimators.* 2001, to appear. Can be downloaded from http://www.math.uconn.edu/~gine/Publications/index.html.

Department of Mathematics, University of Connecticut, Storrs, CT 06269-3009, USA
E-mail address: `gine@uconnvm.uconn.edu`

Department of Mathematics and Statistics, University of New Mexico, Albuquerque, NM 87131-1141, USA
E-mail address: `vlad@math.unm.edu`

Department of Statistics and Probability, Michigan State University, East Lansing, MI 48824-1027, USA
E-mail address: `luda@stt.msu.edu`

Progress in Probability, Vol. 55, 255–292

Estimates of the Rate of Approximation in the CLT for L_1-Norm of Density Estimators

Andrei Yu. Zaitsev

Abstract. We estimate the rate of strong approximation and probabilities of moderate deviations in the CLT for the L_1-norm of the kernel density estimator under no assumptions at all on the density and assuming that the kernel is bounded and has bounded support.

1. Introduction

To fix notation, let X, X_1, $X_2, \dots$ be a sequence of i.i.d. random variables in $\mathbf{R}$ with density f. Further let $\{h_n\}_{n\geq 1}$ be a sequence of positive constants such that $h_n \to 0$ as $n \to \infty$. The classical kernel estimator is defined as

$$f_n(x) \stackrel{\text{def}}{=} \frac{1}{nh_n} \sum_{i=1}^{n} K\left(\frac{x - X_i}{h_n}\right), \quad \text{for } x \in \mathbf{R}, \tag{1.1}$$

where K is a kernel satisfying

$$K(u) = 0, \quad \text{for } |u| > 1/2; \tag{1.2}$$

$$||K||_\infty = \sup_{u\in\mathbf{R}} |K(u)| = \kappa < \infty; \tag{1.3}$$

and

$$\int_{\mathbf{R}} K(u)\, du = 1. \tag{1.4}$$

Let $||\cdot||$ denote the $L_1(\mathbf{R})$-norm. Write $||K^2|| = \int_{\mathbf{R}} K^2(u)\, du$. For any $t \in \mathbf{R}$, set

$$\rho(t) = \rho(t, K) \stackrel{\text{def}}{=} \frac{\int_{\mathbf{R}} K(u)\, K(u+t)\, du}{||K^2||}. \tag{1.5}$$

Received by the editors November 20, 2002.
2000 *Mathematics Subject Classification.* Primary 60F05, 60F15, 60F17; Secondary 62G07.
Key words and phrases. Kernel density function estimator, L_1-norm, central limit theorem, moderate deviations, strong approximation.
Research partially supported by an NSF Grant while visiting the University of Delaware, Russian Foundation of Basic Research (RFBR) Grants 02-01-00265 and 00-15-96019.

Clearly, $\rho(t)$ is a continuous function of t, $|\rho(t)| \leq 1$, $\rho(0) = 1$ and $\rho(t) = 0$ for $|t| \geq 1$. Let Z, Z_1 and Z_2 be independent standard normal random variables and set

$$\sigma^2 = \sigma^2(K) \stackrel{\text{def}}{=} ||K^2|| \int_{-1}^{1} \operatorname{cov}\left(\left|\sqrt{1-\rho^2(t)}\, Z_1 + \rho(t)\, Z_2\right|, |Z_2|\right) dt. \tag{1.6}$$

By definition, any Lebesgue density function f is an element of $L_1(\mathbf{R})$. This reason was used by Devroye and Györfi to justify the assertion that $||f_n - f||$ is the natural distance between a density function f and its estimator f_n. In their book, Devroye and Györfi [6], they posed the question about the asymptotic distribution of $||f_n - f||$.

M. Csörgő and Horváth [4] were the first who proved a Central Limit Theorem (CLT) for $||f_n - f||_p$, the L_p-norm distance, $p \geq 1$. Horváth [9] introduced a Poissonization technique into the study of CLTs for $||f_n - f||_p$. The M. Csörgő and Horváth [4] and Horváth [9] results required some regularity conditions. Beirlant and Mason [1] introduced a general method for deriving the asymptotic normality of the L_p-norm of empirical functionals. Mason (see Theorem 8.9 in Eggermont and LaRiccia [7]) has applied their method to the special case of the L_1-norm of the kernel density estimator and proved Theorem 1.1 below. Giné, Mason and Zaitsev [10] extended the CLT result of Theorem 1.1 to processes indexed by kernels K.

Theorem 1.1 shows that $\|f_n - \mathbf{E}\, f_n\|$ is asymptotically normal under no assumptions at all on the density f. Centering by $\mathbf{E}\, f_n$ is more natural from a probabilistic point of view. The estimation of $\|f - \mathbf{E}\, f_n\|$ (if needed) is a purely analytic problem. The main results of this paper (Theorems 1.3, 1.5 and 1.7) provide estimates of the rate of strong approximation and bounds for probabilities of moderate deviations in the CLT of Theorem 1.1.

Theorem 1.1. *For any Lebesgue density f and for any sequence of positive constants $\{h_n\}_{n\geq 1}$ satisfying $h_n \to 0$ and $nh_n^2 \to \infty$, as $n \to \infty$, we have*

$$\frac{\|f_n - \mathbf{E}\, f_n\| - \mathbf{E}\, \|f_n - \mathbf{E}\, f_n\|}{\sqrt{\text{rm Var}(\|f_n - \mathbf{E}\, f_n\|)}} \to_d Z \tag{1.7}$$

and

$$\lim_{n\to\infty} n \operatorname{Var}(\|f_n - \mathbf{E}\, f_n\|) = \sigma^2. \tag{1.8}$$

The variance σ^2 has an alternate representation. Using the formulas for the absolute moments of a bivariate normal random variable of Nabeya [13], we can write

$$\operatorname{cov}\left(\left|\sqrt{1-\rho^2(t)}\, Z_1 + \rho(t)\, Z_2\right|, |Z_2|\right) = \varphi\left(\rho(t)\right),$$

where

$$\varphi(\rho) \stackrel{\text{def}}{=} \frac{2}{\pi}\left(\rho \arcsin \rho + \sqrt{1-\rho^2} - 1\right), \quad \rho \in [-1, 1]. \tag{1.9}$$

It is easy to see that $\varphi(\rho)$ is strictly positive for $\rho \neq 0$. Therefore $\sigma^2 > 0$. Note that by (1.2), (1.3) and (1.6),

$$\sigma^2 \leq 2\,||K^2|| \leq 2\,\kappa^2. \tag{1.10}$$

In what follows the conditions of Theorem 1.1 are assumed to hold unless stated otherwise. We shall denote by A_j different universal constants. We write A for different constants when we do not fix their numerical values. Throughout the paper, θ symbolizes any quantity not exceeding one in absolute value. The indicator function of a set E will be denoted by $\mathbf{1}_E(\,\cdot\,)$. We write $\log^* b = \max\{e, \log b\}$.

Let η be a Poisson (n) random variable, i.e. a Poisson random variable with mean n, independent of $X, X_1, X_2, \ldots$ and set

$$f_\eta(x) \overset{\text{def}}{=} \frac{1}{nh_n}\sum_{i=1}^{\eta} K\left(\frac{x - X_i}{h_n}\right), \tag{1.11}$$

where the empty sum is defined to be zero. Notice that

$$\mathbf{E}\, f_\eta(x) = \mathbf{E}\, f_n(x) = h_n^{-1}\,\mathbf{E}\, K\left(\frac{x-X}{h_n}\right), \tag{1.12}$$

$$k_n(x) \overset{\text{def}}{=} n\,\mathrm{Var}\,(f_\eta(x)) = h_n^{-2}\,\mathbf{E}\, K^2\left(\frac{x-X}{h_n}\right), \tag{1.13}$$

and

$$n\,\mathrm{Var}\,(f_n(x)) = h_n^{-2}\,\mathbf{E}\, K^2\left(\frac{x-X}{h_n}\right) - \left\{h_n^{-1}\,\mathbf{E}\, K\left(\frac{x-X}{h_n}\right)\right\}^2. \tag{1.14}$$

Define

$$T_\eta(x) \overset{\text{def}}{=} \frac{\sqrt{n}\,\{f_\eta(x) - \mathbf{E}\, f_n(x)\}}{\sqrt{k_n(x)}}. \tag{1.15}$$

Let η_1 be a Poisson random variable with mean 1, independent of $X, X_1, X_2, \ldots$, and set

$$Y_n(x) \overset{\text{def}}{=} \left[\sum_{j \leq \eta_1} K\left(\frac{x - X_j}{h_n}\right) - \mathbf{E}\, K\left(\frac{x-X}{h_n}\right)\right] \Big/ \sqrt{\mathbf{E}\, K^2\left(\frac{x-X}{h_n}\right)}. \tag{1.16}$$

Let $Y_n^{(1)}(x), \ldots, Y_n^{(n)}(x)$ be i.i.d. $Y_n(x)$. Clearly (see (1.11)–(1.13) and (1.15)),

$$T_\eta(x) =_d \frac{\sum_{i=1}^n Y_n^{(i)}(x)}{\sqrt{n}}. \tag{1.17}$$

Set, for any Borel sets B, E,

$$J_n(B) \overset{\text{def}}{=} \sqrt{n}\int_B \{|f_\eta(x) - \mathbf{E}\, f_n(x)| - \mathbf{E}\,|f_\eta(x) - \mathbf{E}\, f_n(x)|\}\,dx, \tag{1.18}$$

$$v_n(B, E) \overset{\text{def}}{=} \mathbf{E}\,[J_n(B)\,J_n(E)], \tag{1.19}$$

$$\sigma_n^2(B) \overset{\text{def}}{=} \mathbf{E}\, J_n^2(B) = v_n(B, B), \tag{1.20}$$

$$\mathbf{P}(B) \stackrel{\text{def}}{=} \int_B f(x)\,dx = \mathbf{P}\{X \in B\}, \tag{1.21}$$

and

$$R_n(B,E) \stackrel{\text{def}}{=} \int_B \left(\int_{-1}^{1} |g_n(x,t,E) - g(x,t,E)|\,dt \right) dx, \tag{1.22}$$

where

$$g(x,t,E) \stackrel{\text{def}}{=} \mathbf{1}_E(x)\,\mathrm{cov}\left(\left| \sqrt{1-\rho^2(t)}\,Z_1 + \rho(t)\,Z_2 \right|, |Z_2| \right) f(x), \tag{1.23}$$

$$g_n(x,t,E) \stackrel{\text{def}}{=} \mathbf{1}_E(x)\mathbf{1}_E(x+th_n)\,\mathbb{C}_n\,(x,x+th_n)\,\sqrt{f(x)\,f(x+th_n)}, \tag{1.24}$$

$$\mathbb{C}_n\,(x,y) \stackrel{\text{def}}{=} \mathrm{cov}\left(\left| \sqrt{1-\rho_{n,x,y}^2}\,Z_1 + \rho_{n,x,y}\,Z_2 \right|, |Z_2| \right), \tag{1.25}$$

Z_1 and Z_2 are independent standard normal random variables and

$$\rho_{n,x,y} \stackrel{\text{def}}{=} \mathbf{E}\,T_\eta(x)\,T_\eta(y) = \mathbf{E}\,Y_n(x)\,Y_n(y) = \frac{\mathbf{E}\left[K\left(\frac{x-X}{h_n}\right)K\left(\frac{y-X}{h_n}\right)\right]}{\sqrt{\mathbf{E}\,K^2\left(\frac{x-X}{h_n}\right)\mathbf{E}\,K^2\left(\frac{y-X}{h_n}\right)}}. \tag{1.26}$$

Note that $\mathbb{C}_n\,(x,y)$ is non-negative and

$$\sup_{x,y\in\mathbf{R}} \mathbb{C}_n\,(x,y) \le 1. \tag{1.27}$$

The following Lemma 1.2 will be proved in Section 2. It is crucial for the formulation of the main results of the paper, Theorems 1.3, 1.5 and 1.7 below.

Lemma 1.2. *Whenever $h_n \to 0$ and $nh_n^2 \to \infty$, as $n \to \infty$, there exist sequences of Borel sets*

$$E_1 \subset E_2 \subset \cdots \subset E_n \subset \cdots \tag{1.28}$$

and constants $\{\beta_n\}_{n=1}^{\infty}$ and $\{D_n\}_{n=1}^{\infty}$ such that the density $f(x)$ is continuous, for $x \in E_n$, $n = 1,2,\ldots$, and relations

$$\phi_n \stackrel{\text{def}}{=} \int_{\mathbf{R}\setminus E_n} f(x)\,dx \to 0, \quad \text{as } n \to \infty, \tag{1.29}$$

$$0 < \beta_n \stackrel{\text{def}}{=} \inf_{y\in E_n} f(y) \le f(x) \le D_n \stackrel{\text{def}}{=} \sup_{y\in E_n} f(y) < \infty, \quad \text{for } x \in E_n, \tag{1.30}$$

and

$$\varepsilon_n \stackrel{\text{def}}{=} \sup_{H\in\mathcal{H}_0}\,\sup_{x\in E_n} |f * H_{h_n}(x) - I(H)\,f(x)| \to 0, \quad \text{as } n \to \infty, \tag{1.31}$$

are valid, where

$$I(H) \stackrel{\text{def}}{=} \int_{\mathbf{R}} H(x)\,dx, \tag{1.32}$$

$$f * H_h(x) \stackrel{\text{def}}{=} h^{-1} \int_{\mathbf{R}} f(z)\,H\left(\frac{x-z}{h}\right) dz, \tag{1.33}$$

$$\mathcal{H}_0 \stackrel{\text{def}}{=} \left\{ K, K^2, |K|^3, \mathbf{1}\{x : |x| \le 1/2\} \right\}. \tag{1.34}$$

Moreover,

$$\frac{D_n^{1/2}}{\beta_n^{1/2}}\left(\frac{1}{(\beta_n\, nh_n)^{1/5}}+\frac{\varepsilon_n}{\beta_n}\right)+R_n(E_n,E_n)+\frac{\lambda(E_n)}{\sqrt{nh_n^2}}+D_n\, h_n$$

$$+\frac{D_n^3\, P_n}{\beta_n^3}+\mathbb{N}_n\sqrt{h_n}\to 0,\quad \textit{as } n\to\infty, \tag{1.35}$$

where $R_n(E_n,E_n)$ *is defined in* (1.22), $\lambda(\,\cdot\,)$ *means the Lebesgue measure,*

$$\mathbb{N}_n \stackrel{\text{def}}{=} \int_{E_n} f^{3/2}(x)\,dx, \tag{1.36}$$

and

$$P_n \stackrel{\text{def}}{=} \max_{x\in\mathbf{R}} \mathbf{P}\left\{[x, x+2\,h_n]\right\}. \tag{1.37}$$

Theorem 1.3. *There exists an absolute constant A such that, whenever $h_n\to 0$ and $nh_n^2\to\infty$, as $n\to\infty$, for any sequence of Borel sets $E_1, E_2, \dots, E_n, \dots$ satisfying* (1.29)–(1.35)*, there exists an $n_0\in\mathbf{N}$ such that, for any fixed $x>0$ and for sufficiently large fixed $n\ge n_0$, one can construct on a probability space a sequence of i.i.d. random variables $X_1, X_2, \dots$ and a standard normal random variable Z such that*

$$\mathbf{P}\left\{\left|\sqrt{n}\,\|f_n-\mathbf{E}\,f_n\|-\sqrt{n}\,\mathbf{E}\,\|f_n-\mathbf{E}\,f_n\|-\sigma Z\right|\ge y_n+z+x\right\}$$

$$\le A\Big(\exp\left\{-A^{-1}\,\sigma^{-1}x/\tau_n^*\right\}+\exp\left\{-A^{-1}\,\kappa^{-1}\,\Omega_n^{-1/2}z\,\log^*\log^*(z/A\,\kappa\,\Omega_n^{1/2})\right\}$$

$$+\mathbf{P}\left\{|\partial_n Z|\ge z/2\right\}\Big),\quad \textit{for any } z>0, \tag{1.38}$$

where

$$\tau_n^* \stackrel{\text{def}}{=} A\,\Psi_n^{3/2}\,(P_n+\psi_n)^{1/2}\to 0,\quad \textit{as } n\to\infty, \tag{1.39}$$

$$y_n \stackrel{\text{def}}{=} \frac{A\,\lambda(E_n)\,\|K^3\|}{\|K^2\|\,\sqrt{nh_n^2}}+\frac{A\,\mathbb{N}_n\sqrt{h_n}}{\sqrt{\|K^2\|}}\to 0,\quad \textit{as } n\to\infty, \tag{1.40}$$

$$\partial_n \stackrel{\text{def}}{=} \frac{A\,\|K^2\|}{\sigma\,h_n}\left(\mathbb{L}_n+\frac{\varepsilon_n\,\mathbb{M}_n}{\|K^2\|}\right) \tag{1.41}$$

$$+A\,\kappa\,\Omega_n^{1/2}+\frac{A}{\sigma}\left(\frac{\|K^3\|\,\lambda(E_n)}{\|K^2\|\,\sqrt{nh_n^2}}\right)^2\to 0,\quad \textit{as } n\to\infty,$$

$$\mathbb{L}_n \stackrel{\text{def}}{=} \int_{E_n}\int_{E_n} \mathbf{1}\{|x-y|\le h_n\}\,\sqrt{f(x)\,f(y)}\;\mathbb{K}_n(x,y)\,dx\,dy, \tag{1.42}$$

$$\mathbb{K}_n(x,y) \stackrel{\text{def}}{=} \min\left\{1-\rho_{n,x,y}^2,\ \frac{\|K^3\|}{\left(1-\rho_{n,x,y}^2\right)^{3/2}\,\|K^2\|^{3/2}\,\sqrt{n\,h_n\,f(x)}}\right\} \tag{1.43}$$

$$\mathbb{M}_n \stackrel{\text{def}}{=} \int_{E_n}\int_{E_n} \mathbf{1}\{|x-y|\le h_n\}\,f^{1/2}(x)\,f^{-1/2}(y)\,dx\,dy, \tag{1.44}$$

$$\Omega_n \stackrel{\text{def}}{=} \alpha_n + 2\,P_n + 2\,\phi_n + \frac{4\,\|K^2\|\,R_n(E_n,E_n)}{\sigma^2} + L(n,\mathbf{R}) \to 0, \quad \textit{as } n \to \infty, \tag{1.45}$$

$$\alpha_n \stackrel{\text{def}}{=} \frac{1296}{5}\,(\tau_n^*)^2 \log \frac{1}{\tau_n^*}, \tag{1.46}$$

$$\Psi_n \stackrel{\text{def}}{=} \|K^2\|\,D_n\,\beta_n^{-1}\,\kappa^2\,\sigma^{-4}, \tag{1.47}$$

$$\psi_n \stackrel{\text{def}}{=} 256\,\kappa^2\,\sigma^{-2}\,\min\{P_n, D_n\,h_n\}, \tag{1.48}$$

$$L(n,\mathbf{R}) \stackrel{\text{def}}{=} \int_{\mathbf{R}} \left|h_n^{-1}\,\mathbf{P}\{X \in [x - h_n/2, x + h_n/2]\} - f(x)\right|\,dx \to 0, \quad \textit{as } n \to \infty. \tag{1.49}$$

Denote by $F\{\cdot\}$ and $\Phi\{\cdot\}$ the probability distributions which correspond to the random variables $\sqrt{n}\,(\|f_n - \mathbf{E}\,f_n\| - \mathbf{E}\,\|f_n - \mathbf{E}\,f_n\|)\,/\sigma$ and Z, respectively. The Prokhorov distance is defined by $\pi(F,\Phi) = \inf\{\varepsilon : \pi(F,\Phi,\varepsilon) \le \varepsilon\}$, where

$$\pi(F,\Phi,\varepsilon) = \sup_X \max\{F\{X\} - \Phi\{X^\varepsilon\},\, \Phi\{X\} - F\{X^\varepsilon\}\}, \quad \varepsilon > 0,$$

and X^ε is the ε-neighborhood of the Borel set X.

Corollary 1.4. *There exists an absolute constant A such that, whenever $h_n \to 0$ and $nh_n^2 \to \infty$, as $n \to \infty$, for any sequence of Borel sets $E_1, E_2, \ldots, E_n, \ldots$ satisfying (1.29)–(1.35), there exists an $n_0 \in \mathbf{N}$ such that, for sufficiently large fixed $n \ge n_0$ and for any $\varepsilon > 0$,*

$$\pi(F,\Phi,2\,\varepsilon + y_n/\sigma)b \le A\,\Big(\exp\Big\{-A^{-1}\,\kappa^{-1}\,\Omega_n^{-1/2}\sigma\,\varepsilon\,\log^*\log^*(\sigma\,\varepsilon/A\,\kappa\,\Omega_n^{1/2})\Big\}$$
$$+\ \exp\left\{-A^{-1}\,\varepsilon/\tau_n^*\right\} + \mathbf{P}\left\{|\partial_n Z| \ge \sigma\,\varepsilon/2\right\}\Big)$$

and

$$\pi(F,\Phi) \le y_n/\sigma + A\,\tau_n^*\,\log^*(1/\tau_n^*)$$
$$+\,A\,\kappa\,\Omega_n^{1/2}\sigma^{-1}\,\log^*\left(\sigma/\kappa\,\Omega_n^{1/2}\right)/\log^*\log^*(\sigma/\kappa\,\Omega_n^{1/2}) + A\,\partial_n\,\sigma^{-1}\sqrt{\log^*(\sigma/\partial_n)},$$

where $\tau_n^*, y_n, \Omega_n, \partial_n$ are defined in (1.39)–(1.49).

Theorem 1.5. *There exists an absolute constant A such that, whenever $h_n \to 0$ and $nh_n^2 \to \infty$, as $n \to \infty$, for any sequence of Borel sets $E_1, E_2, \ldots, E_n, \ldots$ satisfying (1.29)–(1.35), there exists an $n_0 \in \mathbf{N}$ such that, for sufficiently large fixed $n \ge n_0$ and for any fixed b satisfying $\tau_n^* \le A^{-1}b$, $b \le 1$, one can construct on a probability space a sequence of i.i.d. random variables $X_1, X_2, \ldots$ and a standard normal random variable Z such that*

$$\mathbf{P}\Big\{\big|\sqrt{n}\,\|f_n - \mathbf{E}\,f_n\| - \sqrt{n}\,\mathbf{E}\,\|f_n - \mathbf{E}\,f_n\| - \sigma\,Z\big| \tag{1.50}$$
$$\ge A\,\sigma\,\exp\{-b^2/72\,(\tau_n^*)^2\} + y_n + z + x\Big\}$$
$$\le A\Big(\exp\left\{-A^{-1}\,\sigma^{-1}x/\tau_n^*\right\} + \exp\{-A^{-1}\,\kappa^{-1}\,\Omega_n^{-1/2}z\,\log^*\log^*(z/A\,\kappa\,\Omega_n^{1/2})\}$$
$$+\,\mathbf{P}\left\{b\,|Z| > A^{-1}\,\sigma^{-1}x\right\} + \mathbf{P}\left\{|\partial_n Z| \ge z/2\right\}\Big), \quad \textit{for any } x, z > 0,$$

where $\tau_n^*, y_n, \Omega_n, \partial_n$ *are defined in* (1.39)–(1.49).

Remark 1.6. In the formulations of Theorems 1.3 and 1.5 and Corollary 1.4, the numbers n_0 depend on $\{h_n\}_{n\geq 1}$, $\{E_n\}_{n\geq 1}$, f and K.

Comparing Theorems 1.3 and 1.5, we observe that in Theorem 1.3 the probability space depends essentially on x, while in the statement of Theorem 1.5 inequality (1.50) is valid on the same probability space (depending on b) for any $x > 0$. However, (1.50) is weaker than (1.38) for some values of x. The same rate of approximation (as in (1.38)) is contained in (1.50) if $b^2 \geq 72\,(\tau_n^*)^2 \log(1/\tau_n^*)$ and $x \geq b^2\sigma/\tau_n^*$ only.

Denote now by $F(\,\cdot\,)$ and $\Phi(\,\cdot\,)$ the distribution functions of the random variables $\sqrt{n}\,(\|f_n - \mathbf{E}\,f_n\| - \mathbf{E}\,\|f_n - \mathbf{E}\,f_n\|)\,/\sigma$ and Z, respectively. For example, $\Phi(x) = \Phi\,\{(-\infty, x]\}$. The following statement about moderate deviations follows from Theorem 1.3.

Theorem 1.7. *Under the conditions of Theorem* 1.3, *we have* $F(-x)/\Phi(-x) \to 1$ *and* $(1 - F(x))\,/\,(1 - \Phi(x)) \to 1$, *as* $n \to \infty$, *if*

$$0 < x = x_n = o\left(\min\left\{(\tau_n^*)^{-1/3},\ \Omega_n^{-1/6}\left(\log^*\log^*(1/\,\Omega_n)\right)^{1/3},\ y_n^{-1},\ \partial_n^{-1/2}\right\}\right).$$

The choice of sets E_n, which are involved in the formulations of our results, is not unique. Lemma 1.2 ensures that, for *any* density f, there exist sets E_n such that the quantities τ_n^*, y_n, Ω_n and ∂_n tend to zero. The optimization of the choice of E_n is a separate problem. However, for sufficiently regular densities f, it is not difficult to choose E_n so that the rate of approximation is good enough, see the examples below. In our treatment of these examples, we shall use the fact that the function $\varphi(\rho)$ in (1.9) satisfies the Lipschitz condition $|\varphi(\rho_1) - \varphi(\rho_2)| \leq |\rho_1 - \rho_2|$.

Example 1. Consider the density f of the form $f(x) = \sum_{j=1}^m r_j(x)\,\mathbf{1}_{\mathcal{J}_j}(x)$, where functions $r_j(\,\cdot\,) > 0$ satisfy the Lipschitz condition

$$|\,r_j(x) - r_j(y)| \leq C\,|x - y|^\gamma, \quad 0 < \gamma \leq 1, \qquad \text{for } x, y \in \mathcal{J}_j, \quad j = 1, 2, \ldots, m,$$

where constants C and γ are independent of j and $\mathcal{J}_j = [a_j, b_j)$, $a_j < b_j$, $j = 1, 2, \ldots, m$, is a finite collection of disjoint intervals. Assume that the values of functions r_j are separated from zero and infinity:

$$0 < \beta \leq r_j(x) \leq D < \infty \qquad \text{for } x \in \mathcal{J}_j, \quad j = 1, 2, \ldots, m.$$

Choose

$$E_n = \bigcup_{j=1}^m [a_j + h_n/2, b_j - h_n/2].$$

Without loss of generality we assume $a_j + h_n/2 < b_j - h_n/2$ and $h_n \leq 1/4$. Then it is easy to estimate $\phi_n = O\,(h_n)$, $\beta \leq \beta_n \leq D_n \leq D$, $\varepsilon_n = O\,(h_n^\gamma)$, $P_n = O\,(h_n)$, $\Psi_n = O\,(1)$, $\psi_n = O\,(h_n)$, $\lambda(E_n) = O\,(1)$, $\mathrm{N}_n = O\,(1)$, $y_n = O\left(1/\sqrt{nh_n^2} + \sqrt{h_n}\right)$,

$L(n,\mathbf{R}) = O(h_n^\gamma)$, $\tau_n^* = O(\sqrt{h_n})$, $\alpha_n = O\left(h_n \log \frac{1}{h_n}\right)$, $R_n(E_n, E_n) = O(h_n^\gamma)$, $\Omega_n = O\left(h_n \log \frac{1}{h_n} + h_n^\gamma\right)$, $\mathbb{L}_n = O\left(h_n (nh_n)^{-1/5}\right)$, $\mathbb{M}_n = O(h_n)$,

$$\partial_n = O\left(\sqrt{h_n \log \frac{1}{h_n}} + h_n^{\gamma/2} + (nh_n)^{-1/5} + \frac{1}{nh_n^2}\right).$$

Thus, the statement of Theorem 1.7 is valid for

$$0 \;<\; x = x_n = o\left(\min\left\{h_n^{-1/6}\left(\log \frac{1}{h_n}\right)^{-1/6}\left(\log\log \frac{1}{h_n}\right)^{1/3},\right.\right.$$
$$\left.\left. h_n^{-\gamma/6}\left(\log\log \frac{1}{h_n}\right)^{1/3},\ (nh_n)^{1/10},\ (nh_n^2)^{1/2}\right\}\right).$$

Example 2. Consider the standard normal density $f(x) = e^{-x^2/2}/\sqrt{2\pi}$. Choose

$$E_n = \left[-\sqrt{2^{-1}\log \frac{1}{h_n}}, \sqrt{2^{-1}\log \frac{1}{h_n}}\right].$$

Without loss of generality we assume $h_n \le 1/4$. Then $\phi_n = O\left(h_n^{1/4}\right)$, $\beta_n^{-1} = O\left(h_n^{-1/4}\right)$, $D_n = O(1)$, $\varepsilon_n = O(h_n)$, $P_n = O(h_n)$, $\Psi_n = O\left(h_n^{-1/4}\right)$, $\psi_n = O(h_n)$, $L(n,\mathbf{R}) = O(h_n)$, $\tau_n^* = O\left(h_n^{1/8}\right)$, $\alpha_n = O\left(h_n^{1/4}\log \frac{1}{h_n}\right)$, $R_n(E_n, E_n) = O(h_n)$, $\Omega_n = O\left(h_n^{1/4}\log \frac{1}{h_n}\right)$, $\mathbb{L}_n = O\left(h_n (nh_n)^{-1/5}\right)$, $\mathbb{M}_n = O\left(h_n\sqrt{\log \frac{1}{h_n}}\right)$, $\mathbb{N}_n = O(1)$, $\lambda(E_n) = O\left(\sqrt{\log \frac{1}{h_n}}\right)$, $y_n = O\left(\sqrt{\log \frac{1}{h_n}}/\sqrt{nh_n^2} + \sqrt{h_n}\right)$, $\partial_n = O\left(h_n^{1/8}\sqrt{\log \frac{1}{h_n}} + (nh_n)^{-1/5} + \frac{\log \frac{1}{h_n}}{nh_n^2}\right)$. The statement of Theorem 1.7 is valid for

$$0 < x = x_n = o\left(\min\left\{h_n^{-1/24}\left(\log \frac{1}{h_n}\right)^{-1/6}\left(\log\log \frac{1}{h_n}\right)^{1/3},\right.\right.$$
$$\left.\left.(nh_n^2)^{1/2}\left(\log \frac{1}{h_n}\right)^{-1/2}\right\}\right).$$

Example 3. Consider the density

$$f(x) = f_\gamma(x) = \begin{cases} |x|^{-\gamma}(1-\gamma), & 0 < x \le 1, \\ 0, & \text{otherwise}, \end{cases} \qquad 0 < \gamma < 1.$$

Choose $\alpha = \frac{1-\gamma}{1+2\gamma}$ and $E_n = [h_n^\alpha, 1 - h_n]$. Without loss of generality we assume $h_n \le 1/8$. Then it is easy to estimate $\phi_n = O\left(h_n^{(1-\gamma)\alpha}\right)$, $\beta_n^{-1} = O(1)$, $D_n = O(h_n^{-\gamma\alpha})$, $\varepsilon_n = O\left(h_n^{1-(1+\gamma)\alpha}\right)$, $P_n = O(h_n^{1-\gamma})$, $\Psi_n = O(h_n^{-\gamma\alpha})$, $\psi_n = O(h_n^{1-\gamma\alpha})$, $\mathbb{N}_n = O\left(h_n^{(1-3\gamma/2)\alpha} + \mathbf{1}\{\gamma = 2/3\}\log \frac{1}{h_n}\right)$, $\mathbb{L}_n = O\left(h_n (nh_n)^{-1/5}\right)$,

$\mathbb{M}_n = O(h_n)$, $\lambda(E_n) = O(1)$, $y_n = O\left(1/\sqrt{nh_n^2} + h_n^{(1-3\gamma/2)\alpha}\sqrt{h_n}\right)$, $R_n(E_n, E_n) = O(h_n^{1-2\gamma\alpha})$, $\tau_n^* = O(h_n^{(1-\gamma-3\gamma\alpha)/2})$, $\alpha_n = O\left(h_n^{1-\gamma-3\gamma\alpha}\log\frac{1}{h_n}\right)$, $L(n,\mathbf{R}) = O(h_n)$, $\partial_n = O\left(\Omega_n^{1/2} + (nh_n)^{-1/5} + \frac{1}{nh_n^2}\right)$, $\Omega_n = O\left(h_n^{1-\gamma-3\gamma\alpha}\log\frac{1}{h_n} + h_n^{(1-\gamma)\alpha}\right)$. The statement of Theorem 1.7 is valid for

$$0 < x_n = o\Big(\min\Big\{h_n^{-(1-\gamma)^2/6(1+2\gamma)}\left(\log\frac{1}{h_n}\right)^{-1/6}\left(\log\log\frac{1}{h_n}\right)^{1/3},$$
$$(nh_n)^{1/10},\ (nh_n^2)^{1/2}\Big\}\Big). \tag{1.51}$$

Note that the logarithmic factor in (1.51) could be slightly improved by means of a more careful choice of the intervals E_n.

Remark 1.8. When estimating $\mathbb{L}_n$ in the examples, we used the fact that, by (1.30) and (1.43), for $x, y \in E_n$, we have

$$\mathbb{K}_n(x,y) \leq \frac{A\,\|K^3\|^{2/5}}{\|K^2\|^{3/5}\,(nh_n\,f(x))^{1/5}} \leq \frac{A\,\|K^3\|^{2/5}}{\|K^2\|^{3/5}\,(\beta_n\,nh_n)^{1/5}}. \tag{1.52}$$

For some densities f and kernels K, formula (1.42) may give sharper bounds. For example, if $K = f = \mathbf{1}\{x : |x| \leq 1/2\}$ and $E_n = [h_n/2, 1 - h_n/2]$, one can show that $\mathbb{L}_n = O\left(h_n\,(nh_n)^{-2/5}\right)$. This is better than the rates given in Examples 1 and 3.

Studying the examples and analyzing the statements of Theorems 1.3, 1.5 and 1.7, we see that the rates of normal approximation become worse when the density f is non-smooth or has too small or too large values. To show that this is essential, let us consider a scheme of series, where the density f may be depending on n. Namely, let

$$f(x) = \left(2a_n^{-1}\right)\mathbf{1}_{[-a_n, a_n]}(x),$$

where a_n may tend to zero or to infinity as $n \to \infty$. It is not difficult to understand that we can choose a_n tending to infinity so fast that, with probability tending to 1, the intervals $[X_i - h_n/2, X_i - h_n/2]$, $i = 1, 2, \ldots, n$, are disjoint and the distribution of $\sqrt{n}\,\|f_n - \mathbf{E}\,f_n\| - \sqrt{n}\,(\|K\| + 1)$ converges to the degenerate distribution $\mathbb{E}_0$ concentrated at zero. On the other hand, we can choose a_n tending to zero so fast that $\sqrt{n}\,\|f_n - \mathbf{E}\,f_n\|$ converges to the same degenerate distribution $\mathbb{E}_0$ since it behaves as in the case where $\mathbf{P}\{X = 0\} = 1$. Thus, if all non-zero values of f are very large or very small, then the distribution of $\sqrt{n}\,(\|f_n - \mathbf{E}\,f_n\| - \mathbf{E}\,\|f_n - \mathbf{E}\,f_n\|)$ is far from that of $\sigma\,Z$.

Sections 2–5 are devoted to the proof of Theorems 1.3, 1.5 and 1.7. In the proof, we shall use the Poissonization of the sample size, considering integrals $\int_{E_n}\{|f_\eta - \mathbf{E}\,f_n| - \mathbf{E}\,|f_\eta - \mathbf{E}\,f_n|\}$ instead of $\int_{E_n}\{|f_n - \mathbf{E}\,f_n| - \mathbf{E}\,|f_n - \mathbf{E}\,f_n|\}$. This allows us to use independence properties of the Poisson point process $\{X_1, \ldots, X_\eta\}$. In Section 2, we prove Lemma 1.2. Lemma 2.1 provides bounds for variances of integrals over some exceptional sets. Lemma 2.5 gives estimates for variances of integrals over sets of the form $(a,b) \cap E_n$. Lemma 2.6 implies bounds for $\sqrt{n} \int_{E_n} |\mathbf{E}\,|f_\eta - \mathbf{E}\,f_n| - \mathbf{E}\,|f_n - \mathbf{E}\,f_n||$. In Section 3, we replace sets E_n by some sets $C_n \subset E_n$ removing "bad" intervals and tails of small measure. Then we represent the integral over C_n as a sum (in i) S_n of 1-dependent integrals $\delta_{i,n}$ over some sets $I_{i,n}$. Lemma 3.3 provides Bernstein-type bounds for moments of summands $\delta_{i,n}$. Lemma 3.4 contains a bound for the correlation between S_n and some centered and normalized Poisson random variable $U_n = \sum_i u_{i,n}$. The summands $u_{i,n}$ are independent centered and normalized Poisson random variables and the bivariate random vectors $(\delta_{i,n}, u_{i,n})$ are 1-dependent. In Lemma 3.6, using bounds from Lemma 3.3, we prove Bernstein-type bounds for moments of projections of vectors $(\delta_{i,n}, u_{i,n})$ to one-dimensional directions. A result of Heinrich [11], see Lemma 3.5, implies bounds for cumulants of projections of vectors (S_n, U_n). In Lemma 3.8, we use these bounds to show that distribution $\mathcal{L}\left((S_n, U_n)\right) \in \mathcal{A}_2(\tau_n)$ with some $\tau_n \leq \tau_n^*$, where $\mathcal{A}_2(\tau_n)$ is a class of distributions introduced by Zaitsev [19]. In Section 4, we get bounds for exponential moments of integrals over exceptional sets, see Lemma 4.1. These bounds imply exponential inequalities for the tails of the corresponding distributions. Theorems 1.3, 1.5 and 1.7 are proved in Section 5. We use there a result of Zaitsev [23] providing an estimate of the rate of approximation in a de-Poissonization lemma of Beirlant and Mason [1].

2. Preliminary lemmas.

Lemma 2.1 (cf. the proof of Giné, Mason and Zaitsev [10], Lemma 6.2). *Whenever $h_n \to 0$ and $nh_n \to \infty$, as $n \to \infty$, for any Borel subset B of $\mathbf{R}$ and any sequence of functions $a_n \in L_1(\mathbf{R})$,*

$$\mathbf{E}\left(\sqrt{n}\int_B \{|f_n(x) - a_n(x)| - \mathbf{E}\,|f_n(x) - a_n(x)|\}\,dx\right)^2 \tag{2.1}$$

$$\leq d(n,B) \overset{\text{def}}{=} 4\,\|K\|_\infty\,\mathbf{E}\,\frac{1}{h_n}\int_B \left|K\left(\frac{x-X}{h_n}\right)\right|\,dx,$$

$$\mathbf{E}\left(\sqrt{n}\int_B \{|f_\eta(x) - a_n(x)| - \mathbf{E}|f_\eta(x) - a_n(x)|\}\,dx\right)^2 \leq 2\,d(n,B), \tag{2.2}$$

where $d(n,B)$ satisfies

$$d(n,B) \leq 4\,\kappa^2\,\Omega(n,B) \tag{2.3}$$

with

$$\Omega(n,B) \overset{\text{def}}{=} \left(\int_B f(x)\,dx + L(n,B)\right), \tag{2.4}$$

$$L(n,B) \overset{\text{def}}{=} \int_B \left|h_n^{-1}\mathbf{P}\{X \in [x - h_n/2, x + h_n/2]\} - f(x)\right|\,dx \leq L(n,\mathbf{R}) \to 0, \tag{2.5}$$

as $n \to \infty$.

Proof. Applying the main result in Pinelis [15], we get (see (1.2))

$$\begin{aligned}\mathbf{E}\left(\sqrt{n}\int_B\{|f_n(x)-a_n(x)|-\mathbf{E}|f_n(x)-a_n(x)|\}\,dx\right)^2 \\ \le 4\,\mathbf{E}\left(\frac{1}{h_n}\int_B\left|K\left(\frac{x-X}{h_n}\right)\right|dx\right)^2 \\ \le 4\,\|K\|_\infty\,\mathbf{E}\frac{1}{h_n}\int_B\left|K\left(\frac{x-X}{h_n}\right)\right|dx.\end{aligned} \tag{2.6}$$

Similarly, taking into account (1.13) and (1.15)–(1.17), we have

$$\begin{aligned}\mathbf{E}\left(\sqrt{n}\int_B\{|f_\eta(x)-a_n(x)|-\mathbf{E}|f_\eta(x)-a_n(x)|\}\,dx\right)^2 \\ \le 4\,\mathbf{E}\left(\frac{1}{h_n}\int_B\left|\sum_{j\le\eta_1}K\left(\frac{x-X_j}{h_n}\right)\right|dx\right)^2 \\ \le 4\,\|K\|_\infty\,\mathbf{E}\frac{1}{h_n}\int_B\left|K\left(\frac{x-X}{h_n}\right)\right|dx\,\mathbf{E}\,\eta_1^2.\end{aligned}$$

Using (1.2) and (1.3), we obtain

$$\|K\|_\infty\,\mathbf{E}\frac{1}{h_n}\int_B\left|K\left(\frac{x-X}{h_n}\right)\right|dx \le \kappa^2 h_n^{-1}\int_B \mathbf{P}\{X\in[x-h_n/2,x+h_n/2]\}\,dx.$$

Furthermore, $\mathbf{E}\,\eta_1^2=2$ and

$$h_n^{-1}\int_B \mathbf{P}\{X\in[x-h_n/2,x+h_n/2]\}\,dx \le \int_B f(x)\,dx + L(n,B)=\Omega(n,B). \tag{2.7}$$

By a special case of Theorem 1 in Chapter 2 of Devroye and Györfi [6],

$$L(n,\mathbf{R})=\int_{\mathbf{R}}\left|h_n^{-1}\,\mathbf{P}\{X\in[x-h_n/2,x+h_n/2]\}-f(x)\right|dx\to 0,\quad \text{as } n\to\infty,$$

which completes the proof of Lemma 2.1. □

We shall apply Lemma 2.1 in the case where $a_n(x)=\mathbf{E}\,f_n(x)$. Note that in this situation a similar bound may be derived from Theorem 2.1 of de Acosta [3]. Also see Devroye [5], who obtains the bound (2.6) with $a_n(x)=f(x)$. The following standard lemma follows from Theorem 3 in Chapter 2 of Devroye and Györfi [6].

Lemma 2.2 (see Giné, Mason and Zaitsev [10], Lemma 6.1). *Suppose that H is a uniformly bounded real valued function, which is equal to zero off a compact interval. Then*

$$|f*H_h(x)-I(H)\,f(x)|\to 0,\quad \textit{as } h\searrow 0,\quad \textit{for almost all } x\in\mathbf{R}, \tag{2.8}$$

where $I(H)$ and $f*H_h(x)$ are defined in (1.32) and (1.33).

Proof of Lemma 1.2. Applying for each $m \in \mathbf{N}$ and for $\mathcal{H} = \mathcal{H}_0$ Lemma 6.1 from Giné, Mason and Zaitsev [10], we conclude that there exist measurable sets $Q_1, Q_2, \ldots, Q_m, \ldots$ such that

$$\int_{Q_m} f(x)\,dx \geq 1 - 2^{-m}, \tag{2.9}$$

f is continuous, for $x \in Q_m$, $m = 1, 2, \ldots$, and, uniformly in $H \in \mathcal{H}_0$,

$$\sup_{x \in Q_m} |f * H_{h_n}(x) - I(H)\,f(x)| \to 0, \quad \text{as } n \to \infty. \tag{2.10}$$

Write

$$Q_s^* = \bigcup_{m=1}^{s} Q_m. \tag{2.11}$$

By (2.9)–(2.11),

$$\int_{Q_s^*} f(x)\,dx \geq 1 - 2^{-s}, \tag{2.12}$$

$$Q_1^* \subset Q_2^* \subset \cdots \subset Q_s^* \subset \cdots \tag{2.13}$$

and, for $s = 1, 2, \ldots$,

$$\sup_{x \in Q_s^*,\, H \in \mathcal{H}_0} |f * H_{h_n}(x) - I(H)\,f(x)| \to 0, \quad \text{as } n \to \infty. \tag{2.14}$$

Define $m_1 = 1$,

$$m_s = \min\left\{ m > m_{s-1} : \sup_{n \geq m}\ \sup_{x \in Q_s^*,\, H \in \mathcal{H}_0} |f * H_{h_n}(x) - I(H)\,f(x)| < 2^{-s} \right\}, \tag{2.15}$$

for $s = 2, 3, \ldots$, and

$$F_l = Q_s^*, \quad \text{for } m_s \leq l < m_{s+1}. \tag{2.16}$$

By (2.12)–(2.16),

$$\int_{F_l} f(x)\,dx \nearrow 1, \quad \text{as } l \to \infty, \tag{2.17}$$

$$F_1 \subset F_2 \subset \cdots \subset F_l \subset \cdots \tag{2.18}$$

and, for $l = 1, 2, \ldots$,

$$\varepsilon_{l,n}^* \stackrel{\text{def}}{=} \sup_{m \geq n}\ \sup_{x \in F_l,\, H \in \mathcal{H}_0} |f * H_{h_m}(x) - I(H)\,f(x)| \to 0, \quad \text{as } n \to \infty. \tag{2.19}$$

Let sequences $\{\beta_n^*\}_{n=1}^{\infty}$ and $\{D_n^*\}_{n=1}^{\infty}$ satisfy conditions

$$0 < \beta_n^* < D_n^* < \infty; \qquad \beta_n^* \searrow 0, \quad D_n^* \nearrow \infty, \quad \text{as } n \to \infty. \tag{2.20}$$

Define, for $l = 1, 2, \ldots$,

$$G_l = \{x \in F_l : \beta_l^* \leq f(x) \leq D_l^*\} \tag{2.21}$$

Recall that $\mathbb{C}_n(x,y)$ and $\rho_{n,x,y}$ were defined in (1.25) and (1.26). Also observe that

$$\rho_{n,x,x+th_n} = \frac{h_n^{-1}\,\mathbf{E}\left[K\left(\frac{x-X}{h_n}\right)K\left(\frac{x-X}{h_n}+t\right)\right]}{\sqrt{h_n^{-1}\,\mathbf{E}\,K^2\left(\frac{x-X}{h_n}\right)h_n^{-1}\,\mathbf{E}\,K^2\left(\frac{x-X}{h_n}+t\right)}},$$

see (1.26). Applying Lemma 2.2, with $H(u) = K(u)\,K(u+t)$, we get, for each t, that, for almost every $x \in G_l$,

$$h_n^{-1}\,\mathbf{E}\left[K\left(\frac{x-X}{h_n}\right)K\left(\frac{x-X}{h_n}+t\right)\right] \to f(x)\int_{\mathbf{R}} K(u)\,K(u+t)\,du, \quad \text{as } n\to\infty.$$

Moreover, we get with $H(u) = K^2(u)$ and $H(u) = K^2(u+t)$, respectively, for almost every $x \in G_l$, both

$$h_n^{-1}\,\mathbf{E}\,K^2\left(\frac{x-X}{h_n}\right) \to f(x)\,\|K^2\|, \quad \text{and} \quad h_n^{-1}\,\mathbf{E}\,K^2\left(\frac{x-X}{h_n}+t\right) \to f(x)\,\|K^2\|.$$

Thus, for each t and almost every $x \in G_l$,

$$\rho_{n,x,x+th_n} \to \rho(t), \quad \text{as } n\to\infty,$$

and $\mathbb{C}_n(x, x+th_n) \to \operatorname{cov}\left(\left|\sqrt{1-\rho^2(t)}Z_1 + \rho(t)\,Z_2\right|, |Z_2|\right)$. By Lemma 6.4 from Giné, Mason and Zaitsev [10], $\mathbf{1}_{G_l}(x+h_nt)$ converges in measure to $\mathbf{1}_{G_l}(x) = 1$ on $G_l\times[-1,1]$, and $f(x+h_nt)\,\mathbf{1}_{G_l}(x+h_nt)$ converges in measure to $f(x)$ on $G_l\times[-1,1]$ as functions of x and t. Combining these observations, we readily conclude that $g_n(x,t,G_l)$ converges in measure on $G_l\times[-1,1]$ to $g(x,t,G_l)$. By (1.23), (1.24), (1.27) and (2.21), functions $g(x,t,G_l)$ and $g_n(x,t,G_l)$ are uniformly bounded on $G_l\times[-1,1]$. This implies

$$R_n(G_l,G_l) = \int_{G_l}\left(\int_{-1}^{1} |g_n(x,t,G_l) - g(x,t,G_l)|\,dt\right)dx \to 0, \quad \text{as } n\to\infty.$$

It is easy to see that

$$P_n \to 0, \qquad \text{as } n\to\infty. \tag{2.22}$$

Define $j_1 = 1$,

$$\begin{aligned} j_l \;=\; \min\Bigg\{ j > j_{l-1} : \sup_{m\ge j}\Bigg\{ &\frac{\sqrt{D_l^*}}{\sqrt{\beta_l^*}}\left(\frac{1}{(\beta_l^*\,mh_m)^{1/5}} + \frac{\varepsilon_{l,m}^*}{\beta_l^*}\right) + R_m(G_l,G_l) \\ &+ \frac{1}{\beta_l^*\sqrt{mh_m^2}} + \left(\frac{D_l^*}{\beta_l^*}\right)^3 P_m < 2^{-l}\Bigg\}\Bigg\}, \quad \text{for } l = 2,3,\dots, \end{aligned} \tag{2.23}$$

and

$$E_n = G_l = \{x \in F_l : \beta_l^* \le f(x) \le D_l^*\}, \quad \text{for } j_l \le n < j_{l+1}. \tag{2.24}$$

Using (2.17)–(2.24), we obtain

$$\frac{D_n^{1/2}}{\beta_n^{1/2}}\left(\frac{1}{(\beta_n\, nh_n)^{1/5}}+\frac{\varepsilon_n}{\beta_n}\right)+R_n(E_n,E_n)$$
$$+\frac{1}{\beta_n\sqrt{nh_n^2}}+\frac{D_n^3\,P_n}{\beta_n^3}\to 0,\quad \text{as } n\to\infty, \tag{2.25}$$

with

$$\varepsilon_n\le \sup_{m\ge j_l}\varepsilon^*_{l,m},\quad \beta_n\ge\beta_l^*,\quad D_n\le D_l^*,\quad \text{for } j_l\le n<j_{l+1}.$$

It remains to note that, by (1.21), (1.30) and (1.36),

$$\beta_n\,\lambda(B)\le \mathbf{P}(B)\le D_n\,\lambda(B),\quad \text{for any Borel set } B\subset E_n, \tag{2.26}$$

$\mathrm{N}_n\le D_n^{1/2}$ and

$$P_n\ge c_f\,h_n, \tag{2.27}$$

for sufficiently large $n\ge n_0$, where $c_f>0$ depends on density f only. Therefore, (1.30) and (2.25) imply (1.35). □

Remark 2.3. The choice of the sets $E_1,E_2,\dots,E_n,\dots$ depends on the choice of the sequences $\{\beta_n^*\}_{n=1}^{\infty}$ and $\{D_n^*\}_{n=1}^{\infty}$ in the proof of Lemma 1.2.

In the sequel we shall assume that $h_n\to 0$ and $nh_n^2\to\infty$, as $n\to\infty$ and $n\ge n_0$, where n_0 is a positive integer which will be chosen as large as it is necessary for the arguments below to hold. Let $E_1,E_2,\dots,E_n,\dots$ be *any* sequence of Borel sets satisfying (1.29)–(1.35). By (1.30) and (1.35), $\frac{\varepsilon_n}{\beta_n}\to 0$ as $n\to\infty$. Let $n\ge n_0$ be so large that

$$\varepsilon_n\le\beta_n\,\min\{I(H):H\in\mathcal{H}_0\}/2. \tag{2.28}$$

Then, by (1.30), (1.31) and (2.28), for any $x\in E_n$, $H\in\mathcal{H}_0$, we have

$$f(x)\,I(H)/2\le f*H_{h_n}(x)\le 2\,f(x)\,I(H). \tag{2.29}$$

We shall use the following fact that follows from Theorem 1 of Sweeting [18].

Lemma 2.4. *Let $(\omega,\zeta),(\omega_1,\zeta_1),(\omega_2,\zeta_2),\dots,$ be a sequence of i.i.d. bivariate random vectors such that each component has variance* 1*, mean* 0 *and finite moments of the third order. Further, let (Z_1^*,Z_2^*) be bivariate normal vector with mean* 0*, $\mathrm{Var}(Z_1^*)=\mathrm{Var}(Z_2^*)=1$, and with $\mathrm{cov}(Z_1^*,Z_2^*)=\mathrm{cov}(\omega,\zeta)=\rho$. Then there exists a universal positive constant A such that*

$$\left|\mathbf{E}\left|\frac{\sum_{i=1}^n\zeta_i}{\sqrt{n}}\right|-\mathbf{E}\,|Z_1^*|\right|\le\frac{A}{\sqrt{n}}\,\mathbf{E}\,|\zeta|^3 \tag{2.30}$$

and, whenever $\rho^2<1$,

$$\left|\mathbf{E}\left|\frac{\sum_{i=1}^n\omega_i}{\sqrt{n}}\cdot\frac{\sum_{i=1}^n\zeta_i}{\sqrt{n}}\right|-\mathbf{E}\,|Z_1^*Z_2^*|\right|\le\frac{A}{(1-\rho^2)^{3/2}\sqrt{n}}\left(\mathbf{E}\,|\omega|^3+\mathbf{E}\,|\zeta|^3\right) \tag{2.31}$$

and

$$\left|\mathbf{E}\left[\frac{\sum_{i=1}^{n}\omega_i}{\sqrt{n}}\cdot\left|\frac{\sum_{i=1}^{n}\zeta_i}{\sqrt{n}}\right|\right]\right| \le \frac{A}{(1-\rho^2)^{3/2}\sqrt{n}}\left(\mathbf{E}\,|\omega|^3+\mathbf{E}\,|\zeta|^3\right). \tag{2.32}$$

Lemma 2.5. *For sufficiently large $n \ge n_0$ and for arbitrary (possibly depending on n) interval (a,b), $-\infty \le a < b \le \infty$,*

$$\begin{aligned}&\left|\,\sigma_n^2(B) - \mathbf{P}(B)\,\sigma^2\,\right| \qquad (2.33)\\ &\quad\le A\,\mathbf{P}(B)\,\|K^2\|\,D_n^{1/2}\,\beta_n^{-1/2}\left(\frac{\|K^3\|^{2/5}}{\|K^2\|^{3/5}\,(\beta_n\,nh_n)^{1/5}}+\frac{\varepsilon_n}{\|K^2\|\,\beta_n}\right)\\ &\qquad+\|K^2\|\,R_n(B,E_n)+16\,\kappa^2\left(1+\beta_n^{-1}\,\varepsilon_n\right)\,\min\left\{P_n, D_n\,h_n\right\},\end{aligned}$$

where $B = B(n) = (a,b)\cap E_n$. Moreover,

$$\begin{aligned}&\left|\,\sigma_n^2(E_n) - \mathbf{P}(E_n)\,\sigma^2\,\right| \qquad (2.34)\\ &\quad\le A\,h_n^{-1}\,\|K^2\|\left(\mathbb{L}_n+\frac{\varepsilon_n\,\mathbb{M}_n}{\|K^2\|}\right)+\|K^2\|\,R_n(E_n,E_n),\end{aligned}$$

where $\mathbb{L}_n$ and $\mathbb{M}_n$ are defined in (1.42)–(1.44).

Proof. Notice that whenever $|x-y| > h_n$, random variables $|f_\eta(x) - \mathbf{E}\,f_n(x)|$ and $|f_\eta(y)-\mathbf{E}\,f_n(y)|$ are independent. This follows from the fact that they are functions of independent increments of the Poisson process with intensity nf. Therefore (see (1.15), (1.18) and (1.19))

$$\begin{aligned}v_n(B,E_n) &= n\int_B\int_{E_n}\mathbf{E}\left\{|f_\eta(x)-\mathbf{E}\,f_n(x)|\,|f_\eta(y)-\mathbf{E}\,f_n(y)|\right\}dx\,dy \qquad (2.35)\\ &\quad - n\int_B\int_{E_n}\left\{\mathbf{E}\,|f_\eta(x)-\mathbf{E}\,f_n(x)|\,\mathbf{E}\,|f_\eta(y)-\mathbf{E}\,f_n(y)|\right\}dx\,dy\\ &= \int_B\int_{E_n}\mathbf{1}\{|x-y|\le h_n\}\,\mathrm{cov}\left(|T_\eta(x)|,|T_\eta(y)|\right)\sqrt{k_n(x)\,k_n(y)}\,dx\,dy.\end{aligned}$$

According to (1.6) and (1.21)–(1.24), we have, for $x \in E_n$,

$$\int_{-1}^{1} g(x,t,E_n)dt = f(x)\int_{-1}^{1}\mathrm{cov}\left(\left|\sqrt{1-\rho^2(t)}Z_1+\rho(t)\,Z_2\right|,\left|Z_2\right|\right)dt = \frac{f(x)\,\sigma^2}{\|K^2\|}, \tag{2.36}$$

$$\int_B\left(\int_{-1}^{1} g(x,t,E_n)\,dt\right)\,dx = \frac{\mathbf{P}(B)\,\sigma^2}{\|K^2\|} \tag{2.37}$$

and

$$|\varphi_n^2(B) - \mathbf{P}(B)\,\sigma^2| \le \|K^2\|\,R_n(B,E_n), \tag{2.38}$$

where

$$\varphi_n^2(B) = \|K^2\|\int_B\int_{-1}^{1} g_n(x,t,E_n)\,dx\,dt. \tag{2.39}$$

Furthermore, $\mathrm{Var}(Y_n(x)) = 1$ (see (1.12), (1.13) and (1.15)–(1.17)) and

$$\mathbf{E}\left|Y_n(x)\right|^3 \le A \frac{h_n^{-3/2}\,\mathbf{E}\left|K\left(\frac{x-X}{h_n}\right)\right|^3}{\left(h_n^{-1}\,\mathbf{E}\,K^2\left(\frac{x-X}{h_n}\right)\right)^{3/2}}. \tag{2.40}$$

Using (1.30), (1.32)–(1.34), (2.29) and (2.40), we get that, for $n \ge n_0$,

$$\mathbf{E}\left|Y_n(x)\right|^3 \le A \frac{2\left\|K^3\right\|\,h_n^{-1/2}}{\sqrt{f(x)}\,\left(\left\|K^2\right\|/2\right)^{3/2}} \le \frac{A\left\|K^3\right\|}{\sqrt{\beta_n h_n}\,\left\|K^2\right\|^{3/2}}. \tag{2.41}$$

By (1.13), (1.31), (1.32) and (1.34),

$$\sup_{x\in E_n}\left|h_n\,k_n(x) - \left\|K^2\right\|\,f(x)\right| \le \varepsilon_n. \tag{2.42}$$

Assume that $n \ge n_0$ is so large that $\frac{\varepsilon_n}{\|K^2\|\,\beta_n} \le 1/6$, see (1.35). Thus, for $x \in E_n$, we have

$$h_n\,k_n(x) = \left\|K^2\right\|\,f(x)\,\exp\left(\frac{A\,\theta\,\varepsilon_n}{\left\|K^2\right\|\,f(x)}\right), \tag{2.43}$$

where $|\theta| \le 1$. Using (2.43), we see that, for $x, y \in E_n$,

$$\sqrt{k_n(x)\,k_n(y)} = h_n^{-1}\left\|K^2\right\|\,\sqrt{f(x)\,f(y)}\,\exp\left(\frac{A\,\theta\varepsilon_n}{\left\|K^2\right\|}\left(f^{-1}(x)+f^{-1}(y)\right)\right). \tag{2.44}$$

We shall use the elementary fact that if X and Y are mean zero and variance 1 random variables with $\rho = \mathbf{E}\,XY$, then $1 - \mathbf{E}|XY| \le 1 - |\rho| \le 1 - \rho^2$. By an application of Lemma 2.4, keeping (1.17), (1.25), (1.26), (1.35), (1.43), (1.52) and (2.41) in mind, we obtain, for $n \ge n_0$ large enough and $x, y \in E_n$,

$$\begin{aligned}
&\left|\mathrm{cov}\left(|T_\eta(x)|, |T_\eta(y)|\right) - \mathbb{C}_n(x,y)\right| \qquad (2.45)\\
&\le A\min\left\{1-\rho_{n,x,y}^2 + \frac{\mathbf{E}\left|Y_n(x)\right|^3 + \mathbf{E}\left|Y_n(y)\right|^3}{\sqrt{n}}, \frac{\mathbf{E}\left|Y_n(x)\right|^3 + \mathbf{E}\left|Y_n(y)\right|^3}{\left(1-\rho_{n,x,y}^2\right)^{3/2}\sqrt{n}}\right\}\\
&\le A\left(\mathbb{K}_n(x,y) + \mathbb{K}_n(y,x)\right)\\
&\le \frac{A\left\|K^3\right\|^{2/5}\left(f^{-1/5}(x)+f^{-1/5}(y)\right)}{\left\|K^2\right\|^{3/5}\left(nh_n\right)^{1/5}} \le \frac{A\left\|K^3\right\|^{2/5}}{\left\|K^2\right\|^{3/5}\left(\beta_n\,nh_n\right)^{1/5}}.
\end{aligned}$$

Using (1.24), (1.25), (1.27), (1.30), (1.35), (2.35), (2.39), (2.44), (2.45) and the change of variables $y = x + th_n$, we see that, for sufficiently large $n \geq n_0$,

$$\begin{aligned}
&\left|v_n(B, E_n) - \varphi_n^2(B)\right| \\
&\leq A \int_B \int_{E_n} \mathbf{1}\{|x-y| \leq h_n\}\, h_n^{-1} \left\|K^2\right\| \sqrt{f(x)\, f(y)} \\
&\quad \times \left(\mathbb{K}_n(x,y) + \mathbb{K}_n(y,x) + \frac{\varepsilon_n}{\|K^2\|} \left(f^{-1}(x) + f^{-1}(y)\right)\right) dx\, dy \qquad (2.46) \\
&\leq A\, \mathbf{P}(B)\, \|K^2\|\, D_n^{1/2}\, \beta_n^{-1/2} \left(\frac{\left\|K^3\right\|^{2/5}}{\|K^2\|^{3/5} \left(\beta_n n h_n\right)^{1/5}} + \frac{\varepsilon_n}{\|K^2\|\, \beta_n} \right). \qquad (2.47)
\end{aligned}$$

Define

$$\begin{array}{ll} B_1 = (a - h_n, a) \cap E_n, & B_2 = (b, b + h_n) \cap E_n, \\ B_3 = (a, a + h_n) \cap B, & B_4 = (b - h_n, b) \cap B. \end{array} \qquad (2.48)$$

Clearly,

$$B = B_3 \cup B_4 \cup (B \backslash (B_3 \cup B_4)) \qquad (2.49)$$

and

$$\mathbf{E}\, J_n(B)\, J_n(E_n \backslash (B \cup B_1 \cup B_2)) = 0, \qquad (2.50)$$

since $J_n(B)$ and $J_n(E_n \backslash (B \cup B_1 \cup B_2))$ are independent. Similarly, according to (2.48) and (2.49), $\mathbf{E}\, J_n(B)\, J_n(B_1 \cup B_2) = \mathbf{E}\, J_n(B_1)\, J_n(B_3) + \mathbf{E}\, J_n(B_2)\, J_n(B_4)$. Note that, by (1.31)–(1.34), (2.5) and (2.26), we have

$$L(n, B) \leq \lambda(B)\, \varepsilon_n \leq \beta_n^{-1}\, \mathbf{P}(B)\, \varepsilon_n, \quad \text{for any Borel set } B \subset E_n. \qquad (2.51)$$

By (1.18)–(1.21), (1.37), (2.2)–(2.5), (2.26), (2.48), (2.50) and (2.51),

$$\begin{aligned}
\left|\sigma_n^2(B) - v_n(B, E_n)\right| &= \left|\mathbf{E}\, J_n(B)\, J_n(B_1 \cup B_2)\right| \\
&\leq \left|\mathbf{E}\, J_n(B_1)\, J_n(B_3)\right| + \left|\mathbf{E}\, J_n(B_2)\, J_n(B_4)\right| \\
&\leq 4 \max_{1 \leq i \leq 4} d(n, B_i) \qquad (2.52) \\
&\leq 16\, \kappa^2 \max_{1 \leq i \leq 4} \left(\mathbf{P}(B_i) + L(n, B_i)\right) \\
&\leq 16\, \kappa^2 \left(1 + \beta_n^{-1}\, \varepsilon_n\right) \max_{1 \leq i \leq 4} \mathbf{P}(B_i) \\
&\leq 16\, \kappa^2 \left(1 + \beta_n^{-1}\, \varepsilon_n\right) \min\left\{P_n, D_n\, h_n\right\},
\end{aligned}$$

for sufficiently large $n \geq n_0$. Inequalities (2.26), (2.38), (2.47) and (2.52) imply (2.33). Clearly, $\sigma_n^2(E_n) = v_n(E_n, E_n)$, see (1.20). The proof of (2.34) repeats that of (2.33). Instead of (2.47) one should use (2.46) coupled with (1.44). □

Lemma 2.6. *For sufficiently large $n \geq n_0$, we have*

$$\int_{E_n} \left| \sqrt{n}\, \mathbf{E}\, |f_\eta(x) - \mathbf{E}\, f_n(x)| - \mathbf{E}\, |Z| \sqrt{k_n(x)} \right| dx \leq \frac{A\, \lambda(E_n)\, \left\|K^3\right\|}{\|K^2\|\, \sqrt{n h_n^2}} \qquad (2.53)$$

and

$$\int_{E_n} \left| \sqrt{n}\, \mathbf{E}\, |f_n(x) - \mathbf{E}\, f_n(x)| - \mathbf{E}\, |Z| \sqrt{k_n(x)} \right| dx \tag{2.54}$$
$$\leq \frac{A\, \lambda(E_n)\, \|K^3\|}{\|K^2\|\, \sqrt{nh_n^2}} + \frac{A\, \mathbb{N}_n \sqrt{h_n}}{\sqrt{\|K^2\|}},$$

where $\mathbb{N}_n$ *is defined by* (1.36).

Proof. By (1.15), (1.17), (2.30) and (2.41), for $x \in E_n$,

$$\left| \frac{\mathbf{E}\, |\sqrt{n}\, \{f_\eta(x) - \mathbf{E} f_n(x)\}|}{\sqrt{k_n(x)}} - \mathbf{E}|Z| \right| \leq \frac{A}{\sqrt{n}} \mathbf{E}|Y_n(x)|^3 \leq \frac{A\, \|K^3\|}{\sqrt{nf(x)h_n}\, \|K^2\|^{3/2}}. \tag{2.55}$$

Using (1.4), (1.13), (1.14), (1.30), (1.34) and (2.29), we get, for $n \geq n_0$, $x \in E_n$,

$$f(x)\, \|K^2\|\, h_n^{-1}/2 \leq k_n(x) \leq 2\, f(x)\, \|K^2\|\, h_n^{-1} \leq 2\, D_n\, \|K^2\|\, h_n^{-1} \tag{2.56}$$

and

$$\left| \sqrt{k_n(x)} - \sqrt{n \operatorname{Var}(f_n(x))} \right| \leq \frac{(2\, f(x))^2 \sqrt{h_n}}{\sqrt{f(x)\, \|K^2\|/2}} \leq \frac{A\, f^{3/2}(x) \sqrt{h_n}}{\sqrt{\|K^2\|}}. \tag{2.57}$$

Now by (1.30), (1.35), (2.55) and (2.56), we obtain (2.53), for sufficiently large $n \geq n_0$. Similarly one obtains

$$\int_{E_n} \left| \sqrt{n}\, \mathbf{E}\, |f_n(x) - \mathbf{E}\, f_n(x)| - \mathbf{E}\, |Z| \sqrt{n \operatorname{Var}(f_n(x))} \right| dx \leq \frac{A\, \lambda(E_n)\, \|K^3\|}{\|K^2\|\, \sqrt{nh_n^2}},$$

which by (1.36) and (2.57) implies (2.54). □

3. Reduction of the problem to a CLT for 1-dependent random vectors

Let

$$\alpha_n \to 0, \quad \text{as } n \to \infty, \tag{3.1}$$

be a non-increasing sequence of strictly positive numbers. In Section 3, we assume (3.1) only, keeping in mind that α_n will be defined later by (1.46). Using the continuity of our measure, we may find an interval $[-M_n, M_n]$ so that

$$\alpha_n = \int_{|x|>M_n} f(x)\, dx. \tag{3.2}$$

Assume that $n \geq n_0$ is so large that

$$0 < \alpha_n \leq 1/4 \quad \text{and} \quad h_n \leq \min\{M_n/4, 1 - \alpha_n\}. \tag{3.3}$$

Define $m_n = [M_n/h_n] - 1$, $h_n^* = (M_n - h_n)/m_n$, where $[x]$ denotes the integer part of x. Clearly, by (3.3), we have $M_n/2h_n \leq m_n \leq M_n/h_n$. Hence,

$$h_n \leq h_n^* \leq 2h_n. \tag{3.4}$$

Recall that P_n and ψ_n were defined in (1.37) and (1.48). Note that (1.35), (1.48), (3.1) and (3.2) imply that $\mathbf{P}([-M_n + h_n, M_n - h_n]) > \psi_n$, for sufficiently large

$n \geq n_0$. Define, recurrently, integers $l_1 = -m_n$, $l_i \in \mathbf{Z}$, $l_1 < l_2 < \cdots < l_{s_n-1} = m_n$. Let l_{i-1} be constructed. Then if, for some $l \in \mathbf{Z}$, we have $\mathbf{P}([l_{i-1}h_n^*, (l-1)\, h_n^*]) < \psi_n$, $\mathbf{P}([l_{i-1}h_n^*, lh_n^*]) \geq \psi_n$ and $\mathbf{P}([lh_n^*, M_n - h_n]) \geq \psi_n$, we set $l_i = l$. If, for some $l \in \mathbf{Z}$, we have $\mathbf{P}([l_{i-1}h_n^*, (l-1)\, h_n^*]) < \psi_n$, $\mathbf{P}([l_{i-1}h_n^*, lh_n^*]) \geq \psi_n$ and $\mathbf{P}([lh_n^*, M_n - h_n]) < \psi_n$, we set $s_n - 1 = i$ and $l_{s_n-1} = m_n$. Denote

$$z_{0,n} \stackrel{\text{def}}{=} -M_n; \qquad z_{s_n,n} \stackrel{\text{def}}{=} M_n; \qquad z_{i,n} \stackrel{\text{def}}{=} l_i h_n^*, \qquad \text{for } i = 1, \ldots, s_n - 1; \quad (3.5)$$

$$I_{i,n} \stackrel{\text{def}}{=} E_n \cap [z_{i-1,n}, z_{i,n}), \quad p_{i,n} \stackrel{\text{def}}{=} \mathbf{P}(I_{i,n}), \quad q_{i,n} \stackrel{\text{def}}{=} \mathbf{P}([z_{i-1,n}, z_{i,n})), \qquad (3.6)$$

for $i = 1, \ldots, s_n$. Clearly, we have

$$z_{0,n} < z_{1,n} = -M_n + h_n < z_{1,n} < \cdots < z_{s_n-1,n} = M_n - h_n < z_{s_n,n}. \qquad (3.7)$$

Furthermore,

$$P_n = \max_{x \in \mathbf{R}} \mathbf{P}([x, x + 2\, h_n]) \geq \max_{x \in \mathbf{R}} \mathbf{P}([x, x + h_n^*]) \qquad (3.8)$$

(see (3.4)). By (3.6),

$$p_{i,n} \leq q_{i,n}, \qquad i = 1, \ldots, s_n. \qquad (3.9)$$

Clearly, by construction, we have

$$\psi_n \leq q_{i,n} \leq P_n + 2\, \psi_n, \qquad i = 2, \ldots, s_n - 1, \qquad (3.10)$$

and

$$\max\{q_{1,n}, q_{s_n,n}\} \leq P_n, \qquad (3.11)$$

for sufficiently large $n \geq n_0$. Hence, by (1.35), (1.48), (2.22) and (3.9)–(3.11),

$$\max_{1 \leq i \leq s_n} p_{i,n} \leq \max_{1 \leq i \leq s_n} q_{i,n} \to 0, \qquad \text{as } n \to \infty. \qquad (3.12)$$

Introduce sets of indices

$$\Upsilon_1 = \left\{ i = 2, \ldots, s_n - 1 : 4\, \|K^2\|\, R_n(I_{i,n}, E_n) \geq p_{i,n}\, \sigma^2 \right\}, \qquad (3.13)$$

$$\Upsilon_2 = \{ i = 2, \ldots, s_n - 1 : p_{i,n} \leq \mathbf{P}([z_{i-1,n}, z_{i,n}) \backslash I_{i,n}) \}, \qquad (3.14)$$

$$\Upsilon = \Upsilon_1 \cup \Upsilon_2, \quad \Upsilon_3 = \{2, \ldots, s_n - 1\} \backslash \Upsilon. \qquad (3.15)$$

Define

$$C_n = [-M_n + h_n, M_n - h_n] \cap E_n \backslash \bigcup_{i \in \Upsilon} [z_{i-1,n}, z_{i,n}). \qquad (3.16)$$

By construction,

$$C_n = \bigcup_{i \in \Upsilon_3} I_{i,n}, \quad \text{and} \quad I_{i,n} \cap I_{j,n} \text{ are empty}, \quad \text{for } i \neq j. \qquad (3.17)$$

Using (1.22), (1.35), (3.6), (3.7) and (3.13), we obtain

$$\mathbf{P}\left(\bigcup_{i \in \Upsilon_1} I_{i,n} \right) = \sum_{i \in \Upsilon_1} p_{i,n} \leq \frac{4\, \|K^2\|\, R_n(E_n, E_n)}{\sigma^2} \to 0, \quad \text{as } n \to \infty. \qquad (3.18)$$

Furthermore, by (1.29), (3.6), (3.7) and (3.14), we get

$$\mathbf{P}\left(\bigcup_{i\in\Upsilon_2} I_{i,n}\right) = \sum_{i\in\Upsilon_2} p_{i,n} \le \sum_{i\in\Upsilon_2} \mathbf{P}([z_{i-1,n}, z_{i,n})\backslash I_{i,n}) \tag{3.19}$$
$$= \sum_{i\in\Upsilon_2} \mathbf{P}([z_{i-1,n}, z_{i,n})\backslash E_n) \le \mathbf{P}(\mathbf{R}\backslash E_n) = \phi_n \to 0, \quad \text{as } n\to\infty.$$

By (2.4), (2.5), (2.22), (3.1), (3.2), (3.8) and (3.13)–(3.19), we have

$$\Omega(n,\overline{C}_n) \le \alpha_n + 2P_n + 2\phi_n + \frac{4\|K^2\|R_n(E_n,E_n)}{\sigma^2} + L(n,\mathbf{R}) \to 0, \text{ as } n\to\infty, \tag{3.20}$$

where $\overline{C}_n$ denotes the complement of C_n. By Lemma 2.1,

$$\mathbf{E}\left(\sqrt{n}\int_{\overline{C}_n}\{|f_n(x)-\mathbf{E}f_n(x)|-\mathbf{E}|f_n(x)-\mathbf{E}f_n(x)|\}dx\right)^2 \le d_n \overset{\text{def}}{=} d(n,\overline{C}_n), \tag{3.21}$$

and

$$d_n \le 4\,\kappa^2\,\Omega(n,\overline{C}_n) \le 4\,\kappa^2\,\Omega_n, \tag{3.22}$$

where Ω_n is defined in (1.45). Similarly, using (2.2) instead of (2.1), we obtain (see (1.18) and (1.20))

$$\sigma_n^2(E_n\backslash C_n) \le 8\kappa^2\left(\alpha_n + 2P_n + \phi_n + \frac{4\|K^2\|R_n(E_n,E_n)}{\sigma^2} + L(n,\mathbf{R})\right) \to 0, \tag{3.23}$$

as $n\to\infty$. It is easy to see that, by (1.30), (1.42), (1.44) and (1.52),

$$\mathbb{L}_n \le \frac{A\,\|K^3\|^{2/5}\,D_n^{3/10}\,h_n}{\|K^2\|^{3/5}\,(nh_n)^{1/5}\,\beta_n^{1/2}}, \qquad \mathbb{M}_n \le 2\,\beta_n^{-1}\,h_n. \tag{3.24}$$

Clearly, $J_n(E_n) = J_n(C_n)+J_n(E_n\backslash C_n)$. Therefore, applying (1.20), (1.29), (1.30), (1.35), (2.34), (3.16), (3.23), (3.24) and the triangle inequality, we get $\sigma_n^2(C_n) = \sigma^2 + o(1)$ and

$$\frac{1}{2}\,\sigma^2 \le \sigma_n^2(C_n) \le 2\,\sigma^2, \tag{3.25}$$

for sufficiently large $n \ge n_0$.

Denote, for $i = 1,\dots,s_n$,

$$\delta_{i,n} \overset{\text{def}}{=} \frac{\int_{z_{i-1,n}}^{z_{i,n}} \mathbf{1}_{C_n}(x)\,W_\eta(x)\,dx}{\sigma_n(C_n)}, \tag{3.26}$$

where

$$W_\eta(x) \overset{\text{def}}{=} \Delta_\eta(x) - \mathbf{E}\,\Delta_\eta(x) = (|T_\eta(x)| - \mathbf{E}\,|T_\eta(x)|)\,\sqrt{k_n(x)}, \tag{3.27}$$

and

$$\Delta_\eta(x) \overset{\text{def}}{=} \sqrt{n}|f_\eta(x)-\mathbf{E}f_n(x)| = \frac{1}{\sqrt{nh_n}}\left|\sum_{i=1}^{\eta} K\left(\frac{x-X_i}{h_n}\right) - n\mathbf{E}K\left(\frac{x-X}{h_n}\right)\right|. \tag{3.28}$$

Obviously (see (3.5)–(3.7), (3.15) and (3.16)),

$$\delta_{i,n} = 0, \quad \text{for } i \notin \Upsilon_3 \qquad \text{and} \qquad \delta_{i,n} = \frac{\int_{I_{i,n}} W_\eta(x)\,dx}{\sigma_n(C_n)}, \quad \text{for } i \in \Upsilon_3. \tag{3.29}$$

Furthermore, $z_{i,n} - z_{i-1,n} \geq h_n$, for $i = 1, \ldots, s_n$. This implies that the sequence $\delta_{i,n}$, $1 \leq i \leq s_n$, is 1-dependent. We used (1.2), (3.28), (3.29) and that any functions of the Poisson point process $\{X_1, \ldots, X_\eta\}$ restricted to disjoint sets are independent.

The use of the sets C_n has the advantage over the sets E_n in that they permit us to control the variances of the summands $\delta_{i,n}$ from below.

Lemma 3.1. *For sufficiently large $n > n_0$, we have*

$$p_{i,n}\,\sigma^2/4 \leq \sigma_n^2(I_{i,n}) \leq 2\,p_{i,n}\,\sigma^2, \qquad \textit{for } i \in \Upsilon_3.$$

Proof. According to (1.48), (3.6), (3.10), (3.14) and (3.15), we have, for $i \in \Upsilon_3$,

$$p_{i,n} \geq q_{i,n}/2 \geq \psi_n/2 = 128\,\kappa^2\,\sigma^{-2}\min\{P_n, D_n\,h_n\}. \tag{3.30}$$

Hence, by (1.30), (1.35), (2.33), (3.6), (3.13), (3.15) and (3.30), $\beta_n^{-1}\,\varepsilon_n \leq 1$ and

$$\begin{aligned}
\sigma_n^2(I_{i,n}) &\geq p_{i,n}\,\sigma^2 - \left|\sigma_n^2(I_{i,n}) - p_{i,n}\,\sigma^2\right| \\
&\geq \frac{1}{2}\,p_{i,n}\,\sigma^2 - \frac{A\,\|K^2\|\,D_n^{1/2}\,p_{i,n}}{\beta_n^{1/2}}\left(\frac{\|K^3\|^{2/5}}{\|K^2\|^{3/5}\,(\beta_n\,nh_n)^{1/5}} + \frac{\varepsilon_n}{\|K^2\|\,\beta_n}\right) \\
&\geq \frac{1}{4}\,p_{i,n}\,\sigma^2,
\end{aligned}$$

for sufficiently large $n > n_0$. Similarly,

$$\begin{aligned}
\sigma_n^2(I_{i,n}) &\leq p_{i,n}\,\sigma^2 + \left|\sigma_n^2(I_{i,n}) - p_{i,n}\,\sigma^2\right| \\
&\leq \frac{3}{2}\,p_{i,n}\,\sigma^2 + \frac{A\,\|K^2\|\,D_n^{1/2}\,p_{i,n}}{\beta_n^{1/2}}\left(\frac{\|K^3\|^{2/5}}{\|K^2\|^{3/5}\,(\beta_n\,nh_n)^{1/5}} + \frac{\varepsilon_n}{\|K^2\|\,\beta_n}\right) \\
&\leq 2\,p_{i,n}\,\sigma^2,
\end{aligned}$$

for sufficiently large $n > n_0$. □

The following fact will be useful below: if ξ_i are independent centered random variables, then, for every $r \geq 2$,

$$\mathbf{E}\left|\sum_{i=1}^n \xi_i\right|^r \leq 2^{r+1}\,e^{2r}\max\left[r^{r/2}\left(\sum_{i=1}^n \mathbf{E}\,\xi_i^2\right)^{r/2}, \; r^r\sum_{i=1}^n \mathbf{E}\,|\xi_i|^r\right] \tag{3.31}$$

(Pinelis [16], with a unspecified constant A^r; after symmetrization, in the form (3.31), it follows from Latała [12]).

The following Lemma 3.2 gives a Rosenthal-type inequality for Poissonized sums of independent random variables.

Lemma 3.2 (Giné, Mason and Zaitsev [10], Lemma 2.2). *Assume that it is known that for any $n \in \mathbf{N}$, any i.i.d. centered random variables $\xi, \xi_1, \xi_2, \ldots$ for some $r \geq 2$,*

$$\mathbf{E}\left|\sum_{i=1}^{n} \xi_i\right|^r \leq F\left(n\,\mathbf{E}\,\xi^2,\, n\,\mathbf{E}\,|\xi|^r\right), \tag{3.32}$$

where $F(\cdot\,,\cdot)$ is a non-decreasing continuous function of two arguments. Then, for any $\mu > 0$ and any i.i.d. random variables $\zeta, \zeta_1, \zeta_2, \ldots$,

$$\mathbf{E}\left|\sum_{i=1}^{\eta} \zeta_i - \mu\,\mathbf{E}\,\zeta\right|^r \leq F\left(\mu\,\mathbf{E}\,\zeta^2,\, \mu\,\mathbf{E}\,|\zeta|^r\right), \tag{3.33}$$

where η is a Poisson random variable with mean μ, independent of $\zeta_1, \zeta_2, \ldots$.

Lemma 3.3. *We have, uniformly in $i \in \Upsilon_3$, for sufficiently large $n \geq n_0$ and for all integers $r \geq 2$,*

$$\mathbf{E}\,|\delta_{i,n}|^r \leq A^r\, r^r\, p_{i,n}^{r/2-1}\left(\|K^2\|\, D_n\, \beta_n^{-1}\, \kappa^2\, \sigma^{-4}\right)^{r/2} \mathrm{Var}(\delta_{i,n}). \tag{3.34}$$

Proof. By the Hölder and generalized Minkowski inequalities (see, e.g., Folland [8], p. 194), (3.27) and (3.29),

$$\sigma_n^r(C_n)\,\mathbf{E}\,|\delta_{i,n}|^r \leq 2^r\mathbf{E}\left(\int_{I_{i,n}} \Delta_\eta(x)\,dx\right)^r \leq \left(2\int_{I_{i,n}} \left(\mathbf{E}\,\Delta_\eta^r(x)\right)^{1/r} dx\right)^r. \tag{3.35}$$

Write (see (3.28))

$$\mathbf{E}\,\Delta_\eta^r(x) = \frac{1}{\left(\sqrt{n}\,h_n\right)^r}\,\mathbf{E}\left|\sum_{i=1}^{\eta} K\left(\frac{x - X_i}{h_n}\right) - n\,\mathbf{E}\,K\left(\frac{x - X}{h_n}\right)\right|^r. \tag{3.36}$$

Applying Lemma 3.2 coupled with inequality (3.31), we obtain

$$\mathbf{E}\left|\sum_{i=1}^{\eta} K\left(\frac{x - X_i}{h_n}\right) - n\,\mathbf{E}\,K\left(\frac{x - X}{h_n}\right)\right|^r \tag{3.37}$$
$$\leq 2^{r+1} e^{2r} \max\left\{r^{r/2}\left(n\,\mathbf{E}\,K^2\left(\frac{x - X}{h_n}\right)\right)^{r/2},\, r^r\, n\,\mathbf{E}\left|K\left(\frac{x - X}{h_n}\right)\right|^r\right\}.$$

Therefore, using (1.3), (1.32)–(1.34), (2.29), (3.16), (3.36) and (3.37), we see that, for $n \geq n_0$, $x \in C_n$, the moment $\mathbf{E}\,\Delta_\eta^r(x)$ may be estimated from above by

$$\frac{2^{r+1}\, e^{2r}}{\left(\sqrt{n}\,h_n\right)^r} \max\left\{r^{r/2}\left(n\,\mathbf{E}\,K^2\left(\frac{x - X}{h_n}\right)\right)^{r/2},\, r^r\, n\,\mathbf{E}\left|K\left(\frac{x - X}{h_n}\right)\right|^r\right\}$$
$$\leq 2^{r+1} e^{2r} \max\left\{r^{r/2}\left(2\, f(x)\, \|K^2\|\, h_n^{-1}\right)^{r/2}, 2\, r^r n^{1-r/2} \kappa^{r-2} f(x)\, \|K^2\|\, h_n^{1-r}\right\}.$$

Since $\beta_n n h_n \to \infty$, as $n \to \infty$ (see (1.30) and (1.35)), we estimate for sufficiently large $n \geq n_0$, $x \in C_n$,

$$\mathbf{E}\,\Delta_\eta^r(x) \leq 2^{r+1}\, e^{2r}\, r^r \left(2\, f(x)\, \|K^2\|\, h_n^{-1}\right)^{r/2}.$$

Substituting this into (3.35), and using Hölder's inequality, we get

$$\mathbf{E}\,|\delta_{i,n}|^r \le A^r\, r^r \sigma_n^{-r}(C_n)\,\lambda^{r/2}(I_{i,n})\,\left(p_{i,n}\,\|K^2\|\,h_n^{-1}\right)^{r/2}, \tag{3.38}$$

where $p_{i,n}$ is defined in (3.6). By Lemma 3.1,

$$\sigma_n^2(I_{i,n}) \ge p_{i,n}\,\sigma^2/4, \tag{3.39}$$

for sufficiently large $n \ge n_0$. It is easy to see that

$$\mathrm{Var}(\delta_{i,n}) = \frac{\sigma_n^2(I_{i,n})}{\sigma_n^2(C_n)}. \tag{3.40}$$

Each $I_{i,n}$, $i = 2, \ldots, s_n - 1$, can be represented as $I_{i,n} = (J_{i,n} \cup L_{i,n}) \cap E_n$, where $J_{i,n}$ is an interval of length h_n^* and $L_{i,n}$ is a set with $\mathbf{P}(L_{i,n}) \le 2\,\psi_n$ with ψ_n defined in (1.48). Therefore, by (1.10), (1.30), (1.48), (2.26), (3.4) and (3.6),

$$\begin{aligned}\lambda(I_{i,n}) &\le \lambda(J_{i,n} \cap E_n) + \lambda(L_{i,n} \cap E_n) \le 2\,h_n + \beta_n^{-1}\,\mathbf{P}(L_{i,n})\\ &\le \left(2 + 512\,D_n\,\beta_n^{-1}\,\kappa^2\,\sigma^{-2}\right)\,h_n \le A\,D_n\,\beta_n^{-1}\,\kappa^2\,\sigma^{-2} h_n.\end{aligned} \tag{3.41}$$

Substituting (3.39) into (3.38) and using (3.25), (3.40) and (3.41), we obtain inequality (3.34). □

Define

$$S_n = \sum_{i=1}^{s_n} \delta_{i,n} = \sum_{i \in \Upsilon_3} \delta_{i,n} = \frac{\int_{C_n} W_\eta(x)\,dx}{\sigma_n(C_n)} \tag{3.42}$$

(see (3.15), (3.16) and (3.29)),

$$U_n = \frac{1}{\sqrt{n}}\left\{\sum_{j \le \eta} \mathbf{1}\{X_j \in [-M_n, M_n]\} - n\,\mathbf{P}\{X \in [-M_n, M_n]\}\right\} \tag{3.43}$$

and

$$V_n = \frac{1}{\sqrt{n}}\left\{\sum_{j \le \eta} \mathbf{1}\{X_j \notin [-M_n, M_n]\} - n\,\mathbf{P}\{X \notin [-M_n, M_n]\}\right\}. \tag{3.44}$$

Set

$$u_{i,n} = \frac{1}{\sqrt{n}}\left\{\sum_{j \le \eta} \mathbf{1}\left\{X_j \in [z_{i-1,n}, z_{i,n})\right\} - n\,q_{i,n}\right\}, \qquad i = 1, \ldots, s_n. \tag{3.45}$$

It is easy to see that $\sqrt{n}\,u_{i,n}$ is a centered Poisson random variable with

$$\mathrm{Var}(\sqrt{n}\,u_{i,n}) = n\,q_{i,n}, \qquad i = 1, \ldots, s_n. \tag{3.46}$$

Recall that we have $C_n \subset [-M_n + h_n, M_n - h_n]$, see (3.16). Clearly, (S_n, U_n) is a function of the Poisson point process $\{X_1, \ldots, X_\eta\}$ restricted to the set $[-M_n, M_n]$

and V_n is a function of the same process restricted to the set $\mathbf{R}\setminus[-M_n, M_n]$. Therefore, (S_n, U_n) is independent of V_n. Obviously,

$$U_n = \sum_{i=1}^{s_n} u_{i,n}$$

and summands $u_{i,n}$, $i = 1, \dots, s_n$, are independent. Hence,

$$\operatorname{Var}(U_n) = \sum_{i=1}^{s_n} \operatorname{Var}(u_{i,n}) = \sum_{i=1}^{s_n} q_{i,n} = \mathbf{P}\{X \in [-M_n, M_n]\}, \tag{3.47}$$

see (3.5)–(3.7). Observe that

$$\operatorname{Var}(S_n) = 1 \quad \text{and} \quad \operatorname{Var}(U_n) = 1 - \alpha_n, \tag{3.48}$$

where $\alpha_n = \mathbf{P}\{X \notin [-M_n, M_n]\}$, see (3.2).

Lemma 3.4. *For sufficiently large $n \geq n_0$, we have ...*

$$|\operatorname{cov}(S_n, U_n)| \leq \frac{A\,\|K^3\|\,\lambda(E_n)}{\sigma\,\|K^2\|\,\sqrt{nh_n^2}}. \tag{3.49}$$

Moreover,

$$\max_{i\in\Upsilon_3} \frac{|\operatorname{cov}(\delta_{i,n}, u_{i,n})|}{\left(\operatorname{Var}(u_{i,n})\operatorname{Var}(\delta_{i,n})\right)^{1/2}} \to 0, \quad \text{as } n \to \infty. \tag{3.50}$$

Proof. According to (3.6), (3.27), (3.29) and (3.45), we have, for $i \in \Upsilon_3$,

$$\sigma_n(C_n)\operatorname{cov}(\delta_{i,n}, u_{i,n}) = q_{i,n}^{1/2} \int_{I_{i,n}} \left(\mathbf{E}\,|T_\eta(x)|\,u_{i,n}\,q_{i,n}^{-1/2}\right) \sqrt{k_n(x)}\,dx. \tag{3.51}$$

Note that (3.10) and (3.12) imply that

$$\psi_n \leq \min_{2\leq i\leq s_n-1} q_{i,n} \to 0, \quad \text{as } n \to \infty. \tag{3.52}$$

Below we assume that $n \geq n_0$ is sufficiently large. By (2.26), (3.6), (3.39) and (3.40),

$$\lambda(I_{i,n}) \leq \beta_n^{-1}\,p_{i,n} \leq 4\,\beta_n^{-1}\,\sigma^{-2}\sigma_n^2(I_{i,n}) = 4\,\beta_n^{-1}\,\sigma^{-2}\sigma_n^2(C_n)\operatorname{Var}(\delta_{i,n}). \tag{3.53}$$

Note now that

$$\left(T_\eta(x), u_{i,n}\,q_{i,n}^{-1/2}\right) =_d n^{-1/2} \sum_{l=1}^{n} \left(Y_n^{(l)}(x), U^{(l)}\right), \tag{3.54}$$

where $\left(Y_n^{(l)}(x), U^{(l)}\right)$, $l = 1, \dots, n$, are i.i.d. $(Y_n(x), U)$, with $Y_n(x)$ defined in (1.16) and

$$U = q_{i,n}^{-1/2} \left\{ \sum_{j\leq\eta_1} \mathbf{1}\{X_j \in [z_{i-1,n}, z_{i,n})\} - q_{i,n} \right\}, \tag{3.55}$$

η_1 denoting a Poisson random variable with mean 1 from (1.16), which is independent of $X, X_1, X_2, \ldots$. Using (1.2), (1.10), (1.16), (1.32)–(1.34), (1.48), (2.29), (3.10) and (3.55), we see that, for any $x \in C_n$,

$$|\mathrm{cov}(Y_n(x), U)| = \frac{\left|\mathbf{E}\left[K\left(\frac{x-X}{h_n}\right)\mathbf{1}\{X\in[z_{i-1,n}, z_{i,n})\}\right]\right|}{q_{i,n}^{1/2}\left(\mathbf{E}K^2\left(\frac{x-X}{h_n}\right)\right)^{1/2}} \le \frac{2\sqrt{2D_n}\kappa h_n^{1/2}}{q_{i,n}^{1/2}\,\|K^2\|^{1/2}} \le \frac{1}{4}, \quad (3.56)$$

if $\psi_n = 256\,\kappa^2\,\sigma^{-2}\,D_n\,h_n$. Furthermore, using the first equality in (3.56), (1.2), (1.10), (1.37), (3.10) and Hölder's inequality, we get

$$|\mathrm{cov}(Y_n(x), U)| \le q_{i,n}^{-1/2}\mathbf{P}^{1/2}\{X\in[x - h_n/2, x + h_n/2]\} \le \psi_n^{-1/2}P_n^{1/2} \le \frac{1}{8\sqrt{2}}, \quad (3.57)$$

if $\psi_n = 256\,\kappa^2\,\sigma^{-2}\,P_n$.

Applying part (2.32) of Lemma 2.4 and using (2.41), (3.54), (3.56), (3.57) and inequality (3.33) of Lemma 3.2 in the case $\mathbf{P}\{\zeta = 1\} = 1$ together with inequality (3.31), we get

$$\begin{aligned}\mathbf{E}\left[|T_\eta(x)|\; u_{i,n}\, q_{i,n}^{-1/2}\right] &\le \frac{A}{\sqrt{n}}\left(\mathbf{E}\,|Y_n(x)|^3 + \mathbf{E}\,|U|^3\right)\\ &\le \frac{A}{\sqrt{n}}\left(\frac{\|K^3\|}{\|K^2\|^{3/2}\sqrt{f(x)\,h_n}} + \frac{1}{\sqrt{q_{i,n}}}\right). \quad (3.58)\end{aligned}$$

Using (1.30), (1.35), (1.48), (2.27), (2.56), (3.25), (3.41), (3.46), (3.51)–(3.53) and (3.58), we get (3.50):

$$\begin{aligned}&\max_{i\in\Upsilon_3}\frac{|\,\mathrm{cov}(\delta_{i,n}, u_{i,n})|}{(\mathrm{Var}(u_{i,n})\,\mathrm{Var}(\delta_{i,n}))^{1/2}}\\ &\le A\max_{i\in\Upsilon_3}\frac{q_{i,n}^{1/2}\sqrt{\lambda\,(I_{i,n})\;\lambda\,(I_{i,n})}}{\sigma_n(C_n)\,(q_{i,n}\,\mathrm{Var}(\delta_{i,n}))^{1/2}}\\ &\quad\times\max_{x\in E_n}\left\{\left(\frac{\|K^3\|}{\|K^2\|^{3/2}\sqrt{f(x)\,h_n}} + \frac{1}{\sqrt{q_{i,n}}}\right)\sqrt{\frac{f(x)\,\|K^2\|}{n\,h_n}}\right\}\\ &\le \frac{A\,\left(D_n\,\beta_n^{-1}\,\kappa^2\,\sigma^{-2}\right)^{1/2}}{\sigma\,\beta_n^{1/2}\sqrt{n}}\\ &\quad\times\max_{x\in E_n}\left\{\left(\frac{\|K^3\|}{\|K^2\|^{3/2}\sqrt{f(x)\,h_n}} + \frac{1}{\sqrt{\psi_n}}\right)\sqrt{f(x)\,\|K^2\|}\right\} \to 0, \text{ as } n\to\infty.\end{aligned}$$

Similarly,

$$\sigma_n(C_n)\mathrm{cov}(S_n, U_n) = (1-\alpha_n)^{1/2}\int_{C_n}\mathbf{E}\left[|T_\eta(x)|\,U_n(1-\alpha_n)^{-1/2}\right]\sqrt{k_n(x)}\,dx \quad (3.59)$$

and

$$\left|\mathrm{cov}(T_\eta(x), U_n\,(1-\alpha_n)^{-1/2})\right| \le 1/4.$$

Applying part (2.32) of Lemma 2.4 and using (1.17), (2.41) and again inequality (3.33) of Lemma 3.2 in the case $\mathbf{P}\{\zeta = 1\} = 1$ coupled with inequality (3.31), we get

$$\mathbf{E}\left[|T_\eta(x)|\, U_n\,(1-\alpha_n)^{-1/2}\right] \le \frac{A}{\sqrt{n}}\left(\frac{\|K^3\|}{\|K^2\|^{3/2}\,\sqrt{f(x)\,h_n}} + \frac{1}{\sqrt{1-\alpha_n}}\right). \tag{3.60}$$

By (3.16),

$$\lambda(C_n) \le \lambda(E_n). \tag{3.61}$$

Using (1.30), (1.35), (2.56), (3.3), (3.25), (3.59)–(3.61), we get (3.49):

$$\begin{aligned}|\mathrm{cov}(S_n, U_n)| &\le \frac{A(1-\alpha_n)^{1/2}\lambda(E_n)}{\sigma_n(C_n)} \\ &\quad \times \frac{1}{\sqrt{n}}\max_{x\in E_n}\left\{\left(\frac{\|K^3\|}{\|K^2\|^{3/2}\,\sqrt{f(x)\,h_n}} + \frac{1}{\sqrt{1-\alpha_n}}\right)\sqrt{f(x)\,\|K^2\|\,h_n^{-1}}\right\} \\ &\le \frac{A\,\|K^3\|\,\lambda(E_n)}{\sigma\,\|K^2\|\,\sqrt{nh_n^2}}.\end{aligned}$$

□

Below, for $z = (z_1, z_2)$, $u = (u_1, u_2) \in \mathbf{C}^2$, we shall use the notation

$$|z| = |z_1| + |z_2|, \quad \|z\|^2 = |z_1|^2 + |z_2|^2, \quad \langle z, u\rangle = z_1\overline{u_1} + z_2\overline{u_2}.$$

We shall write $\Gamma_r\{\xi\}$ for the k-th cumulant of a random variable ξ. Recall that if, for some $c > 0$, a random variable ξ has finite exponential moments $\mathbf{E}\,e^{z\xi}$, $z \in \mathbf{C}$, $|z| < c$, then (choosing $\log 1 = 0$)

$$\log \mathbf{E}\,e^{z\xi} = \sum_{r=0}^{\infty} \frac{\Gamma_r\{\xi\}\,z^r}{r!} \quad \text{and} \quad \Gamma_r\{\xi\} = \frac{d^r}{dz^r}\log \mathbf{E}\,e^{z\xi}\bigg|_{z=0}. \tag{3.62}$$

Clearly, $\Gamma_0\{\xi\} = 0$, $\Gamma_1\{\xi\} = \mathbf{E}\,\xi$, $\Gamma_2\{\xi\} = \mathrm{Var}(\xi)$,

$$\Gamma_r\{a\xi\} = a^r\Gamma_r\{\xi\}, \quad r = 0, 1, \ldots. \tag{3.63}$$

In the two-dimensional case, when $\xi = (\xi_1, \xi_2)$ is a bivariate random vector, if $\left|\mathbf{E}\,e^{\langle z,\xi\rangle}\right| < \infty$, $z \in \mathbf{C}^2$, $|z| < c$, $c > 0$, then

$$\log \mathbf{E}e^{\langle z,\xi\rangle} = \sum_{r_1,r_2=0}^{\infty} \frac{\Gamma_{r_1,r_2}\{\xi\}\,z_1^{r_1} z_2^{r_2}}{r_1!\,r_2!}, \quad \text{where } \Gamma_{r_1,r_2}\{\xi\} = \frac{\partial^{r_1+r_2}}{\partial z_1^{r_1}\partial z_2^{r_2}}\log \mathbf{E}e^{\langle z,\xi\rangle}\bigg|_{z=0}. \tag{3.64}$$

Lemma 3.5 (a particular case of Heinrich [11], Lemma 5). *Let $\zeta_1, \zeta_2, \ldots, \zeta_m$ be 1-dependent bivariate random vectors with zero means. Let Λ_i^2 be the maximal eigenvalue of the covariance matrix of ζ_i, $i = 1, \ldots, m$. Let λ^2 be the minimal eigenvalue of the covariance matrix $\mathbf{B}$ of $\Xi = \zeta_1 + \zeta_2 + \cdots + \zeta_m$. Set $\Theta = \mathbf{B}^{-1/2}\Xi$.*

Assume that there exists a constant $H \geq 1/2$ *and a real number* γ *such that*

$$18\,H \max_{1\leq i\leq m} \Lambda_i^2 \leq \gamma^2 \tag{3.65}$$

and, for any $t \in \mathbf{R}^2$,

$$|\mathbf{E}\,\langle t, \zeta_i\rangle^r| \leq H\,r!\,\gamma^{r-2}\,|t|^{r-2}\,\mathrm{Var}\,(\langle t, \zeta_i\rangle)\,, \quad i = 1, \ldots, m, \quad r = 3, 4, \ldots. \tag{3.66}$$

Then

$$\sup_{\|t\|=1} |\Gamma_r\,\{\langle t, \Theta\rangle\}| \leq H^*\,(r-2)!\left(8\sqrt{2}\,\gamma/\lambda\right)^{r-2}, \quad r = 2, 3, \ldots, \tag{3.67}$$

where $H^* = 280\,H\,\lambda^{-2}\sum_{i=1}^m \Lambda_i^2$.

Note that (3.66) is automatically satisfied for $r = 2$, since $H \geq 1/2$.

Lemma 3.6. *For sufficiently large* $n \geq n_0$, *we have, uniformly in* $i = 1, \ldots, s_n$,

$$\mathbf{E}\,|t_1\,\delta_{i,n} + t_2\,u_{i,n}|^r \leq A\,r!\,\gamma_n^{r-2}\,\|t\|^{r-2}\,\mathrm{Var}(t_1\,\delta_{i,n} + t_2\,u_{i,n}), \tag{3.68}$$

for all integers $r \geq 2$ *and for all* $t = (t_1, t_2) \in \mathbf{R}^2$, *where*

$$\gamma_n = A\left(\Psi_n^{3/2} \max_{i\in\Upsilon_3} p_{i,n}^{1/2} + \max_{1\leq i\leq s_n} q_{i,n}^{1/2}\right) \to 0, \text{ as } n \to \infty, \tag{3.69}$$

and Ψ_n *is defined in* (1.47). *Moreover, for all integers* $r \geq 3$,

$$\sup_{\|t\|=1} |\Gamma_r\,\{t_1\,S_n + t_2\,U_n\}| \leq (r-2)!\,(A\,\gamma_n)^{r-2}. \tag{3.70}$$

Proof. Let us prove (3.68). Without loss of generality we assume that

$$\|t\| = 1. \tag{3.71}$$

Applying inequality (3.33) of Lemma 3.2 in the case $\mathbf{P}\,\{\zeta = 1\} = 1 - \mathbf{P}\,\{\zeta = 0\} = q_{i,n}$ (see (3.45)) coupled with inequality (3.31), we get, for $i = 1, \ldots, s_n$,

$$\mathbf{E}\,|u_{i,n}|^r \leq A^r\,n^{-r/2}\left(r^{r/2}\,(n\,q_{i,n})^{r/2} + r^r\,n\,q_{i,n}\right). \tag{3.72}$$

Using (3.46) and (3.72), we obtain

$$\mathbf{E}\,|u_{i,n}|^r \leq A^r\,r^r\,\left(q_{i,n} + n^{-1}\right)^{r/2-1}\,\mathrm{Var}(u_{i,n}). \tag{3.73}$$

Relation (3.50) of Lemma 3.4 implies that

$$\begin{aligned}\mathrm{Var}(t_1\,\delta_{i,n} + t_2\,u_{i,n}) &= t_1^2\,\mathrm{Var}(\delta_{i,n}) + t_2^2\,\mathrm{Var}(u_{i,n}) + 2\,t_1 t_2\,\mathrm{cov}(\delta_{i,n}, u_{i,n}) \\ &\geq \frac{1}{2}\,\left(t_1^2\,\mathrm{Var}(\delta_{i,n}) + t_2^2\,\mathrm{Var}(u_{i,n})\right),\end{aligned} \tag{3.74}$$

if $n \geq n_0$ is large enough (for $i \notin \Upsilon_3$ inequality (3.74) is trivial, see (3.29)). Recall that $nh_n^2 \to \infty$, as $n \to \infty$. Therefore, (1.48), (2.27) and (3.10) imply that

$$n^{-1} \leq q_{i,n}, \quad \text{for } i = 2, \ldots, s_n - 1 \tag{3.75}$$

and sufficiently large $n \geq n_0$. Notice that $y \leq (y+1)^{r-2}$, for $y \geq 0$, $r \geq 2$. Moreover, by (1.10), (1.30) and (1.47), we have $\Psi_n \geq 1/4$. Hence, applying Lemma 3.3 together with (1.47), (3.29), (3.69) and (3.71)–(3.74), we get (3.68):

$$\begin{aligned}
\mathbf{E}\,|t_1\,\delta_{i,n} + t_2\,u_{i,n}|^r &\leq 2^r\,\mathbf{E}\,|t_1\,\delta_{i,n}|^r + 2^r\,\mathbf{E}\,|t_2\,u_{i,n}|^r \\
&\leq A^r\,r^r\,\Big(p_{i,n}^{r/2-1}\,\big(\|K^2\|\,D_n\,\beta_n^{-1}\,\kappa^2\,\sigma^{-4}\big)^{r/2}\,t_1^2\,\mathrm{Var}(\delta_{i,n}) \\
&\qquad + \big(q_{i,n} + n^{-1}\big)^{r/2-1}\,t_2^2\,\mathrm{Var}(u_{i,n})\Big) \\
&\leq A\,r!\,\gamma_n^{r-2}\,\big(t_1^2\,\mathrm{Var}(\delta_{i,n}) + t_2^2\,\mathrm{Var}(u_{i,n})\big) \\
&\leq A\,r!\,\gamma_n^{r-2}\,\mathrm{Var}(t_1\,\delta_{i,n} + t_2\,u_{i,n}),
\end{aligned} \tag{3.76}$$

for sufficiently large $n \geq n_0$. Using (3.76) for $r = 4$ and Hölder's inequality, we get

$$(\mathrm{Var}(t_1\,\delta_{i,n} + t_2\,u_{i,n}))^2 \leq \mathbf{E}\,|t_1\,\delta_{i,n} + t_2\,u_{i,n}|^4 \leq A\,\gamma_n^2\,\mathrm{Var}(t_1\,\delta_{i,n} + t_2\,u_{i,n}).$$

Hence,

$$\mathrm{Var}(t_1\,\delta_{i,n} + t_2\,u_{i,n}) \leq A\,\gamma_n^2, \quad \text{for } \|t\| = 1. \tag{3.77}$$

Limit relation (3.69) follows from (1.35), (1.47), (1.48), (3.10) and (3.12).

We shall apply Lemma 3.5 with $m = s_n$,

$$H = A_1, \quad \gamma = A_2\gamma_n, \quad \lambda^2 = \min_{\|t\|=1} \mathrm{Var}(t_1\,S_n + t_2\,U_n), \quad \zeta_i = (\delta_{i,n}, u_{i,n}), \tag{3.78}$$

$$\Lambda_i^2 = \max_{\|t\|=1} \mathrm{Var}(t_1\,\delta_{i,n} + t_2\,u_{i,n}) \leq 2\,\mathrm{Var}(\delta_{i,n}) + 2\,\mathrm{Var}(u_{i,n}), \quad i = 1, \dots, s_n, \tag{3.79}$$

$$H^* = 280\,H\,\lambda^{-2}\sum_{i=1}^{s_n}\Lambda_i^2, \quad \Xi = (S_n, U_n) \in \mathbf{R}^2, \quad \Theta = \mathbf{B}^{-1/2}\Xi, \tag{3.80}$$

where $\mathbf{B}$ is the covariance operator of Ξ. Fixing $A_1 = A$ from (3.68), using (3.77) and (3.79) and choosing A_2 to be large enough, we ensure the validity of the inequality

$$18\,H \max_{1\leq i\leq s_n} \Lambda_i^2 \leq \gamma^2. \tag{3.81}$$

Using (3.16), (3.17), (3.25), (3.29), (3.40) and Lemma 3.1, we obtain (for sufficiently large $n \geq n_0$)

$$\sum_{i=1}^{s_n} \mathrm{Var}(\delta_{i,n}) = \sum_{i\in\Upsilon_3} \frac{\sigma_n^2(I_{i,n})}{\sigma_n^2(C_n)} \leq 4 \sum_{i\in\Upsilon_3} \frac{\mathbf{P}(I_{i,n})\,\sigma^2}{\sigma^2} \leq 4. \tag{3.82}$$

By (3.47) and (3.48),

$$\sum_{i=1}^{s_n} \mathrm{Var}(u_{i,n}) = 1 - \alpha_n. \tag{3.83}$$

Now (3.79), (3.82) and (3.83) imply

$$\sum_{i=1}^{s_n} \Lambda_i^2 \leq 10. \tag{3.84}$$

Furthermore, by (1.35), (3.3), (3.48), (3.78) and inequality (3.49) of Lemma 3.4,

$$\lambda \geq \min\{\mathrm{Var}(S_n), \mathrm{Var}(U_n)\} - 2\,|\mathrm{cov}(S_n, U_n)| \geq 1/2, \tag{3.85}$$

$$\mu \leq \max\{\mathrm{Var}(S_n), \mathrm{Var}(U_n)\} + 2\,|\mathrm{cov}(S_n, U_n)| \leq 2, \tag{3.86}$$

for sufficiently large $n \geq n_0$, where μ is the maximal eigenvalue of the covariance matrix $\mathbf{B}$. Applying Lemma 3.5 and taking into account relations $\Xi = \mathbf{B}^{1/2}\Theta$, (3.63), (3.78)–(3.81) and (3.84)–(3.86), we obtain, for $r \geq 3$, $n \geq n_0$:

$$\begin{aligned}\sup_{\|t\|=1} |\Gamma_r\{t_1 S_n + t_2 U_n\}| &\leq A^r \sup_{\|t\|=1} |\Gamma_r\{\langle t, \Theta\rangle\}| \\ &\leq A^r H^* (r-2)! \left(8\sqrt{2}\,\gamma/\lambda\right)^{r-2} \leq (r-2)!\,(A\,\gamma_n)^{r-2},\end{aligned}$$

proving (3.70). □

The following fact is well known. It may be easily derived from Remark 2 in Rivlin [17], p. 96. It allows us to estimate coefficients of a polynomial via its maximum on an interval.

Lemma 3.7. *Let $\mathbb{P}(x) = a_0 + a_1 x + \cdots + a_r x^r$ be a polynomial of degree not exceeding r. Then*

$$|a_k| \leq \max\left\{\left|t_k^{(r)}\right|, \left|t_k^{(r-1)}\right|\right\} \max_{-1\leq x\leq 1} |\mathbb{P}(x)|,$$

where $t_k^{(r)}$ are coefficients of $\mathbb{T}_r$, the Chebyshev polynomial of order r.

The Chebyshev polynomial

$$\mathbb{T}_r(x) = t_0^{(r)} + t_1^{(r)} x + \cdots + t_r^{(r)} x^r, \quad r = 1, 2, \ldots,$$

is characterized as having the maximal leading coefficient $t_r^{(r)} = 2^{r-1}$ among all polynomials $\mathbb{P}(x)$ with $\max_{-1\leq x\leq 1} |\mathbb{P}(x)| \leq 1$. We have

$$\mathbb{T}_0(x) = 1, \quad \mathbb{T}_1(x) = x, \quad \mathbb{T}_r(x) = 2x\,\mathbb{T}_{r-1}(x) - \mathbb{T}_{r-2}(x), \quad r = 2, 3, \ldots, \tag{3.87}$$

see Rivlin ([17], formulas (1.11), (1.101)). By induction in r, it is easy to derive from (3.87) the rough bound

$$\sum_{k=0}^{r} \left|t_k^{(r)}\right| \leq 3^{r-1}, \quad r = 1, 2, \ldots. \tag{3.88}$$

Let us consider the definition and some useful properties of classes of d-dimensional distributions $\mathcal{A}_d(\tau)$, $\tau \geq 0$, introduced in Zaitsev [19], see as well Zaitsev [20], [21] and [22]. The class $\mathcal{A}_d(\tau)$ (with a fixed $\tau \geq 0$) consists of d-dimensional distributions F for which the function

$$\varphi(z) = \varphi(F, z) = \log \int_{\mathbf{R}^d} e^{\langle z, x\rangle} F\{dx\} \qquad (\varphi(0) = 0)$$

is defined and analytic for $\|z\|\,\tau < 1$, $z \in \mathbf{C}^d$, and

$$\left|d_u d_v^2\, \varphi(z)\right| \leq \|u\|\tau\, \langle \mathbb{D}\,v, v\rangle \qquad \text{for all } u, v \in \mathbf{R}^d \text{ and } \|z\|\,\tau < 1,$$

where $\mathbb{D}$ is the covariance operator corresponding to F, and $d_u\varphi$ denotes the derivative of the function φ in direction u. It is easy to see that $\tau_1 < \tau_2$ implies $\mathcal{A}_d(\tau_1) \subset \mathcal{A}_d(\tau_2)$. Moreover, the class $\mathcal{A}_d(\tau)$ is closed with respect to convolution: if $F_1, F_2 \in \mathcal{A}_d(\tau)$, then $F_1 * F_2 \in \mathcal{A}_d(\tau)$. The class $\mathcal{A}_d(0)$ coincides with the class of all Gaussian distributions in $\mathbf{R}^d$.

Lemma 3.8. *For sufficiently large $n \geq n_0$, we have*

$$G \stackrel{\text{def}}{=} \mathcal{L}((S_n, U_n)) \in \mathcal{A}_2(\tau_n), \qquad \textit{where} \tag{3.89}$$

$$\tau_n = A\gamma_n = A\left(\Psi_n^{3/2} \max_{i\in\Upsilon_3} p_{i,n}^{1/2} + \max_{1\leq i\leq s_n} q_{i,n}^{1/2}\right) \to 0, \ \textit{as}\ n\to\infty, \tag{3.90}$$

with Ψ_n defined in (1.47).

Proof. Comparing formulas (3.62) and (3.64), we see that

$$\frac{\Gamma_r\{t_1 S_n + t_2 U_n\}}{r!} = \sum_{k=0}^{r} \frac{\Gamma_{k,r-k}\{(S_n, U_n)\}\, t_1^k t_2^{r-k}}{k!\,(r-k)!}, \qquad r = 1, 2, \ldots. \tag{3.91}$$

Define polynomials $\mathbb{P}_r(x) = a_0^{(r)} + a_1^{(r)} x + \cdots + a_r^{(r)} x^r$ with

$$a_k^{(r)} = \frac{\Gamma_{k,r-k}\{(S_n, U_n)\}}{k!\,(r-k)!}, \qquad k = 0, 1, \ldots, r. \tag{3.92}$$

By inequality (3.70) of Lemma 3.5, (3.63) and (3.91), we have, for $r = 3, 4, \ldots$:

$$\max_{-1\leq x\leq 1} |\mathbb{P}_r(x)| \leq \frac{1}{r!} \sup_{\|t\|\leq\sqrt{2}} |\Gamma_r\{t_1 S_n + t_2 U_n\}| \leq \frac{2^{r/2}(r-2)!}{r!} (A\gamma_n)^{r-2}, \tag{3.93}$$

if $n \geq n_0$ is sufficiently large. Applying Lemma 3.7 and relations (3.88), (3.92) and (3.93), we get

$$\begin{aligned} |\Gamma_{k,r-k}\{(S_n, U_n)\}| &\leq \frac{3^{r-1} 2^{r/2} (r-2)!\, k!\,(r-k)!}{r!} (A\gamma_n)^{r-2}, \\ &\leq (r-2)!\,(A\gamma_n)^{r-2}, \quad k = 0, 1, \ldots, r, \quad r = 3, 4, \ldots. \end{aligned} \tag{3.94}$$

Further, expanding, for $u = (u_1, u_2) \in \mathbf{R}^2$, $v = (v_1, v_2) \in \mathbf{R}^2$, $w = (w_1, w_2) \in \mathbf{C}^2$,

$$u = u_1 e_1 + u_2 e_2, \quad v = v_1 e_1 + v_2 e_2, \quad w = w_1 e_1 + w_2 e_2,$$

and rewriting $\Gamma_{r_1,r_2}\{(S_n, U_n)\}$ as

$$\Gamma_{r_1,r_2} \stackrel{\text{def}}{=} \Gamma_{r_1,r_2}\{(S_n, U_n)\} = d_{e_1}^{r_1} d_{e_2}^{r_2} \log \mathbf{E} \exp(z_1 S_n + z_2 U_n)\big|_{z=0},$$

we have

$$\begin{aligned} &d_u d_v^2 d_w^r \log \mathbf{E} \exp(z_1 S_n + z_2 U_n)\big|_{z=0} \\ &= \sum_{k=0}^{r} \frac{r!}{k!\,(r-k)!} w_1^k w_2^{r-k} (\Gamma_{k+3,r-k}\, u_1 v_1^2 + \Gamma_{k+2,r+1-k}\, (u_2 v_1^2 + 2\, u_1 v_1 v_2) \\ &\qquad + \Gamma_{k+1,r+2-k}\, (u_1 v_2^2 + 2\, u_2 v_1 v_2) + \Gamma_{k,r+3-k}\, u_2 v_2^2). \end{aligned}$$

Coupled with (3.94), this implies

$$\left|d_u d_v^2 d_w^r \log \mathbf{E} \exp(z_1 S_n + z_2 U_n)\big|_{z=0}\right| \le r!\ \|u\| \cdot \|v\|^2 \cdot \|w\|^r \cdot (A\gamma_n)^{r+1},$$

for $r = 0, 1, \ldots$. By Taylor's formula,

$$d_u d_v^2 \log \mathbf{E} \exp(z_1 S_n + z_2 U_n)\big|_{z=w} = \sum_{r=0}^{\infty} \frac{d_u d_v^2 d_w^r \log \mathbf{E} \exp(z_1 S_n + z_2 U_n)\big|_{z=0}}{r!}.$$

Therefore,

$$\left|d_u d_v^2 \log \mathbf{E} \exp(z_1 S_n + z_2 U_n)\right| \le A\gamma_n\ \|u\| \cdot \|v\|^2, \quad \text{for } \|z\| \cdot A\gamma_n \le 1,$$

for a suitably chosen absolute constant A. It remains to note that, by (1.35), (3.3), (3.48) and (3.49),

$$\operatorname{Var}(v_1 S_n + v_2 U_n) = v_1^2 \operatorname{Var}(S_n) + v_2^2 \operatorname{Var}(U_n) + 2 v_1 v_2 \operatorname{cov}(S_n, U_n) \ge \|v\|^2/2,$$

for sufficiently large $n \ge n_0$. Limit relation (3.90) is a consequence of (3.69). □

4. Exponential bound for the integral over an exceptional set

The proof of the following Lemma 4.1 is similar to the proof of Giné, Mason and Zaitsev [10], Proposition 3.1.

Lemma 4.1. *Let B be a Borel subset of* $\mathbf{R}$,

$$\xi_n = \int_B (\Delta_n(x) - \mathbf{E}\,\Delta_n(x))\,dx \tag{4.1}$$

where

$$\Delta_n(x) = \sqrt{n}\,|f_n(x) - \mathbf{E} f_n(x)| = \frac{1}{h_n\sqrt{n}}\left|\sum_{i=1}^{n}\left\{K\left(\frac{x-X_i}{h_n}\right) - \mathbf{E}K\left(\frac{x-X}{h_n}\right)\right\}\right|. \tag{4.2}$$

Then

$$\mathbf{E}\exp\{\lambda|\xi_n|\} \le 4\exp\left\{\sum_{m=2}^{\infty}\left(\frac{720e\lambda\kappa}{\log m}\right)^m\left(\Omega^{m/2}(n,B) + \frac{1}{n^{m/2-1}}\Omega(n,B)\right)\right\}, \tag{4.3}$$

for all $\lambda \ge 0$.

Proof. Let $X, X_1, X_1', X_2, X_2', \ldots,$ be i.i.d. random variables. Further, we let η be a Poisson random variable with mean n, independent of $X_1, X_1', X_2, X_2', \ldots,$ and set

$$\Delta_\eta(x) = \frac{1}{h_n\sqrt{n}}\left|\sum_{i=1}^{\eta} K\left(\frac{x-X_i}{h_n}\right) - n\,\mathbf{E}\,K\left(\frac{x-X}{h_n}\right)\right|.$$

Define

$$\bar{\xi}_n = \int_B (\Delta_n(x) - \mathbf{E}\,\Delta_\eta(x))\,dx. \tag{4.4}$$

Let $\mathcal{I}_s$, $s = 1, \ldots, 6$, be a partition of the integers $\mathbf{Z}$ such that:

i) if $i \ne j \in \mathcal{I}_s$ then $|i - j| \ge 2$, and

ii) for every $s = 1, \ldots, 6$, $\sum_{i \in \mathcal{I}_s} \mathbf{P}\{X \in ((i-1/2)h_n, (i+3/2)h_n]\} \le 1/2$, and set

$$A_s = \cup_{i \in \mathcal{I}_s} B_{j,n}, \quad s = 1, \ldots, 6, \quad \text{where } B_{j,n} = (ih_n, (i+1)h_n] \cap B,$$

Now, replacing K_1, K_2, η_n and $(ih_n, (i+1)h_n]$ in the proof of inequalities (3.5), (3.7), (3.8) and (3.13) in Giné, Mason and Zaitsev [10] by K, 0, η and $B_{j,n}$, respectively, and using the arguments therein, we obtain

$$\begin{aligned} \mathbf{E}\exp\{\lambda|\xi_n|\} &\le \mathbf{E}\exp\left\{2\lambda|\overline{\xi}_n|\right\} \\ &\le \prod_{s=1}^{6}\left(\mathbf{E}\exp\left\{12\lambda\left|\int_{A_s}(\Delta_n(x) - \mathbf{E}\,\Delta_\eta(x))\,dx\right|\right\}\right)^{1/6} \\ &\le 2\prod_{s=1}^{6}\left(\mathbf{E}\exp\left\{12\lambda\left|\int_{A_s}(\Delta_\eta(x) - \mathbf{E}\,\Delta_\eta(x))\,dx\right|\right\}\right)^{1/6} \end{aligned} \tag{4.5}$$

and

$$\begin{aligned} &\exp\left\{12\lambda\left|\int_{A_s}(\Delta_\eta(x) - \mathbf{E}\,\Delta_\eta(x))\,dx\right|\right\} \\ &\le 2\exp\left\{\sum_{j\in\mathcal{I}_s}\sum_{m=2}^{\infty}\left(\frac{720e\lambda}{\log m}\right)^m\left[\left(\int_{B_{j,n}}\frac{1}{h_n}\mathbf{E}\,K^2\left(\frac{x-X}{h_n}\right)dx\right)^{m/2}\right.\right. \\ &\qquad \left.\left. + \frac{1}{n^{m/2-1}}\int_{B_{j,n}}\frac{1}{h_n}\mathbf{E}\left|K\left(\frac{x-X}{h_n}\right)\right|^m dx\right]\right\}. \end{aligned} \tag{4.6}$$

Furthermore, by a change of variables,

$$\begin{aligned} \sum_{j\in\mathcal{I}_s}\left(\int_{B_{j,n}}\frac{1}{h_n}\mathbf{E}\,K^2\left(\frac{x-X}{h_n}\right)dx\right)^{m/2} &\le \left(\sum_{j\in\mathcal{I}_s}\int_{B_{j,n}}\frac{1}{h_n}\mathbf{E}\,K^2\left(\frac{x-X}{h_n}\right)dx\right)^{m/2} \\ &\le \left(\mathbf{E}\int_B\frac{1}{h_n}K^2\left(\frac{x-X}{h_n}\right)dx\right)^{m/2}. \end{aligned}$$

Using (1.2), (1.3), (2.4), (2.5) and (2.7), we obtain

$$\begin{aligned} \mathbf{E}\frac{1}{h_n}\int_B K^2\left(\frac{x-X}{h_n}\right)dx &\le \kappa^2 h_n^{-1}\int_B \mathbf{P}\{X \in [x - h_n/2, x + h_n/2]\}\,dx \\ &\le \kappa^2\,\Omega(n,B). \end{aligned}$$

Similarly, we have

$$\begin{aligned} &\frac{1}{n^{m/2-1}}\sum_{j\in\mathcal{I}_s}\int_{B_{j,n}}\frac{1}{h_n}\mathbf{E}\left|K\left(\frac{x-X}{h_n}\right)\right|^m dx \\ &\le \frac{\kappa^m h_n^{-1}}{n^{m/2-1}}\int_B \mathbf{P}\{X \in [x - h_n/2, x + h_n/2]\}\,dx \le \frac{\kappa^m}{n^{m/2-1}}\,\Omega(n,B). \end{aligned}$$

Then, combining these estimates with (2.4) and (4.6), we obtain

$$\mathbf{E}\exp\left\{12\lambda\left|\int_{A_s}(\Delta_\eta(x)-\mathbf{E}\,\Delta_\eta(x))\,dx\right|\right\} \tag{4.7}$$
$$\leq 2\exp\left\{\sum_{m=2}^{\infty}\left(\frac{720e\lambda\kappa}{\log m}\right)^m\left(\Omega^{m/2}(n,B)+\frac{1}{n^{m/2-1}}\,\Omega(n,B)\right)\right\}.$$

Inequalities (4.5) and (4.7) imply (4.3). □

5. Proof of Theorems 1.3, 1.5 and 1.7

Note now that for any absolute constant A we have

$$A/\sqrt{n}\leq\tau_n, \tag{5.1}$$

for sufficiently large $n\geq n_0$ (see (3.75) and (3.90)). Therefore, by Example 1.2 in Zaitsev [19],

$$H\overset{\text{def}}{=}\mathcal{L}\left((0,V_n)\right)\in\mathcal{A}_2\left(A/\sqrt{n}\right)\subset\mathcal{A}_2\left(\tau_n\right). \tag{5.2}$$

Hence, by (3.89) and (5.2),

$$Q\overset{\text{def}}{=}\mathcal{L}\left((S_n,U_n)+(0,V_n)\right)\in\mathcal{A}_2\left(\tau_n\right) \tag{5.3}$$

(recall that (S_n,U_n) is independent of V_n).

The following below Lemmas 5.1 and 5.2 are proved in Zaitsev [23]. They provide estimates of the rate of convergence in a lemma of Beirlant and Mason [1], see as well Giné, Mason and Zaitsev [10], Lemma 2.4.

Lemma 5.1. *Let (for each $n\in\mathbf{N}$) $\eta_{1,n}$ and $\eta_{2,n}$ be independent Poisson random variables with $\eta_{1,n}$ being Poisson $(n(1-\alpha_n))$ and $\eta_{2,n}$ being Poisson $(n\alpha_n)$ where $\alpha_n\in(0,1)$. Denote $\eta_n=\eta_{1,n}+\eta_{2,n}$ and set*

$$U_n=\frac{\eta_{1,n}-n(1-\alpha_n)}{\sqrt{n}}\quad\text{and}\quad V_n=\frac{\eta_{2,n}-n\alpha_n}{\sqrt{n}}.$$

Let $\{S_n\}_{n=1}^{\infty}$ be a sequence of random variables such that for each $n\in\mathbf{N}$, the random vector (S_n,U_n) is independent of V_n. Assume that $\operatorname{Var}(S_n)=1$,

$$\mathcal{L}\left((S_n,U_n)+(0,V_n)\right)\in\mathcal{A}_2\left(\tau_n\right), \tag{5.4}$$

and

$$|\chi_n|\leq 1/2, \tag{5.5}$$

where

$$\chi_n=\operatorname{cov}\left(S_n,U_n\right). \tag{5.6}$$

Then there exist absolute constants A_3,A_4,A_5,A_6 such that, for τ_n satisfying the estimates

$$5\alpha_n^{-1}\exp\left\{-5\alpha_n/432\,\tau_n^2\right\}\leq\tau_n, \tag{5.7}$$
$$A_3\,n^{-1/2}\leq\tau_n\leq A_4, \tag{5.8}$$

and for any fixed $n \in \mathbf{N}$ and $y > 0$, one can construct on a probability space random variables ζ_n and Z so that the distribution of ζ_n is the conditional distribution of S_n given $\eta_n = n$, Z is a standard normal random variable and

$$\mathbf{P}\left\{\left|\sqrt{1-\chi_n^2}\, Z - \zeta_n\right| \geq y\right\} \leq A_5 \exp\left\{-A_6\, y/\tau_n\right\}. \tag{5.9}$$

Lemma 5.2. *Let the conditions of Lemma 5.1 be satisfied. Then there exists absolute constants A_7, A_8, A_9, A_{10} such that, for any fixed $n \in \mathbf{N}$ and b satisfying*

$$A_3 n^{-1/2} \leq \tau_n \leq A_7 b, \qquad b \leq 1, \tag{5.10}$$

one can construct on a probability space random variables ζ_n and Z with distributions described in Lemma 5.1 so that, for any $y > 0$,

$$\begin{aligned}\mathbf{P}\left\{\left|\sqrt{1-\chi_n^2}\, Z - \zeta_n\right| \geq A_{10} \exp\left\{-b^2/72\tau_n^2\right\} + y\right\} \\ \leq A_8 \exp\left\{-A_9\, y/\tau_n\right\} + 2\,\mathbf{P}\left\{|\omega| > y/6\right\},\end{aligned} \tag{5.11}$$

where ω have the centered normal distribution with variance b^2.

Comparing Lemmas 5.1 and 5.2, we observe that in Lemma 5.1 the probability space depends essentially on y, while the statement (5.11) of Lemma 5.2 is valid on the same probability space (depending on b) for any $y > 0$. However, (5.11) is weaker than (5.9) for some values of y. The same rate of approximation (as in (5.9)) is contained in (5.11) if $b^2 \geq 72\tau_n^2 \log(1/\tau_n)$ and $y \geq b^2/\tau_n$ only.

Now we return to the estimation and note that, for random variables S_n, U_n and V_n defined in (3.42)–(3.44), the conditions of Lemmas 5.1 and 5.2 are satisfied (with τ_n defined in (3.90) and $\eta_n = \eta$) for $n \geq n_0$. Indeed, by (3.9)–(3.11), we have

$$\tau_n^* \geq \tau_n, \tag{5.12}$$

if the constants A in (1.39) and (3.90) are chosen in a suitable way. Limit relation (1.39) follows from (1.35), (1.47) and (1.48). By (1.39), (1.46) and (5.12), α_n is chosen so that condition (5.7) is satisfied for $n \geq n_0$. Note that by (3.90) and (5.1), condition (5.8) and the first inequality in (5.10) are fulfilled for sufficiently large $n \geq n_0$. Moreover, by (1.35), and (3.49), χ_n (defined in (5.6)) tends to zero, as $n \to \infty$, and condition (5.5) is satisfied for sufficiently large $n \geq n_0$. Thus, we can apply to S_n, U_n, V_n the statements of Lemmas 5.1 and 5.2.

By Lemma 5.1, for sufficiently large fixed $n \geq n_0$ and for any fixed $y > 0$, one can construct on a probability space random variables ζ_n and Z so that the distribution of ζ_n is the conditional distribution of S_n given $\eta = n$,

$$\zeta_n =_d \sigma_n^{-1}(C_n) \int_{C_n} (\Delta_n(x) - \mathbf{E}\,\Delta_\eta(x))\, dx \tag{5.13}$$

(see (3.27), (3.28), (3.42) and (4.2)) and a standard normal random variable Z so that

$$\mathbf{P}\left\{\left|\sqrt{1-\chi_n^2}\, Z - \zeta_n\right| \geq y\right\} \leq A_5 \exp\left\{-A_6\, y/\tau_n\right\}. \tag{5.14}$$

By Lemma 5.2, for sufficiently large fixed $n \geq n_0$ and for any fixed b satisfying

$$\tau_n \leq A_7 b, \qquad b \leq 1, \tag{5.15}$$

one can construct on a probability space a random variable ζ_n with distribution described in (5.13) and a standard normal random variable Z so that, for any $y > 0$,

$$\begin{aligned} \mathbf{P}\left\{\left|\sqrt{1-\chi_n^2}\, Z - \zeta_n\right| \geq A_{10} \exp\left\{-b^2/72\tau_n^2\right\} + y\right\} \\ \leq A_8 \exp\left\{-A_9\, y/\tau_n\right\} + 2\,\mathbf{P}\left\{|\omega| > y/6\right\}, \end{aligned} \tag{5.16}$$

where ω have the centered normal distribution with variance b^2.

In both cases described above we can apply Lemma A of Berkes and Philipp [2] assuming that there exists a sequence of i.i.d. random variables $X_1, X_2, \ldots$ with probability density f and such that

$$\zeta_n = \sigma_n^{-1}(C_n) \int_{C_n} \left(\Delta_n(x) - \mathbf{E}\,\Delta_\eta(x)\right) dx, \tag{5.17}$$

where $\Delta_n(x)$ is defined in (4.2).

By (1.1), (1.11), (1.35), (3.16), (3.28), (4.2) and Lemma 2.6, we have

$$\begin{aligned} &\int_{C_n} \left|\mathbf{E}\,\Delta_n(x) - \mathbf{E}\,\Delta_\eta(x)\right| dx \\ &= \sqrt{n} \int_{C_n} \left|\mathbf{E}\left|f_n(x) - \mathbf{E}\, f_n(x)\right| - \mathbf{E}\left|f_\eta(x) - \mathbf{E}\, f_n(x)\right|\right| dx \\ &\leq \sqrt{n} \int_{E_n} \left|\mathbf{E}\left|f_n(x) - \mathbf{E}\, f_n(x)\right| - \mathbf{E}\left|f_\eta(x) - \mathbf{E}\, f_n(x)\right|\right| dx \\ &\leq \frac{A\,\lambda(E_n)\ \|K^3\|}{\|K^2\|\,\sqrt{nh_n^2}} + \frac{A\,\mathbb{N}_n\sqrt{h_n}}{\sqrt{\|K^2\|}} \stackrel{\text{def}}{=} y_n \end{aligned} \tag{5.18}$$

and $y_n \to \infty$, as $n \to \infty$. Applying Lemma 4.1 for $B = \overline{C}_n$, we see that

$$\mathbf{E}\exp\left\{\lambda|\xi_n|\right\} \leq 4\exp\left\{\sum_{m=2}^{\infty}\left(\frac{720e\lambda\kappa}{\log m}\right)^m \left(\Omega^{m/2}(n,\overline{C}_n) + \frac{1}{n^{m/2-1}}\Omega(n,\overline{C}_n)\right)\right\}, \tag{5.19}$$

for all $\lambda \geq 0$, where

$$\xi_n = \int_{\overline{C}_n} \left(\Delta_n(x) - \mathbf{E}\,\Delta_n(x)\right) dx. \tag{5.20}$$

By (1.45) and (2.27),

$$n^{1/2}\,\Omega_n \to \infty, \qquad \text{as} \quad n \to \infty, \tag{5.21}$$

since we assume $nh_n^2 \to \infty$. Using (1.45), (3.22), (5.19) and (5.21), we obtain that, for sufficiently large $n \geq n_0$,

$$\mathbf{E}\,\exp\left\{\lambda|\xi_n|\right\} \leq 4\,\exp\left\{\sum_{m=2}^{\infty}\left(\frac{A_{11}\,\lambda\,\kappa\,\Omega_n^{1/2}}{\log m}\right)^m\right\}, \tag{5.22}$$

for all $\lambda \geq 0$. It may be shown that there exists an absolute constant A such that

$$\sum_{m=2}^{\infty} \left(\frac{\mu}{\log m} \right)^m \leq A \exp\{\exp\{A\,\mu\}\}, \qquad \text{for all } \mu > 0.$$

Applying the exponential Chebyshev inequality coupled with (5.22), where $\lambda = A_{12}\,\kappa^{-1}\,\Omega_n^{-1/2} \log^* \log^*(z/A_{11}\,\kappa\,\Omega_n^{1/2})$ and A_{12} is sufficiently small, we obtain that

$$\mathbf{P}\{|\xi_n| \geq z\} \leq A \exp\left\{-A^{-1}\kappa^{-1}\Omega_n^{-1/2} z \log^* \log^*(z/A\kappa\Omega_n^{1/2})\right\}, \text{ for any } z > 0. \tag{5.23}$$

Inequalities (3.25), (5.14), (5.18) and (5.23) imply that, for any fixed $n \geq n_0$ and for any fixed $x > 0$, one can construct on a probability space a sequence of i.i.d. random variables $X_1, X_2, \ldots$, and a standard normal random variable Z so that

$$\begin{aligned} &\mathbf{P}\left\{\left|\int_{-\infty}^{\infty} (\Delta_n(x) - \mathbf{E}\,\Delta_n(x))\,dx - \sigma\,Z\right| \geq y_n + z + x\right\} \qquad (5.24) \\ &\leq A\Big(\exp\{-A^{-1}\,\sigma^{-1} x/\tau_n\} + \exp\left\{-A^{-1}\kappa^{-1}\,\Omega_n^{-1/2} z\right\} \\ &\qquad + \mathbf{P}\left\{\left|\left(\sigma - \sigma_n(C_n)\sqrt{1-\chi_n^2}\right) Z\right| \geq z/2\right\}\Big), \text{ for any } z > 0. \end{aligned}$$

Similarly, using (5.16) instead of (5.14), we establish that, for any fixed $n \geq n_0$ and for any fixed b satisfying (5.15), one can construct on a probability space a sequence of i.i.d. random variables $X_1, X_2, \ldots$ and a standard normal random variable Z so that

$$\begin{aligned} &\mathbf{P}\left\{\left|\int_{-\infty}^{\infty} (\Delta_n(x) - \mathbf{E}\,\Delta_n(x))\,dx - \sigma\,Z\right| \geq A\,\sigma\,\exp\{-b^2/72\tau_n^2\} + y_n + z + x\right\} \\ &\leq A\Big(\exp\{-A^{-1}\,\sigma^{-1} x/\tau_n\} + \exp\{-A^{-1}\kappa^{-1}\Omega_n^{-1/2} z \log^* \log^*(z/A\kappa\Omega_n^{1/2})\} \\ &\qquad + \mathbf{P}\left\{\left|\left(\sigma - \sigma_n(C_n)\sqrt{1-\chi_n^2}\right) Z\right| \geq z/2\right\} \\ &\qquad + \mathbf{P}\{b\,|Z| > A^{-1}\,\sigma^{-1} x\}\Big), \text{ for any } x, z > 0. \end{aligned} \tag{5.25}$$

Now, by (1.10), (1.18), (1.20), (1.29), (1.45), (2.34), (3.16), (3.23), (3.49) and (5.6), we have

$$\begin{aligned} &\left|\sigma - \sigma_n(C_n)\sqrt{1-\chi_n^2}\right| \qquad (5.26) \\ &\leq |\sigma - \sigma_n(C_n)| + \sigma\left|\sqrt{1-\chi_n^2} - 1\right| \\ &\leq \sigma\left(1 - \sqrt{\mathbf{P}(E_n)}\right) + \left|\sigma_n(E_n) - \sigma\sqrt{\mathbf{P}(E_n)}\right| + \sigma_n(E_n \backslash C_n) + \sigma\,\chi_n^2 \\ &\leq \frac{A\,\|K^2\|}{\sigma\,h_n}\left(\mathbb{L}_n + \frac{\varepsilon_n\,\mathrm{M}_n}{\|K^2\|}\right) + A\,\kappa\,\Omega_n^{1/2} + \frac{A}{\sigma}\left(\frac{\|K^3\|\,\lambda(E_n)}{\|K^2\|\sqrt{nh_n^2}}\right)^2, \end{aligned}$$

for sufficiently large $n \geq n_0$. Now inequality (1.38) follows from (4.2), (5.12), (5.24) and (5.26). Relations (1.35) and (3.24) imply the limit relation in (1.41). The proof of Theorem 1.5 repeats that of Theorem 1.3. The only difference is that we apply (5.25) instead of (5.24). □

Proof of Theorem 1.7. Without loss of generality, we assume $x \geq 1$. By Theorem 1.3, for any $z > 0$,

$$1 - F(x) \leq 1 - \Phi\left(x - 2z - y_n/\sigma\right) + A\Big(\exp\{-A^{-1}\, z/\tau_n^*\} + \exp\{-A^{-1}\kappa^{-1}\Omega_n^{-1/2}\sigma z \log^* \log^*(\sigma z/A\kappa\Omega_n^{1/2})\} + \mathbf{P}\left\{|\partial_n Z| \geq \sigma z/2\right\}\Big)$$

and

$$1 - F(x) \geq 1 - \Phi\left(x + 2z + y_n/\sigma\right) - A\Big(\exp\{-A^{-1}\, z/\tau_n^*\} + \exp\{-A^{-1}\kappa^{-1}\Omega_n^{-1/2}\sigma z \log^* \log^*(\sigma z/A\kappa\Omega_n^{1/2})\} + \mathbf{P}\left\{|\partial_n Z| \geq \sigma z/2\right\}\Big).$$

Choosing here $z = \max\left\{\sqrt{\tau_n^* x},\ \Omega_n^{1/4}\sqrt{x}\left(\log^* \log^*(1/\Omega_n)\right)^{-1/2}, \sqrt{\partial_n}\right\}$ and using elementary properties of normal distribution function, we get the result. □

Acknowledgment

A big part of the results of this paper was obtained by the author while visiting the Delaware University. He would like to thank Professor David M. Mason for his hospitality and for the statement of the problem. Many steps of proofs of the present paper repeat corresponding steps of a detailed proof of Theorem 1.1 (due to Mason) from the joint paper Giné, Mason and Zaitsev [10]. The author thanks Paul Eggermont, Evarist Giné and David Mason for useful discussions. The author is grateful to the anonymous referee for careful reading the manuscript and useful suggestions.

References

[1] J. Beirlant and D. M. Mason, *On the asymptotic normality of L_p-norms of empirical functionals.* Math. Methods Statist. **4** (1995), 1–19.

[2] I. Berkes and W. Philipp, *Approximation theorems for independent and weakly dependent random vectors.* Ann. Probab. **7** (1979), 29–54.

[3] A. de Acosta, *Inequalities for B-valued random vectors with applications to the strong law of large numbers.* Ann. Probab. **9** (1981), 157–161.

[4] M. Csörgő and L. Horváth, *Central limit theorems for L_p-norms of density estimators.* Z. Wahrsch. Verw. Gebiete **80** (1988), 269–291.

[5] L. Devroye, *Exponential inequalites in nonparametric estimation.* In: Nonparametric Functional Estimation and Related Topics (G. Roussas ed.) 31–44, Kluwer Academic Publishers, Dordrecht, 1991.

[6] L. Devroye and L. Györfi, *Nonparametric Density Estimation: The L_1 View.* Wiley, New York, 1985.

[7] P. P. B. Eggermont and V. N. LaRiccia, *Maximum Penalized Likelihood Estimation, Volume 1; Density Estimation.* Springer, New York, 2001.

[8] G. B. Folland, *Real Analysis.* 2nd Edition, Wiley, New York, 1999.

[9] L. Horváth, *On L_p-norms of multivariate density estimators.* Ann. Statist. **19** (1991), 1933–1949.

[10] E. Giné, D. M. Mason and A. Yu. Zaitsev, *The L_1-norm density estimator process.* Ann. Probab. **31** (2003) (to appear).

[11] L. Heinrich, *Some estimates of the cumulant-generating function of a sum of m-dependent random vectors and their application to large deviations.* Math. Nachrichten **120** (1985), 91–101.

[12] R. Latała, *Estimation of moments of sums of independent random variables.* Ann. Probab. **25** (1997), 1502–1513.

[13] S. Nabeya, *Absolute moments in 2-dimensional normal distributions.* Ann. Inst. Statist. Math. Tokyo. **3** (1951), 2–6.

[14] I. F. Pinelis, *Inequalities for sums of independent random vectors and their application to estimating a density.* Theory Probab. Appl. **35** (1990), 605–607.

[15] I. F. Pinelis, *On a majorization inequality for sums of independent random variables.* Probab. Statist. Letters **19** (1994), 97–99.

[16] I. F. Pinelis, *Optimal bounds for the distributions of martingales in Banach spaces.* Ann. Probab. **22** (1994), 1679–1706.

[17] T. J. Rivlin, *The Chebyshev Polynomials.* Wiley, New York, 1974.

[18] T. J. Sweeting, *Speeds of convergence in the multidimensional central limit theorem.* Ann. Probab. **5** (1977), 28–41.

[19] A. Yu. Zaitsev, *Estimates of the Lévy–Prokhorov distance in the multivariate central limit theorem for random variables with finite exponential moments.* Theor. Probab. Appl. **31** (1986), 203–220.

[20] A. Yu. Zaitsev, *Estimates for quantiles of smooth conditional distributions and multidimensional invariance principle.* Siberian Math. J. **37** (1996), 807–831 (in Russian).

[21] A. Yu. Zaitsev, *Multidimensional version of the results of Komlós, Major and Tusnády for vectors with finite exponential moments.* ESAIM : Probability and Statistics **2** (1998), 41–108.

[22] A. Yu. Zaitsev, *Multidimensional version of the results of Sakhanenko in the invariance principle for vectors with finite exponential moments.* I; II; III, Theor. Probab. Appl. **45** (2000), 718–738; **46** (2001), 535–561; 744–769.

[23] A. Yu. Zaitsev, *Estimates of the rate of approximation in a de-Poissonization lemma.* Ann. Inst. Henri Poincaré, Probab. Statist. **38** (2002), 1071–1086.

Laboratory of Statistical Methods, St. Petersburg Branch of the Steklov Mathematical Institute, 27 Fontanka, St. Petersburg 191011, RUSSIA
E-mail address: zaitsev@pdmi.ras.ru

Part 7

Statistics via Empirical Process Theory

Progress in Probability, Vol. 55, 295–312

Statistical Nearly Universal Glivenko–Cantelli Classes

R. M. Dudley

Abstract. Suppose given a parametric family of densities $f(x,\theta)$. Let $X_1, \ldots, X_n, \ldots$ be i.i.d. observations with a law P not necessarily in the family. If the maximum likelihood estimates of θ exist and converge to some $\theta_0 = \theta_0(P)$, then θ_0 is called the pseudo-true value of the parameter. Analogues of pseudo-true values, here called M-limits, also can occur as limits of M-estimators more general than maximum likelihood estimators, where the sample average of a function $\rho(x,\theta)$ is (approximately) minimized with respect to θ. In this paper, the question is for what ρ do M-limits exist for all probability measures P on the sample space X, or for all P outside some "small" exceptional class?

1. Introduction

Let $(X, \mathcal{B})$ be a measurable space. Let $\mathcal{P}(X, \mathcal{B})$ be the collection of all probability measures on $(X, \mathcal{B})$. For each $P \in \mathcal{P}(X, \mathcal{B})$, we can form the countable product $(\Omega, \mathcal{B}^\infty, \Pr)$ of copies of $(X, \mathcal{B}, P)$. Let $X_1, X_2, \ldots,$ be coordinates on the product. Then X_j will be said to be *strictly i.i.d.* (P). For such X_i let $P_n := (\delta_{X_1} + \cdots + \delta_{X_n})/n$. Then P_n are empirical measures for P and give the "standard model" of such empirical measures. Given a probability space $(\Omega, \mathcal{S}, \Pr)$, having in mind the above countable product, and any set $A \subset \Omega$, the *outer probability* $\Pr^*(A)$ is defined by $\Pr^*(A) := \inf\{\Pr(B) : B \in \mathcal{S}, A \subset B\}$. If f_n are functions, not necessarily measurable, from Ω into S where (S, d) is a metric space, then f_n are said to converge to f_0 *in outer probability* if for every $\varepsilon > 0$, $\Pr^*(d(f_n, f_0) > \varepsilon) \to 0$ as $n \to \infty$. Also, f_n is said to converge to f_0 *almost uniformly* if for every $\varepsilon > 0$, $\Pr^*(d(f_n, f_0) > \varepsilon \text{ for some } n > m) \to 0$ as $m \to \infty$. If $d(f_n, f_0)$ are measurable, e.g. if f_n are measurable and S is separable, then convergence in outer probability (resp. almost uniformly) is equivalent to ordinary convergence in probability (resp. almost sure convergence).

Let $(\Theta, \mathcal{T})$ be a Hausdorff topological space, to be called the *parameter space*. Recall that a function g from Θ into $(-\infty, +\infty]$ is called *lower semicontinuous* iff for each $\theta \in \Theta$, $g(\theta) \le \liminf_{\phi \to \theta} g(\phi)$. Here $g(\theta)$ may be $+\infty$ for some θ, if $g(\phi) \to +\infty$ as $\phi \to \theta$. Lower semicontinuity of g is equivalent to the statement that for each real y, $\{\theta : g(\theta) \le y\}$ is closed.

2000 *Mathematics Subject Classification.* Primary 62G20,60B12; Secondary 62G35.

Key words and phrases. M-estimators, pseudo-true values, location, scale, exponential families.
This work was partially supported by NSF grant DMS-0103821.

For a signed measure ν and function f let $\nu f := \int f d\nu$ (if defined, possibly $\pm\infty$). Let $\mathcal{H}(X,\Theta)$ be the set of all functions ρ from $X \times \Theta$ into $(-\infty,\infty]$ which are jointly measurable, with the Borel σ-algebra generated by the open sets of Θ, and such that $\rho(x,\cdot)$ is lower semicontinuous on Θ for each $x \in X$.

Let $\mathcal{L}^0(X) := \mathcal{L}^0(X,\mathcal{B})$ be the set of measurable real-valued functions on X for $\mathcal{B}$. If $Q \in \mathcal{P}(X,\mathcal{B})$ and $\rho \in \mathcal{H}(X,\Theta)$, then ρ will be called *adjustable* for Q and an $a(\cdot) \in \mathcal{L}^0(X)$ will be called an *adjustment function* for ρ and Q if and only if for

$$h(x,\theta) := \rho(x,\theta) - a(x), \tag{1.1}$$

$-\infty < Qh(\theta) := Qh(\cdot,\theta) \leq +\infty$ for all $\theta \in \Theta$, and $Qh(\phi) < +\infty$ for some $\phi \in \Theta$. For any $a(\cdot) \in \mathcal{L}^0(X)$, if (1.1) holds, clearly $h \in \mathcal{H}(X,\Theta)$ if and only if $\rho \in \mathcal{H}(X,\Theta)$. It is known and not hard to check that if an adjustment function $a(\cdot)$ for ρ and Q exists, then another $\alpha(\cdot) \in \mathcal{L}^0(X)$ is also such an adjustment function if and only if $Q|a - \alpha| < \infty$, and we can take $a(x) \equiv \rho(x,\phi)$ [19], [20, p. 129].

If $a(\cdot)$ is an adjustment function for ρ and Q, then $\theta_1 \in \Theta$ is called the *M-functional* of Q for ρ, $\theta_1 := \theta_1(Q) := \theta_1(Q,\rho)$, if and only if $Qh(\theta)$ achieves a unique minimum at $\theta = \theta_1$. It's easily seen that the M-functional, if it exists, does not depend on the choice of adjustment function.

If Q is an empirical measure P_n, then an *M-estimate* $\hat{\theta}_n$ is a $\theta \in \Theta$ at which $P_n\rho(\theta)$ has its minimum value. For empirical measures, if ρ is real-valued, no adjustment is necessary; we can take $a(x) \equiv 0$, and an M-functional of P_n is the same as a unique M-estimate. P. J. Huber, e.g. [19], [20], defined M-estimation and M-functionals.

If $\rho(x,\theta) = -\log f(x,\theta)$ for a parametric family $f(\cdot,\theta)$ of densities, the M-functional $\theta_1(P)$, if it exists, is called the *pseudo-true value* of θ. Pseudo-true values (without adjustment) have been treated e.g. in [30], [27], [2, p. 47], and [8, p. 10]. Very often in the literature, M-functionals, specifically pseudo-true values, have been assumed to exist. When deciding between two non-nested parametric models by Cox's likelihood ratio test [6], [7], one needs to show that for each law in one model there is a corresponding pseudo-true parameter for the other. Here a *law* is a probability measure defined on the Borel sets.

In this paper an *M-limit* $\theta_0 = \theta_0(P)$ (with respect to ρ) will mean a point of Θ such for every neighborhood U of θ_0, as $n \to \infty$,

$$\Pr{}^*\Big\{ \inf\{P_n\rho(\theta) : \theta \notin U\} \leq \inf\{P_n\rho(\phi) : \phi \in U\}\Big\} \to 0. \tag{1.2}$$

Then, if $\mathcal{T}$ is metrizable and M-estimates $\hat{\theta}_n$ exist for n large enough, they must converge in outer probability to θ_0.

An M-limit $\theta_0 = \theta_0(P)$ with respect to ρ will be called *definite* iff for every neighborhood U of θ_0 there is an $\varepsilon > 0$ such that as $n \to \infty$,

$$\Pr{}^* \{\inf\{P_n\rho(\theta) : \theta \notin U\} \leq \varepsilon + \inf\{P_n\rho(\phi) : \phi \in U\} \to 0. \tag{1.3}$$

If an M-functional $\theta_1(P)$ exists for a given ρ and P, it does not necessarily follow that it is an M-limit. Conversely, in most cases treated in this paper, existing M-limits are M-functionals, but they need not be: for example, let $X = \mathbb{R}$, $\Theta = \{0,1\}$, and $\rho(x,\theta) := x\theta$. Let $p_k := 3/(\pi^2 k^2)$, $t(k) := 2^k$, and $P := \sum_{k=1}^{\infty} p_k(\delta_{-k}+\delta_{t(k)})$. Then $Px = P(\rho(\cdot,1)-\rho(\cdot,0))$ is undefined, so ρ is not adjustable for P and there is no M-functional. But as $n \to \infty$, $P_n[\rho(\cdot,1)-\rho(\cdot,0)] = P_n x \to +\infty$, so almost surely M-estimates $\hat{\theta}_n$ exist for all n large enough and equal 0. Thus the M-limit $\theta_0(P) = 0$.

A class $\mathcal{F} \subset \mathcal{L}^0(X,\mathcal{B})$ is called a *(strong) Glivenko-Cantelli class* for $P \in \mathcal{P}(X,\mathcal{B})$ if, when X_i are strictly i.i.d. (P), we have $\sup_{f\in\mathcal{F}} |(P_n - P)(f)| \to 0$ almost uniformly. Talagrand [32], [33] gave a characterization of Glivenko-Cantelli classes.

A class $\mathcal{F} \subset \mathcal{L}^0(X,\mathcal{B})$ is called a *universal Glivenko-Cantelli class* iff for every $P \in \mathcal{P}(X,\mathcal{B})$, $\mathcal{F}$ is a Glivenko-Cantelli class for P. Each f in such a class must be bounded, and the set of all functions $f - \inf f$, $f \in \mathcal{F}$, must be uniformly bounded [14, Proposition 4].

Among bounded classes of functions, universal Glivenko-Cantelli classes can be rather large, see [14, §3]. For example, if X is a countable set, then for any $M < \infty$, the class of all functions f on X with $|f(x)| \leq M$ for all x is a universal Glivenko-Cantelli class.

A universal Glivenko-Cantelli class $\mathcal{F}$ is called a *uniform* Glivenko-Cantelli class if $\lim_{n\to\infty} \sup_{P\in\mathcal{P}(X,\mathcal{B})} \Pr^*\{\sup\{|\int f d(P_k - P)| : f \in \mathcal{F},\ k \geq n\} = 0$. On such classes see [14, §4].

2. Preliminaries and generalities

Let $\mathcal{Q} \subset \mathcal{P}(X,\mathcal{B})$. Then a $\rho \in \mathcal{H}(X,\Theta)$ will be called *M-consistent for $P \notin \mathcal{Q}$* iff for all $P \notin \mathcal{Q}$, an M-limit $\theta_0(P)$ exists. If $\mathcal{Q} = \emptyset$ then ρ will be called *universally M-consistent*. It will turn out that universal M-consistency is a relatively rare phenomenon.

If $\rho(\cdot,\cdot)$ is universally M-consistent and for a metric d metrizing the topology of Θ, there is a sequence $\varepsilon_n \downarrow 0$ such that for every probability measure Q on $(X,\mathcal{B})$, and $\theta_0 := \theta_0(Q)$,

$$\Pr^*\Big\{ \inf\{Q_n\rho(\theta) : d(\theta,\theta_0) \geq \varepsilon_n\} \leq \inf\{Q_n\rho(\phi) : \phi \in \Theta\}\Big\} \leq \varepsilon_n,$$

then ρ will be called *uniformly M-consistent for d*.

Remark. Of course, a ρ that is not universally M-consistent cannot be uniformly M-consistent. But, as will be seen, there are functions ρ for which Q has an M-limit $\theta_0(Q)$ except for Q in a very small class. One might think that such Q could be neglected, and indeed they can for some purposes. For example, if the exceptional Q all have support in only one or two points, then as soon as three different values of X_j have been observed, one knows one doesn't have such an exceptional Q. But, uniformity will still fail, in the sense that convergence may hold but be slow for Q

close to those for which no M-limit exists. The class of such "close" Q may appear much larger. Moreover, in some situations universally M-consistent functions ρ exist, but none of them is uniformly M-consistent, see Proposition 3.4.

A θ minimizing $P_n\rho(\theta)$, namely an M-estimate, if it exists, is not necessarily unique. Since $\rho(x,\cdot)$ is lower semicontinuous on Θ for each x, each $P_n\rho(\cdot)$ is also lower semicontinuous in θ. Thus the set M_n where it attains its minimum is closed, possibly empty. A measurable function $\hat{\theta}_n$ of $(X_1,\ldots,X_n)$ whose values are M-estimates is called an *M-estimator.* Under mild conditions, based on measurable selection theorems, e.g. Sainte-Beuve [29, Theorems 3,4], if M-estimates exist, M-estimators can be chosen.

In the present paper, it has not seemed necessary to treat approximate, randomized, or non-measurable estimators, partly because the definition of M-limit is not formulated in terms of estimates. On more general estimators see e.g. [17, p. 345], [10].

If $(\Theta,\mathcal{T})$ is locally compact as well as Hausdorff, and if an M-limit $\theta_0(P)$ exists and K is a compact neighborhood of it, then

$$\inf\{P_n\rho(\theta):\ \theta\notin K\}\ >\ \inf\{P_n\rho(\theta):\ \theta\in K\},$$

except with outer probability $\to 0$ as $n\to\infty$ by (1.2). Then, M_n will be a non-empty compact set included in K, so M-estimates exist.

In most, but not all cases in this paper, ρ is defined via log likelihoods, as follows. Suppose we have a family $\{P_\theta,\ \theta\in\Theta\}$ of laws on $(X,\mathcal{B})$, where $\theta\mapsto P_\theta$ is 1-1, and that for some σ-finite measure μ on $(X,\mathcal{B})$, each P_θ is absolutely continuous with respect to μ, with a density $f(x,\theta)$. Let $\rho(x,\theta)\ :=\ -\log f(x,\theta)\le +\infty$. The assumption that $\rho(x,\cdot)$ is lower semicontinuous is equivalent to assuming that $f(x,\cdot)$ is upper semicontinuous for each x. It seems that for the parametric families actually considered in statistics, this assumption is not very restrictive; even if it requires changing the values of $f(x,\theta)$ for some x and θ, and even to $+\infty$, if the set of such x has μ-measure 0 for each θ, then we still have densities of the same laws P_θ. When $\rho(x,\theta)=-\log f(x,\theta)$, an M-estimate is a maximum likelihood estimate.

For any two laws P and Q on $(X,\mathcal{B})$, the Lebesgue decomposition gives a set $A\in\mathcal{B}$ with $P(A^c)=0$ such that Q is absolutely continuous with respect to P on A, with Radon-Nikodym derivative dQ/dP. The *likelihood ratio* $R_{Q/P}$ is defined as dQ/dP on A and $+\infty$ on A^c. The *Kullback-Leibler divergence* (or *information*) [23], [22] satisfies

$$0\ \le\ I(P,Q)\ :=\ -\int \log(R_{Q/P})dP\ \le\ +\infty,$$

where $\log 0\ :=\ -\infty$. It is known that for any $P,\ Q\in\mathcal{P}(X,\mathcal{B})$, $I(P,Q)=0$ if and only if $P=Q$: [26, p. 63] or [22]. In general $I(Q,P)\neq I(P,Q)$, and $I(\cdot,\cdot)$ does not satisfy the triangle inequality; the convergence defined by $Q_{(n)}\ \to P$ if and only if $I(P,Q_{(n)})\to 0$ is convergence for a nonmetrizable Hausdorff topology [10, Theorem 3.8, Corollary 4.5].

If $\{P_\theta,\ \theta \in \Theta\} \subset \mathcal{P}(X,\mathcal{B})$, Θ contains more than one point, each P_θ has a density $f(\cdot,\theta)$ with respect to a σ-finite measure ν on $(X,\mathcal{B})$, $\rho(x,\theta) := -\log f(x,\theta)$, $\rho \in \mathcal{H}(X,\Theta)$, and $Q \in \mathcal{P}(X,\mathcal{B})$, if $I(Q,P_\theta)$ is minimized with respect to θ at a unique $\theta = \theta_1$, then θ_1 is the M-functional at Q for ρ. But an M-functional may exist, via adjustment, even when $I(Q,P_\theta) = +\infty$ for all θ, e.g. when $X = \Theta = \mathbb{R}$, ν is Lebesgue measure, $f(x,\theta) = e^{-|x-\theta|}/2$, and Q is a Cauchy distribution.

A real-valued function f on a Hausdorff topological space will be called *proper-minimal* if it attains its absolute minimum at a point x and for every neighborhood U of x there is an $\varepsilon > 0$ such that $f(y) > f(x)+\varepsilon$ for all $y \notin U$. A function $\rho : \mathbb{R}^d \to \mathbb{R}$ will be called *properly convex* iff it is strictly convex and $\rho(x) \to +\infty$ as $|x| \to +\infty$. Clearly a properly convex function is proper-minimal. Recall that a convex function g defined on an interval J in $\mathbb{R}$ has left and right derivatives everywhere on the interior of J, which are nondecreasing, i.e. $g'(x-) \le g'(x+) \le g'(y-)$ for $x < y$ in J; if g is strictly convex on J then $g'(x+) < g'(y-)$; and a convex function on an open set in $\mathbb{R}^d$ is continuous, e.g. [12, §6.3].

Proposition 2.1. *Let $g : \mathbb{R} \to \mathbb{R}$ be strictly convex. Then the following are equivalent: (a) g is properly convex; (b) For some x and y, $g'(x-) < 0 < g'(y+)$; (c) For some u and v, $g'(u-) \le 0 \le g'(v+)$.*

Proof. (c) $\Rightarrow$ (b) for any $x < u$ and $v < y$. If (b) holds then for any $t < x$, $(g(x) - g(t))/(x-t) \le g'(x-)$. Letting $t \downarrow -\infty$, this implies $g(t) \to +\infty$. Similarly, $g(t) \to +\infty$ as $t \uparrow +\infty$, so (a) holds. Assuming (a), there exist some $v < y$ such that $g(v) < g(y)$, so $g'(y+) \ge g'(y-) > 0$. Similarly, $g'(x-) < 0$ for some x, so (b) and (c) hold, completing the proof. □

Proposition 2.2. *For any measurable space $(X,\mathcal{B})$ and separable metric space (Θ,d), if ρ is a jointly measurable function from $X \times \mathbb{R}^d$ into $\mathbb{R}$ such that for each $x \in X$, $\rho(x,\cdot)$ is proper-minimal and continuous, then $\rho \in \mathcal{H}(X,\Theta)$ and the map $x \mapsto \theta(x)$ such that $\theta(x) = \inf_\theta \rho(x,\theta)$ is measurable.*

Proof. Clearly $\rho \in \mathcal{H}(X,\Theta)$. Let $\{\theta_k\}_{k\ge1}$ be dense in Θ. For each x, by the assumptions, $u(x) := \inf_\theta \rho(x,\theta) = \inf_k \rho(x,\theta_k)$, which is evidently measurable in x. For each $m = 1,2,\ldots$, let $\theta_m(x) = \theta_k$ for the least k such that $\rho(x,\theta_k) < u(x)+(1/m)$. Then for each m, $\theta_m(\cdot)$ is measurable from X into Θ. By definition of proper-minimal, for each x, $\theta_m(x) \to \theta(x)$ as $m \to \infty$. Thus $\theta(\cdot)$ is measurable (e.g. [12, Theorem 4.2.2]). □

A sum of proper-minimal functions may not be proper-minimal, but proper convexity is preserved by sums, giving the following:

Corollary 2.3. *For any measurable space $(X,\mathcal{B})$ and $\Theta = \mathbb{R}^d$, if ρ is a function from $X \times \mathbb{R}^d$ into $\mathbb{R}$ such that $\rho(\cdot,\theta)$ is measurable on X for each $\theta \in \Theta$, and for all $x \in X$, $\rho(x,\cdot)$ is properly convex, then $\rho \in \mathcal{H}(X,\Theta)$ and for any empirical measure P_n, a unique M-estimate $\hat{\theta}_n$ exists and is a measurable function of $(X_1,\ldots,X_n)$.*

Proof. Since $\rho(x,\cdot)$ is continuous on $\mathbb{R}^d$ for each x, $\rho(\cdot,\cdot)$ is jointly measurable, e.g. via [12, Proposition 4.2.6] applied to $\rho_k(x,\theta) := \rho(x, f_k(\theta))$. Thus $P_n\rho(\cdot)$ is jointly measurable on $X^n \times \Theta$. Clearly, for each fixed P_n, $P_n\rho(\cdot)$ is properly convex on $\mathbb{R}^d$ and so continuous and proper-minimal. Thus Proposition 2.2 applies. □

Recall the notion of VC (Vapnik-Chervonenkis) class of sets, e.g. [11, Chapter 4]. A class $\mathcal{F}$ of real-valued functions on a set X is a *VC major* class iff the class of all sets $\{x \in X : f(x) > t\}$ for all $f \in \mathcal{F}$ and $t \in \mathbb{R}$ is a VC class of sets (e.g. [11, §4.7]).

Theorem 2.4. *Let $h(x,\theta)$ be measurable in x for each θ and continuous in $\theta \in \Theta$ for each x where Θ is a Borel set in a complete separable metric space. Suppose that $h(\cdot,\cdot)$ is uniformly bounded and that $\mathcal{F} := \{h(\cdot,\theta) : \theta \in \Theta\}$ is a VC major class of functions. Then $\mathcal{F}$ is a uniform, thus universal, Glivenko-Cantelli class.*

Proof. We apply [14, Theorem 6]; for the Koltchinskii-Pollard entropy properties of uniformly bounded VC major classes of functions see [9]. Sufficient measurability of the class $\mathcal{F}$ follows from the continuity in θ and the Borel measurability of Θ. □

3. The 1-dimensional location case

In this section, ρ will be of the form $\rho(x,\theta) = \rho(x-\theta)$ for a function ρ from $\mathbb{R}$ into $\mathbb{R}$, with $X = \Theta = \mathbb{R}$ having its usual topology and Borel σ-algebra. An M-estimate of location is a θ at which an integral $\int \rho(x-\theta)dP_n(x)$ is minimized. It seems natural to assume that for $n = 2$ the estimate should be the average of the two observations. The next proposition shows that if so and ρ is Lebesgue measurable, then it is properly convex and an even function, $\rho(-x) \equiv \rho(x)$. For $\int \rho(x-\theta) - a(x)dQ(x)$ to be defined (for some θ) whenever Q has a density, where $a(\cdot)$ is Borel measurable, ρ must be Lebesgue measurable (for general Q, even universally measurable).

Proposition 3.1. *Let f be a Lebesgue measurable function from $\mathbb{R}$ into itself. Then f is an even, properly convex function if and only if for every $a \le b$ and $Q := (\delta_a + \delta_b)/2$, $\int f(x-\theta)dQ(x)$ is minimized uniquely at $\theta = (a+b)/2$.*

Proof. If: for any $t \ge 0$ let $a := -t$ and $b = t$. Then $\int f(x-\theta)dQ(x) = (1/2)(f(-t-\theta)+f(t-\theta))$. For this to be minimized uniquely at $\theta = 0$ is equivalent to the inequality, for all real $t, u := -\theta$ with $u \ne 0$,

$$f(-t) + f(t) < f(u+t) + f(u-t). \tag{3.1}$$

Subtracting a constant, we can assume that $f(0) = 0$. Letting $t = 0$, we get $f(u) > 0$ for all $u \ne 0$. Let $g(t) := f(-t) + f(t)$ for all t. Then applying (3.1) to $-u$ in place of u and adding, we get $2g(t) < g(t-u) + g(t+u)$ for all real t, u with $u \ne 0$. Thus g is what is called (strictly) midconvex (or midpoint convex or Jensen convex, e.g. [4]). Being Lebesgue measurable, g is continuous and (strictly) convex by a classical theorem of Sierpiński [31]. Thus g is strictly increasing in $t \geq 0$ and decreasing in $t \leq 0$. It is properly convex.

For $0 < s \leq t$ we have by (3.1) for $u = t$ that

$$f(2t) \;=\; f(2t) + f(0) \;>\; g(t) \;\geq\; g(s) \;>\; f(s). \tag{3.2}$$

Rewriting (3.1) as $f(u+t) - f(t) > f(-t) - f(u-t)$ and setting $y := u - t$ and $x := -u$, we get

$$f(u+t) - f(t) \;>\; f(x+y) - f(y) \;>\; f(-y) - f(x-y) \;=\; f(t-u) - f(t-2u).$$

Iterating, we get for any $k = 1, 2, \ldots$,

$$f(u+t) - f(t) \;>\; f(t - [2k-1]u) - f(t - 2ku).$$

Then if $0 < u < t$ and

$$0 \;<\; t - 2ku \leq \frac{1}{2}\left[t - (2k-1)u\right], \tag{3.3}$$

it follows by (3.2) that $f(u+t) > f(t)$. Clearly, (3.3) is equivalent to $2ku < t \leq (2k+1)u$ or to $t/(2k+1) \leq u < t/(2k)$. Thus $f((1+r)t) > f(t)$ for all $t > 0$ for any r such that $1/(2k+1) \leq r < 1/(2k)$ for some $k = 1, 2, \ldots$. Here $1 + r$ can be replaced by any finite product of such factors $1 + r_j$, say. But it's easily seen that such finite products give an arbitrary real number $v > 1$. Thus f is strictly increasing in $t \geq 0$. Symmetrically, so is h with $h(t) := f(-t)$. Since $f + h = g$ and g is absolutely continuous on bounded sets, so are f and h. Thus for all $x \geq 0$, $f(x) = \int_0^x f'(t)dt$ and likewise for h. Now, for those (Lebesgue almost all) $t > 0$ for which $f'(t)$ and $f'(-t)$ both exist, since $u \mapsto f(t+u) + f(u-t)$ has its absolute minimum at $u = 0$, we have $f'(t) + f'(-t) = 0$. It follows that f' is an odd function (at least in an almost everywhere sense) and f is an even function, so that $f \equiv g/2$, and f is properly convex.

Conversely, if f is even and properly convex, then for all t and all $u \neq 0$, $2f(t) < f(t+u) + f(t-u) = f(u+t) + f(u-t)$, implying (3.1) and proving "only if." □

The following fact is closely related to the theory of robust estimation of location in $\mathbb{R}$ and to the work of Huber [18], [19], [20], even though Huber chose ρ functions that are not strictly convex everywhere, but only on some interval, as in Proposition 3.5 below.

Theorem 3.2. *Let $\rho(x, \theta) \equiv \rho(x - \theta)$ where ρ is real-valued, properly convex, and even, $\rho(-x) = \rho(x)$ for all x. Then the following are equivalent:*

(I) ρ is definitely M-consistent for all laws P on $\mathbb{R}$;

(II) ρ is universally M-consistent;

(III) $\rho(x) = O(|x|)$ as $|x| \to +\infty$.

If these conditions hold, then for every law P on $\mathbb{R}$, an M-functional $\theta_1(P)$ exists, and unique, measurable M-estimators $\hat{\theta}_n = \hat{\theta}_n(X_1, \ldots, X_n)$ exist and converge almost surely to $\theta_0(P) = \theta_1(P)$.

Proof. Clearly (I) implies (II). To show (II) implies (III), we have for all $x < y$, $\rho'(x-) \le \rho'(x+) < \rho'(y-)$. To prove (III) is equivalent to showing that $\psi(x) := \rho'(x+)$ remains bounded as $x \uparrow +\infty$. We have $\psi(0) \ge 0$, so $\psi(x) > 0$ for all $x > 0$ and ψ is strictly increasing. It is continuous except on the at most countable set where it has jumps. Outside that set, $\psi(-x) = -\psi(x)$. Next, the following will be useful.

Lemma 3.3. *Suppose $\{x_k\}_{k\ge1}$ and $\{y_k\}_{k\ge1}$ are sequences of positive numbers with $x_k \le y_k < x_{k+1}$ for all k. Let $c := 3/\pi^2$. Let Q be the probability measure with $Q(x_k) = Q(-y_k) = c/k^2$ for all $k \ge 1$. Let $X_1, X_2, \dots$ be i.i.d. (Q). For $n = 3, 4, \dots$, let A_n be the event that for some $j = 1, \dots, n$, $X_j \ge x_{n+1}$ and for all $i = 1, \dots, n$ with $i \ne j$, we have $|X_i| \le x_{n-2}$. Let B_n be that event with "$X_j \ge x_{n+1}$" replaced by "$X_j \le -y_{n+1}$." Then*

$$\lim_{n\to\infty} Pr(A_n) = \lim_{n\to\infty} Pr(B_n) = ce^{-2c}.$$

Proof. Since $\sum_{k\ge1} 1/k^2 = \pi^2/6 = 1/(2c)$, Q is a probability measure. We have $P(X_1 > y_n) = \sum_{k>n} c/k^2 \sim c/n$ as $n \to \infty$. Similarly, $P(X_1 < -x_n) \sim c/n$. Thus since A_n is a union of n disjoint events for $j = 1, \dots, n$,

$$\Pr(A_n) \sim n\left(\frac{c}{n}\right)\left[1 - \frac{2c}{n-2}\right]^{n-1} \to ce^{-2c}$$

as $n \to \infty$. The proof for B_n is similar. □

Now returning to the proof of Theorem 3.2, suppose ψ is unbounded. Choose $x_0 > 0$ at which $\psi(x_0) > 1$. Recursively choose x_n for $n \ge 1$ such that $x_n \ge 2x_{n-1}$ and $\psi(x_n) \ge 2n\psi(x_{n-1})$ for each n. Let $y_n = x_n$ for all n. Applying Lemma 3.3, suppose $n \ge 3$ and A_n occurs. Then $I_n(\theta) := \int \psi(x-\theta)dQ_n(x)$ is bounded below by its value when $X_j = x_{n+1}$ and $X_i = -x_{n-2}$ for $i = 1, \dots, n$, $i \ne j$. Thus on A_n, if ψ is continuous at $x_{n-2} + \theta$,

$$\begin{aligned} I_n(\theta) &\ge [\psi(x_{n+1} - \theta) + (n-1)\psi(-x_{n-2} - \theta)]/n \\ &= [\psi(x_{n+1} - \theta) - (n-1)\psi(x_{n-2} + \theta)]/n. \end{aligned}$$

If $\theta \le x_{n-2}$ then $x_{n+1} - \theta \ge x_n$ and $x_{n-2} + \theta \le x_{n-1}$. Then

$$I_n(\theta) \ge n^{-1}\psi(x_n) - \psi(x_{n-1}) \ge \psi(x_{n-1}) > 0.$$

Thus $\int \rho(x-\theta)dQ_n(x)$ has negative left and right derivatives with respect to θ at all $\theta < x_{n-2}$ and so is decreasing on that half-line. Since we assume $\theta_0 = \theta_0(Q)$ exists, we have $\theta_0 < x_{n-2}$ for n large enough, and the assumption that $Q_n\rho(\theta)$ has its smallest values in a neighborhood of θ_0 with probability $\to 1$ (1.2) leads to a contradiction. So (II) implies (III).

To prove (III) implies (I), let $M := \sup_x |\psi(x)| < \infty$. Then ρ is a Lipschitz function with $|\rho(x) - \rho(y)| \le M|x-y|$ for all real x and y. Thus $|\rho(x - \theta) - \rho(x)| \le M|\theta|$ for all real x and θ. For any law P on $\mathbb{R}$, the integral $g(\theta) := \int \rho(x-\theta) - \rho(x)dP(x)$ exists. By dominated convergence, it has left derivatives that can be taken under the integral sign, $g'(\theta-) = -\int \psi(x-\theta)dP(x)$. Similarly, right derivatives of g also exist. Since ψ is strictly increasing, g' is strictly increasing in θ. By dominated convergence again, $\lim_{\theta\to-\infty} g'(\theta-) = -M < 0$ and $\lim_{\theta\to+\infty} g'(\theta-) = M > 0$. Thus there is a unique $\theta_1 = \theta_1(P)$ such that $g'(\theta-) < 0$ for all $\theta < \theta_1$ and $g'(\theta-) > 0$ for all $\theta > \theta_1$.

To show that (1.3) holds for $\theta_0(P) = \theta_1(P)$, clearly, g has exactly one relative minimum, which is an absolute minimum, at θ_1. Thus θ_1 is the M-functional at P for ρ. For any neighborhood U of θ_1, there is a $\gamma > 0$ such that $(\theta_1-\gamma, \theta_1+\gamma) \subset U$. Let $\delta := \gamma/2$. Then

$$-\int \psi(x-\theta_1-\delta)dP(x) = g'((\theta_1+\delta)-) > 0$$

$$> -\int \psi(x-\theta_1+\delta)dP(x) = g'((\theta_1-\delta)-).$$

Let $\varepsilon := \min(g'((\theta_1+\delta)-), -g'((\theta_1-\delta)-))/2 > 0$ and $\psi_n(\phi) := \int \psi(x-\phi)dP_n(x)$ for any $\phi \in \mathbb{R}$. Then $\Pr(\psi_n(\theta_1+\delta) > -\varepsilon) \to 0$ and $\Pr(\psi_n(\theta_1-\delta) < \varepsilon) \to 0$ as $n \to \infty$. Since ψ is increasing, ψ_n is decreasing. So for all $\theta \geq \theta_1+\delta$, $\psi_n(\theta) \leq \psi_n(\theta_1+\delta)$, and for all $\theta \leq \theta_1-\delta$, $\psi_n(\theta) \geq \psi_n(\theta_1-\delta)$. Let $E_n := \{\psi_n(\theta_1+\delta) \leq -\varepsilon\} \cap \{\psi_n(\theta_1-\delta) \geq \varepsilon\}$. Then $\Pr(E_n) \to 1$ as $n \to \infty$ and by the strong law of large numbers, almost surely there is an $n_0(\omega)$ such that E_n holds for all $n \geq n_0(\omega)$. On E_n we have $\psi_n(\theta) \leq -\varepsilon$ for all $\theta \geq \theta_1+\delta$ and $\psi_n(\theta) \geq \varepsilon$ for all $\theta \leq \theta_1-\delta$. Let $g_n(\theta) := \int \rho(x-\theta)dP_n(x)$. Then for any $\theta < \phi$ in $\mathbb{R}$, $g_n(\phi) - g_n(\theta) = -\int_\theta^\phi \psi_n(t)dt$. Thus on E_n, $g_n(\theta_1+\gamma) - g_n(\theta_1+\delta) \geq \varepsilon\delta$ and $g_n(\theta) - g_n(\theta_1+\delta) \geq \varepsilon\delta$ for all $\theta \geq \theta_1+\gamma$. Similarly, also on E_n, $g_n(\theta_1-\delta) - g_n(\theta_1-\gamma) \leq -\varepsilon\delta$ and $g_n(\theta) \geq g_n(\theta_1-\delta) + \delta\varepsilon$ for all $\theta \leq \theta_1-\gamma$. It follows that (1.3) holds with $\theta_0 = \theta_1$ and $\delta\varepsilon$ in place of ε, so (I) holds and (I), (II), and (III) are equivalent.

If the conditions hold, then existence of the M-functional $\theta_1(P) = \theta_0(P)$, the M-limit, for any law P was shown in the proof that (III) implies (I). It follows that unique M-estimates $\hat{\theta}_n$ always exist. By Corollary 2.3, $\hat{\theta}_n$ are measurable functions of $(X_1, \ldots, X_n)$. Since the events E_n eventually hold almost surely, $\hat{\theta}_n \to \theta_0$ almost surely and the theorem is proved. □

Remarks. (1) In Theorem 3.2, the functions $x \mapsto \rho(x-\theta)$ never form a universal Glivenko-Cantelli class; in fact they are not integrable for any P such that $\int |x|dP(x) = +\infty$.

(2) The adjusted functions $x \mapsto \rho(x-\theta) - \rho(x)$ are each bounded and so integrable for all P, but they do not form a universal Glivenko-Cantelli class unless θ is restricted to a bounded set.

(3) Still, the proof of (III) implies (I) in Theorem 3.2 is rather easy; it uses a simple form of the technique of one-sided bracketing, apparently due to Huber [19], see also [10].

(4) The functions $x \mapsto \psi(x-\theta)$, being increasing and uniformly bounded, do form a universal and indeed uniform Glivenko-Cantelli class [14] and moreover a uniform Donsker class [15].

(5) For a ρ as in Theorem 3.2, if a law P on $\mathbb{R}$ is symmetric around a point m, meaning that the map $x \mapsto 2m - x$ preserves P, then the proof of Theorem 3.2 shows that $\theta_0(P) = m$.
(6) A simple example of a function ρ satisfying the conditions of Theorem 3.2 is $\rho(x) := \sqrt{1+x^2} - 1$.
(7) There is no familiar classical density f on $\mathbb{R}$ for which $\rho(y) := -\log f(y)$ satisfies the conditions of Theorem 3.2. Perhaps coming closest is the double-exponential density $f(y) \equiv e^{-|y|}/2$, for which ρ is convex, even, and of order $|y|$ at ∞, but not strictly convex.

Next is a negative fact about uniform M-consistency for location.

Proposition 3.4. *Let $\rho(x,\theta) := \rho(x-\theta)$ where $x \in X = \mathbb{R}$ and $\theta \in \Theta = \mathbb{R}$, and ρ is a real-valued, even, properly convex function. Then ρ is never uniformly M-consistent.*

Remark. This fact, on reflection, is not surprising. If one estimates location without simultaneously estimating scale, then the location estimators will naturally not converge uniformly over more dispersed distributions. On estimating location and scale jointly, see section 5.

Proof. First, ρ must be universally M-consistent. Theorem 3.2 tells us that $\rho(x) = O(|x|)$ as $|x| \to \infty$. Thus $\psi(x) := \rho'(x+)$ exists for all x and increases up to some finite limit K as $x \uparrow +\infty$. Let $Q := \frac{1}{2}(\delta_{-M} + \delta_M)$ for $1 < M < \infty$. Then $Q_n = p_n\delta_{-M} + q_n\delta_M$ where $q_n \equiv 1 - p_n$ and $p_n = S_n/n$ where S_n is a binomial $(n, 1/2)$ random variable. By the central limit theorem, there is some $n_0 < \infty$ such that for $n \geq n_0$,

$$\Pr\left(\left|p_n - \frac{1}{2}\right| < \frac{1}{2\sqrt{n}}\right) = \Pr\left(\frac{2}{\sqrt{n}}\left|S_n - \frac{n}{2}\right| < 1\right) < 0.7.$$

Take any $\delta > 0$ with $\delta \leq 1/\sqrt{n_0}$. Then if $n \geq n_0$ and $\Pr(|p_n - \frac{1}{2}| < \delta/2) \geq 3/4$, we have $n \geq 1/\delta^2$. Take $T < \infty$ such that $\psi(T) > K/(1+\delta)$. Take any $M > T + 1$. Recall that the M-estimator $\hat{\theta}_n = \sup\{\theta : Q_n\psi(\cdot - \theta) > 0\} = \inf\{\theta : Q_n\psi(\cdot - \theta) < 0\}$. Since Q is symmetric around 0, $\theta_0(Q) = 0$. Let's bound the probability that $|\hat{\theta}_n| < 1$. For this it's necessary that $q_n\psi(M+1) - p_n\psi(M-1) > 0$ and $q_n\psi(M-1) - p_n\psi(M+1) < 0$, and so

$$1 - \delta < \frac{1}{1+\delta} < \frac{\psi(M-1)}{\psi(M+1)} < \frac{q_n}{p_n} < \frac{\psi(M+1)}{\psi(M-1)} < 1 + \delta.$$

As $M \to +\infty$, it follows that $\delta = \delta_M \to 0$. By choice of T and M, it follows that $|p_n - 1/2| < \delta/2$. Thus, $|\hat{\theta}_n| < 1$ with probability $\geq 3/4$ for $n \geq n_0$ only for $n \geq 1/\delta_M^2$, which depends on M and so on Q. Thus the uniform M-consistency fails. □

There are some well known functions ρ that don't give universal M-consistency but give M-consistency except for rather small families of laws, as follows. For each $r > 0$, let $\rho_r(x) = |x|^2$ for $|x| \le r$ and $2r|x| - r^2$ for $|x| \ge r$. Let $\rho_0(x) := |x|$. Huber [20, pp. 43-55] treats ρ_r. Let $\rho_r(x, \theta) := \rho_r(x - \theta)$. The following is known, being classical for $r = 0$ and noted by Huber for $r > 0$:

Proposition 3.5. *M-consistency holds for ρ_0 and a law Q if and only if Q has a unique median, which then equals $\theta_0(Q)$. The M-estimates for ρ_0 are sample medians, unique if and only if there is a unique sample median.*

For $r > 0$, M-consistency holds for ρ_r for a law Q on $\mathbb{R}$ unless there exists c in $\mathbb{R}$ with $Q((-\infty, c]) = 1/2$ and $Q((c, c + 2r)) = 0$, i.e. the interval of medians of Q has length at least $2r$. M-estimates $\hat{\theta}_n$ for ρ_r are unique unless n is even and the interval of sample medians has length at least $2r$.

4. Multidimensional location M-estimation

First we recall the case $\rho(x) = |x|$, $x \in \mathbb{R}^d$, where the M-estimate, if it exists, is called the *spatial median.*

Proposition 4.1. *Let $\rho(x, \theta) := |x - \theta|$ for all $x, \theta \in \mathbb{R}^d$. Then ρ is M-consistent except for $P \in \mathcal{Q}$ where $\mathcal{Q}$ is the set of all laws Q on $\mathbb{R}^d$ such that for some line L, $Q(L) = 1$ and Q does not have a unique median in L.*

Proof. Let $h(x, \theta) := |x - \theta| - |x|$. Then $|h(x, \theta)| \le |\theta|$ for all x. Haldane [16] proved that a finite sample or empirical measure Q_n has a unique spatial median, minimizing $Q_n\rho(\cdot)$, whenever $Q_n \notin \mathcal{Q}$. Milasevic and Ducharme [25] proved that any law $Q \notin \mathcal{Q}$ has a unique spatial median minimizing $Qh(\cdot)$. It is also known that M-consistency holds for $Q \notin \mathcal{Q}$, e.g. [10, Theorem 6.6]. □

The following for the Huber functions is not claimed as new, but a proof will be indicated.

Proposition 4.2. *For $0 < r < \infty$ let $\rho_r(x, \theta) := \rho_r(|x - \theta|)$ for the Huber function ρ_r of Proposition 3.5 and any $x, \theta \in \mathbb{R}^d$. Then ρ_r is M-consistent for $P \notin \mathcal{Q}$, where $\mathcal{Q}$ is the set of all laws Q on $\mathbb{R}^d$ such that for some line L, $Q(L) = 1$ and the interval of medians of Q in L has length at least $2r$.*

Proof. It's easy to check that ρ_r is a C^1 function. Let $h(x, \theta) := h_r(x, \theta) := \rho_r(x - \theta) - \rho_r(x)$. Then $h_r(x, \theta) \le 2r|\theta|$ for all x. For any law P on $\mathbb{R}^d$, one can show that $Ph(\theta) \to +\infty$ as $|\theta| \to \infty$. Clearly $Ph(\cdot)$ is convex, so it attains its minimum on some nonempty convex compact set A. It will be shown that $Ph(\cdot)$ is proper-minimal for $P \notin \mathcal{Q}$. It is by convexity if A is a singleton. Suppose $\theta, \phi \in A$ with $\theta \ne \phi$. For any $x \in \mathbb{R}^d$ let $g(t) := \rho_r(y_t)$ where $y_t := x - \theta - t(\phi - \theta)$. Then clearly g is convex. One can show that it is strictly convex if x is not on the line L through θ and ϕ, partly as in the previous proposition, [16]. Thus $P(L) = 1$, or there is a contradiction. It follows that $P \in \mathcal{Q}$ as in the 1-dimensional case,

Proposition 3.5. So for $P \notin \mathcal{Q}$, A is a singleton, $Ph(\cdot)$ is proper-minimal, and $P_n h(\cdot)$ is proper-minimal for $P_n \notin \mathcal{Q}$ and so with probability converging to 1. The rest follows from [10, Theorem 6.6]. □

Now, consider another specific ρ function for M-estimation of location in $\mathbb{R}^d$, as follows.

Proposition 4.3. *Let* $\rho(x) := \sqrt{1+|x|^2} - 1$ *and* $h(x,\theta) := \rho(x-\theta) - \rho(x)$ *for* $x \in \mathbb{R}^d$. *Then:*

(a) ρ *is* C^∞ *in* x *and* θ*;*
(b) For each $\theta \in \mathbb{R}^d$, $|h(x,\theta)| \le |\theta|$*;*
(c) For every law Q *on* $\mathbb{R}^d$, $Qh(\cdot)$ *is properly convex and so proper-minimal;*
(d) The class $\mathcal{F}$ *of all* $h(\cdot,\theta)$ *for* $\theta \in \mathbb{R}^d$ *is a VC major class of functions;*
(e) ρ *is universally M-consistent.*

Proof. Parts (a), (b), and (c) are easy to check. To prove (d), for any $\theta \in \mathbb{R}^d$ and $t \in \mathbb{R}$, the set $A(\theta,t) := \{x \in \mathbb{R}^d : h(x,\theta) > t\}$ equals $B(t) \cup C(\theta,t)$ where $B(t) := \{x : \rho(x) + t < 0\}$, which is empty for $t \ge 0$ and for $t < 0$ equals $\{x : \rho(x) < |t|\} = \{x : 1 + |x|^2 < t^2\}$. Next,

$$C(\theta,t) := \{x : \rho(x-\theta) > \rho(x) + t \ge 0\} = D(\theta,t) \setminus B(t)$$

where

$$\begin{aligned} D(\theta,t) &:= \{x : 1 + |x-\theta|^2 > 1 + |x|^2 + t^2 + 2t\sqrt{1+|x|^2}\} \\ &= \{x : -2x\cdot\theta + |\theta|^2 - t^2 > 2t\sqrt{1+|x|^2}\} = E(\theta,t) \cap F(\theta,t), \\ E(\theta,t) &:= \{x : -2x\cdot\theta + |\theta|^2 - t^2 > 0\}, \quad \text{and} \\ F(\theta,t) &:= \{x : (-2x\cdot\theta + |\theta|^2 - t^2)^2 > 4t^2(1+|x|^2)\}. \end{aligned}$$

Thus $A(\theta,t)$ is a Boolean combination, with a fixed bounded number of Boolean operations, of positivity sets of quadratic polynomials on $\mathbb{R}^d$. So these sets belong to a VC class of sets (e.g. [11, Theorems 4.2.1 and 4.2.4]) and $\mathcal{F}$ is a VC major class of functions, as desired.

For part (e), for any $P \in \mathcal{P}(\mathbb{R}^d)$, $Ph(\cdot)$ is properly convex and has a unique minimum at some $\theta_1(P)$. Also, for any bounded neighborhood U of $\theta_1(P)$, $\{h(\cdot,\theta) : \theta \in U\}$ is a universal Glivenko-Cantelli class by Theorem 2.4. The functions $P_n h(\cdot)$ are properly convex and converge to $Ph(\cdot)$, uniformly on U. Then (e) follows. □

5. Location-scale M-estimation in the line

For any probability density f on $\mathbb{R}$ with respect to Lebesgue measure, $\mu \in \mathbb{R}$ and $0 < \sigma < \infty$, setting $\theta := (\mu,\sigma)$, we have the location-scale family of probability densities defined by $f_{\mu,\sigma}(x) := f(x,\theta) := \sigma^{-1} f((x-\mu)/\sigma)$. Let $\rho(y) := -\log f(y)$ as before and

$$\rho(x,\theta) := -\log f(x,\theta) = \log\sigma + \rho((x-\mu)/\sigma). \tag{5.1}$$

It turns out paradoxically that the ρ functions which worked well for location estimation in Theorem 3.2 work poorly for the location-scale case, whereas some ρ functions that work not so well for location alone work better for joint estimation of location and scale. Here is a negative fact about location-scale M-estimation for ρ of power type:

Proposition 5.1. *Let ρ from $\mathbb{R}$ to $\mathbb{R}$ be such that for some positive constants α, c, C, M, and N, for $|y| \geq M$ we have $c|y|^\alpha \leq \rho(y) \leq C|y|^\alpha$, and for $|y| \leq M$ we have $0 \leq \rho(y) \leq N$. Then $\rho(\cdot,\cdot)$ defined by (5.1) has no M-limit (μ_0, σ_0) for any law Q such that $\int |x|^\alpha dQ(x) = +\infty$.*

Remark. The ρ for location in Theorem 3.2 all satisfy the hypotheses of the Proposition with $\alpha = 1$ and suitable $c < C$, N, and M.

Proof. Take a law Q with $\int |x|^\alpha dQ(x) = +\infty$ and $X_1, X_2, \ldots$ i.i.d. (Q). Suppose that an M-limit $(\mu_0, \sigma_0) = \theta_0 = \theta_0(Q)$ exists. Then as $n \to \infty$, with probability $\to 1$, $Q_n\rho(\theta)$ has its smallest values somewhere on the set U of θ where $|\mu - \mu_0| < 1$ and $\sigma_0/2 < \sigma < 2\sigma_0$. Let $\lambda_0 := \log(\sigma_0/2)$ and let Y_j be the indicator of the event $\{|X_j| \geq 2|\mu_0| + 2 + 2M\sigma_0\}$. Let $c_\alpha := c/(4\sigma_0)^\alpha$. Then for $\theta \in U$ we have

$$Q_n\rho(\mu,\sigma) \geq \lambda_0 + n^{-1}\sum_{j=1}^n Y_j\rho((X_j - \mu)/\sigma) \tag{5.2}$$

$$\geq \lambda_0 + \frac{1}{n}\sum_{j=1}^n Y_j c\,||X_j| - |\mu_0| - 1|^\alpha/(2\sigma_0)^\alpha \geq \lambda_0 + \frac{c_\alpha}{n}\sum_{j=1}^n Y_j|X_j|^\alpha.$$

On the other hand, for any $\mu \in \mathbb{R}$ and $\sigma > 0$ we have

$$Q_n\rho(\mu,\sigma) \leq \log\sigma + \sum_{j=1}^n \left(N + C\sigma^{-\alpha}|X_j - \mu|^\alpha\right)/n$$

$$= N + \log\sigma + C\sigma^{-\alpha}n^{-1}\sum_{j=1}^n |X_j - \mu|^\alpha.$$

If $|\mu - \mu_0| < 1$, then for each j, $(1 - Y_j)|X_j - \mu|^\alpha \leq T := (3|\mu_0| + 3 + 2M\sigma_0)^\alpha$, and $Y_j|X_j - \mu|^\alpha \leq 2^\alpha Y_j|X_j|^\alpha$. Thus

$$Q_n\rho(\mu,\sigma) \leq N + \log\sigma + C\sigma^{-\alpha}\left[T + 2^\alpha n^{-1}\textstyle\sum_{j=1}^n Y_j|X_j|^\alpha\right]. \tag{5.3}$$

Let $S_n := \sum_{j=1}^n Y_j|X_j|^\alpha$. Then since $Y_j|X_j|^\alpha$ are i.i.d., $E|X_j|^\alpha = +\infty$ and $(1 - Y_j)|X_j|^\alpha$ are bounded, we have $S_n/n \to +\infty$ a.s. If we set $\sigma = \sigma_n := (C(T + 2^\alpha S_n/n))^{1/\alpha} \sim 2(CS_n/n)^{1/\alpha}$ as $n \to \infty$ in (5.3), we get $Q_n\rho(\mu,\sigma_n) \leq N + 1 + \log\sigma_n \sim \alpha^{-1}\log(S_n/n)$, while in (5.2), $Q_n\rho(\mu,\sigma) \geq \lambda_0 + c_\alpha S_n/n$, which is asymptotically larger, a contradiction. □

Proposition 5.1 suggests that for location-scale estimation we might consider ρ functions that grow as $|x| \to \infty$ more slowly than power functions, specifically, logarithmically. In other words, we might consider longer-tailed densities f. The case of the Cauchy density $f(x) := 1/[\pi(1+x^2)]$, $x \in \mathbb{R}$, has been considered in several papers. Here, up to an unimportant additive constant, $\rho(x) = \log(1+x^2)$. If Q is a law symmetric around 0 and concentrated close to ± 2, then one can check that $\int \rho(x-\theta)dQ(x)$ has a strict relative *maximum* at $\theta = 0$ and has its absolute minimum at some points $\pm\phi$ for $\phi \neq 0$. Such non-uniqueness for estimation of location alone is to be expected from Proposition 3.1 since ρ is not convex. On multiple relative extrema for the Cauchy location case see [28].

With ρ as in (5.1) for the Cauchy density f, Copas [5] and Mäkeläinen, Schmidt and Styan [24] proved that $Q_n\rho(\cdot)$ has a unique relative minimum, which is then a global minimum (maximum likelihood estimator), if $Q_n(\{x\}) < 1/2$ for all x. There are extensions to ρ functions defined by t densities with degrees of freedom $\nu \geq 1$ and to location-scatter estimation in $\mathbb{R}^d$, for Q depending on ν and d [21], and to M-consistency [13].

6. Exponential families

Let μ be a σ-finite measure on the Borel sets of $\mathbb{R}^d$. Let $\Theta := \{\theta \in \mathbb{R}^d : \int e^{\theta\cdot x}d\mu(x) < \infty\}$. Then Θ is convex by Jensen's inequality. For $\theta \in \Theta$ let

$$\kappa(\theta) := \log \int e^{\theta\cdot x}d\mu(x), \quad dP_\theta(x) := e^{\theta\cdot x-\kappa(\theta)}d\mu(x).$$

Then P_θ is a probability measure. Let X be the *support* of μ, that is, the smallest closed set with $\mu(\mathbb{R}^d \setminus X) = 0$. Then X is the support of each law P_θ. We assume that the family $\{P_\theta,\ \theta \in \Theta\}$ is an *exponential family in standard form*, meaning that Θ and the convex hull of X each have non-empty interiors in $\mathbb{R}^d$. Θ is called the *natural parameter space* of the family. Let $f(x,\theta) := e^{\theta\cdot x-\kappa(\theta)}$ and $\rho(x,\theta) := -\log f(x,\theta) = -\theta\cdot x + \kappa(\theta)$. It is known that $\kappa(\cdot)$ is a strictly convex real-valued function on Θ (e.g. [1, Theorem 9.1], [3, pp. 1,19]). For any empirical measure $P_n = n^{-1}\sum_{j=1}^n \delta_{X_j}$, we have $P_n\rho(\theta) = -\theta\cdot\overline{X}_n + \kappa(\theta)$ where $\overline{X}_n$ is the nth sample mean $(X_1+\cdots+X_n)/n \in \mathbb{R}^d$. Thus $P_n\rho(\cdot)$ is also strictly convex on Θ. Clearly the behavior of $P_n\rho$ is entirely determined by that of $\overline{X}_n$. It is well known that for long-tailed distributions P, $\overline{X}_n$ do not converge to a finite limit in any sense. For example, if $X = \mathbb{R}^1$ and P is the Cauchy distribution with density $1/[\pi(1+x^2)]$, then $\overline{X}_n$ have the same (Cauchy) distribution for all n. Call $X \subset \mathbb{R}$ *two-sided unbounded (TSU)* iff $\sup X = -\inf X = +\infty$. If $X \subset \mathbb{R}^1$ is TSU, one can find a P concentrated on X such that $\overline{X}_n$ behave even worse than for the Cauchy distribution:

Proposition 6.1. *If $X \subset \mathbb{R}$ is TSU, then there exist sequences $\{x_k\}_{k\geq 1} \subset X$ and $\{y_k\}_{k\geq 1} \subset -X$ of positive numbers such that $y_k \geq x_k \uparrow +\infty$ as $k \to \infty$ and a probability measure Q on $\{x_k\}_{k\geq 1} \cup \{-y_k\}_{k\geq 1} \subset X$ such that for $X_1, X_2, \ldots$ i.i.d. (Q) and $c := 3/\pi^2$,*

$$\liminf_{n\to\infty} Pr(\overline{X}_n < -x_n) \geq ce^{-2c} \leq \liminf_{n\to\infty} Pr(\overline{X}_n > x_n).$$

Proof. By assumption there exist sequences $\{\xi_i\}_{i\geq 1} \subset X$ and $\{-\eta_j\}_{j\geq 1} \subset X$ with $0 < \xi_i \uparrow +\infty$ and $0 < \eta_j \uparrow +\infty$. Then $\{x_k\}$ will be a subsequence $\{\xi_{i(k)}\}$ and y_k will equal some $\eta_{j(k)}$. Let $i(1) := 1$ and choose $j(1)$ such that $y_1 \geq x_1$. Recursively, for $k \geq 2$ choose indices $i(k)$ and $j(k)$ strictly increasing in k such that $x_k = \xi_{i(k)} \geq 2ky_{k-1}$ and $y_k = \eta_{j(k)} \geq x_k$. Then defining Q as in Lemma 3.3, on the event A_n there, we have

$$\overline{X}_n > n^{-1}x_{n+1} - x_{n-2} \geq 2x_n - x_{n-1} \geq x_n.$$

On the event B_n, $\overline{X}_n < -n^{-1}y_{n+1} + x_{n-2} \leq -2x_n + x_{n-1} \leq -x_n$. Lemma 3.3 then gives the conclusion. □

Now for an exponential family $\{P_\theta,\ \theta \in \Theta\}$ as defined above, let $\mathcal{K}$ be the closure of the convex hull of X. The family is called *regular* if the natural parameter space Θ is open or *steep* if for every $\theta \in \Theta \cap \partial\Theta$, where $\partial\Theta$ is the boundary of Θ, $E_\theta|x| := \int |x| dP_\theta(x) = +\infty$. Thus any regular family is steep. Let $\operatorname{Int} A$ denote the interior of a set A. The following is known (e.g. [3, p. 74]):

Theorem 6.2. *If $\{P_\theta,\ \theta \in \Theta\}$ is a steep exponential family in standard form, then $\theta \mapsto E_\theta x$ is a homeomorphism from $\operatorname{Int}\Theta$ onto $\operatorname{Int}\mathcal{K}$.*

The following must also be known. It is easily proved by dominated convergence and Jensen's inequality. It is more of interest when θ_1 or $\theta_2 \in \partial\Theta$.

Lemma 6.3. *For an exponential family in standard form, if θ_1 and $\theta_2 \in \Theta$, then $\kappa(\cdot)$ is continuous on the closed line segment joining θ_1 to θ_2. If $E_{\theta_j}|x| < \infty$ for $j = 1, 2$, then $\theta \mapsto E_\theta x$ is also continuous on the line segment.*

Here is a negative fact about universal M-consistency for 1-dimensional exponential families. It is noted however that if $\mathcal{K}$ has a finite boundary, M-consistency holds except for point masses on the boundary.

Proposition 6.4. *For an exponential family $dP_\theta(x) = e^{\theta x - \kappa(\theta)}d\mu(x)$ in standard form with $d = 1$, the function $\rho(x, \theta) = -\theta x + \kappa(\theta)$ is never universally M-consistent.*

If X is bounded, with $\mathcal{K} = [c, d]$, then an M-limit $\theta_0(Q)$ exists for all laws Q on the Borel sets of X except for $Q = \delta_c$ or δ_d.

If $\mathcal{K}$ and Θ are both closed half-lines (one bounded above, the other below), then an M-limit $\theta_0(Q)$ exists for all laws Q on the Borel sets of X or $\mathcal{K}$ except $Q = \delta_b$ where b is the finite endpoint of $\mathcal{K}$.

Proof. If X is TSU (unbounded above and below), take the law Q given by Proposition 6.1. The convex hull of X is all of $\mathbb{R}$, hence so is its closure $\mathcal{K}$ and the interior of $\mathcal{K}$. The function $g(\theta) := E_\theta x = \kappa'(\theta)$ is strictly increasing and continuous (in fact, C^∞) on $\operatorname{Int}\Theta = (a, b)$ for some $a < b$ onto some open interval (c, d). The maximum likelihood estimate of θ, if it exists and is in $\operatorname{Int}\Theta$, is $g^{-1}(\overline{X}_n)$. Choose real u, v such that $a < u < v < b$. Apply Proposition 6.1. If $\overline{X}_n > x_n$ for some large enough n, then the smallest values of $Q_n\rho(\theta)$ will be taken for $\theta > v$, but if $\overline{X}_n < -x_n$, they will be taken for $\theta < u$. It follows that no M-limit $\theta_0(Q)$ exists.

If X is bounded, then $\mathcal{K} = [c, d]$ for some $c = \inf X < d = \sup X \in X$, and $\Theta = \mathbb{R}$. Thus the family is regular and steep, and g is an increasing homeomorphism of $\mathbb{R}$ onto (c, d) by Theorem 6.2. Let $Q = \delta_d$. Then $Q_n \equiv \delta_d$ for all n, and $Q_n\rho(\theta)$ is smallest as $\theta \to +\infty$, so no $\theta_0(Q)$ exists. A symmetrical situation occurs if $Q = \delta_c$. But for any law Q on X other than δ_c or δ_d, we have $m_Q := \int x dQ(x) \in (c, d)$, $\overline{X}_n \to m_Q$ almost surely, and a definite M-limit $\theta_0(Q)$ exists and equals $g^{-1}(m_Q)$.

If X is bounded above but not below, then $\mathcal{K} = (-\infty, d]$ for some $d \in X$. For any $\theta \in \Theta$ we have $\phi \in \Theta$ for all $\phi > \theta$, and g is an increasing homeomorphism of $\text{Int}\Theta = (a, \infty)$ for some a, $-\infty \leq a < \infty$, onto (c, d) where $-\infty \leq c < d$. Again if $Q = \delta_d$, then $Q_n\rho(\theta)$ is smallest as $\theta \to +\infty$, so there is no $\theta_0(Q)$. Or if $a = -\infty$ or $a \in \mathbb{R}$ but $a \notin \Theta$, take Q on X with $\int x dQ(x) = -\infty$. Then $\overline{X}_n \to -\infty$ a.s. and $Q_n\rho(\theta)$ is smallest asymptotically for $\theta \downarrow a$, so there is no $\theta_0(Q)$. If $-\infty < a \in \Theta$, either (a) $E_a|x| = +\infty$ (the family is steep) or (b) $E_a|x| < \infty$. In case (b), $E_a x = c$ by Lemma 6.3. In either case, if $-\infty \leq \int x dQ(x) \leq c$ we have an M-limit $\theta_0(Q) = a$. Or if $c < m_Q = \int x dQ(x) < d$ we have $\theta_0(Q) = g^{-1}(m_Q)$. Thus an M-limit $\theta_0(Q)$ exists unless $Q = \delta_d$. The situation where $\mathcal{K} = [c, \infty)$ for some $c \in \mathbb{R}$ is symmetrical. Thus the proposition is proved. □

In dimension $d > 1$, let $\mathcal{Q}$ be the class of laws whose supports have their convex hulls included in the boundary of $\mathcal{K}$. Even if $\mathcal{K}$ and Θ are both closed proper subsets of $\mathbb{R}^d$, however, there can be laws on X not in $\mathcal{Q}$ without M-limits.

References

[1] Barndorff-Nielsen, O. (1978). *Information and Exponential Families in Statistical Theory.* Wiley, Chichester.

[2] Boente, G., and Fraiman, R. (1988). On the asymptotic behaviour of general maximum likelihood estimates for the nonregular case under nonstandard conditions. *Biometrika* **75**, pp. 45-56.

[3] Brown, L. D. (1986). *Fundamentals of Statistical Exponential Families. Inst. Math. Statist. Lect. Notes Monograph Ser.*, **9**.

[4] Chademan, A., and Mirzapour, F. (1999). Midconvex functions in locally compact groups. *Proc. Amer. Math. Soc.* **127**, 2961-2968.

[5] Copas, J. B. (1975). On the unimodality of the likelihood for the Cauchy distribution. *Biometrika* **62**, 701-704.

[6] Cox, D. R. (1961). Tests of separate families of hypotheses. In *Proc. 4th Berkeley Symp. Math. Statist. Prob.* **1**, 105-123.

[7] Cox, D. R. (1962). Further results on tests of separate families of hypotheses. *J. R. Statist. Soc. B* **24**, 406-424.

[8] Dhaene, Geert (1997). *Encompassing: Formulation, Properties and Testing. Lect. Notes in Econ. and Math. Systems* **446**. Springer, Berlin.

[9] Dudley, R. M. (1987). Universal Donsker classes and metric entropy. *Ann. Probab.* **15**, 1306-1326.

[10] Dudley, R. M. (1998). Consistency of M-estimators and one-sided bracketing. In *High Dimensional Probability*, ed. E. Eberlein, M. Hahn, M. Talagrand. Birkhäuser, Basel, pp. 33-58.

[11] Dudley, R. M. (1999). *Uniform Central Limit Theorems.* Cambridge University Press.

[12] Dudley, R. M. (2002). *Real Analysis and Probability.* 2d edition, Cambridge University Press.

[13] Dudley, R. M. (2002). Nearly universal location and scatter parameters. Preprint.

[14] Dudley, R. M., Giné, E., and Zinn, J. (1991). Uniform and universal Glivenko-Cantelli classes. *J. Theoretical Probab.* **4** 485-510.

[15] Giné, E., and Zinn, J. (1991). Gaussian characterization of uniform Donsker classes of functions. *Ann. Probab.* **19**, 758-782.

[16] Haldane, J. B. S. (1948). Note on the median of a multivariate distribution. *Biometrika* **35**, 414-415.

[17] Hoffmann-Jørgensen, J. (1994). *Probability with a View Toward Statistics*, vol. **2**, Chapman and Hall, New York.

[18] Huber, P. J. (1964). Robust estimation of a location parameter. *Ann. Math. Statist.* **35**, 73-101.

[19] Huber, P. J. (1967). The behavior of maximum likelihood estimates under nonstandard conditions. *Proc. Fifth Berkeley Symp. Math. Statist. Probability* **1**, 221-233. University of California Press, Berkeley and Los Angeles.

[20] Huber, P. J. (1981). *Robust Statistics.* Wiley, New York.

[21] Kent, J. T., and Tyler, D. E. (1991). Redescending M-estimates of multivariate location and scatter. *Ann. Statist.* **19**, 2102-2119.

[22] Kullback, S. (1983). Kullback information. In *Encyclopedia of Statistical Sciences*, vol. **4**, pp. 421-425, Eds. S. Kotz, N. L. Johnson. Wiley, New York.

[23] Kullback, S., and Leibler, R. A. (1951). On information and sufficiency. *Ann. Math. Statist.* **22**, 79-86.

[24] Mäkeläinen, T., Schmidt, K., and Styan, G. P. H. (1981). On the existence and uniqueness of the maximum likelihood estimate of a vector-valued parameter in fixed-size samples. *Ann. Statist.* **9**, 758-767.

[25] Milasevic, P. and Ducharme, G. R. (1987). Uniqueness of the spatial median. *Ann. Statist.* **15**, 1332-1333.

[26] Pitman, E. J. G. (1979). *Some Basic Theory for Statistical Inference.* Chapman and Hall, London.

[27] Poskitt, D. S. (1987). Precision, complexity and Bayesian model determination. *J. R. Statist. Soc.* B **49** 199-208.

[28] Reeds, J. A. (1985). Asymptotic number of roots of Cauchy location likelihood equations. *Ann. Statist.* **13**, 775-784.

[29] Sainte-Beuve, M.-F. (1974). On the extension of von Neumann-Aumann's theorem. *J. Functional Analysis* **17**, 112-129.

[30] Sawa, T. (1978). Information criteria for discriminating among alternative regression models. *Econometrica* **46**, 1273-1291.

[31] Sierpiński, W. (1920). Sur les fonctions convexes mesurables. *Fund. Math.* **1**, 125-129.

[32] Talagrand, M. (1987). The Glivenko-Cantelli problem. *Ann. Probab.* **15**, 837-870.

[33] Talagrand, M. (1996). The Glivenko-Cantelli problem, ten years later. *J. Theoret. Probab.* **9**, 371-384.

Room 2-245, MIT, Cambridge, MA 02139, USA
E-mail address: rmd@math.mit.edu

Progress in Probability, Vol. 55, 313–320

Smoothed Empirical Processes and the Bootstrap

Peter Gaenssler and Daniel Rost

Abstract. Based on a uniform functional central limit theorem (FCLT) for unbiased smoothed empirical processes indexed by a class $\mathcal{F}$ of measurable functions defined on a linear metric space we present a consistency theorem for smoothed bootstrapped empirical processes. Our approach and the results are comparable with those in Giné and Zinn [8], and Giné [10], respectively, in the case of empirical processes; especially, our Theorem 2.2 below is comparable with the main result stated as Theorem 2.3 in Giné and Zinn [8].

1. Introduction

Bootstrap resampling methods have become an important tool in statistics since their introduction by Efron [2, 3]. Bickel and Freedman [1] carried out an extensive asymptotic analysis of Efron's nonparametric bootstrap for i.i.d. real-valued data. An extension of the bootstrap for general empirical measures based on data in arbitrary sample spaces and indexed by Vapnik-Chervonenkis classes of sets was presented for the first time in Gaenssler [4] (as acknowledged in Wellner [15] and Giné [9]).

As in Gaenssler and Rost [6] the present context is as follows: Let $(X, \mathcal{X})$ be an arbitrary measurable space (sample space) and let $\eta_j, j \in \mathbb{N}$, be independent identically distributed (i.i.d.) random elements (r.e.) in $(X, \mathcal{X})$ with law $\mathcal{L}\{\eta_j\} \equiv \nu$ on $\mathcal{X}$, viewed as coordinate mappings on the product probability space $(\Omega, \mathcal{A}, \mathbb{P}) := (X^{\mathbb{N}}, \mathcal{X}^{\mathbb{N}}, \nu^{\mathbb{N}})$. Let ν_n be the empirical measure based on $\eta_1, ..., \eta_n$, i.e.

$$\nu_n := n^{-1} \sum_{j \leq n} \delta_{\eta_j},$$

where δ_x denotes the Dirac measure in $x \in X$. Given a class $\mathcal{F}$ be of $\mathcal{X}$-measurable functions $f : X \longrightarrow \mathbb{R}$ with measurable envelope $F : X \longrightarrow \mathbb{R}$ (i.e. $\sup_{f \in \mathcal{F}} |f(x)| \leq F(x)$ for all $x \in X$) put

$$\nu_n(f) := \int_X f \, d\nu_n,$$

Received by the editors December 1, 2002. Revised March 2003.
2000 *Mathematics Subject Classification.* Primary 60F05, 62F40; Secondary 60G15.
Key words and phrases. Smoothed Empirical Process, Bootstrap.

and let

$$\mathbb{G}_n^\nu := \sqrt{n}(\nu_n - \nu) \equiv \left(\sqrt{n}(\nu_n(f) - \nu(f))\right)_{f\in\mathcal{F}}$$

be the *empirical process* (of sample size n). FCLT's for $\mathbb{G}_n^\nu$, i.e. weak convergence of $\mathbb{G}_n^\nu$to a mean-zero Gaussian process $\mathbb{G}_\nu = (G_\nu(f))_{f\in\mathcal{F}}$ denoted by $\mathbb{G}_n^\nu \xrightarrow[sep]{\mathcal{L}} \mathbb{G}_\nu$ in the space $l^\infty(\mathcal{F}) := \{x : \mathcal{F} \longrightarrow \mathbb{R} \ : \ ||x||_\mathcal{F} := \sup_{f\in\mathcal{F}} |x(f)| < \infty\}$, are well known through the work of Dudley, Giné-Zinn, Pollard and others (cf. e.g. van der Vaart and Wellner [14] and the references cited there). Here $\xrightarrow[sep]{\mathcal{L}}$ denotes weak convergence in the sense of Hoffmann-Jørgensen (see e.g. van der Vaart and Wellner [14], Section 1.3 for the definition) where the limiting process $\mathbb{G}_\nu$ (with covariance structure given by $cov(G_\nu(f), G_\nu(g)) = \nu(fg) - \nu(f)\cdot\nu(g)\ , f, g \in \mathcal{F})$ has all its sample paths in a *separable* subspace S_0 of $(l^\infty(\mathcal{F}), ||\cdot||_\mathcal{F})$; usually $S_0 = U^b(\mathcal{F}, d_\nu^{(2)})$, the space of all bounded, uniformly $d_\nu^{(2)}$-continuous functions on $\mathcal{F}$ will serve for this, where

$$d_\nu^{(2)}(f,g) := \left(\nu(f-g)^2\right)^{1/2} \equiv \left(\int_X (f-g)^2 d\nu\right)^{1/2} \quad , f, g \in \mathcal{F},$$

since the usual conditions on $\mathcal{F}$ (cf. below) imply that $(\mathcal{F}, d_\nu^{(2)})$ is totally bounded, being equivalent with the separability of $U^b(\mathcal{F}, d_\nu^{(2)})$ as a subspace of $(l^\infty(\mathcal{F}), ||\cdot||_\mathcal{F})$. Equivalently, with the law $\mathcal{L}\{\mathbb{G}_\nu\}$ of $\mathbb{G}_\nu$ on the Borel σ-field $\mathcal{B}(l^\infty(\mathcal{F}))$ in $(l^\infty(\mathcal{F}), ||\cdot||_\mathcal{F})$ being tight, saying also that the limiting process $\mathbb{G}_\nu$ is tight, we have (cf. van der Vaart and Wellner [14], Section 1.12, and Giné [10, Corollary 1.5]; see also Gaenssler and Rost [5, pp.79 -])

$$\mathbb{G}_n^\nu \xrightarrow[sep]{\mathcal{L}} \mathbb{G}_\nu \quad \text{in } l^\infty(\mathcal{F}) \quad \Longleftrightarrow \quad d_{BL}(\mathbb{G}_n^\nu, \mathbb{G}_\nu) \longrightarrow 0. \tag{1.1}$$

Here, given stochastic processes $S_n, n \in \mathbb{N}$, and S indexed by $\mathcal{F}$, where S_n and S have all its sample paths in $l^\infty(\mathcal{F})$, but S being tight, the *bounded Lipschitz distance* $d_{BL}(S_n, S)$ is defined by (with $\mathbb{E}^*$ denoting outer expectation)

$$d_{BL}(S_n, S) := \sup\{\left|\mathbb{E}^*(H(S_n)) - \mathbb{E}(H(S))\right| : H \in BL_1(l^\infty(\mathcal{F}))\},$$

with

$$BL_1(l^\infty(\mathcal{F})) := \Big\{H : l^\infty(\mathcal{F}) \longrightarrow \mathbb{R} \ : H \text{ bounded by 1, and}$$

$$\sup_{x_1,x_2\in l^\infty(\mathcal{F}), x_1\neq x_2} |H(x_1) - H(x_2)|/||x_1 - x_2||_\mathcal{F} \leq 1\Big\}.$$

In case of (1.1) $\mathcal{F}$ is called a *ν-Donsker class.*

Now, in view of bootstrap results for empirical processes it is important to know whether $\mathcal{F}$ is also a uniform Donsker class, i.e. whether

$$\sup_{\nu\in\mathcal{M}^1(X)} d_{BL}(\mathbb{G}_n^\nu, \mathbb{G}_\nu) \longrightarrow 0 \text{ (and } (\mathcal{F}, d_\nu^{(2)})\text{is totally bounded uniformly in } \nu),$$

where $\mathcal{M}^1(X)$ denotes the class of all p-measures on $\mathcal{X}$.

The following result is known (see Giné [10, Theorem 5.3], together with Ledoux and Talagrand [11], Sections 11.1 and 12.1, and Giné and Zinn [8, Proposition 3.1]). For this we assume that $\mathcal{F}$ has *uniformly integrable L_2-entropy* which means that (cf. van der Vaart and Wellner [14, condition (2.5.1), p. 127])

$$\int_0^\infty \sup_Q \sqrt{\log N\big(\tau \cdot (Q(F^2))^{\frac{1}{2}}, \mathcal{F}, d_Q^{(2)}\big)}\, d\tau < \infty \tag{1.2}$$

where the supremum is taken over all finite measures Q on $\mathcal{X}$ with $0 < Q(F^2) < \infty$. We also assume $\mathcal{M}^1(X)$ to be equipped with the sup-norm $\|\cdot\|_{\mathcal{F}\mathcal{F}}$ where

$$\mathcal{F}\mathcal{F} := \{f, f^2 : f \in \mathcal{F}\}.$$

Note that for each $\nu \in \mathcal{M}^1(X)$ $\quad \|\nu\|_{\mathcal{F}\mathcal{F}} := \sup_{h\in\mathcal{F}\mathcal{F}} |\nu(h)| < \infty$, provided $\mathcal{F}$ is uniformly bounded as will be assumed in the following, i.e. it is assumed (w.l.o.g.) that $\mathcal{F}$ has $F \equiv 1$ as its envelope.

Theorem 1.1. *Let $\mathcal{F}$ be uniformly bounded having uniformly integrable L_2-entropy. Then*

(a) $\sup_{\nu\in\mathcal{M}^1(X)} d_{BL}(\mathbb{G}_n^\nu, \mathbb{G}_\nu) \longrightarrow 0;$

(b) *the mapping from $(\mathcal{M}^1(X), \|\cdot\|_{\mathcal{F}\mathcal{F}})$ into $(\mathcal{R}, d_{BL})$, defined by $\nu \longmapsto \mathcal{L}\{\mathbb{G}_\nu\}$, is uniformly continuous, where $\mathcal{R}$ denotes the space of all tight p-measures on $\mathcal{B}(l^\infty(\mathcal{F}))$.*

Concerning the bootstrapped empirical processes

$$\mathbb{G}_n^{\nu_n} := \sqrt{n}\Big(n^{-1}\sum_{i\le n}\delta_{\eta_i^*} - \nu_n\Big)$$

based on i.i.d. r.e. $\eta_j^*, j \in \mathbb{N}$, in $(X, \mathcal{X})$ with $\mathcal{L}\{\eta_j^*\} = \nu_n$, Theorem 1.1 yields the following consistency result (an *in probability FCLT*) for $\mathbb{G}_n^{\nu_n} = (G_n^{\nu_n}(f))_{f\in\mathcal{F}}$ (Giné and Zinn [8, Corollary 2.7]; see also Sheehy and Wellner [13, Theorem 4.1]), where $\xrightarrow{\mathbb{P}}$ denotes convergence in probability:

Theorem 1.2. *Let $\mathcal{F}$ be uniformly bounded having uniformly integrable L_2-entropy. Then*

$$d_{BL}(\mathbb{G}_n^{\nu_n}, \mathbb{G}_\nu)^* \xrightarrow{\mathbb{P}} 0. \tag{1.3}$$

(Note that $d_{BL}(\mathbb{G}_n^{\nu_n}, \mathbb{G}_\nu)$ is random (defined on the basic p-space $(\Omega, \mathcal{A}, \mathbb{P})$ through the empirical measure ν_n based on the original data $\eta_1, ..., \eta_n$), being not necessarily measurable and therefore one works with the *minimal measurable cover function* of $d_{BL}(\mathbb{G}_n^{\nu_n}, \mathbb{G}_\nu)$, denoted by $d_{BL}(\mathbb{G}_n^{\nu_n}, \mathbb{G}_\nu)^*$ (cf. van der Vaart and Wellner [14, Chap. 1.2]).

Proof. The proof of Theorem 1.2 is very simple and relies on Theorem 1.1: In fact, we have that

$$d_{BL}(\mathbb{G}_n^{\nu_n}, \mathbb{G}_\nu) \le d_{BL}(\mathbb{G}_n^{\nu_n}, \mathbb{G}_{\nu_n}) + d_{BL}(\mathbb{G}_{\nu_n}, \mathbb{G}_\nu),$$

where Theorem 1.1 (a) yields $d_{BL}(\mathbb{G}_n^{\nu_n}, \mathbb{G}_{\nu_n})^* \xrightarrow{\mathbb{P}} 0$.

Since $\mathcal{F}$ has uniformly integrable L_2-entropy by assumption, the same holds for $\mathcal{FF}$; note that for $f, g \in \mathcal{F}$ (with $F \equiv 1$ as envelope of $\mathcal{F}$)

$$\nu((f^2 - g^2)^2) = \nu((f-g)^2(f+g)^2) \leq 4\nu((f-g)^2),$$

whence for all $\tau > 0$ $N(2\tau, \mathcal{FF}, d_\nu^{(2)}) \leq 2\,N(\tau, \mathcal{F}, d_\nu^{(2)})$.
On the other hand, since $\mathcal{FF}$ has uniformly integrable L_2-entropy, we have for each $p \geq 1$

$$\sup_{h \in \mathcal{FF}} |\nu_n(h) - \nu(h)| \xrightarrow{L_p} 0 \tag{1.4}$$

(cf. Gaenssler and Rost [6, Theorem 2.2] in the special case of uniformly bounded $\mathcal{F}$ and $\mu_n \equiv 0$ together with the fact that $N(\cdot, \cdot, d_\nu^{(1)}) \leq N(\cdot, \cdot, d_\nu^{(2)})$ since $d_\nu^{(1)} \leq d_\nu^{(2)}$, where $d_\nu^{(1)}(f,g) := \nu(|f-g|)$).
Finally, according to Theorem 1.1 (b) we have that for each $\varepsilon > 0$ there is a $\delta > 0$ such that

$$\|\nu_n - \nu\|_{\mathcal{FF}} < \delta \quad \text{implies} \quad d_{BL}(\mathbb{G}_{\nu_n}, \mathbb{G}_\nu) < \varepsilon.$$

Therefore (note that d_{BL} is bounded by 2)

$$d_{BL}(\mathbb{G}_{\nu_n}, \mathbb{G}_\nu) \leq \varepsilon + 2I(\|\nu_n - \nu\|_{\mathcal{FF}} \geq \delta),$$

the last term converging to 0 in probability by (1.4). Thus (1.3) is shown. □

2. Main results

Now, considering instead of ν_n *smoothed empirical measures* $\tilde{\nu}_n$ as in Gaenssler and Rost [6] and Rost [12] which arise though convolution of ν_n with given probability measures μ_n on $\mathcal{X}$ ($\tilde{\nu}_n \equiv \nu_n \star \mu_n$), where now X is a *linear metric space* endowed with its Borel σ-field $\mathcal{X} \equiv \mathcal{B}(X)$, it turns out that (1.3) in case of separable X is also valid for the processes $\mathbb{G}_n^{\tilde{\nu}_n}$ (indexed by $\mathcal{F}$) instead of $\mathbb{G}_n^{\nu_n}$, i.e. for

$$\mathbb{G}_n^{\tilde{\nu}_n} := \sqrt{n}\big(n^{-1} \sum_{j \leq n} \delta_{\tilde{\eta}_j^*} - \tilde{\nu}_n\big)$$

based on i.i.d. r.e. $\tilde{\eta}_j^*$ in $(X, \mathcal{X})$ with $\mathcal{L}\{\tilde{\eta}_j^*\} = \tilde{\nu}_n$, under the additional assumption (cf. (2.5) in Gaenssler and Rost [6])

$$\sup_{h \in \mathcal{FF}} |\nu \star \mu_n(h) - \nu(h)| \longrightarrow 0, \tag{2.1}$$

where as in Section 1 $\mathcal{F}$ is assumed to be uniformly bounded (with $F \equiv 1$ as its envelope).
Note that for this one has only to check the analogue to (1.4), i.e. whether a ULLN holds for smoothed empirical measures $\tilde{\nu}_n = \nu_n \star \mu_n$ indexed by $\mathcal{FF}$.

But if $\mathcal{F}$ has uniformly integrable L_2-entropy, so has $\mathcal{FF}$ and in consequence $\left(N(\tau,\mathcal{FF},d^{(1)}_{\tilde{\nu}_n})\right)_{n\in\mathbb{N}}$ is stochastically bounded for all $\tau>0$ and thus the analogue to (1.4) for $\tilde{\nu}_n$, i.e. for each $p\geq 1$

$$\sup_{h\in\mathcal{FF}}|\tilde{\nu}_n(h)-\nu(h)| \xrightarrow{L_p} 0$$

holds according to Theorem 2.2 in Gaenssler and Rost [6].

Especially any *Vapnik-Chervonenkis subgraph class* (*VCSG-class*) $\mathcal{F}$ and so $\mathcal{FF}$ has uniformly integrable L_2-entropy; this, along with the proof of Theorem 2.4 in Gaenssler and Rost [6] (according to which for separable X one can deduce (2.1) from $\mu_n \longrightarrow \delta_0$ weakly), yields the following result (noticing that with $\mathcal{F}$ also $\mathcal{FF}$ is uniformly bounded and equicontinuous):

Theorem 2.1. *Let X be a separable linear metric space and let $\mathcal{F}$ be a uniformly bounded equicontinuous VCSG-class. Suppose $\mu_n \longrightarrow \delta_0$ weakly. Then*

$$d_{BL}(\mathbb{G}_n^{\tilde{\nu}_n},\mathbb{G}_\nu)^* \xrightarrow{\mathbb{P}} 0. \tag{2.2}$$

Next, recalling the definition of $\tilde{\nu}_n$ indexed by $\mathcal{F}$, i.e.

$$\tilde{\nu}_n(f)\equiv\nu_n\star\mu_n(f):=\int_X\int_X f(x+y)\nu_n(dx)\mu_n(dy)=\frac{1}{n}\sum_{j\leq n}\int_X f(\eta_j+y)\mu_n(dy)$$

for $f\in\mathcal{F}$, where $\mathbb{E}(\tilde{\nu}_n(f))=\nu\star\mu_n(f)$, the unbiased smoothed empirical process $\tilde{\mathbb{G}}_n^\nu$ is given by

$$\tilde{\mathbb{G}}_n^\nu=\sqrt{n}(\tilde{\nu}_n-\nu\star\mu_n)$$

and so in view of Theorem 1.1(a) it is natural to ask whether this assertion still holds true when replacing $\mathbb{G}_n^\nu$ by $\tilde{\mathbb{G}}_n^\nu$.

Combining the results in Rost [12] with the approach as described in Section 1 in the special case of empirical processes (i.e. with $\mu_n\equiv\delta_0$) yields the following *uniform FCLT for unbiased smoothed empirical processes* $\tilde{\mathbb{G}}_n^\nu$ (indexed by $\mathcal{F}$):

Theorem 2.2. *Let X be a linear metric space and $\mathcal{F}$ be uniformly bounded. Suppose that for every $\nu\in\mathcal{M}^1(X)$ there is a tight mean-zero Gaussian process $\mathbb{G}_\nu=(G_\nu(h))_{h\in\mathcal{F}\cup\mathcal{G}}$ with tight law $\mathcal{L}\{\mathbb{G}_\nu\}$ on $\mathcal{B}(l^\infty(\mathcal{F}\cup\mathcal{G}))$ in $(l^\infty(\mathcal{F}\cup\mathcal{G}),||\cdot||_{\mathcal{F}\cup\mathcal{G}})$ where*

$$\mathcal{G}:=\{g_{f,\mu_n}:f\in\mathcal{F},n\in\mathbb{N}\}\qquad \textit{with } g_{f,\mu_n}(x):=\int_X f(x+y)\mu_n(dy)\,,\ x\in X.$$

Assume that the following conditions (2.3)-(2.5) are satisfied:

$$\sup_{\nu\in\mathcal{M}^1(X)} \mathbb{E}(\|\mathbb{G}_\nu\|_{\mathcal{F}}) < \infty \tag{2.3}$$

$$\lim_{\delta\to 0} \sup_{\nu\in\mathcal{M}^1(X)} \mathbb{E}\Big(\sup_{\substack{h_1, h_2 \in \mathcal{F}\cup\mathcal{G} \\ d_\nu^{(2)}(h_1,h_2)\le\delta}} \big|G_\nu(h_1) - G_\nu(h_2)\big|\Big) = 0 \tag{2.4}$$

$$\sup_{\nu\in\mathcal{M}^1(X)} \sup_{f\in\mathcal{F}} \int_X \Big[\int_X (f(x+y) - f(x))\mu_n(dy)\Big]^2 \nu(dx) \longrightarrow 0. \tag{2.5}$$

Then

$$\sup_{\nu\in\mathcal{M}^1(X)} d_{BL}(\tilde{\mathbb{G}}_n^\nu, \mathbb{G}_\nu) \longrightarrow 0. \tag{2.6}$$

In (2.6) not only $\tilde{\mathbb{G}}_n^\nu$ but also $\mathbb{G}_\nu$ (restricted to the index set $\mathcal{F}$) is considered as a process with sample paths in $l^\infty(\mathcal{F})$ whose law $\mathcal{L}\{\mathbb{G}_\nu\}$ is tight on $\mathcal{B}(l^\infty(\mathcal{F}))$; so d_{BL} in (2.6) stands for the bounded Lipschitz distance based on $l^\infty(\mathcal{F})$ (and not on $l^\infty(\mathcal{F}\cup\mathcal{G})$). The proof of (2.6) is contained in Gaenssler and Rost [5, see Theorem 7.5.5] following the lines of proof of (a) in Theorem 1.1 as given in Giné [10].

Additionally, from van der Vaart and Wellner [14, Theorem 2.8.3] we have that (2.3) and (2.4) are fulfilled if $\mathcal{F}\cup\mathcal{G}$ has uniformly integrable L_2-entropy, which in turn is implied by $\mathcal{F}\cup\mathcal{G}$ being a VCSG-class.
So, Theorem 2.2 yields

Theorem 2.3. *Let X be a linear metric space and let $\mathcal{F}$ be uniformly bounded. Suppose that $\mathcal{F}\cup\mathcal{G}$ has uniformly integrable L_2-entropy and that (2.5) is fulfilled. Then*

$$\sup_{\nu\in\mathcal{M}^1(X)} d_{BL}(\tilde{\mathbb{G}}_n^\nu, \mathbb{G}_\nu) \longrightarrow 0.$$

Now we are in the position of deriving the consistency theorem for smoothed bootstrapped empirical processes indexed by $\mathcal{F}$: For this, returning to the bootstrapped empirical process $\mathbb{G}_n^{\nu_n}$ based on the i.i.d. bootstrapped r.e. η_j^* in $(X,\mathcal{X})$ with $\mathcal{L}\{\eta_j^*\} = \nu_n$, smoothing this process with μ_n gives the *smoothed bootstrapped empirical process*

$$\sqrt{n}\Big(n^{-1}\sum_{j\le n}(\delta_{\eta_j^*}\star\mu_n) - \nu_n\star\mu_n\Big). \tag{2.7}$$

Note that this process is different from the process $\mathbb{G}_n^{\tilde{\nu}_n}$ as considered at the beginning of this section. To obtain a consistency result for the process (2.7), comparable with (1.3), it is important to note that the process (2.7) arises through bootstrapping the unbiased smoothed empirical process $\tilde{\mathbb{G}}_n^\nu$: In fact, given

$$\begin{aligned}\tilde{\mathbb{G}}_n^\nu &= \sqrt{n}(\tilde{\nu}_n - \nu\star\mu_n) = \sqrt{n}(\nu_n\star\mu_n - \nu\star\mu_n)\\ &= \sqrt{n}\Big(\big(n^{-1}\sum_{j\le n}\delta_{\eta_j}\big)\star\mu_n - \nu\star\mu_n\Big) = \sqrt{n}\Big(n^{-1}\sum_{j\le n}(\delta_{\eta_j}\star\mu_n) - \nu\star\mu_n\Big)\end{aligned}$$

based on i.i.d. r.e. η_j in $(X, \mathcal{X})$ with law $\mathcal{L}\{\eta_j\} = \nu$, and then replacing ν by ν_n and η_j by η_j^* with $\mathcal{L}\{\eta_j^*\} = \nu_n$ we get the process (2.7) which will be denoted by $\tilde{\mathbb{G}}_n^{\nu_n}$ (and also called the *bootstrapped smoothed empirical process*).
To this process we can apply Theorem 2.3 to obtain

Theorem 2.4. *Let the conditions of Theorem 2.3 be satisfied. Then*

$$d_{BL}(\tilde{\mathbb{G}}_n^{\nu_n}, \mathbb{G}_\nu)^* \xrightarrow{\mathbb{P}} 0.$$

Proof. We follow the argument given for proving Theorem 1.2: We have

$$d_{BL}(\tilde{\mathbb{G}}_n^{\nu_n}, \mathbb{G}_\nu) \leq d_{BL}(\tilde{\mathbb{G}}_n^{\nu_n}, \mathbb{G}_{\nu_n}) + d_{BL}(\mathbb{G}_{\nu_n}, \mathbb{G}_\nu),$$

where by Theorem 2.3 $d_{BL}(\tilde{\mathbb{G}}_n^{\nu_n}, \mathbb{G}_{\nu_n})^* \xrightarrow{\mathbb{P}} 0$ and $d_{BL}(\mathbb{G}_{\nu_n}, \mathbb{G}_\nu)^* \xrightarrow{\mathbb{P}} 0$ as shown in the proof of (1.3). □

References

[1] P.J. Bickel and D.A. A. Freedman, *Some asypmptotic theory for the bootstrap.* Ann. Statist. **9** (1981), 1196-1217.

[2] B. Efron, *Bootstrap methods: another look at the jackknife.* Ann. Statist. **7** (1979), 1-26.

[3] B. Efron, *The Jackknife, the Bootstrap, and Other Resampling Plans.* CBMS-NSF Regional Conference Series in Applied Mathematics, **38**, Society for Industrial and Applied Mathematics, Philadelphia, 1982.

[4] P. Gaenssler, *Bootstraping empirical measures indexed by Vapnik-Chervonenkis classes of sets.* Proceedings IV Vilnius Conference, Prob. Theory and Math. Statist., Vol. **1** (1986), 467-481.

[5] P. Gaenssler and D. Rost, *Empirical and partial-sum processes; revisited as random measure processes.* MaPhySto, Centre for Mathematical Physics and Stochastics, Aarhus, Lecture Notes no. **5** (1999), 112 pp.

[6] P. Gaenssler and D. Rost, *On Uniform Laws of Large Numbers for Smoothed Empirical Measures.* In: High Dimensional Probability II, Evarist Giné, David Mason, Jon A. Wellner (Editors), Progress in Probability, Birkhäuser, Vol.**47** (2000), 79-87.

[7] E. Giné and J. Zinn, *Bootstrapping general empirical measures.* Ann. Probab. **18** (1990), 851-869.

[8] E. Giné and J. Zinn, *Gaussian characterization of uniform Donsker classes of functions.* Ann. Probab. **19** (1991), 758 - 782.

[9] E. Giné, *Empirical processes and applications: an overview.* Bernoulli **2** (1) (1996), 1-28.

[10] E. Giné, *Lectures on some aspects of the bootstrap.* École d'Été de Calcul de Probabilités de Saint-Flour XXVI. In: Giné, Grimmett and Saloff-Coste: Lectures on Probability Theory and Statistics. Springer Lecture Notes in Math. **1665**, 1997.

[11] M. Ledoux and M. Talagrand, *Probability in Banach spaces.* Springer, 1991.

[12] D. Rost, *Limit Theorems for Smoothed Empirical Processes.* In: High Dimensional Probability II, Evarist Giné, David Mason, Jon A. Wellner (Editors), Progress in Probability, Birkhäuser, Vol. **47** (2000), 107-113.

[13] A. Sheehy and J.A. Wellner, *Uniform Donsker classes of functions.* Ann. Probab. **20** (1992), 1983 - 2030.

[14] A.W. van der Vaart and J.A. Wellner, *Weak convergence and empirical processes.* Springer Series in Statistics, Springer, 1996.

[15] J.A. Wellner, *Bootstrap Limit Theorems: A Partial Survey.* In: Nonparametric Statistics and Related Topics; A.K.Md.E. Saleh (Editor), Elsevier Science Publishers B.V., 1992.

Mathematical Institute, University of Munich, Theresienstr. 39, D-80333 Munich, Germany
E-mail address: gaensler@mathematik.uni-muenchen.de
E-mail address: rost@mathematik.uni-muenchen.de

Progress in Probability, Vol. 55, 321–332

A Note on the Asymptotic Distribution of Berk–Jones Type Statistics under the Null Hypothesis

Jon A. Wellner and Vladimir Koltchinskii

Abstract. Proofs are given of the limiting null distributions of the statistics of Berk and Jones (1979) and of Einmahl and McKeague (2002).

1. Introduction

Suppose that $X_1, \ldots, X_n$ are i.i.d. with distribution function F on R and we want to test

$$H : F = F_0 \quad \text{versus} \quad K : F \neq F_0 \tag{1.1}$$

where F_0 is continuous. Without loss of generality we can take $F_0(x) = (x \vee 0) \wedge 1$, the uniform distribution on $[0,1]$. The test of Berk and Jones (1979) is defined in terms of the empirical distribution function $\mathbb{F}_n$ given, as usual, for $x \in R$ by

$$\mathbb{F}_n(x) = n^{-1} \sum_{i=1}^{n} 1_{[X_i \le x]} \, .$$

For each fixed x the random variable $n\mathbb{F}_n(x) \sim \text{Binomial}(n, F(x))$, and hence the likelihood ratio statistic for testing

$$H_x : F(x) = F_0(x) \quad \text{versus} \quad K_x : F(x) \neq F_0(x)$$

is given by

$$\begin{aligned}
\lambda_n(x) &= \frac{\sup_{F(x)} L_n(F(x))}{L_n(F_0(x))} = \frac{L_n(\mathbb{F}_n(x))}{L_n(F_0(x))} \\
&= \frac{\mathbb{F}_n(x)^{n\mathbb{F}_n(x)} (1 - \mathbb{F}_n(x))^{n(1-\mathbb{F}_n(x))}}{F_0(x)^{n\mathbb{F}_n(x)} (1 - F_0(x))^{n(1-\mathbb{F}_n(x))}} \\
&= \left(\frac{\mathbb{F}_n(x)}{F_0(x)} \right)^{n\mathbb{F}_n(x)} \left(\frac{1 - \mathbb{F}_n(x)}{1 - F_0(x)} \right)^{n(1-\mathbb{F}_n(x))} .
\end{aligned}$$

Received by the editors December 11, 2002.
2000 *Mathematics Subject Classification.* Primary 62G30; Secondary 62E20, 62E17.
Key words and phrases. Goodness of fit, null hypothesis, asymptotic distribution, centering.
We owe thanks to Bob Berk for encouragement to complete this study.

By defining

$$K(x,y) \equiv x\log\left(\frac{x}{y}\right) + (1-x)\log\left(\frac{1-x}{1-y}\right)$$

for $x, y \in [0,1]$, the log-likelihood ratio statistic is

$$\log \lambda_n(x) = nK(\mathbb{F}_n(x), F_0(x)).$$

Note that $K(x,y)$ is the Kullback-Leibler "distance" between two Bernoulli distributions P_x and P_y: $K(x,y) = K(P_x, P_y)$ where the "second K" is the Kullback-Leibler distance (or relative entropy)

$$K(P,Q) = E_P\{\log dP/dQ\} \qquad \text{if } P \ll Q$$

and $P_x(X=k) = x^k(1-x)^{1-k}$ for $k \in \{0,1\}$. It follows that $K(x,y) \geq 0$ with equality if and only if $x = y$ (which can also be proved directly).

The statistic R_n studied by Berk and Jones (1979) is simply the supremum of these "pointwise likelihood ratio" test statistics:

$$R_n = \sup_{0\leq x\leq 1} n^{-1}\log\lambda_n(x) = \sup_{0\leq x\leq 1} K(\mathbb{F}_n(x), F_0(x)).$$

Einmahl and McKeague (2003) propose an integral statistic T_n defined by

$$T_n = 2\int_0^1 n^{-1}\log\lambda_n(x)\,dF_0(x) = 2\int_0^1 K(\mathbb{F}_n(x), F_0(x))\,dF_0(x).$$

Einmahl and McKeague (2003) extend this integral type of test statistic to several other testing problems.

Our goal in this note is to give complete proofs of the following theorems of Berk and Jones (1979) and Einmahl and McKeague (2003). Note that neither of these papers provide proofs of these results since in both cases they are primarily concerned with other questions. Moreover, although the brief sketch of a proof of Theorem 1.1 given by Berk and Jones (1979) is heuristically on target, the first displayed formula of section 6 of Berk and Jones (1979) does not seem to be correct. The formula there asserts that

$$R_n = \frac{1}{2}\left\{\sup_{0<x<1}\frac{(\mathbb{F}_n(x)-x)^2}{x(1-x)}\right\}\left\{1 + O_p(n^{-1/2})\right\},$$

but our proof in section 2 suggests that only a weaker relationship holds.

Theorem 1.1. *Suppose that $F = F_0$; i.e. the null hypothesis H holds. Then:*

$$nR_n - d_n \to_d Y_4 \sim E_v^4$$

where $E_v^4(x) = \exp(-4\exp(-x)) = P(Y_4 \leq x)$, and

$$d_n = \log_2 n + \frac{1}{2}\log_3 n - \frac{1}{2}\log(4\pi)$$

where $\log_2 n \equiv \log(\log n)$ and $\log_3 n \equiv \log(\log_2 n)$. In other words, for all $x \in R$

$$\lim_{n\to\infty} P\left(nR_n - d_n \leq x\right) = \exp\left(-4e^{-x}\right)$$

Theorem 1.2. *Suppose that $F = F_0$; i.e. the null hypothesis H holds. Then:*

$$nT_n \to_d \int_0^1 \frac{\mathbb{U}^2(s)}{s(1-s)}\,ds \equiv A^2$$

where $\mathbb{U}$ denotes a standard Brownian bridge process on $[0,1]$. In other words, for all $t \in R$ and

$$\lim_{n\to\infty} P\left(nT_n \le t\right) = P\left(\int_0^1 \frac{\mathbb{U}^2(s)}{s(1-s)}ds \le t\right).$$

Remark 1. (Centering of nR_n). Set

$$\begin{aligned} c_n &= 2\log_2 n + (1/2)\log_3 n - (1/2)\log(4\pi)\,, \\ b_n &= (2\log_2 n)^{1/2}\,. \end{aligned}$$

As will be seen in the next section, the proof of Theorem 1.1 actually suggests that the appropriate centering of nR_n is given by $c_n^2/(2b_n^2)$. Although

$$\frac{1}{2}\frac{c_n^2}{b_n^2} = \log_2 n + (1/2)\log_3 n - (1/2)\log(4\pi) + o(1) = d_n + o(1) \qquad (1.2)$$

so the two centerings $c_n^2/(2b_n^2)$ and d_n are asymptotically equivalent, simulations and comparison with the finite sample results of Owen (1995) suggest that the centering $c_n^2/(2b_n^2)$ yields a better approximation for finite sample sizes. We will provide evidence in support of this in Section 3.

Remark 2. Theorem 1.2 says that the asymptotic distribution of nT_n under the null hypothesis is the same as that of the classical Anderson - Darling statistic; see e.g. Shorack and Wellner (1986), 148 and 224-227.

2. Proofs of Theorems 1.1 and 1.2

We first prove Theorem 1.1.

Proof of Theorem 1.1. First note that

$$\frac{\partial}{\partial x}K(x,y)\Big|_{x=y} = \log\left(\frac{x}{y}\right) - \log\left(\frac{1-x}{1-y}\right)\Big|_{x=y} = 0\,,$$

and

$$\frac{\partial^2}{\partial x^2}K(x,y) = \frac{1}{x} + \frac{1}{1-x} = \frac{1}{x(1-x)}\,.$$

Hence it follows that

$$\begin{aligned} K(x,y) &= K(y,y) + \frac{\partial}{\partial x}K(x,y)\Big|_{x=y}(y-x) + \frac{1}{2}\frac{\partial^2}{\partial x^2}K(x,y)\Big|_{x=y^*}(y-x)^2 \\ &= 0 + 0 + \frac{1}{2}\frac{(y-x)^2}{y^*(1-y^*)} \\ &= \frac{1}{2}\frac{(y-x)^2}{y^*(1-y^*)} \end{aligned}$$

for some y^* satisfying $|y^* - x| \le |y - x|$. This yields

$$K(\mathbb{F}_n(x), x) = \frac{1}{2} \frac{(\mathbb{F}_n(x) - x)^2}{\mathbb{F}_n^*(x)(1 - \mathbb{F}_n^*(x))} \tag{2.1}$$

for $0 < x < 1$ where $|\mathbb{F}_n^*(x) - x| \le |\mathbb{F}_n(x) - x|$; i.e. $x \le \mathbb{F}_n^*(x) \le \mathbb{F}_n(x)$ on the event $x \le \mathbb{F}_n(x)$ and $\mathbb{F}_n(x) \le \mathbb{F}_n^*(x) \le x$ on the event $\mathbb{F}_n(x) \le x$.

We can write (2.1) as

$$K(\mathbb{F}_n(x), x) = \frac{1}{2} \frac{(\mathbb{F}_n(x) - x)^2}{x(1-x)} \left\{ 1 + \frac{x(1-x)}{\mathbb{F}_n^*(x)(1 - \mathbb{F}_n^*(x))} - 1 \right\}$$

where

$$\begin{aligned} |\mathrm{Rem}_n(x)| &\equiv \left| \frac{x(1-x)}{\mathbb{F}_n^*(x)(1 - \mathbb{F}_n^*(x))} - 1 \right| \\ &= \left| \frac{x - x^2 - (\mathbb{F}_n^*(x) - \mathbb{F}_n^*(x)^2)}{\mathbb{F}_n^*(x)(1 - \mathbb{F}_n^*(x))} \right| \\ &= \left| \frac{(x - \mathbb{F}_n^*(x))(1 - (x + \mathbb{F}_n^*(x)))}{\mathbb{F}_n^*(x)(1 - \mathbb{F}_n^*(x))} \right| \\ &\le 3 \left| \frac{(x - \mathbb{F}_n(x))}{\mathbb{F}_n^*(x)(1 - \mathbb{F}_n^*(x))} \right| . \end{aligned}$$

Fix $\delta \in (0, 1/2)$. Now for $x \in [\delta, 1-\delta]$, $\mathbb{F}_n(x) \in [\delta/2, 1 - \delta/2]$ a.s. for $n \ge N_\omega$, so

$$\sup_{\delta \le x \le 1-\delta} |\mathrm{Rem}_n(x)| \le \frac{3}{(\delta/2)(1 - \delta/2)} \sup_{\delta \le x \le 1-\delta} |\mathbb{F}_n(x) - x| = O_p(n^{-1/2}) .$$

For $0 < x \le 1/2$ the function $g(x) = x(1-x)$ is $\nearrow$, so

$$g(\mathbb{F}_n^*(x)) \ge g(x) \wedge g(\mathbb{F}_n(x))$$

on the set $\{\mathbb{F}_n(x) < 1/2\}$. Since $P(\mathbb{F}_n(\delta) \ge 1/2) \to 0$, we get

$$\begin{aligned} \sup_{X_{(1)} \le x \le \delta} |\mathrm{Rem}_n(x)| &\le 3 \left(\sup_{X_{(1)} \le x \le \delta} \frac{|\mathbb{F}_n(x) - x|}{g(\mathbb{F}_n(x))} \vee \sup_{0 < x \le \delta} \frac{|\mathbb{F}_n(x) - x|}{g(x)} \right) \\ &= O_p(1) \end{aligned} \tag{2.2}$$

where $0 \le X_{(1)} \le \ldots \le X_{(n)} \le 1$ denote the order statistics of the sample. Here the $O_p(1)$ holds by virtue of Daniels (1945), Robbins (1954), and Chang (1955); see Theorem 2, Shorack and Wellner (1986), page 345, and Inequality 1, Shorack and Wellner (1986) page 415. Now note that $K(\mathbb{F}_n(x), x) = 0$ for $x < X_{(1)}$ and $x > X_{(n)}$. Therefore

$$\begin{aligned} R_n &= \sup_{0 < x < 1} K(\mathbb{F}_n(x), x) = \sup_{X_{(1)} \le x \le X_{(n)}} K(\mathbb{F}_n(x), x) \\ &= \frac{1}{2} \sup_{X_{(1)} \le x \le X_{(n)}} \frac{(\mathbb{F}_n(x) - x)^2}{\mathbb{F}_n^*(x)(1 - \mathbb{F}_n^*(x))} \end{aligned}$$

by (2.1). Fix $\delta \in (0, 1/2)$ and define

$$
\begin{aligned}
R_n(I) &= \frac{1}{2} \sup_{\delta \le x \le 1-\delta} \frac{(\mathbb{F}_n(x) - x)^2}{\mathbb{F}_n^*(x)(1 - \mathbb{F}_n^*(x))}, \\
R_n(II) &= \frac{1}{2} \sup_{X_{(1)} \le x \le \delta} \frac{(\mathbb{F}_n(x) - x)^2}{\mathbb{F}_n^*(x)(1 - \mathbb{F}_n^*(x))}, \\
R_n(III) &= \frac{1}{2} \sup_{1-\delta \le x \le X_{(n)}} \frac{(\mathbb{F}_n(x) - x)^2}{\mathbb{F}_n^*(x)(1 - \mathbb{F}_n^*(x))}.
\end{aligned}
$$

Thus $R_n = \max\{R_n(I), R_n(II), R_n(III)\}$. For $R_n(I)$,

$$
\begin{aligned}
R_n(I) &= \frac{1}{2} \sup_{\delta \le x \le 1-\delta} \frac{(\mathbb{F}_n(x) - x)^2}{x(1-x)} \left\{1 + O_p(n^{-1/2})\right\} \\
&\le \frac{1}{2} \sup_{\delta \le x \le 1-\delta} \left\{ \frac{(\mathbb{F}_n(x) - x)^2}{x(1-x)} \vee \frac{(\mathbb{F}_n(x) - x)^2}{\mathbb{F}_n(x)(1 - \mathbb{F}_n(x))} \right\} \left\{1 + O_p(n^{-1/2})\right\}.
\end{aligned}
$$

In the second region the argument above leading to (2.2) yields

$$
\begin{aligned}
R_n(II) &\le \frac{1}{2} \sup_{X_{(1)} \le x \le \delta} \left\{ \frac{(\mathbb{F}_n(x) - x)^2}{x(1-x)} \vee \frac{(\mathbb{F}_n(x) - x)^2}{\mathbb{F}_n(x)(1 - \mathbb{F}_n(x))} \right\} \\
&\ge \frac{1}{2} \sup_{X_{(1)} \le x \le \delta} \left\{ \frac{(\mathbb{F}_n(x) - x)^2}{x(1-x)} \wedge \frac{(\mathbb{F}_n(x) - x)^2}{\mathbb{F}_n(x)(1 - \mathbb{F}_n(x))} \right\},
\end{aligned}
$$

and similarly for $R_n(III)$. It follows that

$$
\begin{aligned}
R_n &\le \frac{1}{2} \sup_{X_{(1)} \le x \le X_{(n)}} \left\{ \frac{(\mathbb{F}_n(x) - x)^2}{x(1-x)} \vee \frac{(\mathbb{F}_n(x) - x)^2}{\mathbb{F}_n(x)(1 - \mathbb{F}_n(x))} \right\} \\
&\qquad \times \left\{1 + O_p(n^{-1/2})\right\} \\
&= \frac{1}{2} \sup_{X_{(1)} \le x \le X_{(n)}} \left\{ \frac{(\mathbb{F}_n(x) - x)^2}{x(1-x)} \left(1 \vee \frac{x(1-x)}{\mathbb{F}_n(x)(1 - \mathbb{F}_n(x))}\right) \right\} \\
&\qquad \times \left\{1 + O_p(n^{-1/2})\right\}, \qquad (2.3)
\end{aligned}
$$

and, on the other hand,

$$
\begin{aligned}
R_n &\ge \frac{1}{2} \sup_{X_{(1)} \le x \le X_{(n)}} \left\{ \frac{(\mathbb{F}_n(x) - x)^2}{x(1-x)} \wedge \frac{(\mathbb{F}_n(x) - x)^2}{\mathbb{F}_n(x)(1 - \mathbb{F}_n(x))} \right\} \\
&\qquad \times \left\{1 + O_p(n^{-1/2})\right\} \\
&= \frac{1}{2} \sup_{X_{(1)} \le x \le X_{(n)}} \left\{ \frac{(\mathbb{F}_n(x) - x)^2}{x(1-x)} \left(1 \wedge \frac{x(1-x)}{\mathbb{F}_n(x)(1 - \mathbb{F}_n(x))}\right) \right\} \\
&\qquad \times \left\{1 + O_p(n^{-1/2})\right\}. \qquad (2.4)
\end{aligned}
$$

Now we break the suprema into the regions $[X_{(1)}, d_n]$, $[d_n, 1-d_n]$, and $[1-d_n, X_{(n)}]$ with $d_n = (\log n)^k/n$ for any $k \geq 1$. Then we have

$$n \sup_{X_{(1)} \leq x \leq d_n} \frac{(\mathbb{F}_n(x) - x)^2}{x(1-x)} = o_p(b_n^2)$$

where $b_n = \sqrt{2 \log_2 n}$; see Shorack and Wellner (1986), (26), page 602. Moreover,

$$\sup_{X_{(1)} \leq x \leq d_n} \left| \frac{x(1-x)}{\mathbb{F}_n(x)(1-\mathbb{F}_n(x))} \right| = O_p(1),$$

so

$$n \sup_{X_{(1)} \leq x \leq d_n} \frac{(\mathbb{F}_n(x) - x)^2}{x(1-x)} \left(1 \,\#\, \frac{x(1-x)}{\mathbb{F}_n(x)(1-\mathbb{F}_n(x))} \right) = o_p(b_n^2) \tag{2.5}$$

for $\# = \wedge$ or $\# = \vee$, and similarly for the region $[1-d_n, X_{(n)}]$. On the other hand if we define

$$Z_n \equiv \sup_{d_n \leq x \leq 1-d_n} \frac{\sqrt{n}|\mathbb{F}_n(x) - x|}{\sqrt{x(1-x)}}, \tag{2.6}$$

then, for $k \geq 5$

$$\frac{Z_n}{b_n} \to_p 1, \tag{2.7}$$

and

$$b_n Z_n - c_n \to_d E_v^4 \tag{2.8}$$

where $c_n = 2 \log_2 n + (1/2) \log_3 n - (1/2) \log(4\pi)$ (see e.g. Shorack and Wellner (1986), page 600, (16.1.20)) and (16.1.17)). Furthermore,

$$\left\| \frac{\mathbb{F}_n(x) - x}{x} \right\|_{d_n}^1 = O(r_n) \tag{2.9}$$

almost surely where

$$r_n^2 \equiv \frac{\log_2 n}{n d_n} = \frac{\log_2 n}{(\log n)^k} \to 0;$$

see Shorack and Wellner (1986), page 424, (4.5.10) and (4.5.11). It follows from (2.3), (2.4), (2.7), (2.8), and (2.9) that

$$\begin{aligned} nR_n &= \frac{1}{2} \left\{ \sup_{d_n \leq x \leq 1-d_n} \frac{n(\mathbb{F}_n(x) - x)^2}{x(1-x)} (1 + O_p(r_n)) \bigvee o_p(b_n^2) \right\} \\ &\qquad \times \left\{ 1 + O_p(n^{-1/2}) \right\} \\ &= \frac{1}{2} \left\{ Z_n^2 \bigvee o_p(b_n^2) \right\} + o_p(1). \end{aligned} \tag{2.10}$$

Hence we can write

$$\begin{aligned} \frac{1}{2} Z_n^2 &= \frac{1}{2}(Z_n - c_n/b_n)(Z_n + c_n/b_n) + \frac{1}{2}\frac{c_n^2}{b_n^2} \\ &= \frac{1}{2} b_n (Z_n - c_n/b_n) \frac{Z_n + c_n/b_n}{b_n} + \frac{1}{2}\frac{c_n^2}{b_n^2} \end{aligned}$$

It follows that

$$\begin{aligned} nR_n - \frac{1}{2}\frac{c_n^2}{b_n^2} &= b_n(Z_n - c_n/b_n)\frac{Z_n + c_n/b_n}{2b_n} \bigvee \left(o_p(b_n^2) - \frac{1}{2}\frac{c_n^2}{b_n^2}\right) + o_p(1) \\ &= b_n(Z_n - c_n/b_n)\frac{Z_n/b_n + c_n/b_n^2}{2} \bigvee (o_p(1) - 1/2)\, b_n^2 + o_p(1) \\ &\to_d Y_4 \frac{1+1}{2} \bigvee \{-\infty\} = Y_4 \,; \end{aligned} \tag{2.11}$$

here we used $c_n^2/b_n^2 \sim b_n^2$ in the second equality. Since

$$\frac{1}{2}\frac{c_n^2}{b_n^2} = \log_2 n + (1/2)\log_3 n - (1/2)\log(4\pi) + o(1) = d_n + o(1) \tag{2.12}$$

this yields

$$P(nR_n - d_n \le x) \quad \to \quad \exp(-4\exp(-x))\,, \tag{2.13}$$

and completes the proof of Theorem 1. Note that the centering $c_n^2/(2b_n^2)$ emerges naturally in the course of this proof. □

Proof of Theorem 2. Let $\alpha \in (1/2, 1)$, set $a_n = n^{-\alpha}$, and write

$$\begin{aligned} nT_n &= \left(\int_0^{a_n} + \int_{a_n}^{1-a_n} + \int_{1-a_n}^{1}\right) 2nK(\mathbb{F}_n(x), x)\, dx \\ &\equiv I_n + II_n + III_n\,. \end{aligned}$$

Now by (2.1) it follows that

$$II_n = \int_{a_n}^{1-a_n} \frac{n(\mathbb{F}_n(x) - x)^2}{\mathbb{F}_n^*(x)(1 - \mathbb{F}_n^*(x))}\, dx \to_d \int_0^1 \frac{\mathbb{U}^2(x)}{x(1-x)}\, dx$$

since

$$\left\| \frac{x(1-x)}{\mathbb{F}_n(x)(1-\mathbb{F}_n(x))} - 1 \right\|_{a_n}^{1-a_n} \to_p 0$$

by Wellner (1978) Theorem 0, page 77, and hence also with $\mathbb{F}_n$ replaced by $\mathbb{F}_n^*$, and

$$\int_{a_n}^{1-a_n} \frac{n(\mathbb{F}_n(x) - x)^2}{x(1-x)}\, dx \to_d \int_0^1 \frac{\mathbb{U}^2(x)}{x(1-x)}\, dx$$

as is well-know. Hence it remains only to show that $I_n \to_p 0$ (and then $III_n \to_p 0$ also by symmetry). To show this, fix $\epsilon > 0$ and choose $\lambda = \lambda_\epsilon$ so large that

$$P(\|\mathbb{F}_n(x)/x\|_0^1 > \lambda) = \lambda^{-1} < \epsilon\,.$$

On the event $\|\mathbb{F}_n(x)/x\|_0^1 \le \lambda$ we have

$$\begin{aligned} I_n &= \int_0^{a_n} 2nK(\mathbb{F}_n(x), x)\, dx \\ &= \int_0^{a_n} 2n\mathbb{F}_n(x) \log \frac{\mathbb{F}_n(x)}{x}\, dx + o(1) \qquad \text{a.s.} \\ &\le \int_0^{a_n} 2n\lambda x \log \lambda\, dx \\ &= \lambda \log \lambda\, n a_n^2 \to 0 \end{aligned}$$

since $\alpha > 1/2$. □

3. On the centering: finite sample approximations

Now we present some graphical evidence in favor of centering the statistic nR_n of Berk and Jones (1979) by $c_n^2/(2b_n^2)$ rather than the asymptotically equivalent form given by d_n. Figures 1 and 2 below give empirical distributions of 5000 Monte Carlo replications of $nR_n - d_n$ for sample sizes $n = 100$ and $n = 1000$ respectively, together with the limit distribution function. From Figure 1 it is apparent that the distribution of $nR_n - d_n$ with $n = 100$ is shifted to the right from the limit distribution in the middle of its range. From Figure 2 we see that this continues to be the case for $n = 1000$, although the shift is somewhat less. Figures 3 and 4 give empirical distributions of 5000 Monte Carlo replications of $nR_n - c_n^2/(2b_n^2)$ for samples sizes $n = 100$ and $n = 1000$ respectively, together with the limit distribution function. It is clear that the distribution of $nR_n - c_n^2/(2b_n^2)$ nearly coincides with the limit distribution in the middle of its range. Furthermore, Figure 4 shows improved agreement with the limit distribution using the centering $c_n^2/(2b_n^2)$ for $n = 1000$.

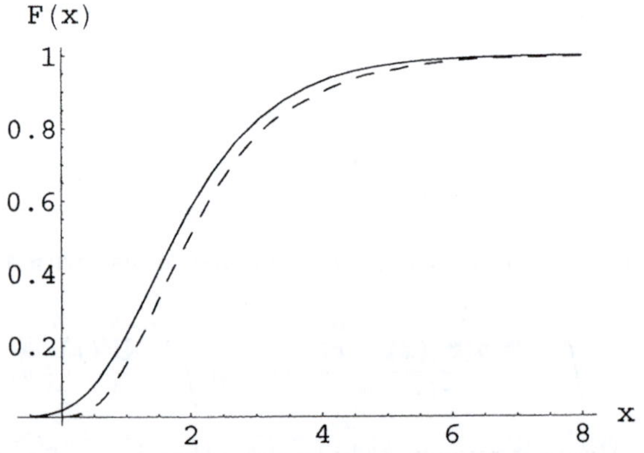

Figure 1: solid line: limit distribution $F(x) = \exp(-4\exp(-x))$; dashed line: empirical distribution function of 5000 replications of $nR_n - d_n$, $n = 100$.

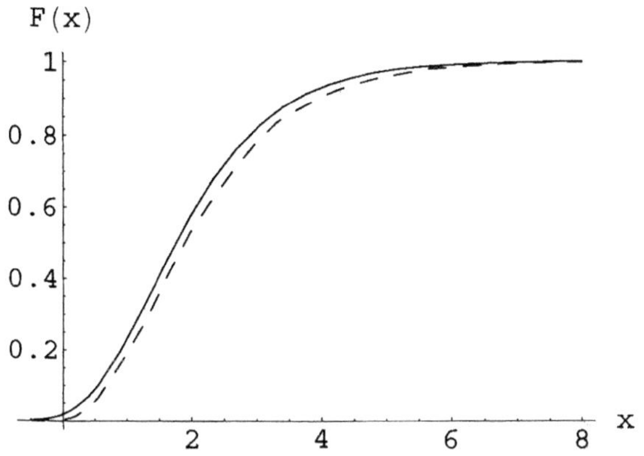

Figure 2: solid line: limit distribution $F(x) = \exp(-4\exp(-x))$; dashed line: empirical distribution function of 5000 replications of $nR_n - d_n$, $n = 1000$.

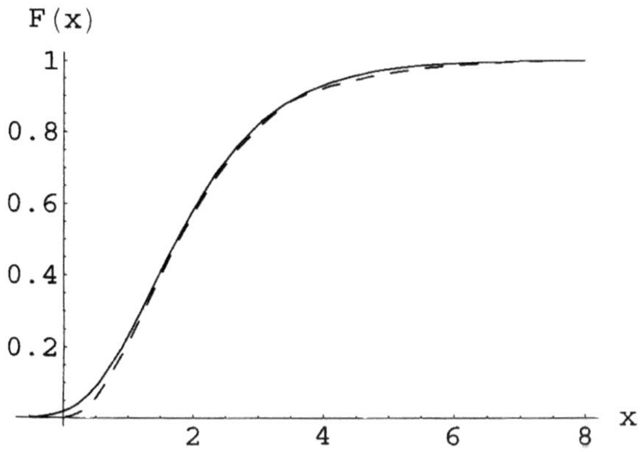

Figure 3: solid line: limit distribution $F(x) = \exp(-4\exp(-x))$; dashed line: empirical distribution function of 5000 replications of $nR_n - c_n^2/(2b_n^2)$, $n = 100$.

Owen (1995) used the recursions of Noe (1972) to find finite sample formulas for the quantiles of nR_n. For example, Owen (1995) reports that approximate .95 and .99 quantiles of nR_n for $1 < n \leq 1000$ are given as follows:

$$\lambda(n,.95) \approx \begin{cases} 3.0123 + .4835\log n - .00957(\log n)^2 - .001488(\log n)^3, & 1 < n \leq 100 \\ 3.0806 + .4894\log n - .02086(\log n)^2, & 100 < n \leq 1000, \end{cases}$$

while

$$\lambda(n,.99) \approx \begin{cases} 4.626 + .541\log n - .0242(\log n)^2, & 1 < n \leq 100, \\ 4.710 + .512\log n - .0219(\log n)^2, & 100 < n \leq 1000. \end{cases} \tag{3.1}$$

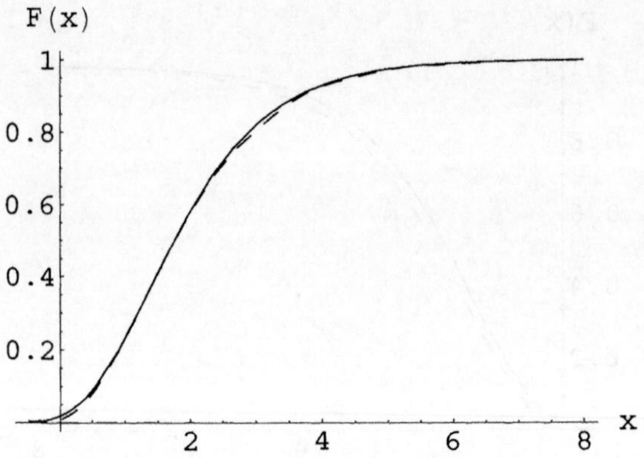

Figure 4: solid line: limit distribution $F(x) = \exp(-4\exp(-x))$; dashed line: empirical distribution function of 5000 replications of $nR_n - c_n^2/(2b_n^2)$, $n = 1000$.

The signs in (3.1) are from Owen (2001), page 159, Table 7.1, and differ by a sign from Owen (1995) formulas (12) and (13); it seems clear that this is a typo in Owen (1995).

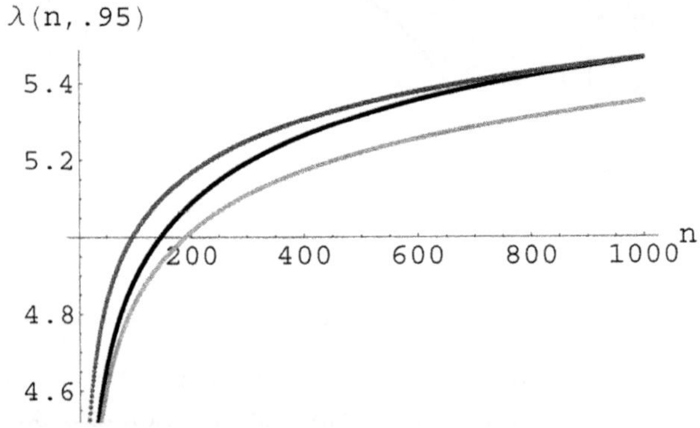

Figure 5: Black: Owen's finite-sample quantiles, $\lambda(n, .95)$.
Medium Gray: approximate quantiles $\lambda(n, .95; c_n^2/(2b_n^2))$
Light Gray: approximate quantiles $\lambda(n, .95; d_n)$

Figures 5 and 6 give plots of these approximations (in blue) from Owen (1995) together with the corresponding quantiles resulting from Theorem 1 with the two asymptotically equivalent centerings $c_n^2/(2b_n^2)$ and d_n. Figure 5 shows that centering by $c_n^2/(2b_n^2)$ gives excellent correspondence with Owen's formula for $1 - \alpha = .95$, and we therefore propose using the resulting formula, namely

$$\lambda(n, .95; c_n^2/(2b_n^2)) = \frac{c_n^2}{2b_n^2} - \log\{(1/4)\log(1/(1-.05))\}$$

rather than

$$\lambda(n, .95; d_n) = d_n - \log\{(1/4)\log(1/(1-.05))\}$$

for $n \geq 800$ (and certainly for $n \geq 1000$). Figure 6 gives a less clear picture for $1-\alpha = .99$. Although the approximation using the centering $c_n^2/(2b_n^2)$ is closer to Owen's finite sample formula, there is still not a clear agreement for $n \geq 1000$, so further work is needed here.

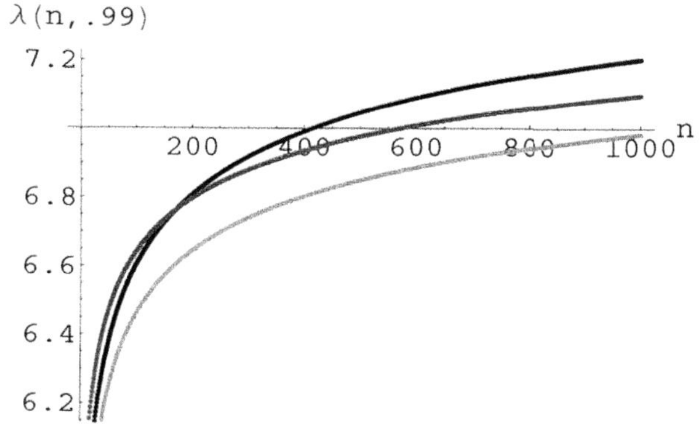

Figure 6: Black: Owen's finite-sample quantiles, $\lambda(n, .99)$.
Medium Gray: approximate quantiles $\lambda(n, .99; c_n^2/(2b_n^2))$
Light Gray: approximate quantiles $\lambda(n, .99; d_n)$

Acknowledgements: We owe thanks to Bob Berk for encouragement to complete this study.

References

[1] Berk, R. H. and Jones, D. H. (1978). Relatively optimal combinations of test statistics. *Scand. Journal of Statistics* **5**, 158 - 162.

[2] Berk, R. H. and Jones, D. H. (1979). Goodness-of-fit statistics that dominate the Kolmogorov statistics. *Z. Wahrsch. verw. Gebiete* **47**, 47 - 59.

[3] Chang, Li-Chien (1955). On the ratio of the empirical distribution to the theoretical distribution function. *Acata Math. Sinica* **5**, 347 - 368. [English Tranlation in *Selected Tranls. Math. Statist. Prob.* **4**, 17 - 38 (1964).]

[4] Daniels, H. E. (1945). The statistical theory of the strength of bundles of thread. *Proc. Roy. Soc. London, Ser. A* **183**, 4-5 - 435.

[5] Darling, D. A. and Erdös, P. (1956). A limit theorem for the maximum of normalized sums of independent random variables. *Duke Math. J.* **23**, 143-155.

[6] Donoho, D. and Jin, J. (2002). Higher criticism for detecting sparse heterogeneous mixtures. Technical Report 2002-12, Department of Statistics, Stanford University.

[7] Eicker, F. (1979). The asymptotic distribution of the suprema of the standardized empirical process. *Ann. Statist.* **7**, 116 - 138.

[8] Einmahl, J.H.J. and McKeague, I. (2003). Empirical likelihood based hypothesis testing. *Bernoulli* **9**, 267 - 290.

[9] Groeneboom, P. and Shorack, G. R. (1981). Large deviations of goodness of fit statistics and linear combinations of order statistics. *Ann. Probab.* **9**, 971 - 987.

[10] Jaeschke, D. (1979). The asymptotic distribution of the supremum of the standardized empirical distibution function on subintervals. *Ann. Statist.* **7**, 108 - 115.

[11] Mason, D. M. and Schuenemeyer, J. H. (1983). A modified Kolmogorov-Smirnov test sensitive to tail alternatives. *Ann. Statist.* **11**, 933-946.

[12] Noe, M. (1972). The calculation of distributions of two-sided Kolmogorov-Smirnov type statistics. *Ann. Math. Statist.* **43**, 58 - 64.

[13] Owen, A. B. (1995). Nonparametric likelihood confidence bands for a distribution function. *J. Amer. Statist. Assoc.* **90**, 516 - 521.

[14] Owen, A. B. (2001). *Empirical Likelihood.* Chapman and Hall/CRC, Boca Raton.

[15] Robbins, H. (1954). A one-sided confidence interval for an unknown distribution function. *Ann. Math. Statist.* **25**, 409.

[16] Shorack, G. R. and Wellner, J. A. (1978). Linear bounds on the empirical distribution function. *Ann. Prob.* **6**, 349 - 353.

[17] Shorack, G. R. and Wellner, J. A. (1986). *Empirical Processes with Applications to Statistics.* Wiley, New York.

[18] Wellner, J. A. (1978). Limit theorems for the ratio of the empirical distribution function to the true distribution function. *Z. Wahrsch. verw. Geb.* **45**, 73 - 88.

University of Washington, Statistics, Box 354322
Seattle, Washington 98195-4322, U.S.A.
E-mail address: `jaw@stat.washington.edu`

Department of Mathematics and Statistics, University of New Mexico
Albuquerque, NM 87131-1141, U.S.A.
E-mail address: `vlad@math.unm.edu`

Progress in Probability, Vol. 55, 333–346

A Note on the Smoothed Bootstrap

Dragan Radulović

Abstract. In this paper we treat the smoothed bootstrap based on histogram induced empirical measure. We demonstrate the superiority of this type of bootstrap in a very general sense. Moreover, we show that this bootstrap can effectively estimate the bias inherent from the histogram density estimation.

1. Introduction

Let $\{X_i\}_{i>0}$ be i.i.d. P random variables, and let $H_n(X_1, ..., X_n, P)$ be a statistic of interest. Classical methods for estimating the distribution of H, have been enhanced by a very intuitive resampling method, the bootstrap. Introduced by Efron [5], the bootstrap considers $\{X_i^*\}_{i=1}^n$, a triangular array such that for every fixed n the X_i^*'s are i.i.d. with distribution $P_n = n^{-1}\sum_{i=1}^n \delta_{X_i}$, and constructs the bootstrap version of H, i.e., $H_n^*(X_1^*, ..., X_n^*, P_n)$. Since $P \approx P_n$ one could hope that in some cases H_n and H_n^* are close in distributions.

A great number of research papers have been devoted to this subject (for a recent overview we refer to books by Giné [10], Efron [7] and Shao and Tu [18]). The two main research areas are: *correctness* - How close are H_n and H_n^*? and *generalization* - For what classes of statistics do we have this "closeness"?

In the case of P absolutely continuous with smooth density p, it is natural to consider the bootstrap random variables resampled from the smooth empirical measure $\widetilde{P}_n$ (an empirical probability measure based on histogram or kernel density estimation) instead of P_n. These methods are usually called "the smoothed bootstrap".

While research on the naive bootstrap (based on P_n) has been more or less completed, the work on the smoothed bootstrap is far from over. For one thing, it is not exactly clear how to use the smoothed bootstrap. In the true spirit of the bootstrap one could replace P with $\widetilde{P}_n$ and construct the statistic $\widetilde{H}_n^*(\widetilde{X}_1^*, ..., \widetilde{X}_n^*, \widetilde{P}_n)$ as an estimator of $H_n(X_1, ..., X_n, P)$.

However, at this moment it is not clear what are the benefits. The naive bootstrap works reasonably well and it seems that there is very little room for improvement. For example, with regard to *correctness*, it is well known that in the

Received by the editors December 20, 2002.
2000 *Mathematics Subject Classification.* Primary 60F05, 62F40; Secondary 60G15.
Key words and phrases. Smoothed Empirical Process, Bootstrap.

case of H, a smooth function of the mean, the naive bootstrap is already second order correct (see Hall [12]). As for *generality* we refer to Giné - Zinn's [9] result. They showed that the empirical process

$$Z_n(f)_{f\in\mathcal{F}} := \sqrt{n}(P_n - P)(f)_{f\in\mathcal{F}} \tag{1.1}$$

converges weakly if and only if its bootstrap counterpart

$$Z_n^*(f)_{f\in\mathcal{F}} := \sqrt{n}(P_n^* - P_n)(f)_{f\in\mathcal{F}} \tag{1.2}$$

does. Results related to the smooth bootstrap are in agreement with the above assertion, for they failed to demonstrate the smoothed bootstrap superiority. They typically show that in some isolated cases the smoothed bootstrap works better (Efron [6], Csörgő [3], Silverman and Young [19]), or they establish that in some general sense the smoothed bootstrap is not worse than the standard bootstrap (Cuevas and Romo [4], Wang [22], Hall, DiCiccio and Romano [11]). For an overview we refer to Shao and Tu [18] (Chapter 3).

In this paper we treat only the bootstrap results related to histogram density estimation. To start with, one has to show that, under some condition on P, the smooth empirical measure $\widetilde{P}_n$ is superior to the classical empirical measure P_n. Recent results by Radulović and Wegkamp [15] established exactly this. Namely, in the case of histogram density estimation, the smoothed empirical process

$$\widetilde{Z}_n(f)_{f\in\mathcal{F}} := \sqrt{n}(\widetilde{P}_n - P)(f)_{f\in\mathcal{F}} \tag{1.3}$$

converges weakly for classes of function $\mathcal{F}$, that are more general than in the case of the usual empirical process (1.1).

Motivation for the bootstrap in this setting is twofold.

First, the CLT for empirical processes (1.1) induces weak convergence for a very large and interesting class of statistics H (see van der Vaart and Wellner [20]). Since the smoothed empirical process works in greater generality, the class of statistics derived from the weak convergence of (1.3) is therefore richer. Theorems 1 and 2 below (see Remark 2), establish the bootstrap version of (1.3), and therefore verify the bootstrap for all the statistics induced by (1.2) (i.e. the statistics for which the usual bootstrap works) as well as for a large class of new statistics not covered by (1.1) and (1.2) (i.e. the statistics for which the usual bootstrap does not apply).

Second, the well known fact that $\widetilde{P}_n \to P$ in total variation, coupled with the results of Radulović and Wegkamp [15], clearly demonstrate the superiority of the smoothed empirical distribution. The only obstacle, from a statistical point of view, is the *bias*. In fact, it is the bias that produces the discrepancy in the optimal bandwidth size. For LLN optimal $h = O(n^{-1/5})$ and for CLT we need $h = o(n^{-1/2})$. The bootstrap procedure, introduced in this paper, not only capture the weak convergence of $\widetilde{Z}_n$ but also estimates its bias (Remarks 1 and 3 below).

Finally we would like to mention another interesting consequence of these results, which is related to the statistical analysis of massive data sets - usually referred as "data mining". Unlike the kernel density estimators or empirical measures P_n, the histograms do not require the recording of the whole data set. In fact, since the bandwidths are typically $h = n^{-\rho}$ and $0 < \rho < 1$, in case of large data sets it is not uncommon to have hundred-fold data reduction. The surprise is that this reduction comes with no cost. Actually, the results in this paper coupled with the work by Radulović and Wegkamp [15], demonstrate that by using histograms we can have both the data reduction and the better estimator than P_n !

This paper is organized as follows. Statements of results are in section 2, applications in section 3, while the proofs are in section 4. In order to keep the proofs clean we state the results under conditions stronger than needed. Section 5 deals with the generalization.

2. Notation and statements

At this point we will slightly simplify the notation introduced in the previous section. We will drop $\tilde{}$ from $\tilde{Z}_n$, $\tilde{P}_n$, and $\tilde{X}_i^*$, and use Z_n ,$P_{n,h}$ and X_i^* instead.

Let $\{X_i\}_{i>0}$ be i.i.d. P random variables, where P is absolutely continuous with density function p. We shall impose the following condition on p.

Condition A. There exists a positive constant C such that $p(x) = 0$ outside $[-C, C]$, and $|p(x) - p(y)| \leq C|x - y|$ for all x and y.

Consider an equidistant partition of the real line $....a^n_{-2} < a^n_{-1} < a^n_0 < a^n_1...$ such that $a_i = ih_n$, and histogram density estimator

$$p_h(x) := \left(\frac{1}{nh_n} \sum_{j=1}^{n} 1_{X_j \in (a_i^n, a_{i+1}^n]} \right) 1_{x \in (a_i^n, a_{i+1}^n]} \tag{2.1}$$

with associated empirical probability measure

$$P_{n,h}(f) := \int f(x) p_h(x) dx. \tag{2.2}$$

We will assume $h_n \to 0$ and $nh_n \to \infty$. *We are not assuming $n^{1/2}h_n \to 0$!*

Let $\mathcal{F}$ be a class of measurable functions such that $|f| \leq 1$. The boundedness condition could be replaced by an envelope condition at the cost of more technical proofs. Finally, the histogram based empirical process indexed by $\mathcal{F}$ is denoted by Z_n, i.e.

$$Z_n(f)_{f \in \mathcal{F}} := \sqrt{n}(P_{n,h} - P)(f)_{f \in \mathcal{F}} \tag{2.3}$$

Given a sample $\{X_i\}_{i=1}^n$,the smooth bootstrap random variables $\{X_i^*\}_{i=1}^n$ are i.i.d. $P_{n,h}$, and analogous to (2.1), (2.2) and (2.3) we define

$$p_h^*(x) := \left(\frac{1}{nh_n} \sum_{j=1}^{n} 1_{X_j^* \in (a_i^n, a_{i+1}^n]} \right) 1_{x \in (a_i^n, a_{i+1}^n]}$$

$$P_{n,h}^*(f) := \int f(x) p_h^*(x) dx$$

and

$$Z_n^*(f) := \sqrt{n}(P_{n,h}^* - P_{n,hl})(f).$$

In order to mimic the process Z_n, which is bias, the bootstrap process above is not centered with $P_{n,h}$ but with $P_{n,hl}$, the histogram based probability measure defined with respect to the bandwidth $h_n l_n$. Typically $l_n \to 0$.

We need the following definitions before stating the theorems. For probability measures P and Q on the real line, define the metric $d_3(P,Q)$ by

$$d_3(P,Q) = \sup\{|\int Hd(P-Q)| : \left\|H^{(i)}\right\|_\infty \leq 1, i \leq 3\}$$

It is well known that this metric metrizes weak convergence of probability measures on $\mathbf{R}$. Another metric with this property is the bounded Lipschitz metric, defined by

$$d_{BL}(P,Q) = \sup\{|\int Hd(P-Q)| : \|H\|_\infty \leq 1,\ |H(x) - H(y)| \leq \|x-y\|\}.$$

The metric above naturally extends to P and Q defined on Banach spaces.

Theorem 1. *Let p satisfy condition A and let f be a bounded function. Then*

$$d_3(\mathcal{L}(Z_n(f)), \mathcal{L}(Z_n^*(f))|X_1, ..., X_n)) = O_P\left(n^{-1/2} + n^{1/2}hl\right) + o_P(1)$$

Theorem 2. *Let p satisfy condition A,* let $h = n^{-\delta}$ *and let $\mathcal{F}$ be a class of bounded functions such that there exists a Gaussian process G for which*

$$\sqrt{n}(P_{n,h} - EP_{n,h})(f)_{f\in\mathcal{F}} \stackrel{W}{\to} G(f)_{f\in\mathcal{F}} \tag{2.4}$$

for every $\delta \subset (0,1)$. Then

$$d_{BL}(\mathcal{L}(Z_n(f)_{f\in\mathcal{F}}), \mathcal{L}(Z_n^*(f)_{f\in\mathcal{F}}|X_1, ..., X_n)) = o_P(1) + O_P\left(n^{1/2}hl\right)$$

Remark 1. The typical bias for Z_n is of order $n^{1/2}h$ which is obviously larger then $n^{1/2}hl$. For the cases $h \geq n^{-1/2}$ (say $h = n^{-1/3}$) the process Z_n diverges while the bootstrap still works (for appropriate l). Actually we can even bootstrap $n^{-1/5}$, which is the optimal bandwidth. Another interesting consequence is the possibility of using histograms in $\mathbf{R}^2$. Namely, here one must have $h > n^{-1/2}$, and since Theorems 1 and 2, still apply, we can bootstrap in $\mathbf{R}^2$ even though the process Z_n always diverges!

Remark 2. It should be observed that the assumptions of Theorem 2 ask for the weak convergence of *centered* version of the process Z_n. For this we need finite dimensional convergence (this is trivially satisfied for any $h = n^{-\delta}$, $\delta \in (0,1)$) and stochastic equicontinuity. In a recent paper Radulović and Wegkamp [15] treated the weak convergence of Z_n. Their results assumed $h = o(n^{-1/2})$, but only in order to control the bias of the process. A careful inspection of their proof (see proof of

Theorem 2.1 in [15]) reveals that stochastic equicontinuity for the centered version of the process Z_n follows for any $h = n^{-\delta}$, $\delta \in (0,1)$. These authors demonstrated that the process Z_n converges for essentially all the classes for which the classical empirical process (i.e. $\sqrt{n}(P_n - P)(f)$) does. More importantly, they were able to show that there are classes of functions for which Z_n converges but for which the classical empirical process does not.

Theorem 2 therefore demonstrates that there exist a whole class of statistics H_n (statistics that can be derived from the process Z_n and class $\mathcal{F}$) for which the smooth bootstrap applies and the "naive" bootstrap does not.

Let us define

$$\phi(x,h) := \left(h^{-1}\int_{ih}^{(i+1)h} f(t)dt\right) 1_{ih \le x < (i+1)h}$$

and

$$L(f,h) := \left(\int (f(x) - \phi(x,h))^2 p(x)dx\right)^{1/2}.$$

A straightforward modification of the proof of Theorem 1 (the expression (4.9) is of the order $O(L(f,h)^2)$) yields the following bias reduction.

Corollary 1 *Let p satisfy condition A and let $f(x)$ be a bounded function. Then*

$$d_3(\mathcal{L}(Z_n(f)), \mathcal{L}(Z_n^*(f))|X_1, ..., X_n)) = O_P\left(n^{-1/2} + n^{1/2}hl + L(f,h)\right).$$

Remark 3. Clearly $L(f,h)) \to 0$ for any $L_2(P)$ function, but in many cases we can get the rate of convergence. For example if f is α-Lipschitz the rate is h^α, hence we can bootstrap very irregular functions and still get reasonable rates.

If f is "almost locally Lipschitz", meaning: for all but fixed number of i's

$$\sup_{x \ne y, x,y \in (ih,(i+1)h)} \frac{|f(x) - f(y)|}{|x - y|} \le C$$

then $L(f,h)) = O(h)$. This class of functions include indicators of a finite union of intervals, many VC classes of sets (see [20] chapter 2.6) and truncated smooth functions.

3. Applications

Consider the following scheme. Let $\mathcal{F}$ be a class of functions and let $\widehat{f} \in \mathcal{F}$ be such that

$$E\widehat{f}(X) \le Ef(X) \text{ for all } f \in \mathcal{F} \tag{3.1}$$

If $\mathcal{F}$ is of the form $\{f_\xi : \xi \in \mathbf{R}^d\}$ than $\widehat{f} = f_{\widehat{\xi}}$ and $\widehat{\xi}$ is called an M-estimator. With a clever choice of class $\mathcal{F}$ one can, using this approach, recover many standard estimators. For example mean, median and sample variance can be obtain by the following three classes respectively $\{f_\xi(x) = (\xi - x)^2 - x^2\}$, $\{f_\xi(x) = |\xi - x| - |x|\}$ and $\{f_\xi(x,y) = (\xi - (x-y)^2/2)^2 - (x-y)^4/2\}$.

This approach is very flexible and can be used for a variety of non standard statistics and real life applications. Many problems in engineering, actuarial and financial mathematics reduce to the statements very similar to (3.1). Namely one has to minimize the risk (or maximize the profit) of a particular process or device over a large class of incoming functions. For more on M-estimation we refer to Van der Geer [21].

Since we typically do not know $P = \mathcal{L}(X)$, in order to use the above scheme one proceeds as follows:

(i) Replace unknown $P = \mathcal{L}(X)$ with its estimate $P_n = 1/n \sum_{i=1}^n \delta_{X_i}$ and compute $\widehat{f}_n$ such that $P_n \widehat{f}_n(X) \leq P_n f(X)$ for all $f \in \mathcal{F}$.

(ii) Prove consistency: $P_n \widehat{f}_n(X) \to Pf(X)$ and $\widehat{f}_n \to f$.

(iii) Prove CLT: $H_n := \sqrt{n}(P_n \widehat{f}_n(X) - P\widehat{f}(X))$ converges in distribution.

(iv) Prove that the Bootstrap version $H_n^* := \sqrt{n}(P_n^* \widehat{f}_n(X^*) - P_n \widehat{f}_n(X))$ is close in distribution to H_n.

For the technical details we refer to Arcones-Gine [2], but would like to comment here that the approach is justified (modulo technical details) by Uniform Limit Theorems. For example: It is well known that for fixed $\widehat{f}$, we have $P_n \widehat{f} \to P\widehat{f}$. However, our estimation of $\widehat{f}$ depends on n, so for consistency we would need that $|P_n \widehat{f}_n - P\widehat{f}_n| \to 0$ which follows from

$$\sup_{f \in \mathcal{F}} |P_n f - Pf| \to 0 \qquad \text{in probability.} \tag{3.2}$$

Similarly, (with some additional assumptions and technical details), **iii)** and **iv)** reduce to the Uniform CLT (1.1) and Uniform Bootstrap CLT (1.2). The power of the method is therefore intrinsically related to the following question: "For how large a class of functions do we have Uniform Limit Theorems?"

Although we seldom know $\mathcal{L}(X)$, it is almost always possible to determine if the law is continuous or discrete. In the former case it is reasonable to replace P_n with its smoothed version $P_{n,h}$. The approach now is identical to **i)-iv)** and the method is valid provided that we have appropriate Uniform Limit Theorems. The results in this paper, Theorem 2 and Remark 2 coupled with (1.3) therefore justify this approach for much larger classes of functions than the usual case using P_n.

To be more specific: It is well known that the smoothed empirical measure $P_{n,h}$, converges to P in total variation, which in our case easily implies

$$\sup_{|f| \leq 1} |P_{n,h} f - Pf| \to 0 \text{ in probability,}$$

and essentially guaranties consistency **ii)** for any class of uniformly bounded functions (actually for all $f : |f| \leq F$ and $PF < \infty$). This class is far larger that the classes for which we have the classical Uniform LLN (3.2). Actually, it is easy to construct relevant classes of functions for which (1.3) as well as Theorem 2 in this paper, are valid but for which (3.2) does not hold.

In other words, using smoothed empirical measures we could get the CLT and Bootstrap CLT for classes which are too large even for simple consistency requirement for classical empirical measure P_n. For more details about the constructions of these classes we refer to Radulovic and Wegkamp [15] and [16].

4. Proofs

In order to simplify the computation we will assume $h_n = 1/M_n$ and $l_n = 1/L_n$ for M_n and L_n integer sequences. Aloso, we will typically write $Z_n, Z_n^*,..$ instead of $Z_n(f), Z_n^*(f),...$ We introduce some notation. Let $.\xi_i = \xi_{i,n}$ such that $\xi_i \in (a_i, a_{i+1})$ and $p(\xi_i) = \frac{1}{h}\int_{a_i}^{a_{i+1}} p(x)dx$. Let Q_n and Q_n^* be two sequences of discrete probability measures supported on $\{\xi_{i,n}\}$ and such that

$$Q_n(\xi_i) = \int_{a_i}^{a_{i+1}} p(x)dx \text{ and } Q_n^*(\xi_i) = \int_{a_i}^{a_{i+1}} p_h(x)dx \tag{4.1}$$

Let us also define

$$\widetilde{f}(\xi_i) := \frac{1}{h}\int_{a_i}^{a_{i+1}} f(x)dx \ , \ Y_j := \xi_i 1_{X_j \in (a_i, a_{i+1})} \ , \ Y_j^* := \xi_i 1_{X_j^* \in (a_i, a_{i+1})}$$

and

$$Q_{n,E}(\widetilde{f}) := \frac{1}{n}\sum_{j=1}^{n} \widetilde{f}(Y_j) \text{ and } Q_{n,E}^*(\widetilde{f}) := \frac{1}{n}\sum_{j=1}^{n} \widetilde{f}(Y_j^*) \tag{4.2}$$

It is easy to see that $Q_{n,E}(\widetilde{f}) = P_{n,h}(f)$, $Q_{n,E}^*(\widetilde{f}) = P_{n,h}^*(f)$ and $Q_n(\widetilde{f}) = E(P_{n,h}(f))$. Using this notation we can decompose Z_n into

$$\begin{aligned} Z_n(f) &= \sqrt{n}(P_{n,h} - P)(f) \\ &= \frac{1}{\sqrt{n}}\sum_{j=1}^{n}(\widetilde{f}(Y_j) - Q_n(\widetilde{f})) + \sqrt{n}(EP_{n,h} - P)(f) = Z_{n,C}(f) + B_{n,h}(f). \end{aligned}$$

In the same way,

$$\begin{aligned} Z_n^*(f) &= \sqrt{n}(P_{n,h}^* - P_{n,hl})(f) \\ &= \frac{1}{\sqrt{n}}\sum_{j=1}^{n}(\widetilde{f}(Y_j^*) - Q_{n,E}(\widetilde{f})) + \sqrt{n}(P_{n,h} - P_{n,hl})(f) = Z_{n,C}^*(f) + B_{h,hl}^*(f) \end{aligned}$$

Proof for Theorem 1. Since the bias parts (i.e. $B_{n,h}$ and $B_{n,hl}^*$) are independent from centered processes $Z_{n,C}$ and $Z_{n,C}^*$ (the latter, conditioned on the sample) we have the following inequality (for example see Araujo and Giné [1])

$$d_3(\mathcal{L}(Z_n), \mathcal{L}(Z_n^*)) \le d_3(\mathcal{L}(Z_{n,C}), \mathcal{L}(Z_{n,C}^*)) + d_3(\mathcal{L}(B_h), \mathcal{L}(B_{n,hl}^*)) = I_n + II_n \tag{4.3}$$

The result follows from Lemma 1 and Lemma 2 below. □

Lemma 1. *For* $|f| \leq 1$,

$$d_3(\mathcal{L}(Z_{n,C}), \mathcal{L}(Z^*_{n,C})) = O_P(n^{-1/2})$$

Proof. First we observe that although the bootstrap r.v.'s Y_i^* are not defined by the usual resampling from $\{Y_1, ..., Y_n\}$, we can treat them as if they were. To see this we define the "usual" bootstrap r.v. $\{\widehat{Y}_i^*\}$ by letting $\widehat{Y}_i^* = Y_j$ with probability $1/n$, and observe that both Y_i^*'s and $\widehat{Y}_i^*$'s are i.i.d., and $\mathcal{L}(\widehat{Y}_i^*) = \mathcal{L}(Y_i^*)$. Namely

$$\begin{aligned} P^*(\widehat{Y}_i^* = \xi_k) &= \frac{1}{n}\sum_{j=1}^n 1_{Y_j=\xi_k} = \frac{1}{n}\sum_{j=1}^n 1_{X_j \in (a_k, a_{k+1})} \\ &= \int_{a_k}^{a_{k+1}} p_h(x)dx = P^*(X_i^* \in (a_k, a_{k+1})) = P^*(Y_i^* = \xi_k). \end{aligned}$$

Next we let $W_j := n^{-1/2}(\widetilde{f}(Y_j) - Q_n(\widetilde{f}))$ and $W_j^* := n^{-1/2}(\widetilde{f}(Y_j^*) - Q_{n,E}(\widetilde{f}))$. Clearly $Z_{n,C} = \sum_{j=1}^n W_j$, $Z^*_{n,C} = \sum_{j=1}^n W_j^*$ and $EW_j = E^*W_j^* = 0$. In order to avoid possible dependency problems between W_j and W_j^* we define the former with respect to a different sequence $\{X_i'\}$, which is i.i.d. P, but independent from $\{X_i\}$.

We proceed using an argument very similar to the classical (Lindeberg's) proof of the CLT (see Pollard [13] pg. 51-52). Let $S_j := W_1^* + ... + W_{j-1}^* + W_{j+1} + ... + W_n$. Then, for fixed H and by virtue of adding and subtracting the appropriate terms we have the following:

$$|EH(Z_{n,C}) - E^*H(Z^*_{n,C})| \leq \sum_{j=1}^n |E_{X'}E^*(H(S_j + W_j)) - E_{X'}E^*(H(S_j + W_j^*))|. \quad (4.4)$$

In the above expression $E_{X'}$ indicate the expectation with respect to the sequence $\{X_i'\}$. We now develop H as a second degree Taylor polynomial. Since the following estimate will not depend on j we will restrict ourselves to the case $j = 1$. Therefore:

$$\begin{aligned} &E_{X'}E^*(H(S_1 + W_1)) - E_{X'}E^*(H(S_1 + W_1^*)) \\ &\quad = E_{X'}E^*(H(S_1)) + E_{X'}E^*(H'(S_1)W_1) \\ &\qquad + E_{X'}E^*(H''(S_1)(W_1)^2/2!) + E_{X'}E^*(H'''(\rho)(W_1)^3/3!) \\ &\qquad - E_{X'}E^*(H(S_1)) - E_{X'}E^*(H'(S_1)W_1^*) \\ &\qquad - E_{X'}E^*(H''(S_1)(W_1^*)^2/2!) - E_{X'}E^*(H'''(v)(W_1^*)^3/3!), \end{aligned}$$

where H', H'', H''' stand for first, second and third derivatives respectively, $\rho \in [S_1, S_1 + W_1]$ and $v \in [S_1, S_1 + W_1^*]$.

The terms containing H cancel. Due to the fact that S_1 and W_1 as well as S_1 and W_1^* are independent (and since W_1 and W_1^* are centered), the terms with H' equal zero. The terms containing H''' are of order $O(n^{-3/2})$. Finally we need to estimate

$$|E_{X'}E^*(H''(S_1)(W_1)^2) - E_{X'}E^*(H''(S_1)(W_1^*)^2)|$$

(Since S_1 and W_1 as well as S_1 and W_1^* are independent)

$$= |E_{X'}E^*(H''(S_1)\left(E_{X'}E^*(W_1)^2 - E_{X'}E^*(W_1^*)^2\right)|$$

(Since $|H''| \leq 1$ and by definition of W_1 and W_i^*)

$$\leq \frac{1}{n}|E(\widetilde{f}(Y_1))^2 - E^*(\widetilde{f}(Y_1^*))^2| + \frac{1}{n}|(E\widetilde{f}(Y_1))^2 - (E^*\widetilde{f}(Y_1^*))^2|.$$

The first expression is equal to

$$\frac{1}{n}\left|\frac{1}{n}\sum_{j=1}^{n}\left((\widetilde{f}(Y_j))^2 - E(\widetilde{f}(Y_j))^2\right)\right| = O_P(n^{-3/2})$$

and the second expression is bounded by

$$\frac{2}{n}\left|\frac{1}{n}\sum_{j=1}^{n}\left((\widetilde{f}(Y_j)) - E(\widetilde{f}(Y_j))\right)\right| = O_P(n^{-3/2}).$$

Going back to (4.4) we observe that there are n terms, each of order $O_P(n^{-3/2})$. This proves Lemma 1. □

Lemma 2. *Under the conditions of Theorem 1,*

$$d_3(\mathcal{L}(B_{n,h}), \mathcal{L}(B_{n,hl}^*)) = O_P\left(n^{1/2}hl\right) + o_P(1)$$

Proof. Notice that, for all ω's

$$\begin{aligned} d_3(\mathcal{L}(B_{n,h}), \mathcal{L}(B_{n,hl}^*)) &\leq |B_{n,h} - B_{n,hl}^*| \\ &= \sqrt{n}\,|(EP_{n,h} - P)(f) - (P_{n,h} - P_{n,hl})(f)| \end{aligned}$$

(By adding and subtracting $(EP_{n,h}(f) - EP_{n,hl}(f))$

$$\leq \sqrt{n}\,|(EP_{n,hl} - P)(f)| + \sqrt{n}\,|(P_{n,h} - P_{n,hl})(f) - (EP_{n,h}(f) - EP_{n,hl}(f))|. \quad (4.5)$$

For the first part we observe

$$\sqrt{n}\,|(EP_{n,hl} - P)(f)| \leq \sqrt{n}\int |Ep_{hl}(x) - p(x)|dx = O(n^{1/2}hl) \quad (4.6)$$

For the second part of (4.5) we have to set up some notation. As before $a_{i+1} - a_i = h_n = 1/M_n$ and we define the equidistant subpartition $c_{i,j}$ such that $a_i = c_{i,1} < c_{i,2} < \dots < c_{i,L_n} = a_{i+1}$ and $c_{i,j+1} - c_{i,j} = 1/(M_nL_n)$ where $l_n = 1/L_n$. In order to simplify the proof we will assume that density function $p(x)$ is supported on the unit interval $[0,1]$. With this notation we have

$$(P_{n,h}-P_{n,hl})(f)=\sum_{i=0}^{M_n}\left(\left(\int_{a_i}^{a_{i+1}}f(x)dx\right)M_n\frac{1}{n}\sum_{j=1}^{n}1_{X_j\in(a_i,a_{i+1})}\right)$$
$$-\sum_{i=0}^{M_n}\sum_{k=0}^{L_n}\left(\left(\int_{c_{i,k}}^{c_{i,k+1}}f(x)dx\right)L_nM_n\frac{1}{n}\sum_{j=1}^{n}1_{X_j\in(c_{i,k},c_{i,k+1})}\right)$$

Let $A^i:=\frac{1}{n}\sum_{j=1}^{n}1_{X_j\in(a_i,a_{i+1})}$, $A^i_k:=\frac{1}{n}\sum_{j=1}^{n}1_{X_j\in(c_{i,k},c_{i,k+1})}$. Clearly A^i and A^i_k are random and $\sum_{k=1}^{L_n}A^i_k=A^i$ and $\sum_{i=1}^{M_n}A^i=1$. With this notation the above is equal to

$$\sum_{i=0}^{M_n}A^i\frac{1}{h_n}\int_{a_i}^{a_{i+1}}f(x)dx-\sum_{i=0}^{M_n}\sum_{k=0}^{L_n}\left(\left(\frac{1}{h_nl_n}\int_{c_{i,k}}^{c_{i,k+1}}f(x)dx\right)A^i_k\right)$$

(By letting $\widetilde{f}_{i,k}:=\frac{1}{h_nl_n}\int_{c_{i,k}}^{c_{i,k+1}}f(x)dx$ and $\widetilde{f}_i:=\frac{1}{h_n}\int_{a_i}^{a_{i+1}}f(x)dx$)

$$=\sum_{i=0}^{M_n}A^i\widetilde{f}_i-\sum_{i=0}^{M_n}\sum_{k=0}^{L_n}A^i_k\widetilde{f}_{i,k}$$

(Since $\sum_{k=1}^{L_n}A^i_k=A^i$)

$$=\sum_{i=0}^{M_n}\sum_{k=0}^{L_n}A^i_k\left(\widetilde{f}_i-\widetilde{f}_{i,k}\right).$$

Therefore the second part of (4.5) is equal to

$$\sqrt{n}\left|\sum_{i=0}^{M_n}\sum_{k=0}^{L_n}(A^i_k-EA^i_k)\left(\widetilde{f}_i-\widetilde{f}_{i,k}\right)\right|=B_n$$

Letting $\lambda_{i,k}:=\widetilde{f}_i-\widetilde{f}_{i,k}$, we trivially have

$$EB_n^2=n\sum_{i=0}^{M_n}\sum_{k=0}^{L_n}\sum_{ii=0}^{M_n}\sum_{kk=0}^{L_n}\lambda_{ii,kk}\lambda_{i,k}E\left((A^i_k-EA^i_k)(A^{ii}_{kk}-EA^{ii}_{kk})\right)\tag{4.7}$$

and

$$E\left((A^i_k-EA^i_k)(A^{ii}_{kk}-EA^{ii}_{kk})\right)=EA^i_kA^{ii}_{kk}-p_{i,k}p_{ii,kk}\tag{4.8}$$

where $p_{i,k}=\int_{c_{i,k}}^{c_{i,k+1}}p(x)dx$. Adopting notation $I_{i,k}:=(c_{i,k},c_{i,k+1})$, for the cases $i=ii$ and $k=kk$ the expression (4.8) is equal to

$$E(A^i_k)^2-p^2_{i,k}=E\left(\frac{1}{n}\sum_{j=1}^{n}1_{X_j\in I_{i,k}}\right)^2-p^2_{i,k}$$
$$=\frac{1}{n^2}\sum_{j-1}^{n}\sum_{jj-1}^{n}E1_{X_j\in I_{i,k}}1_{X_{jj}\in I_{i,k}}-p^2_{i,k}=\frac{1}{n}(p_{i,k}-p^2_{i,k})$$

Since for the cases $i \neq ii$ or $k \neq kk$ the intervals $I_{i,k}$ and $I_{ii,kk}$ are disjoint we have the following

$$EA_k^i A_{kk}^{ii} - p_{i,k}p_{ii.kk} = \frac{1}{n^2}\sum_{\substack{j,jj=1\\ j\neq jj}}^{n} E1_{X_j\in I_{i,k}}E1_{X_{jj}\in I_{ii,kk}} - p_{i,k}p_{ii.kk} = \frac{-1}{n}p_{i,k}p_{ii,kk}\,.$$

Therefore (4.7) is equal to

$$\sum_{i=0}^{M_n}\sum_{k=0}^{L_n}\lambda_{i,k}^2(p_{i,k}-p_{i,k}^2) - \sum_{\substack{i,ii=0,\\ i\neq ii}}^{M_n}\sum_{\substack{k,kk=0\\ k\neq kk}}^{L_n}\lambda_{ii,kk}\lambda_{i,k}p_{i,k}p_{ii,kk}$$

thus implying

$$\begin{aligned} EB_n &\leq (EB_n^2)^{1/2} \\ &\leq \left(\sum_{i=0}^{M_n}\sum_{k=0}^{L_n}(\lambda_{i,k})^2 p_{i,k}\right)^{\frac{1}{2}} + \left(\left(\sum_{i=0}^{M_n}\sum_{k=0}^{L_n}|\lambda_{i,k}|p_{i,k}\right)\left(\sum_{ii=0}^{M_n}\sum_{kk=0}^{L_n}|\lambda_{ii,kk}|p_{ii,kk}\right)\right)^{\frac{1}{2}} \\ &= \left(\sum_{i=0}^{M_n}\sum_{k=0}^{L_n}\lambda_{i,k}^2 p_{i,k}\right)^{\frac{1}{2}} + \left(\sum_{i=0}^{M_n}\sum_{k=0}^{L_n}|\lambda_{i,k}|p_{i,k}\right) \leq 2\left(\sum_{i=0}^{M_n}\sum_{k=0}^{L_n}\lambda_{i,k}^2 p_{i,k}\right)^{\frac{1}{2}}. \end{aligned}$$

Now

$$\sum_{i=0}^{M_n}\sum_{k=0}^{L_n}\lambda_{i,k}^2 p_{i,k} = \sum_{i=0}^{M_n}\sum_{k=0}^{L_n}\lambda_{i,k}^2\int_{c_{i,k}}^{c_{i,k+1}}p(x)dx = \int(\widetilde{f}_i - \widetilde{f}_{i,k})^2 1_{x\in I_{i,k}}p(x)dx. \quad (4.9)$$

It is well known (see Rudin [17] - Lebesgue points) that $|\widetilde{f}_i - \widetilde{f}_{i,k}|1_{x\in I_{i,k}} \to 0$ for almost all x. Since $|\widetilde{f}| \leq 1$, we can use the dominated convergence theorem. This proves Lemma 2. □

Proof for Theorem 2. Using the very same discretization argument as in the proof of Theorem 1 we can write

$$\begin{aligned} Z_n(f)_{f\in\mathcal{F}} &= Z_{n,C}(f)_{f\in\mathcal{F}} + B_{n,h}(f)_{f\in\mathcal{F}} \quad \text{and} \\ Z_n^*(f)_{f\in\mathcal{F}} &= Z_{n,C}^*(f)_{f\in\mathcal{F}} + B_{n,hl}^*(f)_{f\in\mathcal{F}}\,. \end{aligned}$$

Let the set $\{g_1, ..., g_{K_\delta}\}$ be a $\delta-$cover for the class $\mathcal{F}$ under $L_2(P)$ distance (i.e. for every $f \in \mathcal{F}$ there exists a g_k such that $\|f - g_k\|_{L_2(P)} \leq \delta$). For given $f \in \mathcal{F}$, g_f will denote the closest g_k from the cover. For any $H : l^\infty(\mathcal{F}) \to \mathbf{R}$, bounded and Lipschitz, by adding and subtracting the appropriate terms we have

$$H(Z_n(f)) - H(Z_n^*(f)) = A_1 + A_2 + A_3 + A_4,$$

where

$$A_1 = H(Z_{n,C}(f) + B_{n,h}(f)) - H(Z_{n,C}(g_f) + B_{n,h}(f))$$
$$A_2 = H(Z_{n,C}(g_f) + B_{n,h}(f)) - H(Z^*_{n,C}(g_f) + B_{n,h}(f))$$
$$A_3 = H(Z^*_{n,C}(g_f) + B_{n,h}(f)) - H(Z^*_{n,C}(g_f) + B^*_{n,hl}(f))$$
$$A_4 = H(Z^*_{n,C}(g_f) + B^*_{n,hl}(f)) - H(Z^*_{n,C}(f) + B^*_{n,hl}(f))$$

It is sufficient to prove that $E|A_i| = o_P(1) + O_P(n^{1/2}hl)$. For A_3 we have

$$E|A_3| \leq \sup_{f\in\mathcal{F}} |B_{n,h}(f) - B^*_{n,hl}(f)|$$

(By the same decomposition as in (4.5))

$$\leq \sqrt{n} \sup_{f\in\mathcal{F}} |(EP_{n,hl} - P)(f)| + \sqrt{n} \sup_{f\in\mathcal{F}} |(P_{n,h} - P_{n,hl})(f) - E(P_{n,h} - P_{n,hl})(f)| \quad (4.10)$$

By (4.6) the first expression is of the order $O(n^{1/2}hl)$. Th proof of Lemma 2 implies that the process

$$\{\sqrt{n}((P_{n,h} - P_{n,hl})(f_k) - E(P_{n,h} - P_{n,hl})(f_k))\}_{k=1}^m \quad (4.11)$$

converges to 0 for any finite collection $\{f_1, ..., f_m\}$. On the other hand the condition (2.4) implies the stochastic equicontinuity of the process (4.11) . It is well known (see Vaart Wellner [20]) that this implies that the second part of (4.10) is of order $o_P(1)$. Hence $A_3 = O_P(n^{1/2}hl) + o_P(1)$.

The A_2 term reduces to finite dimensional convergence which is of the order $O_p(n^{-1/2})$ by Lemma 1. The fourth term is of the order $o_P(1)$ provided that

$$\lim_{\delta\to 0} \limsup_{n\to\infty} P^* \left(\sup_{Pf^2\leq\delta} \left| \frac{1}{\sqrt{n}} \sum_{j=1}^n \left(\widetilde{f}(Y_j^*) - E^*(\widetilde{f}(Y_1^*))\right) \right| > \varepsilon \right) = 0 \quad (4.12)$$

in probability. For the "in probability" in the above statement we mean a.s. convergence along subsequences. The $\{Y_j^*\}_{j=1}^n$ are resampled (with replacement) from $\{Y_j\}_{j=1}^n$ and we can use the result of Giné-Zinn [9]. The fact that in our case Y_j are actually Y_j^n, thus a triangular array (with Y_j^n , $j = 1, ..., n$ i.i.d. for every fixed n) does not affect the proof. Therefore, (4.12) follows from

$$\lim_{\delta\to 0} \limsup_{n\to\infty} P \left(\sup_{Pf^2\leq\delta} \left| \frac{1}{\sqrt{n}} \sum_{j=1}^n \left(\widetilde{f}(Y_j) - E(\widetilde{f}(Y_1))\right) \right| > \varepsilon \right) = 0 . \quad (4.13)$$

The requirement above also implies that the term A_1 is of order $o(1)$. Finally, by the notation introduced in (4.1) and (4.2) and by the fact that weak convergence of the empirical process implies the stochastic equicontinuity (see Giné and Zinn [8]), we have that condition (2.4) implies (4.13). This proves Theorem 2. □

5. Appendix

In order not to cloud the main idea of the proofs with technical details, all the results in previous sections are stated under conditions much stronger than actually needed. In this section we address some of the straightforward generalizations.

Bounded functions. The very strong condition $|f| \leq 1$ for all $f \in \mathcal{F}$, can be easily replaced with the moment assumption on the envelope function $F(x) = \sup_{f\in\mathcal{F}} |f(x)|$. Namely for $E|F(X)|^3 < \infty$, Lemma 1 remains true. In order to control the bias (see expression (4.6), we would need an additional smoothness condition on the envelope function F:

$$\int F(x)|p(x) - p(x+\lambda)|dx = O(\lambda).$$

It is possible to prove consistency even under the minimal assumption $E(F(X))^2 < \infty$. In this case one has to modify the proof of Lemma 1 (truncation) and the rate is not $O_P(n^{-1/2})$ but $o_P(1)$.

Compact support . Condition A asks for $p(x)$ Lipschitz and compactly supported. Both conditions are too strong. All we needed is the above condition (5) and $p(x)$ continuous. It should be noted that for "nice" $F(x)$ the condition (5) is satisfied for essentially all densities of interest.

Second order correctness. For suitable small h's (i.e. $h = o(n^{-1/2})$) and for function f such that $Ef(X)^4 < \infty$ and $L(f,h) = O(h)$ (see Remark 3) we can get second order correctness for the smoothed bootstrap procedure. Namely, for $\widehat{Z}_n := Z_n/\sigma_f$ and $\widehat{Z}_n^* := Z_n^*/\sigma_f^*$ (with $\sigma_f^2 = Var(Z_n f)$ and $(\sigma_f^*)^2 = Var^*(Z_n^* f)$) we get

$$d_3(\mathcal{L}(\widehat{Z}_n), \mathcal{L}(\widehat{Z}_n^*)|X_1, ..., X_n)) = o_P(n^{-1/2}),$$

even though both $\widehat{Z}_n$ and $\widehat{Z}_n^*$ have bias of order $n^{1/2}h$! The proof for the statement above reduces to a modification of Lemma 1 (see Radulović [14], Remark 3).

Acknowledgment

I would like to thank Marten Wegkamp for useful discussions and interactions.

References

[1] Araujo, A. and Giné, E. (1989). *The Central limit Theorem and Banach Valued Random Variables* , Wiley, New York.

[2] Arcones, M. and Giné, E. (1992). On the Bootstrap of M-Estimators and Other Statistical Functional. *Exploring the Limits of The Bootstrap*, John Wiley & Sons.

[3] Csörgő, M. (1983). *Quantile Processes With Statistical Applications.* SIAM, Philadelphia.

[4] Cuevas, A. and Romo, J. (1997). Differentiable functionals and smoothed bootstrap. *Ann. Inst. Statist. Math.* **49**, 355-370.

[5] Efron, B. (1979). Bootstrap Methods: Another Look at the Jackknife, *Ann. Statist.* **7**, 1-26.

[6] Efron, B. (1982). *The Jackknife, the Bootstrap, and Other Resampling Plans*, SIAM, Philadelphia.

[7] Efron, B. and Tibshirani, R.J. (1993). *An Introduction to the Bootstrap.* Chapman and Hall.

[8] Giné, E. and Zinn, J. (1986). Lectures on the central limit theorem for empirical processes. *Probability and Banach Spaces. Lecture Notes in Math.* **1221**, 50-113. Springer, Berlin.

[9] Giné, E. and Zinn, J. (1990). Bootstrapping general empirical measures. *Ann. Probab.* **18**, 851-869.

[10] Giné, E. (1997). Lectures on some aspects of the bootstrap. *Lectures on Probability Theory and Statistics, Lecture Notes in Mathematics* **1665**, Springer, Berlin.

[11] Hall, P., DiCiccio, T. and Romano, J. (1989). On Smoothing and the Bootstrap, *Ann. Statist.* **17**, 693 - 704.

[12] Hall, P. (1992). *The Bootstrap and Edgeworth Expansion*, Springer-Verlag, New York.

[13] Pollard, D. (1984). *Convergence of Stochastic Processes.* Springer, New York.

[14] Radulović, D. (1998). Can We Bootstrap Even if CLT Fails?, *Journal of Theoretical Probability* **11**, 813 - 830.

[15] Radulović, D. and Wegkamp, M. (2003). Necessary and sufficient conditions for weak convergency of smoothed empirical processes. *Stat. Probab. Letters* **61**, 321-336.

[16] Radulović, D. and Wegkamp, M. (2000). Weak convergence of smoothed empirical processes: beyond Donsker classes. *High Dimenisonal Probability II*, Birkhauser, Boston, 89-106.

[17] Rudin, W. (1966). *Real and Complex Analysis.* McGraw-Hill, New York.

[18] Shao, J and Tu, D. (1995). *The Jackknife and Bootstrap.* Springer, New York.

[19] Silverman, B. W. and Young, G. A. (1987). The bootstrap: To smooth or not to smooth. *Biometrika* **74**, 469 - 479.

[20] Van der Vaart, A. W. and Wellner J.A. (1996). *Weak Convergence and Empirical Processes.* Springer, New York.

[21] Van de Geer, S. (2000). *Empirical Processes in M-Estimation*, Cambridge University Press, Cambridge.

[22] Wang, S. (1989). On the bootstrap and smoothed bootstrap, *Comm. Statist. A,* **18**, 3949-3962.

Department of Statistics, Yale University
24 Hillhause Avenue, New Haven CT, USA
E-mail address: `dragan.radulovic@yale.edu`

GPSR Compliance

The European Union's (EU) General Product Safety Regulation (GPSR) is a set of rules that requires consumer products to be safe and our obligations to ensure this.

If you have any concerns about our products, you can contact us on ProductSafety@springernature.com

In case Publisher is established outside the EU, the EU authorized representative is:

Springer Nature Customer Service Center GmbH
Europaplatz 3
69115 Heidelberg, Germany

Batch number: 10732312

Printed by Printforce, the Netherlands